Geometric Quantum Mechanics

Geometric Quantum Mechanics

Michel van Veenendaal
Northern Illinois University
Argonne National Laboratory
IL, USA

WILEY

Published by John Wiley & Sons, Inc., Hoboken, New Jersey.
Published simultaneously in Canada.

Library of Congress Cataloging-in-Publication Data applied for:
Hardback ISBN: 9781119913214; ePDF: 9781119913221; epub: 9781119913238

Cover Image: © zf L/Getty Images
Cover Design: Wiley

Set in 9.5/12.5pt CMR10 by Straive, Chennai, India

Contents

Preface

The introduction of quantum mechanics has settled into a standard approach, well known from a wide variety of textbooks. These methods are generally heavily based on differential equations and matrices. The drawback of this approach is that many concepts are introduced as uniquely quantum mechanical and often simply postulated. This is often reinforced by the fact that many mathematical techniques are only encountered by students for the first time in a quantum-mechanics course. Take complex operators as an example. Many might view this as quintessential quantum mechanics. However, similar operators can already be obtained by simply rotating a vector in two dimensions. A vector consists of unit vectors and coefficients. A rotation can be obtained by manipulation of the unit vectors or by performing an operation on the coefficients. The manipulation of the unit vectors is done using geometric algebra techniques. This approach is generally avoided in introductory physics books, restricting themselves to the imaginary scalar $i = \sqrt{-1}$. The presence of additional imaginary units is generally postponed until the introduction of gamma and Pauli matrices. This gives the impression that this is something unique to relativistic quantum mechanics, which is not. The rotation of the unit vector can also be obtained by performing an operation (taking the derivative) on the coefficients. The complex operators give rise to eigenvectors since the operation on the coefficients and the manipulation of the unit vectors using imaginary units are constructed such that they cancel each other.

These concepts can be extended to higher order vectors, where the coefficients now become a function space. Generally, the focus of introductory quantum-mechanics text books is on deriving the function space. For example, for spherical harmonics, the Schrödinger equation is set up in spherical symmetry. The differential equation of the angular part is solved by a series solution leading to the spherical harmonics. These functions appear to have all kinds of interesting transformation properties, which are then derived using the commutation properties of the angular momentum operators. Here, we start from the recognition that we are dealing with higher order vectors. The function space is obtained by making an exponential expansion of unit vectors. By writing the unit vectors in terms of Schwinger bosons (comparable to step operators for the quantum-harmonic oscillator), the correct higher order unit vectors are also obtained. These unit vectors contain all the transformation properties of the spherical harmonics and are related to the $3j$ symbols and Clebsch-Gordan coefficients.

Another advantage of using geometric-algebra techniques is that they are often more insightful and easier to use than the Pauli or gamma matrices. The matrix techniques just present one way of incorporating the properties of unit vectors in space and spacetime. Even weird quantities such as spinors with half angles can be understood from straightforward geometric principles.

The stronger emphasis on the unit vector aspect of the Hilbert space compared to the function part leads to a more natural connection to classical mechanics and electromagnetism which are usually expressed in terms of vector operations. It also leads to a better insight into symmetry aspects, which becomes important for understanding the basis of particle physics and many-body systems.

The second chapter deals with relativistic quantum mechanics and its nonrelativistic limit. This adds a time component to our considerations leading to rotations in spacetime or Lorentz boosts. Since photons move at the speed of light, electromagnetism is a relativistic theory, and its incorporation with particles is much more straightforward in the relativistic limit. Connections are made to classical mechanics and electromagnetism via the derivation of the Lorentz and Maxwell equations. The scalar Schrödinger equation is obtained by removing the vector part of the four-momentum. However, some conceptually confusing vector terms cannot be removed.

Chapter 3 looks at a variety of applications of the Schrödinger equation. For the more advanced single-particle problems, the focus is on function spaces that can be "constructed," such as quantum-harmonic oscillator problems in different dimensions. Other sections also focus on how the understanding of the symmetry can provide insight into the solutions, such as translational symmetry for solids.

The last two chapters cover a very wide range of topics in condensed matter, nuclear, and particle physics. The goal is not to provide an in-depth understanding but to demonstrate the quantum-mechanical ideas that underlie these concepts. Chapter 4 focuses on how many-body systems of interacting particles can show different behavior from a system of many independent particles. To understand the electron–electron interaction on atoms, the transformation properties for spherical symmetry are used. This shows how the Coulomb interaction splits the degenerate states into distinct groups. However, the same ideas can be used to understand the structure of mesons and baryons, except that we are now dealing with the strong interaction, and the symmetry has changed. The concepts of the quantum-harmonic oscillator are used to understand the structure of the nucleus.

Chapter 5 looks into more detail on the appearance of new phenomena as a result of mutual interactions. Again, the focus is on the ideas that allowed us to solve these problems and less on the phenomenology. For example, both superconductivity and superfluidity reduce to complicated perturbed quantum-harmonic oscillator problems after applying the appropriate approximations. It addresses how ideas from magnetism and superconductivity affected the issue of mass generation in particle physics. It explains how things that we often consider inherent properties, such as the mass of a particle and the reduced velocity of light in a medium leading to refraction, are actually effective models for much more complicated physics.

June 2022 *Michel van Veenendaal*
DeKalb, Illinois

1

Space

Introduction

The goal of this chapter is to understand motion in free space, *i.e.* in the absence of any potential. For quantum particles, this already turns out to be rather complex. In order to follow the thought process, it can be useful to think of the concepts in terms of pairs. To understand the motion, we initially look at vectors and then extend it to higher order vectors. There are two different types of motion: translation and rotation. Translation changes the length of a vector, but not its direction. Rotation changes the direction of a vector, but not its length. We can construct spaces with translational or rotational symmetry. This leads to conserved quantities, momentum, and angular momentum, respectively, that form a vector space by themselves. This is known as a dual space. Vectors are ideal in tracking rigid objects or point particles, where one only tracks the position in space. Vectors consist of two parts: basis vectors and coefficients. Motion can be achieved in two ways: by manipulation of the basis vectors or by performing an operation on the coefficients. Whichever method you use, the end result should be the same. Quantum particles are not rigid objects. The shape of the particle adjusts itself depending on the geometry (*e.g.* the potential landscape) of the problem. We therefore need to consider the entire particle, which is described by a function. The goal is to obtain functions that correspond to the conserved quantities (momentum or angular momentum). In free space, this can be done by taking the vectors and raising them to an arbitrary positive integer power. This leads to higher order vectors, which again are expressed in terms of basis vectors and coefficients. The coefficients are now called functions, and all the coefficients together form a function space. Motion can still be obtained by either manipulating the basis vectors or performing an operation on the functions.

Along the way, try to follow what part of the following pairs we are trying to describe: translation or rotation; vectors or higher order vectors (function spaces); obtaining motion by manipulating the basis vectors or by performing an operation on the coefficients/functions; and are we looking at the problem in real space or in its dual space (*i.e.* the space of the conserved quantities, momentum and angular momentum). The approach followed here is rather different from most quantum mechanics textbooks, where generally the Schrödinger equation and other concepts are postulated, and function spaces are produced for a wide variety of problems. Here, we start very simple and discover that many of the concepts that are often considered quintessential quantum mechanics simply arise from the manipulation of vectors or our description of extended nonrigid objects in space.

Geometric Quantum Mechanics, First Edition. Michel van Veenendaal.

Since the approach is different from most textbooks on quantum mechanics, we start with a brief outline of the sections and their goals.

1.1 The exponential function is crucial for the understanding of rotation and translation. This section discusses the exponential function in terms of the repeated application of an infinitesimally small action.
1.2 In order to better understand how to manipulate vectors, we start with the relatively simple example of the rotation of a unit vector in two dimensions. Since a vector consists of unit vectors and coefficients, it is important to realize that this operation can be achieved by either manipulating the unit vectors or by performing an operation on the coefficients. In this section, the former is achieved using anticommuting unit vectors. This is known as Clifford or geometric algebra.
1.3 This section looks again at the rotation of a unit vector, but now by performing an operation on the coefficients. This approach directly leads to differential calculus. The operator approach is generally more emphasized in introductory quantum mechanics courses that focus on wave functions. The geometric approach becomes more prevalent when symmetry and geometry dominate the physics. Despite only looking at the rotation of a vector in two dimensions, an operator similar to the quantum angular momentum operator is obtained.
1.4 Although a vector can indicate a position in space, it does not allow the description of each point in space independently. This is necessary if we want to describe functions or fields in space. For the unit circle, a function space can be obtained by raising the unit vector to all integer powers. This function space allows us to describe any function on the unit circle.
1.5 In this section, the anticommuting algebra in two dimensions is extended to three dimensions. It is shown that this allows the multiplication of vectors without the need to split it into an inner and an outer product.
1.6 In physics, the machinery of anticommuting algebra is often incorporated using the properties of matrices. Contrary to what one might expect, three-dimensional space is described in 2×2 and not 3×3 matrices. This reduction is possible since imaginary units give us an additional degree of freedom. This section introduces spinors, which form the basis of the matrices. Spinors are conceptually complicated since they are effectively the square root of the rotation axis.
1.7 In this section, the Pauli matrices are derived. The three 2×2 matrices effectively form a set of unit vectors, with the same properties as the algebraic anticommuting unit vectors.
1.8 The Pauli matrices can be directly related to imaginary units or bivectors that allow us to rotate vectors in three dimensions. The eigenvectors of the rotation matrices are related to the spinors described earlier.
1.9 The relationship between spinors and vectors is described in terms of the Pauli matrices. Additionally, it is shown how Pauli matrices can be described in terms of spinors.
1.10 The function space for rotation in Section 1.4 is adapted to describe one-dimensional translational systems and is then extended to three dimensions. Rotation was treated before translation, since translation is effectively treated like a rotation, which leads to some conceptual complexities that do not occur for rotation.

1.11 The function space from the preceding section allows the description of any arbitrary function in space. For complex exponentials, this procedure is generally called a Fourier transform. Since the function space consists of eigenfunctions of the momentum operator, it allows for the translation of functions in space.

1.12 The concepts for a homogeneous space are modified for a distorted and discrete space. An example of such a space is the periodic lattice of a solid. Here, we encounter a conceptual problem with momentum space. In free space, the translation in the x-, y-, or z-direction is directly related to the momentum or wavevector in the same direction. However, in a distorted space, this intuitive concept falls apart, and we have to rethink what constitutes an orthogonal space. This problem did not occur for rotation, where the rotation occurs in a plane, and the conserved quantity, angular momentum, is perpendicular to this plane, *i.e.* parallel to the axis of rotation.

1.13 The connection between obtaining the reciprocal/momentum space and linear algebra techniques such as matrix inversion is demonstrated. In many ways, these operations are related to vector division.

1.14 This section returns to rotations. Although the rotation of vectors is relatively straightforward, the function space for rotations is rather complex. The function space for the unit circle was obtained by taking integer powers of the unit vector. This is extended to three dimension by studying the unit vector in spherical polar coordinates and expressing it in spinors. To create higher order unit vectors, the spinors are taken as indistinguishable quantities.

1.15 The function space for rotations in three dimensions is derived by raising the unit vector to an integer arbitrary power. This leads to the function space of spherical harmonics and the associated unit vectors.

1.16 In order to perform operations on higher order unit vectors, the concepts of the Pauli matrices are extended to arbitrary rank. It is shown that the transformation properties of the angular momentum operators are equivalent to those of the unit vectors.

1.17 This section demonstrates that the use of operators on the functions leads to results equivalent to the manipulation of the unit vectors. This establishes the intimate connection between the operator and the transformation techniques.

1.1 The Exponential Function

Introduction. – A recapitulation of the exponential function might seem like a peculiar start of a quantum mechanics textbook. Although most readers will be familiar with the exponential function, it is important to understand the physical meaning of the exponential function as the repeated application of an infinitesimally small action on an object, particle, or any other kind of quantity.

The multiple applications of a particular action were first considered by Jacob Bernoulli in 1683 when studying compound interest. Given an interest rate x, the capital M increases as $M' = (1 + x)M$, whenever the interest is paid (let us say, once a year). However, if the same interest rate is divided over n periods, the capital increases as

$$M' = \left(1 + \frac{x}{n}\right)^n M. \tag{1.1}$$

Suppose the interest rate is 6% or $x = 0.06$. If the interest is paid once a year, then $M' = 1.06M$. If the interest is paid monthly ($n = 12$), then $M' = 1.06168M$ for an interest rate of 0.5% per month. If interest is paid daily, then $M' = 1.06183M$. So, the results are apparently not the same. Let us now for convenience assume that the yearly interest rate is 100% or $x = 1$. If it is paid at the end of the year, the capital is doubled, $M' = 2M$. However, if an interest rate of $x/n = 1/n$ is paid n times, the final capital is

$$M' = e_n M \quad \text{with} \quad e_n = \left(1 + \frac{1}{n}\right)^n, \tag{1.2}$$

which gives $e_1 = 2$ for $n = 1$. The value of e_n is plotted in Figure 1.1. This number is always larger than 2 since we also receive interest on the previous interest, which is known as compound interest. If the interest was paid monthly, we have $e_{12} = 2.61304$, which is definitely larger than a single interest payment. Figure 1.1 also shows that e_n approaches an asymptotic value, which can be determined to be the mathematical constant

$$e \equiv e_\infty = \lim_{n\to\infty}\left(1 + \frac{1}{n}\right)^n = 2.71828, \tag{1.3}$$

also known as Euler's constant. This is the multiplication factor if the interest was compounded continuously. For an arbitrary interest rate x, the continuous compound rate becomes

$$\lim_{n\to\infty}\left(1 + \frac{x}{n}\right)^n = \lim_{n'\to\infty}\left(1 + \frac{1}{n'}\right)^{n'x} = \left(\lim_{n'\to\infty}\left(1 + \frac{1}{n'}\right)^{n'}\right)^x = e^x, \tag{1.4}$$

taking $n' = n/x$. The above expression is the definition of the exponential function. Note that in the expression there is no need for n to be an integer. Although compound interest is a useful topic, for physics, the importance of the exponential function is the repeated application of an infinitesimally small change. This change could be many things: a translation, a rotation, propagation in time, growth, decay, etc.

The product on the left-hand side of Eq. (1.4) can also be expanded, for integer n, using Newton's binomial expansion

$$e^x = \lim_{n\to\infty}\sum_{k=0}^{n}\begin{pmatrix} n \\ k \end{pmatrix}\left(\frac{x}{n}\right)^k \cong \lim_{n\to\infty}\sum_{k=0}^{n}\frac{n!}{k!(n-k)!n^k}x^k. \tag{1.5}$$

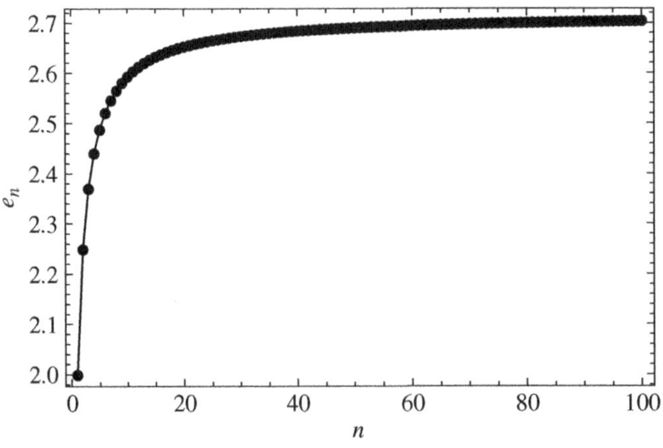

Figure 1.1 The value of e_n given in Eq. (1.2) as a function of n.

In the limit $n \to \infty$, $x/n \ll 1$, and the terms become smaller with increasing n. In that limit, the series becomes

$$e^x = \lim_{n\to\infty} \sum_{k=0}^{n} \frac{n(n-1)(n-2)\cdots(n-k+1)}{k!n^k} x^k = \sum_{k=0}^{\infty} \frac{x^k}{k!}, \tag{1.6}$$

assuming that only the terms with $k \ll n$ contribute to the series. Equation (1.6) expands the exponential function into a power series. The result can also be obtained using a Taylor series.

Conceptual summary. – In this section, we looked at the exponential function. The most important aspect is to view the exponential function as an infinitesimally small action repeated an infinite number of times. This is how objects move in space. This is also how particles interact with a field, giving rise to rotations and Lorentz boosts. In the following section, this approach will be applied to the rotation of a vector. A rotation over a finite angle is divided into an infinite number of infinitesimally small rotations. First, the vector is rotated over an infinitesimally small angle. Then, we take that vector and rotate it again over an infinitesimally small angle, and so on. The rotation over a finite angle is then given by an exponential.

Problems

1 Using Eq. (1.6), show that for an infinitesimally small change in time *dt* the increase in $N(t) = N_0 e^{\alpha t}$ is directly proportional to $N(t)$ and α.

1.2 Rotation in a Plane

Introduction. – The following two sections describe the rotation of a unit vector in two dimensions. This might appear disconnected from quantum mechanics where particles are described by wave functions and not by vectors. However, as we will see, we can construct the appropriate functions for free space by raising vectors to an arbitrary power. By studying the rotation of vectors, we discover several important properties of functions without the additional complexities of an infinitely large function space. Additionally, when we are only looking at rotations, there is no need to consider all the functions at once (we will also see this for atomic orbitals: a p orbital never turns into an s or a d orbital under rotation).

A unit vector therefore contains many of the necessary ingredients of a function space, albeit in its simplest form. It contains unit vectors and functions. The latter are usually called coefficients in a vector. A vector can be rotated in two different ways. One can manipulate the unit vectors, which will be discussed in this section, or one can perform an operation on the coefficients/functions, which is treated in the following section. For function spaces relevant to quantum mechanics, we encounter both techniques.

The starting point is a Cartesian coordinate system defined by two orthonormal unit vectors $\mathbf{e}_x$ and $\mathbf{e}_y$. A unit vector in an arbitrary direction, see Figure 1.2, can be written as

$$\hat{\mathbf{r}} = \hat{\mathbf{r}}(\varphi) = \cos\varphi\, \mathbf{e}_x + \sin\varphi\, \mathbf{e}_y, \tag{1.7}$$

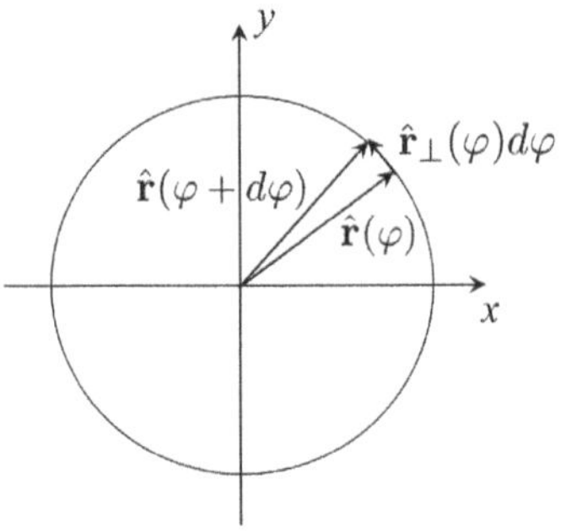

Figure 1.2 The change in the vector $\hat{\mathbf{r}}(\varphi)$ to $\hat{\mathbf{r}}(\varphi + d\varphi)$ when increasing the angle by a small change $d\varphi$. The distance between the points on the unit circle is given by the arc length $d\varphi$. The direction of the displacement is given by $\hat{\mathbf{r}}_\perp(\varphi)$, which is the unit vector $\hat{\mathbf{r}}(\varphi)$ rotated counterclockwise by 90°.

where the hat indicates a unit vector. Obviously, most readers are familiar with sines and cosines, but for the moment let us forget what trigonometric functions are, and assume that the $\cos\varphi$ and $\sin\varphi$ give the length of the projections onto the x- and y-axis, respectively, for a given angle φ between the unit vector $\hat{\mathbf{r}}$ and the x-axis. The norm $|\hat{\mathbf{r}}|$ of the vector is given by

$$|\hat{\mathbf{r}}|^2 = \cos^2\varphi + \sin^2\varphi = 1. \tag{1.8}$$

Therefore, all the unit vectors together describe a circle of radius 1 around the origin.

Let us take a closer look at the unit vectors. The norm of the unit vectors is 1. The product of two equal unit vectors is now given by

$$\mathbf{e}_i^2 = \mathbf{e}_i\mathbf{e}_i = 1 \text{ for } i = x, y. \tag{1.9}$$

Note that the product of two vectors is used and not the inner product. We come back to the distinction in more detail below.

We want to rotate the unit vector $\hat{\mathbf{r}}$. From Figure 1.2, we see that a rotation over a small angle is a displacement of the vector in a direction perpendicular to the vector. Any displacement parallel to $\hat{\mathbf{r}}$ changes the norm and is therefore not a rotation. We would now like to find a vector operation that causes a 90° counterclockwise rotation to obtain a vector perpendicular to $\hat{\mathbf{r}}$. Let us denote this operator as $\boldsymbol{\imath}_{yx}$. First, let us operate with $\boldsymbol{\imath}_{yx}$ on the unit vectors. A 90° counterclockwise rotation transforms $\mathbf{e}_x$ into $\mathbf{e}_y$,

$$\boldsymbol{\imath}_{yx}\mathbf{e}_x = \mathbf{e}_y \quad \Rightarrow \quad \boldsymbol{\imath}_{yx} = \mathbf{e}_y\mathbf{e}_x, \tag{1.10}$$

where the equation has been multiplied by $\mathbf{e}_x$ from the right and using the normalization $\mathbf{e}_x^2 = 1$. A 90° counterclockwise rotation transforms $\mathbf{e}_y$ into $-\mathbf{e}_x$, giving

$$\boldsymbol{\imath}_{yx}\mathbf{e}_y = -\mathbf{e}_x \quad \Rightarrow \quad \boldsymbol{\imath}_{yx} = -\mathbf{e}_x\mathbf{e}_y, \tag{1.11}$$

where the final equation is obtained by multiplying the first equation by $\mathbf{e}_y$ from the right. The quantity defining the rotation is therefore given by the product of two vectors, also known as a bivector,

$$\boldsymbol{\imath}_{yx} = \mathbf{e}_y\mathbf{e}_x = -\mathbf{e}_x\mathbf{e}_y. \tag{1.12}$$

However, this directly implies that vectors do not commute with each other. While scalars (numbers) commute with each other, *i.e.* for two scalars a and b, one has $ab = ba$, the unit vectors follow the anticommutative property,

$$\mathbf{e}_x\mathbf{e}_y + \mathbf{e}_y\mathbf{e}_x = 0. \tag{1.13}$$

The use of anticommuting algebraic unit vectors is known as Clifford or geometric algebra. Using the anticommutative property and the norm of the unit vectors allows us to find the norm of $\boldsymbol{i}_{yx}$

$$\boldsymbol{i}_{yx}^2 = \mathbf{e}_y\mathbf{e}_x\mathbf{e}_y\mathbf{e}_x = -\mathbf{e}_y\mathbf{e}_y\mathbf{e}_x\mathbf{e}_x = -1. \tag{1.14}$$

This implies that $\boldsymbol{i}_{yx} = \sqrt{-1}$ is an imaginary unit. However, there is nothing mysterious about this, since two successive 90° counterclockwise rotations equal a 180° rotation, which simply changes the direction of a vector.

Obviously, this is reminiscent of the imaginary scalar i. In the complex plane, the axes are determined by 1 and i. The multiplication $i \times 1 = i$ is a transition from the real to the imaginary axis. Multiplying again $i \times i = -1$ brings one to the negative real axis. Note that the real and imaginary axes are not essentially different from the x and y axes. Since i is a scalar, it simply commutes with the vectors, for example $i\mathbf{e}_x = \mathbf{e}_x i$. As we will see, in many applications in mathematics and physics, one often avoids the use of anticommuting quantities, or the effects of anticommutation are incorporated by the use of matrix multiplication. In that case, the bivector $\boldsymbol{i}_{yx}$ is replaced by the complex scalar i. However, this needs to be done with some care, since there are two bivectors $\boldsymbol{i}_{yx}$ and $\boldsymbol{i}_{xy} = \mathbf{e}_x\mathbf{e}_y = -\boldsymbol{i}_{yx}$. The imaginary unit $\boldsymbol{i}_{xy}$ causes the same rotation when multiplying the vector from the right $\mathbf{e}_x\boldsymbol{i}_{xy} = \mathbf{e}_y$ and $\mathbf{e}_y\boldsymbol{i}_{xy} = -\mathbf{e}_x$. For vectors in real space, this is often the convention. Therefore, it is the bivector $\boldsymbol{i}_{xy}$ that should be replaced by the imaginary scalar i. In physics, however, there is a strong preference of having operators act to the right. One of the reasons is that derivatives generally act to the right (obviously, this can also be overcome by defining operators such as $f\overleftarrow{\partial}_x = df/dx$ and $\overrightarrow{\partial}_x f = df/dx$. However, we will not go this route). For a bivector $\boldsymbol{i}_{yx}$ acting to the right, the subscript needs to be read from the right to the left, *i.e.* it is a rotation from the x-axis to the y-axis. For a bivector operating to the left, the subscript needs to be read from the left to the right, which makes $\boldsymbol{i}_{xy}$ also a counterclockwise rotation from the x- to the y-axis. However, if we act with $\boldsymbol{i}_{xy}$ to the right, then it is a clockwise rotation from the y- to the x-axis, with $\boldsymbol{i}_{xy}\mathbf{e}_x = -\mathbf{e}_y$ and $\boldsymbol{i}_{xy}\mathbf{e}_y = \mathbf{e}_x$.

The unit circle described by all the unit vectors $\hat{\mathbf{r}}$ can also be obtained by rotation. Rotating a unit vector $\hat{\mathbf{r}}(\varphi)$ over a small angle $d\varphi$ gives a new unit vector

$$\hat{\mathbf{r}}(\varphi + d\varphi) = \hat{\mathbf{r}}(\varphi) + \hat{\mathbf{r}}_\perp(\varphi)d\varphi, \tag{1.15}$$

see Figure 1.2. The angle $d\varphi$ must be expressed in radians and is essentially the length of the path along the unit circle. The unit vector $\hat{\mathbf{r}}_\perp(\varphi)$ is perpendicular to $\hat{\mathbf{r}}$ and can be obtained by multiplying with $\boldsymbol{i}_{yx}$,

$$\hat{\mathbf{r}}_\perp(\varphi) = \boldsymbol{i}_{yx}\mathbf{r}(\varphi) = -\sin\varphi\,\mathbf{e}_x + \cos\varphi\,\mathbf{e}_y. \tag{1.16}$$

It is easily checked that this vector is indeed perpendicular to $\mathbf{r}(\varphi)$. Therefore, we can also express Eq. (1.15) as

$$\hat{\mathbf{r}}(\varphi + d\varphi) = (1 + \boldsymbol{i}_{yx}d\varphi)\hat{\mathbf{r}}(\varphi). \tag{1.17}$$

A rotation over an arbitrary angle $\Delta\varphi$ can be obtained by n successive rotations over an infinitesimally small angle $\Delta\varphi/n$ with $n \to \infty$. This gives

$$\hat{\mathbf{r}}(\varphi + \Delta\varphi) = \lim_{n\to\infty}\left(1 + \frac{\boldsymbol{i}_{yx}\Delta\varphi}{n}\right)^n \hat{\mathbf{r}}(\varphi). \tag{1.18}$$

However, this expression is comparable to Eq. (1.4) and can be rewritten as

$$\hat{\mathbf{r}}(\varphi + \Delta\varphi) = e^{\boldsymbol{i}_{yx}\Delta\varphi}\hat{\mathbf{r}}(\varphi). \tag{1.19}$$

The unit vectors $\hat{\mathbf{r}}(\varphi)$ can now be expressed as rotations with respect to a reference axis. Conventionally, in polar coordinates, the angle φ is defined with respect to the x-axis.

$$\hat{\boldsymbol{r}} = \hat{\boldsymbol{r}}(\varphi) = e^{\boldsymbol{i}_{yx}\varphi} \quad \Rightarrow \quad \hat{\mathbf{r}} = \hat{\boldsymbol{r}}\mathbf{e}_x. \tag{1.20}$$

Note the difference in notation for the unit vector $\hat{\mathbf{r}}$ and the complex exponential $\hat{\boldsymbol{r}}$. The vector $\hat{\mathbf{r}}$ can be viewed as a vector in a fixed basis $\mathbf{e}_x$ and $\mathbf{e}_y$. The exponential $\hat{\boldsymbol{r}}$ is a relative vector. We only know its direction once the reference axis is defined. The relative vector $\hat{\boldsymbol{r}}$ is defined in an orthogonal basis of vector quantities 1 and $\boldsymbol{i}_{yx}$, also known as the complex plane. This basis has the nice property that its unit vectors commute. Therefore, while the vectors $\hat{\mathbf{r}}$ in the fixed basis anticommute, the relative vectors $\hat{\boldsymbol{r}}$ commute. However, while for $\hat{\mathbf{r}}$, the unit vectors are $\mathbf{e}_i$ with $\mathbf{e}_i^2 = 1$ for $i = x, y$, for $\hat{\boldsymbol{r}}$, the unit vectors are 1 and $\boldsymbol{i}_{yx}$ with norms 1 and -1, respectively. Therefore, the unit vectors are not equivalent. A similar notation is used for $\hat{\boldsymbol{r}}$ and $\hat{\mathbf{r}}$, because both generally refer to the same quantity, since the reference axis is usually fixed.

We have now found two different ways to express a unit vector in two dimensions, see Eqs. (1.7) and (1.20). Obviously, these results need to be equivalent to each other. Using the expansion in Eq. (1.6) and taking $\boldsymbol{i}_{yx}\varphi$ as the argument, the complex exponential can be written as

$$e^{\boldsymbol{i}_{yx}\varphi} = 1 + \boldsymbol{i}_{yx}\varphi + \frac{1}{2!}(\boldsymbol{i}_{yx}\varphi)^2 + \frac{1}{3!}(\boldsymbol{i}_{yx}\varphi)^3 + \cdots. \tag{1.21}$$

The powers of the imaginary unit can be rewritten as

$$(\boldsymbol{i}_{yx}\varphi)^2 = -\varphi^2 \text{ and } (\boldsymbol{i}_{yx}\varphi)^3 = -\boldsymbol{i}_{yx}\varphi^3, \text{ etc.} \tag{1.22}$$

The exponential can then be split into a real and imaginary part

$$\begin{aligned} e^{\boldsymbol{i}_{yx}\varphi} &= \left(1 - \frac{\varphi^2}{2!} + \frac{\varphi^4}{4!} + \cdots\right) + \boldsymbol{i}_{yx}\left(\varphi - \frac{\varphi^3}{3!} + \frac{\varphi^5}{5!} + \cdots\right) \\ &\equiv \cos\varphi + \boldsymbol{i}_{yx}\sin\varphi, \end{aligned} \tag{1.23}$$

which is Euler's formula. The two series effectively define the cosine and sine functions,

$$\begin{aligned} \cos\varphi &= 1 - \frac{\varphi^2}{2!} + \frac{\varphi^4}{4!} + \cdots = \sum_{n=0}^{\infty}\frac{(-1)^n}{(2n)!}\varphi^{2n}, \\ \sin\varphi &= \varphi - \frac{\varphi^3}{3!} + \frac{\varphi^5}{5!} + \cdots = \sum_{n=0}^{\infty}\frac{(-1)^n}{(2n+1)!}\varphi^{2n+1}. \end{aligned} \tag{1.24}$$

The cosine and sine contain all the even and odd powers, respectively. Therefore, changing the sign of the angle makes the sine negative, $\sin(-\varphi) = -\sin\varphi$, but leaves the value of the cosine unchanged, $\cos(-\varphi) = \cos\varphi$. We can also define the relative vector with a negative angle

$$\hat{\boldsymbol{r}}^* = (e^{\boldsymbol{i}_{yx}\varphi})^* = e^{-\boldsymbol{i}_{yx}\varphi} = \cos\varphi - \boldsymbol{i}_{yx}\sin\varphi. \tag{1.25}$$

This is also known as the conjugate (indicated by the asterisk *). Instead of changing the sign of φ, we can also think of this as a substitution $\boldsymbol{i}_{yx} \rightarrow \boldsymbol{i}_{xy} = -\boldsymbol{i}_{yx}$. This corresponds to changing the direction of the rotation since $\boldsymbol{i}_{xy}$ corresponds to a clockwise rotation over 90° since $\boldsymbol{i}_{xy}\mathbf{e}_x = -\mathbf{e}_y$ and $\boldsymbol{i}_{xy}\mathbf{e}_y = \mathbf{e}_x$. Note that this is not a simple sign change, since $\hat{\boldsymbol{r}}^*$ is a function that is independent of $\hat{\boldsymbol{r}}$. So, instead of having two functions $\cos\varphi$ and $\sin\varphi$ to describe the unit circle, we can also use $e^{\boldsymbol{i}_{yx}\varphi}$ and $e^{-\boldsymbol{i}_{yx}\varphi}$. Using the conjugate of the exponential, the cosine and sine functions can also be written as

$$\cos\varphi = \frac{1}{2}(e^{\boldsymbol{i}_{yx}\varphi} + e^{-\boldsymbol{i}_{yx}\varphi}) \quad \text{and} \quad \sin\varphi = \frac{1}{2\boldsymbol{i}_{yx}}(e^{\boldsymbol{i}_{yx}\varphi} - e^{-\boldsymbol{i}_{yx}\varphi}). \tag{1.26}$$

A complex basis requires that one takes properly into account that one of the axes is imaginary. For example, calculating the norm of a vector in a complex basis involves taking the conjugate

$$\hat{\mathbf{r}}^2 = \hat{\boldsymbol{r}}\mathbf{e}_x\hat{\boldsymbol{r}}\mathbf{e}_x = \hat{\boldsymbol{r}}\hat{\boldsymbol{r}}^*\mathbf{e}_x\mathbf{e}_x = \hat{\boldsymbol{r}}\hat{\boldsymbol{r}}^* = e^{\boldsymbol{i}_{yx}\varphi}e^{-\boldsymbol{i}_{yx}\varphi} = 1 \equiv |\hat{\boldsymbol{r}}|^2, \tag{1.27}$$

using that $\mathbf{e}_x\boldsymbol{i}_{yx} = \mathbf{e}_x\mathbf{e}_y\mathbf{e}_x = -\mathbf{e}_y\mathbf{e}_x\mathbf{e}_x = -\boldsymbol{i}_{yx}\mathbf{e}_x$, which changes the sign in the exponential.

We have now obtained all possible combinations of the unit vectors $\mathbf{e}_x$ and $\mathbf{e}_y$ in two dimensions

$$(1+\mathbf{e}_x)(1+\mathbf{e}_y) = 1 + \mathbf{e}_x + \mathbf{e}_y + \mathbf{e}_x\mathbf{e}_y, \tag{1.28}$$

or in table form

multivector	rank	number	name
1	0	1	scalar
$\mathbf{e}_x, \mathbf{e}_y$	1	2	vector
$\boldsymbol{i}_{yx} \equiv \mathbf{e}_y\mathbf{e}_x$	2	1	bivector = pseudoscalar.

(1.29)

The rank indicates how many unit vectors appear in the multivector. The total number of multivectors is $2^d = 2^2 = 4$, where d is the dimension. Note that the number of multivectors increases to 8 and 16 for three-dimensional space and spacetime, respectively.

Example 1.1 ***Addition of Rotations***

Let us consider $\hat{\boldsymbol{r}} = e^{\boldsymbol{i}_{yx}\varphi}$ and $\hat{\boldsymbol{r}}' = e^{\boldsymbol{i}_{yx}\varphi'}$. Write the consecutive operation of $\hat{\boldsymbol{r}}$ and $\hat{\boldsymbol{r}}'$ in trigonometric functions in terms of φ and φ' and in terms of $\varphi + \varphi'$. Use this result to derive the trigonometric angle addition formula.

Solution

The great advantage of a complex basis is the ease of performing rotations, since each relative vector $\boldsymbol{r}$ is essentially a rotation. The angle with the x-axis can easily be changed by multiplication of complex exponentials

$$\hat{\boldsymbol{r}}'\hat{\boldsymbol{r}} = e^{\boldsymbol{i}_{yx}\varphi'}e^{\boldsymbol{i}_{yx}\varphi} = e^{\boldsymbol{i}_{yx}(\varphi'+\varphi)} = \cos(\varphi'+\varphi) + \boldsymbol{i}_{yx}\sin(\varphi'+\varphi). \tag{1.30}$$

Note that this provides a way to derive trigonometric addition formulas since the product can also be expanded as

$$\begin{aligned}\hat{\boldsymbol{r}}'\hat{\boldsymbol{r}} &= (\cos\varphi' + \boldsymbol{i}_{yx}\sin\varphi')(\cos\varphi + \boldsymbol{i}_{yx}\sin\varphi) \\ &= \cos\varphi'\cos\varphi - \sin\varphi'\sin\varphi + \boldsymbol{i}_{yx}(\sin\varphi'\cos\varphi + \cos\varphi'\sin\varphi),\end{aligned} \tag{1.31}$$

giving the identities

$$\begin{aligned}&\cos\varphi'\cos\varphi \mp \sin\varphi'\sin\varphi = \cos(\varphi' \pm \varphi), \\ &\sin\varphi'\cos\varphi \pm \cos\varphi'\sin\varphi = \sin(\varphi' \pm \varphi),\end{aligned} \tag{1.32}$$

where the angle difference identity can be obtained in the same fashion or by substituting $\varphi \to -\varphi$. For $\varphi' = \varphi$, the angle addition formulas become

$$\cos^2\varphi - \sin^2\varphi = \cos 2\varphi \quad \text{and} \quad 2\sin\varphi\cos\varphi = \sin 2\varphi \tag{1.33}$$

Conceptual summary. – In this section, we looked at the rotation of a vector in two dimensions. A rotation is the motion of a vector without changing its length. This can be achieved by adding a vector perpendicular to it. A perpendicular vector can be obtained by multiplying it with the imaginary unit $\boldsymbol{i}_{yx} = \mathbf{e}_y\mathbf{e}_x$. The quantity $\boldsymbol{i}_{yx}$ is a bivector, *i.e.* the product of two unit vectors. The total rotation is expressed as an exponential $e^{\boldsymbol{i}_{yx}\varphi}$. In order for the rotation to work, unit vectors need to be treated as anticommuting algebraic quantities (sometimes also called quaternions). This is not the way vectors are generally taught in introductory courses. However, it is conceptually a lot closer to quantum mechanics. We therefore have to dispense with the following ideas that you might have:

- Vectors cannot be simply multiplied, and their product always needs to be split into an inner product $\mathbf{r}' \cdot \mathbf{r}$ (a scalar) and an outer product $\mathbf{r}' \times \mathbf{r}$ (a vector). However, there is nothing wrong with the product $\mathbf{r}'\mathbf{r}$ as long as we take into account the anticommuting behavior of the unit vectors.
- There are only scalars and vectors. In this section, however, we saw that the rotation was directly related to the bivector $\mathbf{e}_y\mathbf{e}_x$. This is often avoided, and one focuses on the rotation axis (here, this would be the z-axis, which is somewhat problematic since the system is two-dimensional). However, it is the bivector that actually performs the rotation and not the rotation axis.
- Scalars and vectors do not mix. Although, an expression such as $5 + \mathbf{r}$ might look uncomfortable, there is nothing wrong with it.
- There is only one imaginary unit: the pseudoscalar i with $i^2 = -1$. In higher dimensions, we encounter several quantities that square to -1. In this section, we found the bivector $\boldsymbol{i}_{yx} = \mathbf{e}_y\mathbf{e}_x$, which is of course directly related to i. Note that there is nothing peculiar about squaring to -1. The action of $\boldsymbol{i}_{yx}$ is a 90° counterclockwise rotation. Performing this twice gives a 180° rotation, which is equivalent to an inversion of the vector.

In quantum mechanics, the physics of anticommuting unit vectors is generally incorporated using matrices, such as the Pauli matrices for space and the γ matrices for spacetime. Although computationally convenient, the use of matrices is often less intuitive.

Problems

1 Show that the norm of a unit vector can also be written as the product of the unit vector with itself, *i.e.* $|\hat{\mathbf{r}}|^2 \equiv \hat{\mathbf{r}}\hat{\mathbf{r}}$.

2 Show that $\boldsymbol{i}_{yx}$ and $\mathbf{e}_i$ with $i = x, y$ anticommute.

3 Derive the commutation relation between $\mathbf{e}_i$ and $e^{\boldsymbol{i}_{yx}\varphi}$.

4 Evaluate the product $\mathbf{r}'\mathbf{r}$ with $\mathbf{r} = r(\cos\varphi\ \mathbf{e}_x + \sin\varphi\ \mathbf{e}_y)$.

5 The properties of anticommuting unit vectors can also be expressed in terms of (Pauli) matrices, the approach typically being followed in quantum mechanics. Let us consider the matrices

$$\mathbf{e}_z = \begin{pmatrix} 1 & 0 \\ 0 & -1 \end{pmatrix} \text{ and } \mathbf{e}_x = \begin{pmatrix} 0 & 1 \\ 1 & 0 \end{pmatrix}, \tag{1.34}$$

representing unit vectors in the zx-plane (avoiding the complex matrix conventionally associated with the y-direction).

a) Show that $\mathbf{e}_z$ and $\mathbf{e}_x$ anticommute.

b) Show that the product $\mathbf{e}_z\mathbf{e}_x$ is an imaginary unit, *i.e.* it squares to -1 (expressed in matrix form as $-\mathbb{1}_2$, where $\mathbb{1}_2$ is the 2×2 identity matrix).

6 Although $e^{\boldsymbol{i}_{yx}\varphi}\hat{\mathbf{r}}$ causes a rotation of the unit vector, it also adds a complex exponential when working on a scalar $e^{\boldsymbol{i}_{yx}\varphi} \times 1$. Show that the transformation $e^{\boldsymbol{i}_{yx}\frac{\varphi}{2}}\boldsymbol{A}e^{-\boldsymbol{i}_{yx}\frac{\varphi}{2}}$ still causes a rotation when $\boldsymbol{A} = \hat{\mathbf{r}}$ is a vector but does not affect a scalar $\boldsymbol{A} = 1$.

1.3 Calculus and Operators

Introduction. – In the previous section, a rotation was obtained by manipulation of the unit vectors with the imaginary unit $\boldsymbol{i}_{yx}$. However, there are two aspects to the unit vector in Eq. (1.7): the unit vectors $\mathbf{e}_x$ and $\mathbf{e}_y$ and the coefficients $\cos\varphi$ and $\sin\varphi$. The rotation can also be obtained by performing an operation on the coefficients/functions. These concepts are crucial for quantum mechanics, which is dominated by operator techniques working on wave functions.

The result of applying the imaginary init $\boldsymbol{i}_{yx}$ on $\hat{\mathbf{r}}(\varphi)$ was a vector perpendicular to it. Rewriting Eq. (1.15) gives for the change in the unit vector under rotation

$$\hat{\mathbf{r}}_\perp(\varphi)d\varphi = \hat{\mathbf{r}}(\varphi + d\varphi) - \hat{\mathbf{r}}(\varphi). \tag{1.35}$$

This expression is only valid in the limit that $d\varphi$ approaches zero, giving

$$\hat{\mathbf{r}}_\perp(\varphi) = \lim_{d\varphi\to 0}\frac{\mathbf{r}(\varphi + d\varphi) - \mathbf{r}(\varphi)}{d\varphi} = \frac{d\mathbf{r}(\varphi)}{d\varphi}. \tag{1.36}$$

This expression is Newton's difference quotient, albeit in vector format. This limit is the definition of the derivative given on the right-hand side. However, we are interested in the components, which are scalar quantities. Equation (1.24) shows that the coefficients can be written as a power series in φ^n. For a particular term in the series, the difference quotient is

$$\begin{aligned}\frac{d\varphi^n}{d\varphi} &= \lim_{d\varphi\to 0}\frac{(\varphi + d\varphi)^n - \varphi^n}{d\varphi} \\ &= \lim_{d\varphi\to 0}\frac{\varphi^n + n\varphi^{n-1}d\varphi + \cdots - \varphi^n}{d\varphi} = n\varphi^{n-1},\end{aligned} \tag{1.37}$$

neglecting all terms $(d\varphi)^n$ with $n > 1$ in Newton's binomial expansion. Obviously, the result is the well-known derivative of φ^n. This result can be used to obtain the derivatives of the coefficients. From Eq. (1.24),

$$\frac{d\sin\varphi}{d\varphi} = \sum_{n=0}^{\infty}\frac{(-1)^n}{(2n+1)!}(2n+1)\varphi^{2n} = \sum_{n=0}^{\infty}\frac{(-1)^n}{(2n)!}\varphi^{2n} = \cos\varphi, \tag{1.38}$$

and for the cosine

$$\frac{d\cos\varphi}{d\varphi} = \sum_{n=1}^{\infty}\frac{(-1)^n}{(2n-1)!}\varphi^{2n-1} = \sum_{n'=0}^{\infty}\frac{(-1)^{n'+1}}{(2n'+1)!}\varphi^{2n'+1} = -\sin\varphi, \tag{1.39}$$

substituting $n = n' + 1$. This gives the derivatives of the sine and the cosine. Note that the effect of the derivative is comparable to $\boldsymbol{i}_{yx}\mathbf{e}_i$: the coefficients are interchanged and one obtains a minus sign.

In Eq. (1.16), a 90° counterclockwise rotation was obtained by multiplying the vector by the imaginary unit $\hat{\mathbf{r}}_\perp(\varphi) = \boldsymbol{i}_{yx}\mathbf{r}$. The same result can be obtained using the operator

$$\hat{\imath}_{yx} = \frac{d}{d\varphi}, \tag{1.40}$$

where the hat here indicates that it is an operator (unfortunately, convention uses hats for both operators and unit vectors). This gives

$$\hat{\mathbf{r}}_\perp(\varphi) = \hat{\imath}_{yx}\mathbf{r} = (\hat{\imath}_{yx}\cos\varphi)\mathbf{e}_x + (\hat{\imath}_{yx}\sin\varphi)\mathbf{e}_y = -\sin\varphi\mathbf{e}_x + \cos\varphi\mathbf{e}_y, \tag{1.41}$$

using the derivatives obtained above. This gives the same result as $\boldsymbol{i}_{yx}\mathbf{r}$. This can also be expressed in terms of the complex exponential in Eq. (1.20). The derivative is then given by

$$\hat{\imath}_{yx}\hat{\mathbf{r}} = \frac{d\hat{\mathbf{r}}}{d\varphi} = \frac{d}{d\varphi}(e^{\boldsymbol{i}_{yx}\varphi}\mathbf{e}_x) = \boldsymbol{i}_{yx}e^{\boldsymbol{i}_{yx}\varphi}\mathbf{e}_x = \boldsymbol{i}_{yx}\hat{\mathbf{r}}. \tag{1.42}$$

Therefore, the operator $\hat{\imath}_{yx} = d/d\varphi$ performs the same operation on a unit vector as the imaginary unit $\boldsymbol{i}_{yx} = \mathbf{e}_y\mathbf{e}_x$. The above could also be directly derived from the power series of the exponential in Eq. (1.21). Therefore, while the components $\cos\varphi$ and $\sin\varphi$ are interchanged under the derivative with respect to φ, the unit vector $\hat{\mathbf{r}}$ also returns to itself, albeit with an imaginary factor.

The operator $\hat{\imath}_{yx}$ is generally not used in physics. This is a result of the preference of eigenfunctions and eigenvectors, *i.e.* functions and vectors that remain unchanged under a certain operation apart from a multiplication factor. Additionally, in physics, one generally requires the multiplication factor, known as the eigenvalue, to be real for physical operators. Therefore, in Eq. (1.42), although the unit vector $\hat{\mathbf{r}}$ is an eigenfunction, the operator $\hat{\imath}_{yx} = d/d\varphi$ is not considered a proper physical operator since its eigenvalue $\boldsymbol{i}_{yx}$ is not a real scalar. However, although a rotation over 90° seems like a perfectly normal operation, no vector in a plane is unchanged under a 90° rotation (with the exception of $\mathbf{r} = 0$). So, $\hat{\imath}_{yx}$ has no eigenvectors with real eigenvalues. Therefore, a different operator is used, which mixes the operations on vectors and coefficients. A clockwise rotation simply produces the vector opposite to that of an anticlockwise rotation. This operation is therefore given by $-\boldsymbol{i}_{yx}\mathbf{r}$ or $-\hat{\imath}_{yx}\mathbf{r}$. Therefore, to obtain an operator that produces eigenvectors, the product is taken of a counterclockwise and clockwise rotation

$$\hat{L} = (-\boldsymbol{i}_{yx})\hat{\imath}_{yx} = -\boldsymbol{i}_{yx}\frac{d}{d\varphi}. \tag{1.43}$$

Operating this on a vector gives

$$\hat{L}\hat{\mathbf{r}} = -\boldsymbol{i}_{yx}\hat{\imath}_{yx}\,\hat{\mathbf{r}} = -\boldsymbol{i}_{yx}\hat{\mathbf{r}}_\perp = \hat{\mathbf{r}}, \tag{1.44}$$

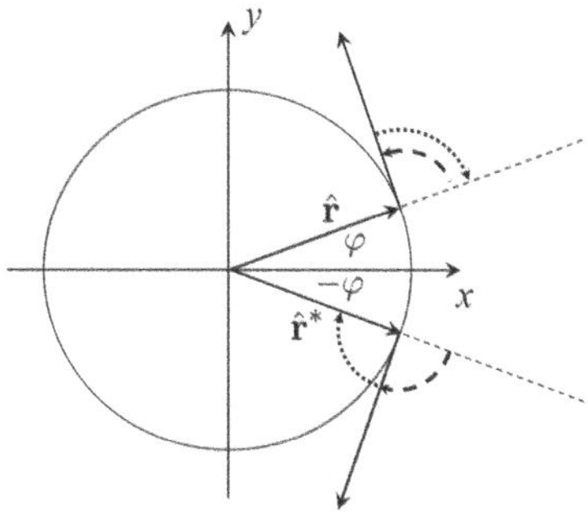

Figure 1.3 The effect of the operation of $-\boldsymbol{\imath}_{yx}\hat{\imath}_{yx}$ on the unit vector $\hat{\mathbf{r}}$ and its conjugate. Applying $\hat{\imath}_{yx}$ on $\hat{\mathbf{r}}$ causes a 90° counterclockwise rotation (dashed arrow); the effect of $-\boldsymbol{\imath}_{yx}$ is a 90° clockwise rotation (dotted arrow) canceling the effect of $\hat{\imath}_{yx}$. Applying $\hat{\imath}_{yx}$ on $\hat{\mathbf{r}}^*$, on the other hand, causes a 90° clockwise rotation. Subsequent application of $-\boldsymbol{\imath}_{yx}$ produces another 90° clockwise rotation, leading to the vector $-\hat{\mathbf{r}}^*$.

using Eq. (1.42). Obviously, this result is not really surprising since a 90° counterclockwise rotation followed by a 90° clockwise rotation should leave a vector unchanged, see Figure 1.3. Another way of thinking about the same operation is to write it as $(-1)\boldsymbol{\imath}_{yx}\hat{\imath}_{yx}$. These are two 90° counterclockwise rotations, producing a 180° rotation, which changes the direction of the vector. Multiplying by -1 again leaves the vector unchanged. Those with some familiarity to quantum mechanics might recognize that the operator $\hat{L}$ is equivalent to the z component of the quantum angular momentum operator in spherical polar coordinates (with $\hbar \to 1$ and $\boldsymbol{\imath}_{yx}$ instead of i).

However, this is not the only eigenvector of $\hat{L}$. In Eq. (1.25), the conjugate vector was introduced. Operating with $\hat{\imath}_{yx}$ on $\hat{\mathbf{r}}^* \equiv \hat{\boldsymbol{r}}^*\mathbf{e}_x = \hat{\mathbf{r}}(-\varphi)$ gives

$$\hat{\imath}_{yx}\hat{\mathbf{r}}^* = \frac{d}{d\varphi}(\cos\varphi\mathbf{e}_x - \sin\varphi\mathbf{e}_y) = -\sin\varphi\mathbf{e}_x - \cos\varphi\mathbf{e}_y. \tag{1.45}$$

This is a rotation in the clockwise direction, see Figure 1.3. This can be understood by noting that the derivative with respect to φ produces a vector that indicates the direction where the vector is going when changing the argument. Since $\hat{\mathbf{r}}^*(\varphi) = \hat{\mathbf{r}}(-\varphi)$, this vector rotates in the clockwise direction when increasing the value of φ. The next operation $-\boldsymbol{\imath}_{yx}$ again produces a 90° clockwise rotation regardless of the vector it is working on, giving the total result, see Figure 1.3,

$$\hat{L}\hat{\mathbf{r}}^* = -\boldsymbol{\imath}_{yx}\hat{\imath}_{yx}\,\hat{\mathbf{r}}^* = -\hat{\mathbf{r}}^*. \tag{1.46}$$

So, we have found a second vector that is an eigenvector of $\hat{L}$, although this time with an eigenvalue -1.

It is not necessary to combine $\boldsymbol{\imath}_{yx}$ and $\hat{\imath}_{yx}$ to obtain these solutions. One can also define the operator purely in terms of derivatives

$$\hat{L} = -\hat{\imath}_{yx}^2 = -\frac{d^2}{d\varphi^2}. \tag{1.47}$$

The product $\hat{\imath}_{yx}^2$ corresponds to two 90° rotations. Since this equals a 180° rotation, it does not matter whether the rotations are clockwise or counterclockwise. Since $\hat{\imath}_{yx}^2$ changes the direction of the vector, the operation $-\hat{\imath}_{yx}^2$ should leave the vector unchanged, *i.e.*

$$\hat{L}\hat{\boldsymbol{r}} = \hat{\boldsymbol{r}}, \tag{1.48}$$

leaving out the reference vector $\mathbf{e}_x$. This leads to the differential equation

$$\frac{d^2\hat{\boldsymbol{r}}}{d\varphi^2} + \hat{\boldsymbol{r}} = 0. \tag{1.49}$$

This is essentially a wave equation. This equation can be rewritten as

$$\left(\frac{d}{d\varphi} + \boldsymbol{i}_{yx}\right)\left(\frac{d}{d\varphi} - \boldsymbol{i}_{yx}\right)\hat{\boldsymbol{r}} = 0. \tag{1.50}$$

The solutions are again $\hat{\boldsymbol{r}} = e^{\pm \boldsymbol{i}_{yx}\varphi}$.

The third way to combine 90° counterclockwise and clockwise rotations is using only vector operators. This gives $\boldsymbol{L} = -\boldsymbol{i}_{yx}^2 = 1$. However, this is the identity operator and does not provide a way to find expressions for the coefficients.

Complex basis. – In the preceding paragraph, we have found that there are two complex solutions $\hat{\boldsymbol{r}} = e^{\pm \boldsymbol{i}_{yx}\varphi}$ that are eigenfunctions of the angular momentum operator $\hat{L}$. We want to find a pair of unit vectors for these functions. For the Cartesian basis, the unit vectors and the coefficients behave in the same fashion under rotation. The bivector $\boldsymbol{i}_{yx}$ interchanges the unit vectors $\mathbf{e}_x$ and $\mathbf{e}_y$, and one obtains a negative sign. The operator $\hat{\imath}_{yx} = d/d\varphi$ also interchanges the coefficients $\cos\varphi$ and $\sin\varphi$, and one obtains a negative sign. This is obviously not a coincidence, since $\boldsymbol{i}_{yx}$ and $\hat{\imath}_{yx}$ were both designed to produce a 90° counterclockwise rotation of the unit vector.

We now want to find unit vectors that behave in the same fashion under rotation as the functions $\hat{\boldsymbol{r}} = e^{\pm \boldsymbol{i}_{yx}\varphi}$. However, we are running into a problem since the complex exponentials are eigenfunctions of $\hat{L}$, and there are no vectors in two dimensions that are eigenvectors of rotation (except for special angles, such as 0° and 180°). This is because the conserved quantity for rotation is the rotation axis that lies out of the plane of rotation. The proper eigenfunctions for rotation are spinors, which are related to the rotation axis and also lie outside of the plane of rotation. Spinors will be discussed in Section 1.6. Surprisingly, there is a way to obtain conserved quantities that lie in the plane of rotation, namely by the introduction of complex vectors. Although the solutions are somewhat abstract, they are very commonly used.

Let us associate the complex exponentials with the unit vectors $\mathbf{e}_{\pm 1}$ and expand the complex exponentials

$$\hat{\mathbf{r}} = \frac{e^{-i\varphi}}{\sqrt{2}}\mathbf{e}_1 + \frac{e^{i\varphi}}{\sqrt{2}}\mathbf{e}_{-1} = \frac{1}{\sqrt{2}}(\mathbf{e}_1 + \mathbf{e}_{-1})\cos\varphi + \frac{i}{\sqrt{2}}(-\mathbf{e}_1 + \mathbf{e}_{-1})\sin\varphi, \tag{1.51}$$

where the square roots enforce proper normalization. Note that the complex scalar i is used instead of $\boldsymbol{i}_{yx}$. Since the above expression has to be equal to $\hat{\mathbf{r}}$, this implies

$$\mathbf{e}_x = \frac{1}{\sqrt{2}}(\mathbf{e}_1 + \mathbf{e}_{-1}) \quad \text{and} \quad \mathbf{e}_y = \frac{i}{\sqrt{2}}(-\mathbf{e}_1 + \mathbf{e}_{-1}). \tag{1.52}$$

The unit vectors in front of the complex exponentials are therefore the complex vectors

$$\mathbf{e}_{\pm 1} = \frac{1}{\sqrt{2}}(\mathbf{e}_x \pm i\mathbf{e}_y). \tag{1.53}$$

For the experts, no Condon–Shortley phase factors have been introduced which will cause additional sign changes. One can visualize $\mathbf{e}_{\pm 1}$ as vectors rotating in the xy-plane in the counterclockwise and clockwise direction, respectively, which looks comparable to the complex exponentials $e^{\pm i\varphi}$.

However, the obtained unit vectors are peculiar when we try to interpret the imaginary unit i as a 90° rotation. Substituting $i \rightarrow \boldsymbol{\imath}_{xy}$ gives $\mathbf{e}_1 = \sqrt{2}\mathbf{e}_x$ and $\mathbf{e}_{-1} = 0$. This gives $\hat{\mathbf{r}} = e^{-\boldsymbol{\imath}_{xy}\varphi}\mathbf{e}_x = e^{\boldsymbol{\imath}_{yx}\varphi}\mathbf{e}_x$. This is still correct, but we no longer have two unit vectors. These complex unit vectors are used for two reasons. First, even though spinors, which are related to the axis of rotation, make more sense as eigenvectors of the rotation, they are also not entirely intuitive from a physical point of view either. Second, even though the complex vectors might look a little weird, they have the same properties under rotation as the spinors, and this is often the only thing we care about when it comes to unit vectors. For example, they are eigenvectors of the imaginary unit

$$\boldsymbol{\imath}_{yx}\mathbf{e}_{\pm 1} = \frac{1}{\sqrt{2}}(\mathbf{e}_y \mp i\mathbf{e}_x) = \mp i\frac{1}{\sqrt{2}}(\mathbf{e}_x \pm i\mathbf{e}_y) = \mp i\mathbf{e}_{\pm 1}. \tag{1.54}$$

Since we modified the exponentials, let us check how they behave under $\hat{\imath}_{yx}$. The coefficients are defined as

$$r^{\pm 1} = \frac{e^{\mp i\varphi}}{\sqrt{2}} \quad \Rightarrow \quad \hat{\imath}_{yx}r^{\pm 1} = \mp i r^{\pm 1}, \tag{1.55}$$

which has the same transformation as the unit vector. Again, the same operation can be obtained by manipulating the appropriate unit vectors or by performing an operation on the coefficients. Therefore, the same transformation properties are contained in the unit vectors and the coefficients.

One might wonder why the sign in the exponential of $r^{\pm 1}$ is opposite to the index, and why superscripts are used instead of subscripts. This is unfortunately the scourge of complex bases. Note that the unit vectors $\mathbf{e}_{\pm 1}$ are counterclockwise, but the coefficients $r^{\pm 1}$ are clockwise. However, a unit vector can also be constructed using unit vectors and coefficients where the direction is reversed

$$\mathbf{e}^{\pm 1} = \mathbf{e}^*_{\pm 1} = \mathbf{e}_{\mp 1} = \frac{1}{\sqrt{2}}(\mathbf{e}_x \mp i\mathbf{e}_y) \tag{1.56}$$

and

$$r_{\pm 1} = (r^{\pm 1})^* = r^{\mp 1} = \frac{e^{\pm i\varphi}}{\sqrt{2}}. \tag{1.57}$$

These problems are related to the covariance and contravariance of vectors. Since these problems do not occur for Cartesian coordinate systems, they are quite often first encountered in detail in quantum mechanics. However, similar issues also occur when discussing spacetime.

Our deep familiarity with Cartesian coordinates often makes us take things for granted that are actually a result of the fact that this basis is real and orthonormal. For example, most readers probably had little issue with the normalization of the unit vectors $\mathbf{e}_i^2 = 1$, with $i = x, y$. However, trying the same thing with the unit vectors in Eq. (1.53) gives $\mathbf{e}_m^2 = 0$, with $m = \pm 1$. Since this is obviously not a proper norm, this is not the correct way to do it.

For a two-dimensional space, we have found two sets of equally valid unit vectors. Let us introduce the bra-ket notation which is commonly used in quantum mechanics because of its convenience in dealing with matrices,

$$\langle m| = \mathbf{e}^m \quad \text{and} \quad |m\rangle = \mathbf{e}_m, \tag{1.58}$$

with $m = 1, -1$. There are some differences in expressing it as a vector or in bra-ket notation, but that does not need to concern us at the moment. The norm of a vector is effectively the projection of the bra $\langle m|$ onto the ket $|m\rangle$ (the name arises from the use of the brackets $\langle$ and $\rangle$),

$$\langle m|m\rangle = \mathbf{e}^m \mathbf{e}_m = 1, \tag{1.59}$$

which gives the desired result. The result of the projection is a scalar or number. Therefore, in order to obtain the norm, we need to combine the different unit vectors. This problem does not occur for Cartesian coordinates since $\mathbf{e}^i \equiv \mathbf{e}_i$ with $i = x, y$.

A unit vector can now be written in two different ways by projecting it onto the bra or the ket basis

$$\hat{\mathbf{r}} = (\hat{\mathbf{r}}| = \sum_{m=1,-1} (\hat{\mathbf{r}}|m\rangle\langle m| = r_1 \mathbf{e}^1 + r_{-1}\mathbf{e}^{-1}, \tag{1.60}$$

$$\hat{\mathbf{r}} = |\hat{\mathbf{r}}) = \sum_{m=1,-1} |m\rangle\langle m|\hat{\mathbf{r}}) = \mathbf{e}_1 r^1 + \mathbf{e}_{-1} r^{-1}, \tag{1.61}$$

where $\sum_{m=1,-1} |m\rangle\langle m| = \mathbb{1}_2$ spans the entire vector space in two dimensions (we will see that the space $|m\rangle$ is actually infinite, hence the brackets $|\cdot\rangle$. However, in two dimensions, a vector, indicated with parentheses $(\cdot|$, only projects onto $m = 1, -1$). We encounter $\langle\hat{\mathbf{r}}|$ and $|\hat{\mathbf{r}}\rangle$ with brackets when dealing with function spaces; $(\hat{\mathbf{r}}|$ and $|\hat{\mathbf{r}})$ with parentheses refer to vectors. The difference will become apparent in the following sections. The coefficients of the vector are the projections of $\hat{\mathbf{r}}$ onto the unit vectors

$$(\hat{\mathbf{r}}|m\rangle = r_m \quad \text{and} \quad \langle m|\hat{\mathbf{r}}) = r^m. \tag{1.62}$$

We can express a vector in the same way in Cartesian coordinates

$$\hat{\mathbf{r}} = (\hat{\mathbf{r}}| = \sum_{i=x,y} (\hat{\mathbf{r}}|i\rangle\langle i| = (\hat{\mathbf{r}}|x\rangle\langle x| + (\hat{\mathbf{r}}|y\rangle\langle y| = \cos\varphi\, \mathbf{e}_x + \sin\varphi\, \mathbf{e}_y, \tag{1.63}$$

with $\langle i| = \mathbf{e}^i \equiv \mathbf{e}_i$, with $i = x, y$. Note that $|\hat{\mathbf{r}})$ produces the same result.

For complex unit vectors, the inner product of a vector with itself is given by

$$(\hat{\mathbf{r}}|\hat{\mathbf{r}}) = \sum_{mm'} \mathbf{e}^{m'} \mathbf{e}_m r_{m'} r^m = \sum_m r_m r^m = \sum_{m=\pm 1} \frac{e^{im\varphi}}{\sqrt{2}} \frac{e^{-im\varphi}}{\sqrt{2}} = 1, \tag{1.64}$$

using that $\mathbf{e}^{m'}\mathbf{e}_m = \delta_{mm'}$. This explains the need for the square roots in $r_{\pm 1}$ and $r^{\pm 1}$. The coefficients for a vector are normalized by summing over the different directions. As we will see, in quantum mechanics, it is more common to normalize the functions by integrating over the variables (in this case the angle φ).

Example 1.2 ***Solving the Wave Equation***
Solve the differential equation in Eq. (1.49).

Solution
Equation (1.49) contains a real and an imaginary part that we can solve separately, leading to the differential equation $d^2 f/d\varphi^2 + f = 0$. Obviously, this is a well-known differential equation, and its solutions are probably familiar to most readers. However, suppose we were not aware of the solutions (say, we were considering the differential equation for the Legendre, Hermite, or Laguerre polynomials), then how do we approach it? One way is to solve it numerically. However, often we prefer analytical solutions if we can obtain them. Since many functions can be expressed in terms of a Taylor series, let us

suggest a solution of the type $f(\varphi) = \sum_{n=0}^{\infty} c_n \varphi^n$, where c_n are coefficients. The second derivative is then given by

$$\frac{d^2 f(\varphi)}{d\varphi^2} = \sum_{n=0}^{\infty} n(n-1)c_n \varphi^{n-2}. \tag{1.65}$$

In order to have equal powers in the series, the index of the other term in the differential equation is shifted $n \to n-2$. The differential equation can then be written as

$$\sum_{n=2}^{\infty} \left(n(n-1)c_n + c_{n-2}\right) \varphi^{n-2} = 0. \tag{1.66}$$

Since each φ^{n-2} is an independent function, the factors in front of them have to be zero. This gives

$$c_n = -\frac{1}{n(n-1)} c_{n-2}. \tag{1.67}$$

Since the difference between the indices is 2, there are two independent solutions with even and odd powers. For the even powers, we find taking $n = 2n'$

$$c_{2n'} = -\frac{1}{2n'(2n'-1)} c_{2n'-2} = \frac{1}{2n'(2n'-1)(2n'-2)(2n'-3)} c_{2n'-4} = \frac{(-1)^{n'}}{(2n')!} c_0.$$

Conventionally, we take $f(0) \equiv 1$ for the even-powered series since the trigonometric functions are defined for a unit circle. This gives $c_0 = 1$, and the values of $c_{2n'}$ are the same as those for the expansion of the cosine in Eq. (1.24). The coefficients for the sine can be found in a similar fashion, which is left to the reader. The complex exponentials can be found by making complex combinations of sine and cosines.

Conceptual summary. – The rotation of a vector in two dimensions directly led to the introduction of differential calculus and allowed us to find expression of the coefficients $\cos\varphi$ and $\sin\varphi$ in terms of the polar angle φ. Note that the derivatives $d\cos\varphi/d\varphi = -\sin\varphi$ and $d\sin\varphi/d\varphi = \cos\varphi$ interchange the coefficients, and one obtains a negative sign. This is the same effect as was found for the bivector operation $\boldsymbol{i}_{yx}\mathbf{e}_x = \mathbf{e}_y$ and $\boldsymbol{i}_{yx}\mathbf{e}_y = -\mathbf{e}_x$, where the unit vectors are interchanged, and one obtains a minus sign. The effect is the same since taking the derivative with respect to φ and multiplying by $\boldsymbol{i}_{yx}$ perform the same operation, *i.e.* a 90° counterclockwise rotation. Although this is rather clear when looking at it from the point of view of a rotation of the unit vectors, it is not at all obvious why the derivatives of the sine and the cosine should behave this way when they are derived in a calculus course. We will see this again later. For function spaces, there are often very nice relations between the functions, which are generally hard to understand from an operator or calculus point of view but make a lot more sense when understanding the underlying geometrical purpose of the functions.

Since mathematics and physics have a predilection for eigenfunctions and eigenvectors (which often correspond to conserved quantities), a new operator was introduced $\hat{L} = -\boldsymbol{i}_{yx} d/d\varphi$. Since the derivative and $-\boldsymbol{i}_{yx}$ cause a 90° counterclockwise and clockwise rotations, respectively, it leaves the unit vectors unchanged, and they are therefore eigenvectors of $\hat{L}$. Note that we have derived a complex operator equivalent to the z component of the quantum angular momentum operator in spherical polar coordinates from purely classical considerations. The underlying reason is that the rotation of a function is not necessarily a quantum-mechanical operation. Note that the eigenfunctions of the angular momentum operator can also be found using a differential equation very similar to the wave equation.

The derivative operator $\hat{\imath}_{yx} = d/d\varphi$ has complex exponentials as eigenfunctions. When trying to find unit vectors that behave in the same fashion under the bivector $\boldsymbol{i}_{yx}$, things get rather abstract. Complex unit vectors $\mathbf{e}_{\pm 1} = \frac{1}{\sqrt{2}}(\mathbf{e}_x \pm i\mathbf{e}_y)$ are found, since no real vector is conserved under rotation in a plane. The real eigenvectors in fact lie out of the plane of rotation and are related to the rotation axis, which is a conserved quantity. Although its interpretation is not entirely intuitive, the complex unit vectors are used often since they satisfy the proper transformation properties under rotation.

We have now seen two ways to solve the problem of a rotation of a unit vector in two dimensions: by manipulation of the unit vectors and by applying an operator on the coefficients/functions. Although both approaches are valid, the emphasis in quantum-mechanics courses often lies on the latter. The underlying reason is that differential equations lend themselves well to brute-force approaches that enable the treatment of problems where the space is not as simple. The use of unit vectors gives us greater insights into the geometry of the space and how functions can be manipulated. It also makes connections to classical mechanics, electromagnetism, and relativity easier.

Problems

1 Although there are no eigenvectors of the rotation operator (except for rotations over 0° and 180°), there is a pair of unit vectors that transform into each other under a 90° rotation. Find the general pair of vectors that transform into each other, and how they transform into each other.

2 Show that the unit vector in the complex basis in Eq. (1.60) is an eigenvector of $\hat{L}$ in Eq. (1.43). Show that $\hat{\mathbf{r}}(-\varphi)$ is the conjugate vector.

3 The series solution for the second-order differential equation in Eq. (1.65) splits into two parts. Find the series solution for the first-order differential equation

$$\frac{df(\varphi)}{d\varphi} = if(\varphi). \tag{1.68}$$

1.4 Function Space for Rotation in a Plane

Introduction. – In the previous sections, the rotation of a unit vector in two dimensions was considered. Although this is an important problem on its own, it was also an introduction to function spaces since vectors are part of a function space. The two-dimensional unit vectors $\hat{\mathbf{r}}$ describe a unit circle. The vector $\hat{\mathbf{r}}$ can indicate any point on the unit circle in terms of the polar angle φ (there is only one variable, since the unit circle is a one-dimensional space). However, that is not all we generally want to do. In many applications, we would like to assign a value to a point in space in order to describe, for example a field, a potential, or a wave function.

One can consider the following analogy. The GPS coordinates (a vector) require only two numbers (latitude and longitude) to describe each position on Earth. However, to describe all the altitudes on Earth (comparable to a wave function or a field) each coordinate has to be associated with a height. Note that, in principle, this requires one to store an infinite number of altitudes. Functions are needed since quantum particles cannot be treated as rigid objects. In many classical mechanical problems, one only tracks the motion of the center of mass and vectors suffice. However, quantum particles adjust themselves to the potential landscape. Even when moving in the absence of a potential, the shape of the particle is not fixed when the system is not in an eigenstate. Since the energetics depends on the shape of the quantum particle, the description cannot be limited to the center of mass.

We continue with the example of rotation on the unit circle. Since the mathematics is relatively easy, we can focus on the concepts. Additionally, the results can be extended to translational motion in one dimension, which corresponds to an infinitely large circle. It is then straightforward to generalize the result to three dimensions. To describe a function on the unit circle, an infinite number of values needs to be stored, since there are an infinite number of points on the unit circle. Each point on the unit circle needs to be associated with a value. Obviously, one can simply define a function $f(\varphi) = f(\hat{\boldsymbol{r}})$ that returns a certain value for each point on the unit circle. However, when dealing with rotation, it is advantageous to expand the function in terms of eigenfunctions of the angular momentum operator $\hat{L}$. The reason is that we generally do not know how a function changes when it is freely rotating. Since we understand the behavior of eigenfunctions under rotation, expressing a function in eigenfunctions allows one to describe the behavior of the function while it is rotating. So far, we have found the eigenfunctions $e^{\pm i\varphi}$. However, if the space contains an infinite number of points, the related function space contains an infinite number of functions.

A key ingredient in describing each point in space independently is that we obtain points on the unit circle that are orthogonal to each other. The unit vectors $\hat{\mathbf{r}}$ do not satisfy this property, since the inner product between two unit vectors is

$$\hat{\mathbf{r}}' \cdot \hat{\mathbf{r}} = (\hat{\mathbf{r}}'|\hat{\mathbf{r}}) = \cos\varphi' \cos\varphi + \sin\varphi' \sin\varphi = \cos(\varphi' - \varphi), \tag{1.69}$$

using the notation from Eq. (1.58). Therefore, the unit vectors are only orthogonal when $\varphi' - \varphi = \frac{\pi}{2}$. It is important to realize that vectors only indicate the location of a point in space, they do not represent that point in space. A point in space in bra-ket notation is indicated as $|\hat{\mathbf{r}}\rangle$ or $|\hat{\boldsymbol{r}}\rangle$. These new vectors are orthonormal to each other

$$\langle \hat{\boldsymbol{r}}'|\hat{\boldsymbol{r}}\rangle = \delta(\hat{\boldsymbol{r}}' - \hat{\boldsymbol{r}}), \tag{1.70}$$

where $\langle \hat{\boldsymbol{r}}| = (|\hat{\boldsymbol{r}}\rangle)^*$ is the conjugate of $|\hat{\boldsymbol{r}}\rangle$. The δ-function is defined as

$$\delta(\hat{\boldsymbol{r}} - \hat{\boldsymbol{r}}') = \begin{cases} \infty & \hat{\boldsymbol{r}} = \hat{\boldsymbol{r}}', \\ 0 & \text{elsewhere.} \end{cases} \tag{1.71}$$

Note that we can also use $\delta(\varphi - \varphi')$, since $\hat{\boldsymbol{r}} = \hat{\boldsymbol{r}}'$ implies that $\varphi = \varphi'$. Equation (1.70) means that all points on the unit circle are orthogonal to each other, meaning that each point is completely independent from another point. So, while $(\hat{\mathbf{r}}|$ indicates the position of a point in space, $\langle \hat{\boldsymbol{r}}|$ *is* that point in space. This allows us to assign a value to each point on the unit circle.

Another important property of space is completeness

$$\int d\hat{\boldsymbol{r}}\, |\hat{\boldsymbol{r}}\rangle\langle\hat{\boldsymbol{r}}| = \mathbb{1}_{\infty}, \tag{1.72}$$

where $\mathbb{1}_{\infty}$ is an infinite-dimensional identity. This means that the vectors $|\hat{\boldsymbol{r}}\rangle$ represent all possible points on the unit circle. Since completeness is equal to the identity, it should not do anything. Applying it to a point gives

$$|\hat{\boldsymbol{r}}'\rangle = \mathbb{1}_{\infty}|\hat{\boldsymbol{r}}'\rangle = \int d\hat{\boldsymbol{r}}\, |\hat{\boldsymbol{r}}\rangle\langle\hat{\boldsymbol{r}}|\hat{\boldsymbol{r}}'\rangle = \int d\hat{\boldsymbol{r}}\, |\hat{\boldsymbol{r}}\rangle\delta(\hat{\boldsymbol{r}}' - \hat{\boldsymbol{r}}) = |\hat{\boldsymbol{r}}'\rangle, \tag{1.73}$$

using Eq. (1.70). In the completeness relation (from right to left) $\langle\hat{\boldsymbol{r}}|$ effectively removes the point $|\hat{\boldsymbol{r}}'\rangle$; however, $|\hat{\boldsymbol{r}}\rangle$ directly creates the same point. Therefore, the completeness relation does not do anything (it is an identity) as long as the integral goes over all positions $|\hat{\boldsymbol{r}}\rangle$. Although this appears like an incredibly boring relation, it becomes important for switching between real and function spaces.

Additional eigenfunctions of the angular momentum can be found by generalizing the eigenvalue equation (1.48),

$$\hat{\boldsymbol{L}}\hat{\boldsymbol{r}}_m = m\hat{\boldsymbol{r}}_m \quad \Rightarrow \quad \hat{\boldsymbol{r}}_m = \frac{1}{\sqrt{2\pi}} e^{\boldsymbol{i}m\varphi}. \tag{1.74}$$

We use here a general imaginary unit $\boldsymbol{i}$ with $\boldsymbol{i}^2 = -1$. For the xy-plane, $\boldsymbol{i} \to \boldsymbol{i}_{yx}$, but the plane can be oriented in any direction in space. The prefactor ensures a proper normalization when integrating over φ from 0 to 2π, as we will see below. Following Eq. (1.49), these functions are also solutions of the wave equation

$$\frac{d^2\hat{\boldsymbol{r}}}{d\varphi^2} + m^2\hat{\boldsymbol{r}} = 0. \tag{1.75}$$

We want the functions to be single valued, *i.e.*

$$\hat{\boldsymbol{r}}_m(\varphi + 2\pi) = \frac{1}{\sqrt{2\pi}} e^{\boldsymbol{i}m(\varphi+2\pi)} = \hat{\boldsymbol{r}}_m(\varphi). \tag{1.76}$$

In order to obtain single-valued functions, $m = \cdots, -2, -1, 0, 1, 2, \ldots$ has to be an integer. This implies that the function space is discrete, which is directly related to the periodicity of the unit circle. A similar effect occurs when the system is confined in space. This effect is at the origin of quantization in quantum mechanics. Although the function space is discrete, the number of functions is still infinite. When the space is infinitely large, the function space becomes continuous as we will see. The functions can be written as

$$\hat{\boldsymbol{r}}_m = \frac{1}{\sqrt{2\pi}}(\hat{\boldsymbol{r}})^m = \frac{1}{\sqrt{2\pi}} e^{\boldsymbol{i}m\varphi} \equiv \langle\hat{\boldsymbol{r}}|m\rangle. \tag{1.77}$$

Just as the points $\langle\hat{\boldsymbol{r}}|$ are a basis for the entire space, these functions are also known as basis functions. In addition to real space vectors, a vector space of functions $|m\rangle$ can also be defined. The interpretation of the inner product $\langle\hat{\boldsymbol{r}}|m\rangle$ is that the function labeled by m is projected onto the site $\hat{\boldsymbol{r}}$, giving $e^{\boldsymbol{i}m\varphi}/\sqrt{2\pi}$, *i.e.* the value of the function at that site. Since the functions are complex, the conjugate is given by

$$\langle m|\hat{\boldsymbol{r}}\rangle = \hat{\boldsymbol{r}}_m^* = \frac{1}{\sqrt{2\pi}} e^{-\boldsymbol{i}m\varphi}. \tag{1.78}$$

Just like the position vectors in Eq. (1.70), the function space of complex exponentials is orthogonal,

$$\langle m'|m\rangle = \langle m'|\mathbb{1}_\infty|m\rangle = \int d\hat{\boldsymbol{r}}\, \langle m'|\hat{\boldsymbol{r}}\rangle\langle\hat{\boldsymbol{r}}|m\rangle, \tag{1.79}$$

where the completeness relation of real space has been inserted. Inserting Eqs. (1.77) and (1.78) into (1.79) gives

$$\langle m'|m\rangle = \frac{1}{2\pi}\int d\hat{\boldsymbol{r}}\, e^{i(m-m')\varphi} = \frac{1}{2\pi}\int_0^{2\pi} d\varphi\, e^{i(m-m')\varphi} = \delta_{mm'}, \tag{1.80}$$

where $\delta_{mm'}$ is a Kronecker delta function, which is 1 for $m' = m$, and 0 otherwise. Due to the inclusion of the factor $1/\sqrt{2\pi}$ in the wave function, the basis functions are properly normalized when integrating over the square of the function.

The complex exponentials can also be used to satisfy the orthogonality of the real space vectors in Eq. (1.70). Let us define a function that is the sum over all complex exponentials

$$\Xi(\varphi) = \frac{1}{2\pi}\sum_{m=-\infty}^{\infty} e^{-im\varphi}. \tag{1.81}$$

For $\varphi = 0$, the exponent is 1, and the summation goes to infinity. The oscillating nature of the exponentials makes the $\varphi \neq 0$ value small with respect to the $\varphi = 0$ value. This satisfies the criteria for a δ-function. However, the above function is periodic since the exponential is also 1 for $\varphi = 2\pi n$ with n an integer. The function $\Xi(\varphi)$ is therefore an impulse train known as the Dirac comb

$$\Xi(\varphi) = \sum_{n=-\infty}^{\infty} \delta(\varphi - 2\pi n). \tag{1.82}$$

However, when we restrict φ to values between 0 and 2π, the periodicity is irrelevant, and we can replace $\delta(\hat{\boldsymbol{r}}' - \hat{\boldsymbol{r}})$ in Eq. (1.70) by $\Xi(\varphi' - \varphi)$. The orthonormality from Eq. (1.70) can now be written as

$$\langle\hat{\boldsymbol{r}}'|\hat{\boldsymbol{r}}\rangle = \Xi(\varphi' - \varphi) = \frac{1}{2\pi}\sum_{m=-\infty}^{\infty} e^{im(\varphi'-\varphi)} = \sum_{m=-\infty}^{\infty} \langle\hat{\boldsymbol{r}}'|m\rangle\langle m|\hat{\boldsymbol{r}}\rangle, \tag{1.83}$$

expressing the exponentials in bra-ket notation using Eq. (1.77) with $(\langle m|\hat{\boldsymbol{r}}\rangle)^* = \langle\hat{\boldsymbol{r}}|m\rangle$. In order for the left- and right-hand sides of Eq. (1.83) to be equal, there must also be a completeness relation for the function space $|m\rangle$

$$\sum_{m=-\infty}^{\infty} |m\rangle\langle m| = \mathbb{1}_\infty. \tag{1.84}$$

So, both the real and the function spaces are orthogonal and complete.

Fourier transform. – We now want to look at an arbitrary function $|f\rangle$ on the unit circle. Note that $|f\rangle$ is the function, but we have to find a way to express it. Commonly, the function is expressed by its values at a particular position. We can do this by projecting the function onto a particular position $\langle\hat{\boldsymbol{r}}|$,

$$f(\hat{\boldsymbol{r}}) = \langle\hat{\boldsymbol{r}}|f\rangle. \tag{1.85}$$

One special type of function is that of a point on the unit circle. Let us consider an arbitrary point $|\hat{\boldsymbol{a}}\rangle$. From Eq. (1.85), the function associated with this position is

$$\hat{\boldsymbol{a}}(\hat{\boldsymbol{r}}) = \langle\hat{\boldsymbol{r}}|\hat{\boldsymbol{a}}\rangle = \delta(\hat{\boldsymbol{r}} - \hat{\boldsymbol{a}}), \tag{1.86}$$

using Eq. (1.70). Therefore, the point $|\hat{\boldsymbol{a}}\rangle$ on the unit circle is described by a δ-function at $\hat{\boldsymbol{a}}$. Note that this is different from the relative unit vector

$$\hat{\boldsymbol{a}} = e^{i_{yx}\varphi_a}, \tag{1.87}$$

taking the vector in the xy-plane with φ_a the polar angle with respect to the x-axis. To find the actual unit vector $\hat{\mathbf{a}}$, the relative unit vector has to be taken with respect to the reference axis,

$$\hat{\mathbf{a}} = \hat{\boldsymbol{a}}\mathbf{e}_x = e^{i_{yx}\varphi_a}\mathbf{e}_x = \cos\varphi_a\ \mathbf{e}_x + \sin\varphi_a\mathbf{e}_y, \tag{1.88}$$

Therefore, although the vector $\hat{\boldsymbol{a}}$ and the point $|\hat{\boldsymbol{a}}\rangle$ look very similar, they represent different quantities. The vector $\hat{\boldsymbol{a}}$ only indicates the location, whereas $|\hat{\boldsymbol{a}}\rangle$ actually represents that point in space.

A function does not necessarily have to be expressed in spatial coordinates. If a set of basis functions is complete, any function can be expressed in terms of those functions, in this case the complex exponentials $e^{im\varphi}$. The amount of a certain complex exponential is given by the projection of $|f\rangle$ onto the exponential $\langle m|$,

$$f_m = \langle m|f\rangle. \tag{1.89}$$

Since the function space is discrete, a subscript is used. There are advantages in expressing a function in terms of basis functions. The behavior of $|f\rangle$ under rotation is unknown. However, the behavior of the functions $|m\rangle$ under rotation is known, since they are eigenfunctions of the angular momentum, see Eq. (1.74). By expressing $|f\rangle$ in terms of $|m\rangle$, we can understand its behavior under rotation.

Any function given in terms of the position $\hat{\boldsymbol{r}}$ can be expressed in terms of complex exponentials by inserting the completeness relation in Eq. (1.84),

$$f(\hat{\boldsymbol{r}}) = \langle\hat{\boldsymbol{r}}|f\rangle = \langle\hat{\boldsymbol{r}}|\mathbb{1}_\infty|f\rangle = \sum_{m=-\infty}^{\infty}\langle\hat{\boldsymbol{r}}|m\rangle\langle m|f\rangle. \tag{1.90}$$

With Eqs. (1.77) and (1.89), this becomes

$$f(\hat{\boldsymbol{r}}) = \frac{1}{\sqrt{2\pi}}\sum_{m=-\infty}^{\infty} e^{im\varphi} f_m. \tag{1.91}$$

This expression is the (inverse) Fourier transform of the function $f(\hat{\boldsymbol{r}})$. The coefficient f_m is the Fourier component, *i.e.* the projection of the function onto the function space of complex exponentials $|m\rangle$. The coefficient f_m giving the amount of $e^{im\varphi}$ in the function can be calculated using

$$f_m = \langle m|f\rangle = \langle m|\mathbb{1}_\infty|f\rangle = \int d\hat{\boldsymbol{r}}\ \langle m|\hat{\boldsymbol{r}}\rangle\langle\hat{\boldsymbol{r}}|f\rangle, \tag{1.92}$$

where the completeness relation of real space from Eq. (1.72) has been inserted. The Fourier coefficient can be written as

$$f_m = \frac{1}{\sqrt{2\pi}}\int_0^{2\pi} d\varphi\ e^{-im\varphi} f(\hat{\boldsymbol{r}}). \tag{1.93}$$

Note the change in sign in the complex exponential in Eqs. (1.91) and (1.93).

Example 1.3 ***Sawtooth Function***

a) Express the function $f(\varphi) = \varphi$ for $0 \leq \varphi \leq 2\pi$ in terms of the eigenfunctions of the unit circle.
b) Plot $f(\varphi)$ for various ranges of m.

Solution

a) The function can be expressed in the basis functions as

$$f(\varphi) = \langle \varphi | f \rangle = \sum_{m=-\infty}^{\infty} \langle \varphi | m \rangle \langle m | f \rangle = \frac{1}{\sqrt{2\pi}} \sum_{m=-\infty}^{\infty} e^{im\varphi} f_m. \tag{1.94}$$

Note that $f(\varphi) \equiv f(\hat{\boldsymbol{r}})$. The coefficient f_m can be obtained from

$$f_m = \langle m | f \rangle = \int_0^{2\pi} d\varphi \, \langle m | \varphi \rangle \langle \varphi | f \rangle = \frac{1}{\sqrt{2\pi}} \int_0^{2\pi} d\varphi \, e^{-im\varphi} f(\varphi). \tag{1.95}$$

For $m \neq 0$, the integral can be integrated by parts as follows:

$$\begin{aligned} f_m &= \frac{1}{\sqrt{2\pi}} \int_0^{2\pi} d\varphi \, e^{-im\varphi} \varphi = \frac{1}{\sqrt{2\pi}} \left[\frac{e^{-im\varphi}}{-im} \varphi \right]_0^{2\pi} - \frac{1}{\sqrt{2\pi}} \int_0^{2\pi} d\varphi \, \frac{e^{-im\varphi}}{-im} \\ &= \frac{1}{\sqrt{2\pi}} \left(\frac{2\pi i}{m} - \left[\frac{e^{-im\varphi}}{-m^2} \right]_0^{2\pi} \right) = \frac{\sqrt{2\pi} i}{m} \end{aligned} \tag{1.96}$$

For $m = 0$, we have $f_0 = \sqrt{2}\pi^{\frac{3}{2}}$.

b) Figure 1.4 shows the function $f(\varphi)$ from Eq. (1.94) for different ranges of $m = -m_{\max}, -m_{\max} + 1, \ldots, m_{\max} - 1, m_{\max}$ with $m_{\max} = 5, 15, 100$. Since $f_m \sim 1/m$, the weight of the exponentials decreases with increasing m. For $m_{\max} = 5$, the basic shape of the sawtooth is reasonably well represented, but the agreement is less at the discontinuity in the function. With increasing $m_{\max}$, the Fourier transform approaches $f(\varphi) = \varphi$ more closely. Note that the transform is periodic with a period 2π. The oscillations near the discontinuity become sharper but do not disappear. This is known as Gibbs phenomenon. Although the basis functions $e^{im\varphi}$ look rather different from $f(\varphi)$, it is still possible to express the function in terms of complex exponentials.

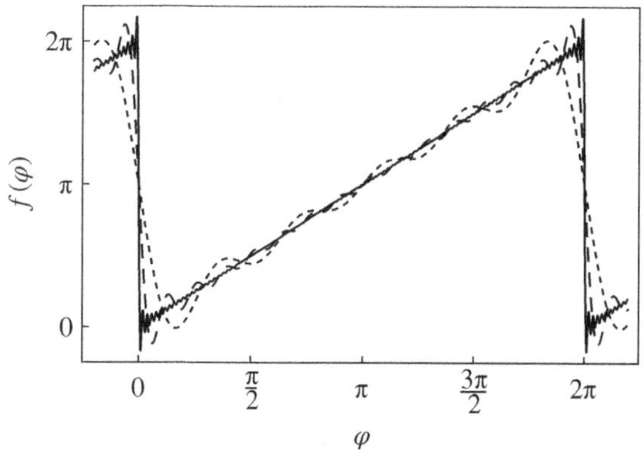

Figure 1.4 The function $f(\varphi) = \varphi$ plotted using different ranges of complex exponentials where the maximum absolute m value included is 5 (dotted), 15 (dashed), and 100 (solid).

Conceptual summary. – In this section, we derived the function space for the unit circle. It is important to understand the difference between $\hat{\boldsymbol{r}}$ and $|\hat{\boldsymbol{r}}\rangle$. The vectors $\hat{\boldsymbol{r}}$ allow us to indicate a position on the unit circle, whereas $|\hat{\boldsymbol{r}}\rangle$ represents an actual point on the unit circle. Unlike the vectors, the points $|\hat{\boldsymbol{r}}\rangle$ are orthogonal and complete. In the previous sections, we saw that $\hat{\boldsymbol{r}}$ could be described in terms of complex functions. The same can be done for $|\hat{\boldsymbol{r}}\rangle$. However, since there are now an infinite number of independent points, we also need an infinite number of functions. For a unit circle these basis functions are given by the vectors $\hat{\boldsymbol{r}}$ raised to an integer power m, *i.e.* $\hat{\boldsymbol{r}}^m = e^{im\varphi}$. This function space is also complete and orthogonal. Any function $f(\hat{\boldsymbol{r}})$ on the unit circle can be expressed in terms of the basis functions. For complex exponentials, this is known as a Fourier series. The advantage of expressing $f(\hat{\boldsymbol{r}})$ in terms of the basis functions is that we know how the basis functions transform under rotation, whereas this is not known for $f(\hat{\boldsymbol{r}})$.

Since a unit circle is essentially a line, the ideas from this section are easily modified to any one-dimensional space. Additionally, the concepts in this section can be extended to higher dimensions and spacetime. Note that the functions depend on the geometry of space. When the spatial properties change, a different function space is needed. Obtaining the appropriate function space in any arbitrary system is one of the key aspects of quantum mechanics. The unit circle is a one-dimensional system defined by rotational symmetry. Even though the system is relatively simple, the mathematics is not entirely trivial. The complexity only increases when looking at higher dimensions and systems with less symmetry, for example due to the presence of a potential. Note that obtaining the function space is not only a mathematical exercise, it is directly related to the physical properties. For example, the orbitals of an atom are the function space for that particular system.

Before extending the results to three dimensions, let us take some time to appreciate the large number of important mathematical concepts that appear in the relatively simple problem of a unit circle: anticommuting quantities, imaginary units, bivectors, exponentials and trigonometric functions, series expansions, differential calculus, complex operators, wave equations, function spaces, and Fourier series. Many of these topics are often taught in completely separate courses, so it is important to see how these approaches interconnect with each other. Additionally, although many physics students encounter some of these concepts for the first time in a quantum mechanics course, the only thing we have done so far is rotating a vector in two dimensions and describing functions on a unit circle.

Problems

1 Consider the following function on the unit circle:

$$f(\varphi) = \begin{cases} 1 & \dfrac{\pi}{2} \le \varphi \le \pi, \\ 0 & \text{elsewhere.} \end{cases} \tag{1.97}$$

a) Find the Fourier transform of $f(\varphi)$.
b) Plot the real part of the Fourier transform.
c) Use the Fourier transform to calculate $f(\varphi)$ for different ranges $[-m_{\max}, m_{\max}]$ for m.

1.5 Three-Dimensional Space

Introduction. – In this section, the results from Sections 1.2 and 1.3 are extended to three dimensions. Therefore, we return to looking at vectors $\mathbf{r}$ that point to positions in space and how to transform these vectors. Therefore, the vectors in this section only give the location of a point in space but do not represent points in space. This will be described in Section 1.10. The properties of the vectors are described using anticommuting algebraic unit vectors, known as geometric or Clifford algebra. In quantum mechanics, these properties are often incorporated using 2×2 matrices, known as Pauli matrices. We will see in the coming sections that this is essentially equivalent.

The starting point is a Cartesian coordinate system with three orthonormal unit vectors $\mathbf{e}_i$, with $i = x, y, z$. A vector in three dimensions can then be expressed as

$$\mathbf{r} = x\mathbf{e}_x + y\mathbf{e}_y + z\mathbf{e}_z. \tag{1.98}$$

When x, y, and z are real numbers, $\mathbf{r}$ can point to any position in three-dimensional space. For a proper Cartesian space, the unit vectors need to satisfy several conditions. First, the unit vectors are normalized

$$\mathbf{e}_i^2 = \mathbf{e}_i\mathbf{e}_i = 1 \quad \text{for} \quad i = x, y, z. \tag{1.99}$$

Second, following Eq. (1.13), unequal unit vectors are assumed to anticommute,

$$\mathbf{e}_i\mathbf{e}_j + \mathbf{e}_j\mathbf{e}_i = 0 \quad \text{for } i \neq j. \tag{1.100}$$

This can be cast in one equation as

$$\{\mathbf{e}_i, \mathbf{e}_j\} = 2\delta_{ij}, \tag{1.101}$$

where the anticommutator is defined as

$$\{A, B\} = AB + BA. \tag{1.102}$$

Equation (1.101) can be viewed as the metric for Cartesian unit vectors.

In addition to the vectors, there are several multivectors in three dimensions

$$(1 + \mathbf{e}_x)(1 + \mathbf{e}_y)(1 + \mathbf{e}_z) = 1 + \mathbf{e}_x + \mathbf{e}_y + \mathbf{e}_z + \mathbf{e}_x\mathbf{e}_y + \mathbf{e}_y\mathbf{e}_z + \mathbf{e}_x\mathbf{e}_z + \mathbf{e}_x\mathbf{e}_y\mathbf{e}_z.$$

In two dimensions, the imaginary unit $\boldsymbol{i}_{yx} = \mathbf{e}_y\mathbf{e}_x$ is a pseudoscalar, *i.e.* the highest order element in two dimensions. This is no longer the case in three dimensions, where there are three distinct bivectors

$$\boldsymbol{i}_{ij} = \mathbf{e}_i\mathbf{e}_j \quad \text{with} \quad ij = yx, zy, xz. \tag{1.103}$$

These quantities are all imaginary units since

$$\boldsymbol{i}_{ij}^2 = \mathbf{e}_i\mathbf{e}_j\mathbf{e}_i\mathbf{e}_j = -\mathbf{e}_i\mathbf{e}_j\mathbf{e}_j\mathbf{e}_i = -1. \tag{1.104}$$

As in two dimensions, these bivectors can be associated with rotations, but now around different axes of rotation. Therefore, unlike the impression often created in many courses, there are more imaginary units than the complex scalar i. The pseudoscalar, the highest order element in three dimensions, is now the product of all three unit vectors

$$\boldsymbol{i} = \mathbf{e}_x\mathbf{e}_y\mathbf{e}_z. \tag{1.105}$$

This is still an imaginary unit since

$$\boldsymbol{i}^2 = \mathbf{e}_x\mathbf{e}_y\mathbf{e}_z\mathbf{e}_x\mathbf{e}_y\mathbf{e}_z = \mathbf{e}_x\mathbf{e}_x\mathbf{e}_y\mathbf{e}_z\mathbf{e}_y\mathbf{e}_z = 1 \times (-1) = -1. \tag{1.106}$$

The pseudoscalar provides a direct relationship between vectors and bivectors

$$\boldsymbol{i}_{ij} = \underline{\mathrm{e}}_k = \boldsymbol{i}\mathbf{e}_k \quad \text{for} \quad k = x, y, z \quad \text{and} \quad ij = yz, zx, xy. \tag{1.107}$$

The notation $\underline{\mathrm{e}}_k$ indicates that $\boldsymbol{i}_{ij}$ is the full space represented by the pseudoscalar $\boldsymbol{i}$ containing all the unit vectors, but with the vector $\mathbf{e}_k$ removed. The vectors and bivectors form a dual space. The bivectors $\boldsymbol{i}_{ij}$ are related to a 90° counterclockwise rotation in the ji-plane. The unit vector $\mathbf{e}_k$ corresponds to the axis of rotation perpendicular to that plane.

In summary, in three dimensions there are $2^3 = 8$ different independent multivectors

multivector	rank	number	name
1	0	1	scalar
$\mathbf{e}_i$	1	3	vector
$\boldsymbol{i}_{ij} = \underline{\mathrm{e}}_k$	2	3	bivector = pseudovector
$\boldsymbol{i} \equiv \mathbf{e}_x\mathbf{e}_y\mathbf{e}_z$	3	1	trivector = pseudoscalar

(1.108)

with $ij = yz, zx, xy$ for $k = x, y, z$.

In analogy to scalars, vectors can be combined by different operations, such as addition, subtraction, multiplication, and division. The operation comparable to division is more complex and will be postponed to a later section. Most of the readers will be familiar with vector addition/subtraction, which can be accomplished by adding/subtracting the coefficients in each of the directions of the Cartesian coordinate system,

$$\mathbf{r} \pm \mathbf{r}' = (x \pm x')\mathbf{e}_x + (y \pm y')\mathbf{e}_y + (z \pm z')\mathbf{e}_z. \tag{1.109}$$

This can be done because each direction is independent due to the orthogonality of the unit vectors $\mathbf{e}_i$.

Multiplication of vectors is more complicated. It is generally taught by splitting the multiplication into two entirely separate operations, namely the inner and outer products. However, there is nothing that prevents the direct multiplication of vectors. The direct product of two vectors splits into a scalar and a bivector term,

$$\begin{aligned} \mathbf{r}\mathbf{r}' &= (x\mathbf{e}_x + y\mathbf{e}_y + z\mathbf{e}_z)(x'\mathbf{e}_x + y'\mathbf{e}_y + z'\mathbf{e}_z) \\ &= \mathbf{r} \cdot \mathbf{r}' + \mathbf{r} \wedge \mathbf{r}'. \end{aligned} \tag{1.110}$$

The scalar part is the inner product

$$\mathbf{r} \cdot \mathbf{r}' = xx' + yy' + zz'. \tag{1.111}$$

The bivector is called the wedge product

$$\mathbf{r} \wedge \mathbf{r}' = \mathbf{e}_x\mathbf{e}_y(xy' - yx') + \mathbf{e}_y\mathbf{e}_z(yz' - zy') + \mathbf{e}_z\mathbf{e}_x(zx' - xz'). \tag{1.112}$$

Using the relation between the vectors and the bivectors in Eq. (1.107), the wedge product is directly related to the outer product

$$\mathbf{r} \wedge \mathbf{r}' = \boldsymbol{i}\mathbf{r} \times \mathbf{r}'. \tag{1.113}$$

Both definitions are valid in three dimensions. The bivector defines the plane of rotation; the outer product gives the axis of rotation which is perpendicular to the plane. In introductory physics classes, there is a strong tendency to remove all aspects of bivectors and express everything in terms of vectors. Although this is possible, it does leave out important insights between rotations and bivectors, as introduced in Section 1.2.

The product in Eq. (1.110) can also be reversed, which only changes the sign of the wedge product

$$\mathbf{r}'\mathbf{r} = \mathbf{r} \cdot \mathbf{r}' - \mathbf{r} \wedge \mathbf{r}'. \tag{1.114}$$

Combining the two equations gives

$$\mathbf{r} \cdot \mathbf{r}' = \frac{1}{2}(\mathbf{r}\mathbf{r}' + \mathbf{r}'\mathbf{r}) \tag{1.115}$$

and

$$\mathbf{r} \wedge \mathbf{r}' = \frac{1}{2}(\mathbf{r}\mathbf{r}' - \mathbf{r}'\mathbf{r}). \tag{1.116}$$

Every action involving two vectors is effectively a two-dimensional problem, since the two vectors form a plane. One of the vectors can be taken as the x-axis (either vector will do). Aligning the x-axis with $\mathbf{r}$ allows us to write $\mathbf{r} = r\mathbf{e}_x$, where $r = |\mathbf{r}|$ is the norm of the vector. The y-axis is perpendicular to $\mathbf{e}_x$ and also in the plane defined by $\mathbf{r}$ and $\mathbf{r}'$. The second vector can then be written as $\mathbf{r}' = r'(\cos\alpha\mathbf{e}_x + \sin\alpha\mathbf{e}_y)$, where α is the angle between $\mathbf{r}$ and $\mathbf{r}'$. The product of the two vectors can now be written as

$$\mathbf{r}'\mathbf{r} = rr'\cos\alpha + rr'\sin\alpha\ \mathbf{e}_y\mathbf{e}_x. \tag{1.117}$$

The inner product is the scalar part

$$\mathbf{r}' \cdot \mathbf{r} = rr'\cos\alpha. \tag{1.118}$$

The norm of the wedge product and outer product is

$$|\mathbf{r}' \times \mathbf{r}| = |\mathbf{r}' \wedge \mathbf{r}| = rr'\sin\alpha. \tag{1.119}$$

Since the norm, which is a scalar, is invariant of the choice of coordinate system, these expressions are valid regardless of the axes system.

Conceptual summary. – Although many of the concepts in three dimensions are similar to two dimensions, it leads to alternative ways of looking at rotations. The bivectors $\boldsymbol{i}_{ij}$ are the quantities that produce rotations in the ji-plane. However, the additional dimension allows the definition of an axis of rotation perpendicular to the plane of rotation. Vectors parallel to the axis of rotation are conserved quantities, and rotations are often defined in terms of the rotation axis.

Another important concept that is introduced in this section is the mixing of multivectors of different ranks. For example, many readers might consider an expression such as $7 + \mathbf{r}$ incorrect, since it mixes a scalar and a vector. This is why the product of two vectors is generally split into an inner and an outer product. The idea that these two can mix is often only encountered when discussing the vector properties of matrices. However, there is nothing that prevents us from introducing these concepts at the algebraic level.

Problems

1 Show that the pseudoscalar $\boldsymbol{i}$ and the unit vectors $\mathbf{e}_i$, with $i = x, y, z$, commute.

2 Find the product of $3\mathbf{e}_x + \mathbf{e}_y$ and $\mathbf{e}_x - 2\mathbf{e}_y$.

3 Show that the norm of the wedge product is equal to area of the parallelogram spanned by the two vectors.

4 What is the vector normal to the plane defined by the bivector $4\boldsymbol{i}_{xy} - 3\boldsymbol{i}_{xz}$?

1.6 Spinors

Introduction. – In this section, we return to a rotation in a plane, but now oriented in a three-dimensional space. The eigenfunctions for rotation in two dimensions were $r_{\pm 1} \sim e^{\pm \boldsymbol{i}_{yx}\varphi}$, see Section 1.3. The coefficients are eigenfunctions of the angular momentum operator, $\hat{L}r_{\pm 1} = \pm r_{\pm 1}$. The coefficients raised to an integer power created the function space for the unit circle. These results still hold in three dimensions, except that the plane of rotation must be oriented in space.

These results were all very satisfactory. However, we ran into issues when trying to find a vector quantity to associate with the coefficients. The goal is to describe a space in terms of vectors that comprise unit vectors and coefficients, *i.e.* the eigenfunctions of $\hat{L}$. Ideally, we want to be able to perform a certain action, say a translation or a rotation, in two different ways: by manipulation of the unit vectors or by performing an operation on the coefficients, as was shown for a unit vector in two dimensions in Sections 1.2 and 1.3, respectively. In physics, we have a preference for eigenfunctions, since they teach us valuable information about the system. In the xy-plane, coefficients could be obtained that were eigenfunctions of the angular momentum operator. The eigenfunctions could be found using $\hat{L} = -\boldsymbol{i}_{yx}\hat{\imath}_{yx} = -\boldsymbol{i}_{yx}d/d\varphi$ or $\hat{L} = -\hat{\imath}_{yx}^2 = -d^2/d\varphi^2$ from Eqs. (1.43) and (1.47), respectively. Now, we want the unit vectors to be eigenvectors of an operator written in terms of multivectors related to the rotation. However, trying to find a vector operator in a similar fashion as the angular momentum operator leads nowhere since $-\boldsymbol{i}_{yx}^2 = 1$. Every vector in space is an eigenvector of the identity operator, so this is not useful. We are therefore left with looking at the bivector $\boldsymbol{i}_{yx}$. Now, in principle, no vector in the xy-plane is conserved under $\boldsymbol{i}_{yx}$, since $\boldsymbol{i}_{yx}$ causes a 90° counterclockwise rotation. We got around this in Section 1.3 by introducing complex vectors. This works formally but is somewhat unsatisfactory.

In three dimensions, the situation is different, since there is a vector that is clearly conserved under rotation, namely the axis of rotation. This corresponds to the classical expectation that the vector associated with angular momentum is perpendicular to the plane of rotation. Unfortunately, our problems do not end here. We want to find eigenvectors of a vector operation. We will see that in order to write the rotation axis in eigenvector form, we effectively need to take the square root of a vector. The resulting quantity is known as a spinor.

In two dimensions, see Section 1.2, a rotation is determined by the bivector given by the product of two orthonormal unit vectors in the plane. For a typical Cartesian coordinate system, this is written as $\mathbf{e}_y\mathbf{e}_x = \boldsymbol{i}_{yx}$. In three dimensions, the situation is very similar except that the plane can be randomly oriented in space. The direction of the plane of rotation is given by the rotation axis which can be defined by a unit vector $\hat{\mathbf{r}}$. The bivector associated with the rotation axis is

$$\boldsymbol{i}_{\underline{\hat{\mathbf{r}}}} = \boldsymbol{i}\hat{\mathbf{r}}, \tag{1.120}$$

where $\boldsymbol{i} = \mathbf{e}_x\mathbf{e}_y\mathbf{e}_z$ is the pseudoscalar, see Eq. (1.105). The rotation bivector and the axis of rotation are therefore dual vectors, see Eq. (1.107). The quantity $\boldsymbol{i}_{\underline{\hat{\mathbf{r}}}}$ is the total space $\boldsymbol{i}$ but with the vector $\hat{\mathbf{r}}$ removed. For example, for rotations around the z-axis, the bivector is given by $\boldsymbol{i}\mathbf{e}_z = \mathbf{e}_x\mathbf{e}_y\mathbf{e}_z\mathbf{e}_z = \mathbf{e}_x\mathbf{e}_y = \boldsymbol{i}_{xy}$, see Section 1.2. The bivector associated with counterclockwise rotations when working on a vector to the right is $-\boldsymbol{i}_{\underline{\hat{\mathbf{r}}}}$. The eigenvectors are $\hat{\boldsymbol{r}} \sim e^{\mp \boldsymbol{i}_{\underline{\hat{\mathbf{r}}}}\varphi}$.

Let us now approach the problem using the axis of rotation, which is clearly a conserved quantity when rotating around it. Since the axis is a one-dimensional space, there are only two distinct unit vectors, parallel, and antiparallel to the unit vector $\hat{\mathbf{r}}$ indicating the axis of rotation

$$\hat{\mathbf{r}}\hat{\mathbf{r}}_\pm = \pm 1 \quad \Rightarrow \quad \hat{\mathbf{r}}_\pm = \pm\hat{\mathbf{r}}, \tag{1.121}$$

where $\pm$ corresponds to counterclockwise/clockwise rotations with respect to the rotation axis using the right-hand rule. This usage is very common in classical physics to indicate the direction of rotation. Note that working with $\hat{\mathbf{r}}$ on $\hat{\mathbf{r}}_\pm$ gives the same ± 1 as the eigenvalues of $\hat{L}$ in Eqs. (1.44) and (1.46). However, $\hat{\mathbf{r}}_\pm$ are not eigenvectors of the rotation axis, since the vectors $\hat{\mathbf{r}}_\pm$ do not appear on the right-hand side. This is problematic since often we want to do successive interactions. However, if the result reduces to a scalar, there is no vector left to act on. Using $\hat{\mathbf{r}}_\pm$ has the additional disadvantage that the vectors are not orthogonal to each other and are therefore not conducive to building a vector space. The above equation can also be rewritten in terms of the bivector by multiplying it by $-\boldsymbol{i}$,

$$-\boldsymbol{i}_{\underline{\hat{\mathbf{r}}}}\hat{\mathbf{r}}_\pm = -\boldsymbol{i}\hat{\mathbf{r}}\hat{\mathbf{r}}_\pm = \mp\boldsymbol{i}, \tag{1.122}$$

using Eq. (1.120). Again, the values $\mp\boldsymbol{i}$ are the same as the complex eigenvalues found in Eqs. (1.54) and (1.55) for counterclockwise rotations. However, again, Eq. (1.122) is not an eigenvalue equation.

Before continuing, it is convenient to change the coordinate system for the vectors. Just as the polar coordinates used in Section 1.2, the use of angles with respect to the coordinate system is more suited when discussing rotations. Additionally, since the length of a vector is unchanged under rotations, the discussion is limited to unit vectors. An arbitrary unit vector can be described in terms of two rotations. Taking the z-axis as our reference, the first rotation is around the y'-axis or in the zx'-plane around an angle θ, see Figure 1.5,

$$\hat{\mathbf{r}} = e^{\boldsymbol{i}_{x'z}\theta}\mathbf{e}_z = (\cos\theta + \sin\theta\ \boldsymbol{i}_{x'z})\mathbf{e}_z = \cos\theta\ \mathbf{e}_z + \sin\theta\ \mathbf{e}_{x'}. \tag{1.123}$$

The x'-axis can be obtained by rotating the original x-axis over an angle φ

$$\hat{\mathbf{r}} = \cos\theta\mathbf{e}_z + \sin\theta e^{\boldsymbol{i}_{yx}\varphi}\mathbf{e}_x = \cos\theta\mathbf{e}_z + \sin\theta(\cos\varphi + \boldsymbol{i}_{yx}\sin\varphi)\mathbf{e}_x. \tag{1.124}$$

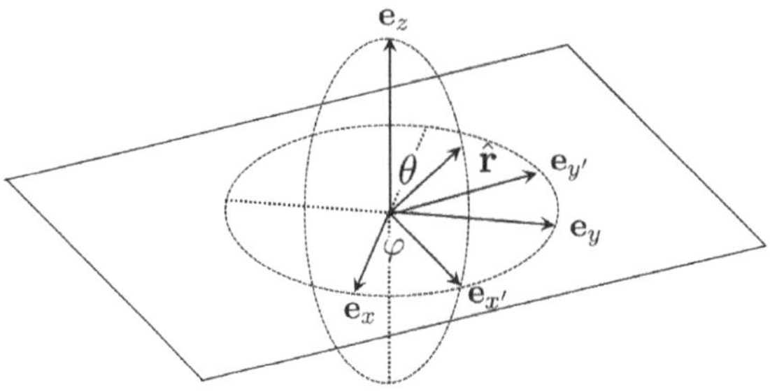

Figure 1.5 A unit vector $\hat{\mathbf{r}}$ is in an arbitrary direction in a Cartesian coordinate system and forms a plane with the reference axis $\mathbf{e}_z$. The angle between $\hat{\mathbf{r}}$ and $\mathbf{e}_z$ is θ. This plane can also be described by the two orthonormal unit vectors $\mathbf{e}_z$ and $\mathbf{e}_{x'}$. The unit vector $\mathbf{e}_{x'}$ is obtained by rotating the unit vector $\mathbf{e}_x$ of our reference system by an angle φ around the z-axis.

In order to obtain a vector with a finite length, the unit vector $\hat{\mathbf{r}}$ is multiplied by its norm r

$$\mathbf{r} = r\hat{\mathbf{r}} = r\sin\theta\cos\varphi\mathbf{e}_x + r\sin\theta\sin\varphi\mathbf{e}_y + r\cos\theta\mathbf{e}_z. \tag{1.125}$$

This is the well-known expression for a vector defined in a spherical polar coordinate system projected onto a Cartesian basis, see Figure 1.5. The coefficients can be shortened by defining

$$x = r\sin\theta\cos\varphi, \quad y = r\sin\theta\sin\varphi, \quad z = r\cos\theta, \tag{1.126}$$

which reproduces the vector in the Cartesian basis in Eq. (1.98).

To turn Eq. (1.121) into an eigenvalue equation, the same vector quantity needs to be present on both sides of the equation. This can be achieved by effectively taking the square root of $\hat{\mathbf{r}}_\pm$

$$\hat{\mathbf{r}}_\pm = \pm\hat{\mathbf{r}} = \hat{\boldsymbol{r}}_{\pm\frac{1}{2}}\hat{\boldsymbol{r}}_{\pm\frac{1}{2}}\mathbf{e}_z, \tag{1.127}$$

where $\hat{\boldsymbol{r}}_{\pm\frac{1}{2}}$ is a complex exponential in the zx'-plane. Comparison with Eq. (1.123) directly shows that

$$\hat{\boldsymbol{r}}_{\frac{1}{2}} = \hat{\boldsymbol{r}}_\uparrow = e^{\boldsymbol{i}_{x'z}\frac{\theta}{2}} \text{ and } \hat{\boldsymbol{r}}_{-\frac{1}{2}} = \hat{\boldsymbol{r}}_\downarrow = \boldsymbol{i}_{x'z}e^{\boldsymbol{i}_{x'z}\frac{\theta}{2}}, \tag{1.128}$$

where the imaginary unit $\boldsymbol{i}_{x'z}$ takes care of the minus sign needed for $\hat{\mathbf{r}}_-$. The arrows are the conventional physics notation for the half integers, $\uparrow, \downarrow = \frac{1}{2}, -\frac{1}{2}$. The vectors $\hat{\mathbf{r}}_{\pm\frac{1}{2}} = \hat{\boldsymbol{r}}_{\pm\frac{1}{2}}\mathbf{e}_z$ are known as up and down spinors.

By writing the vector as a square in Eq. (1.121), the rotations can be rearranged

$$\hat{\mathbf{r}}\hat{\boldsymbol{r}}_{\pm\frac{1}{2}}\hat{\boldsymbol{r}}_{\pm\frac{1}{2}}\mathbf{e}_z = \hat{\mathbf{r}}\hat{\boldsymbol{r}}_{\pm\frac{1}{2}}\mathbf{e}_z\hat{\boldsymbol{r}}^*_{\pm\frac{1}{2}} = \pm 1, \tag{1.129}$$

where the conjugate occurs due to the fact that $\mathbf{e}_z$ and $\boldsymbol{i}_{x'z}$ in the exponential anticommute with each other, $\boldsymbol{i}_{x'z}\mathbf{e}_z = \mathbf{e}_{x'}\mathbf{e}_z\mathbf{e}_z = -\mathbf{e}_z\mathbf{e}_{x'}\mathbf{e}_z = -\mathbf{e}_z\boldsymbol{i}_{x'z}$. This causes a conjugation of all bivectors $\boldsymbol{i}_{x'z}$. By multiplying the above equation by $\hat{\boldsymbol{r}}_{\pm\frac{1}{2}}$ from the right, an eigenvector equation is obtained

$$\hat{\mathbf{r}}\hat{\mathbf{r}}_{\pm\frac{1}{2}} = \pm\hat{\boldsymbol{r}}_{\pm\frac{1}{2}} \quad \text{or} \quad \hat{\mathbf{r}}\hat{\boldsymbol{r}}_{\pm\frac{1}{2}} = \pm\hat{\mathbf{r}}_{\pm\frac{1}{2}}, \tag{1.130}$$

with $\hat{\mathbf{r}}_{\pm\frac{1}{2}} = \hat{\boldsymbol{r}}_{\pm\frac{1}{2}}\mathbf{e}_z$. Apart from the unit vector associated with the reference axis, Eq. (1.130) is an eigenvector equation. For $\hat{\mathbf{r}}_\pm$, we can write

$$\hat{\mathbf{r}}\hat{\mathbf{r}}_\pm = \hat{\mathbf{r}}\hat{\boldsymbol{r}}_{\pm\frac{1}{2}}\hat{\boldsymbol{r}}_{\pm\frac{1}{2}}\mathbf{e}_z = \pm\hat{\boldsymbol{r}}_{\pm\frac{1}{2}}\mathbf{e}_z\hat{\boldsymbol{r}}_{\pm\frac{1}{2}}\mathbf{e}_z = \pm\hat{\boldsymbol{r}}_{\pm\frac{1}{2}}\hat{\boldsymbol{r}}^*_{\pm\frac{1}{2}}\mathbf{e}_z\mathbf{e}_z = \pm 1, \tag{1.131}$$

reproducing Eq. (1.121). Note that the rotation axis only operates on one of the $\hat{\boldsymbol{r}}_{\pm\frac{1}{2}}$.

These eigenvectors of the rotation axis are the well-known spinors. They can be associated with vectors in real space. The $\hat{\boldsymbol{r}}_{\uparrow}$ solution can be written as

$$\hat{\boldsymbol{r}}_{\uparrow} = e^{\boldsymbol{i}_{x'z}\frac{\theta}{2}} = \cos\frac{\theta}{2} + \boldsymbol{i}_{x'z}\sin\frac{\theta}{2}. \tag{1.132}$$

Rotating the reference x-axis over an angle φ to obtain the x'-axis, see Eq. (1.124) and Figure 1.5, gives

$$\hat{\mathbf{r}}_{\uparrow} = \hat{\boldsymbol{r}}_{\uparrow}\mathbf{e}_z = \cos\frac{\theta}{2}\mathbf{e}_z + e^{\boldsymbol{i}_{yx}\varphi}\sin\frac{\theta}{2}\,\mathbf{e}_x. \tag{1.133}$$

The other solution is rotated by 90° in the zx'-plane, see Eq. (1.128),

$$\hat{\mathbf{r}}_{\downarrow} = \hat{\boldsymbol{r}}_{\downarrow}\mathbf{e}_z = \boldsymbol{i}_{x'z}\hat{\mathbf{r}}_{\uparrow} = -\sin\frac{\theta}{2}\mathbf{e}_z + e^{\boldsymbol{i}_{yx}\varphi}\cos\frac{\theta}{2}\,\mathbf{e}_x. \tag{1.134}$$

The vectors $\hat{\boldsymbol{r}}_{\pm\frac{1}{2}}$ are also known as the Pauli spinors. Due to the relation in Eq. (1.120), they are also directly eigenvectors of the imaginary unit. Multiplying Eq. (1.130) by $\boldsymbol{i}$ gives

$$-\boldsymbol{i}_{\underline{\hat{\mathbf{r}}}}\hat{\mathbf{r}}_{\pm\frac{1}{2}} = \mp\boldsymbol{i}\hat{\boldsymbol{r}}_{\pm\frac{1}{2}}. \tag{1.135}$$

Note that the complex eigenvalues $\mp\boldsymbol{i}$ from Eq. (1.122) now appear in the form of an eigenvalue equation.

We can now generalize the expression for a unit vector in two dimensions in Eq. (1.51) for a plane randomly oriented in a three-dimensional space

$$|\varphi) = \sum_{\sigma=\uparrow,\downarrow}|\sigma\rangle\langle\sigma|\varphi) = \hat{\mathbf{r}}_{\uparrow}\frac{e^{-\boldsymbol{i}_{\underline{\hat{\mathbf{r}}}}\varphi}}{\sqrt{2}} + \hat{\mathbf{r}}_{\downarrow}\frac{e^{\boldsymbol{i}_{\underline{\hat{\mathbf{r}}}}\varphi}}{\sqrt{2}} \tag{1.136}$$

Note that $\hat{\mathbf{r}}$ is here the normal vector of the plane (the rotation axis) and not a unit vector in the plane. The function space $e^{\pm\boldsymbol{i}_{\underline{\hat{\mathbf{r}}}}\varphi}/\sqrt{2}$ associated with counterclockwise and clockwise rotations around $\hat{\mathbf{r}}$ consists of, what we can call, oriented complex exponentials. The spinors are effective unit vectors that contain the essential transformation properties of the functions. The spinors $|\sigma\rangle = \hat{\boldsymbol{r}}_{\pm\frac{1}{2}}$, which lie out of the plane of rotation, take over the role of the complex unit vectors $\mathbf{e}_{\pm 1}$, see Eq. (1.51), that lie in the plane of rotation. It might seem strange that we have different unit vectors for rotation. However, often we are only interested in how unit vectors transform into each under a certain operation. For rotation, $\mathbf{e}_{\pm 1}$ and $\hat{\boldsymbol{r}}_{\pm\frac{1}{2}}$ have the same transformation properties.

Example 1.4 ***Rotation Around the z-Axis***

a) Determine the spinors for a rotation around the z-axis.
b) Demonstrate that the squares of the spinors are parallel to the z-axis.

Solution

a) For the z-axis, $\theta = 0$. Additionally, we take $\varphi = 0$, although we can, in principle, choose any value of φ. The spinors are $\hat{\mathbf{r}}_{\uparrow} = \mathbf{e}_z$ and $\hat{\mathbf{r}}_{\downarrow} = \mathbf{e}_x$. These form an orthogonal set, as opposed to the rotation axes $\hat{\mathbf{r}}_{\pm} = \pm\mathbf{e}_z$.
b) The rotation axis can be obtained using $\hat{\boldsymbol{r}}_{\uparrow} = \hat{\mathbf{r}}_{\uparrow}\mathbf{e}_z = 1$ and $\hat{\boldsymbol{r}}_{\downarrow} = \hat{\mathbf{r}}_{\downarrow}\mathbf{e}_z = \boldsymbol{i}_{xz}$, giving $\hat{\mathbf{r}}_{+} = \hat{\boldsymbol{r}}_{\uparrow}\hat{\boldsymbol{r}}_{\uparrow}\mathbf{e}_z = 1 \times 1 \times \mathbf{e}_z = \mathbf{e}_z$ and $\hat{\mathbf{r}}_{-} = \hat{\boldsymbol{r}}_{\downarrow}\hat{\boldsymbol{r}}_{\downarrow}\mathbf{e}_z = \boldsymbol{i}_{xz} \times \boldsymbol{i}_{xz} \times \mathbf{e}_z = -\mathbf{e}_z$. This gives the unit vectors describing the counterclockwise and clockwise rotation axes.

Conceptual summary. – In this section, we obtained unit vectors that are eigenvectors of a vector operation associated with the rotation. The vector operation is the rotation axis $\hat{\mathbf{r}}$. The eigenvectors are the spinors $\hat{\boldsymbol{r}}_{\pm\frac{1}{2}}$ with eigenvalues ± 1. The real physical quantities are related to $\hat{\boldsymbol{r}}_{\pm\frac{1}{2}}\hat{\boldsymbol{r}}_{\pm\frac{1}{2}}\mathbf{e}_z = \pm\hat{\mathbf{r}}$, which are parallel/antiparallel to the rotation axis corresponding to counterclockwise and clockwise rotations, respectively. The spinors are therefore effectively the square roots of $\pm\hat{\mathbf{r}}$.

From the example, we saw that the spinors for rotations around the z-axis correspond to $\hat{\mathbf{r}}_\uparrow = \mathbf{e}_z$ and $\hat{\mathbf{r}}_\downarrow = \mathbf{e}_x$. Looking purely at the spinors, it is not at all obvious that the down spinor is associated with clockwise rotations around the z-axis. However, squaring the spinors gives $\pm\mathbf{e}_z$, which corresponds to the two directions of rotation around the z-axis. This is the typical basis chosen in the construction of the Pauli matrices, as we will see in the following section.

Although spinors are often considered a purely quantum-mechanical phenomenon, they are derived here using only geometric considerations. Obviously, taking the square root of a vector is a rather unusual operation with peculiar consequences. For example, since the spinors rotate only by half the angle, it takes four spin reversals to make a full circle. All of this might appear a little esoteric. However, the spinors will be used to construct the Pauli matrices. They will be generalized and used to build unit vectors for the spherical harmonics. Since the resulting unit vectors have the same transformation properties as the spherical harmonics, they behave the same under rotation. Therefore, instead of performing complicated operations on spherical harmonics, we can understand their behavior under rotation from the unit vectors, *i.e.* the spinors. Additionally, spinors are important for the description of spin arising from relativity.

Problems

1 Let us consider a rotation around the $\mathbf{e}_x$.
   a) Find the spinors $\boldsymbol{e}_{x,\pm\frac{1}{2}}$ and show that they produce $\pm\mathbf{e}_x$.
   b) Show that the spinors are eigenvectors of the rotation axis.
   c) Show that the spinors are also eigenvectors of the imaginary unit $\boldsymbol{i}_{zy}$.

2 Although exponentials are convenient, show that the rotation axis can also be obtained using the expression for $\hat{\boldsymbol{r}}_\uparrow$ in Eq. (1.132) in terms of trigonometric functions.

3 In Eq. (1.53), we found complex unit vectors that had the same transformation properties under the imaginary unit $\boldsymbol{i}_{yx}$ as the spinors found in this section. Find an appropriate imaginary unit that turns the complex unit vectors into real spinors. Show the transformation properties under rotation axis $\mathbf{e}_z$ and discuss the consequences of the results.

1.7 Pauli Matrices

Introduction. – In the preceding sections, the behavior of vectors was described by treating them as anticommuting algebraic quantities. In quantum mechanics, similar behavior is usually obtained with 2×2 matrices, known as the Pauli matrices. Similar to the unit vectors, there are three Pauli matrices in three dimensions. Although they are only 2×2 matrices, they work in an effective three-dimensional space, with the imaginary unit providing an additional degree of freedom. Additionally, the basis of the Pauli matrices is the same as those of the spinors for rotations around the z-axis as derived in the previous section. Many of the properties from the previous sections can be derived for Pauli matrices.

The spinors in Eqs. (1.133) and (1.134) are expressed in a basis of two unit vectors $\mathbf{e}_z$ and $\mathbf{e}_x$. Note that this basis corresponds to the two spinors for rotation around the z-axis, as derived in the example in the previous section. Therefore, the z-axis is the reference axis of the spinors. The component of the spinor in the x-direction contains the imaginary unit $\boldsymbol{i}_{yx}$ to indicate rotations in the xy-plane. So far, the properties of vectors were described using geometric algebra. However, these properties can also be expressed in terms of matrices. To this end, we first need to define a basis. The unit vectors can be defined as

$$|\mathbf{e}_z\rangle = \begin{pmatrix} 1 \\ 0 \end{pmatrix} \text{ and } |\mathbf{e}_x\rangle = \begin{pmatrix} 0 \\ 1 \end{pmatrix}, \tag{1.137}$$

using bra-ket notation to indicate a vector. This only contains two directions z and x. The y-axis will be included using an imaginary unit. Therefore, the coefficients of the vector can be complex. Since the z-axis is effectively our reference axis, the associated Pauli matrix will be diagonal. Unlike the algebraic approach, where each direction is treated in an equivalent fashion, the Pauli matrices are different for each direction due to the fact that we need to pick a basis for the matrices.

The conjugates of the unit vectors in Eq. (1.137) are given by

$$\langle \mathbf{e}_z| = \begin{pmatrix} 1 & 0 \end{pmatrix} \text{ and } \langle \mathbf{e}_x| = \begin{pmatrix} 0 & 1 \end{pmatrix}. \tag{1.138}$$

The inner product is given by the product of a bra and a ket. For example,

$$\langle \mathbf{e}_z|\mathbf{e}_z\rangle = \begin{pmatrix} 1 & 0 \end{pmatrix}\begin{pmatrix} 1 \\ 0 \end{pmatrix} = 1, \tag{1.139}$$

where the product of a 1×2 times a 2×1 matrix is a 1×1 matrix, *i.e.* a scalar (the parentheses are usually omitted for the scalar). The product of the ket and the bra is given by a matrix, for example

$$|\mathbf{e}_z\rangle\langle \mathbf{e}_z| = \begin{pmatrix} 1 \\ 0 \end{pmatrix}\begin{pmatrix} 1 & 0 \end{pmatrix} = \begin{pmatrix} 1 & 0 \\ 0 & 0 \end{pmatrix}, \tag{1.140}$$

where the product of a 2×1 times a 1×2 matrix is a 2×2 matrix. This allows us to write the identity matrix as

$$\mathbb{1}_2 = \begin{pmatrix} 1 & 0 \\ 0 & 1 \end{pmatrix} = |\mathbf{e}_z\rangle\langle \mathbf{e}_z| + |\mathbf{e}_x\rangle\langle \mathbf{e}_x|. \tag{1.141}$$

We now want to set up 2×2 matrices that describe the operation of vectors on other vectors. Since the Pauli matrices are an alternative way of representing the properties of a three-dimensional Cartesian space, the properties of the anticommuting algebraic unit vectors can be used to find the matrix elements of the Pauli matrices. The general expression for obtaining the matrix elements is written as

$$\mathbf{e}_k\mathbf{e}_j = \mathbf{e}_i\mathbf{e}_z(\mathbf{e}_k)_{ij}, \tag{1.142}$$

where $\mathbf{e}_k$, with $k = x, y, z$, is the vector operating on a basis vector; the vector $\mathbf{e}_k$ transforms $\mathbf{e}_j$ into $\mathbf{e}_i$, with $i, j = z, x$; the vector $\mathbf{e}_z$ on the right-hand side is included since the matrices need to be defined with respect to a certain axis, and the z-axis is the conventional choice. The coefficient $(\mathbf{e}_k)_{ij}$ necessary to make the left- and right-hand side equal is the resulting matrix element. We now want to write all the transformation into the format of Eq. (1.142).

For $k = z$, this gives

$$\mathbf{e}_z\mathbf{e}_z \quad \text{and} \quad \mathbf{e}_z\mathbf{e}_x = \mathbf{e}_x\mathbf{e}_z(-1) \tag{1.143}$$

for $j = x, z$, respectively. When necessary, expressions are rewritten in the form $\mathbf{e}_i\mathbf{e}_z(\mathbf{e}_k)_{ij}$. Both $\mathbf{e}_z$ and $\mathbf{e}_x$ end up in themselves under the operation of $\mathbf{e}_z$. However, they have a different matrix element, namely $(\mathbf{e}_z)_{zz} = 1$ and $(\mathbf{e}_z)_{xx} = -1$. From the previous section, we know that $\hat{\mathbf{r}}_{\frac{1}{2}} = \mathbf{e}_z$ and $\hat{\mathbf{r}}_{-\frac{1}{2}} = \mathbf{e}_x$ are the spinors for counterclockwise and clockwise rotations around the z-axis, which explains the eigenvalues of ± 1, see Eq. (1.130). This can be written in matrix form as

$$\mathbf{e}_z = \begin{pmatrix} 1 & 0 \\ 0 & -1 \end{pmatrix}. \tag{1.144}$$

This is the Pauli matrix for the z-direction.

Operating $\mathbf{e}_x$ on the unit vectors $\mathbf{e}_z$ and $\mathbf{e}_x$ that form the basis gives

$$\mathbf{e}_x\mathbf{e}_z \quad \text{and} \quad \mathbf{e}_x\mathbf{e}_x = 1 = \mathbf{e}_z\mathbf{e}_z, \tag{1.145}$$

where the reference axis $\mathbf{e}_z$ always needs to be present on the right-hand side. Again, the end result of the operation is written in the form $\mathbf{e}_i\mathbf{e}_z(\mathbf{e}_k)_{ij}$. Operating with $\mathbf{e}_x$ transforms the unit vectors into each other. In matrix form, this gives

$$\mathbf{e}_x = \begin{pmatrix} 0 & 1 \\ 1 & 0 \end{pmatrix}. \tag{1.146}$$

For $k = y$, one finds

$$\begin{aligned} \mathbf{e}_y\mathbf{e}_z &= -\mathbf{e}_z\mathbf{e}_y = -\mathbf{e}_z\mathbf{e}_x\mathbf{e}_x\mathbf{e}_y = \mathbf{e}_x\mathbf{e}_z\boldsymbol{\imath}_{xy}, \\ \mathbf{e}_y\mathbf{e}_x &= \mathbf{e}_z\mathbf{e}_z\mathbf{e}_y\mathbf{e}_x = -\mathbf{e}_z\mathbf{e}_z\boldsymbol{\imath}_{xy}, \end{aligned} \tag{1.147}$$

giving $(\mathbf{e}_y)_{xz} = \boldsymbol{\imath}_{xy}$ and $(\mathbf{e}_y)_{zx} = -\boldsymbol{\imath}_{xy}$. Again, the vectors are transformed into each other, but this time with the imaginary unit as coefficient

$$\mathbf{e}_y = \begin{pmatrix} 0 & -\boldsymbol{\imath}_{xy} \\ \boldsymbol{\imath}_{xy} & 0 \end{pmatrix} \rightarrow \begin{pmatrix} 0 & -i \\ i & 0 \end{pmatrix}. \tag{1.148}$$

In the conventional notation, the bivector $\boldsymbol{\imath}_{xy}$ is replaced by the scalar imaginary unit i, since it is the only imaginary unit left in the 2×2 matrices.

In summary, we have found the following matrices:

$$\mathbf{e}_x = \begin{pmatrix} 0 & 1 \\ 1 & 0 \end{pmatrix}, \quad \mathbf{e}_y = \begin{pmatrix} 0 & -i \\ i & 0 \end{pmatrix}, \quad \mathbf{e}_z = \begin{pmatrix} 1 & 0 \\ 0 & -1 \end{pmatrix}. \tag{1.149}$$

These are the well-known Pauli matrices. They are usually denoted as σ_k, with $k = x, y, z$. Here, they are indicated as $\mathbf{e}_k$, since they behave similarly to the anticommuting unit vectors from Section 1.5. First, they satisfy normalization

$$\mathbf{e}_i^2 = \mathbb{1}_2, \quad \text{with } i = x, y, z, \tag{1.150}$$

where the identity matrix is given in Eq. (1.141). In addition, the Pauli matrices satisfy the anticommutation relation (1.101)

$$\mathbf{e}_i\mathbf{e}_j + \mathbf{e}_j\mathbf{e}_i = 0, \quad \text{for } i \neq j. \tag{1.151}$$

Therefore, the Pauli matrices behave in exactly the same way as the anticommuting unit vectors described in Section 1.5, which can be checked by straightforward matrix multiplication.

Since the Pauli matrices $\mathbf{e}_i$ correspond to unit vectors, a general vector can be written as

$$\mathbf{r} = x\mathbf{e}_x + y\mathbf{e}_y + z\mathbf{e}_z = \begin{pmatrix} z & x - iy \\ x + iy & -z \end{pmatrix}. \tag{1.152}$$

The norm of the vector is given by the equation

$$\mathbf{r} = r\mathbb{1}_2 \quad \Rightarrow \quad \begin{pmatrix} z - r & x - iy \\ x + iy & -z - r \end{pmatrix} = 0. \tag{1.153}$$

Note that this equation mixes the vector and the scalar terms. This possibility was already anticipated using the algebraic unit vectors. This equation can be solved by finding the determinant of the matrix

$$\begin{aligned} \begin{vmatrix} z - r & x - iy \\ x + iy & -z - r \end{vmatrix} &= (z - r)(-z - r) - (x + iy)(x - iy) \\ &= r^2 - z^2 - x^2 - y^2 = 0. \end{aligned} \tag{1.154}$$

This gives two solutions

$$r = \pm\sqrt{x^2 + y^2 + z^2}. \tag{1.155}$$

At first, it might seem strange that the norm has positive and negative values. However, this corresponds to the earlier discussion, where a vector effectively is a rotation axis, and there are two solutions $\mathbf{r}_\pm = \pm\mathbf{r}$ corresponding to the two directions of rotation around the axis. The norm can also be obtained using geometric algebra methods,

$$\mathbf{r} = r \quad \Rightarrow \quad \mathbf{r}^2 = r^2 \quad \Rightarrow \quad r = \pm\sqrt{\mathbf{r}^2} = \pm\sqrt{x^2 + y^2 + z^2}, \tag{1.156}$$

which gives the same result without having to solve a determinant.

Example 1.5 ***Two Coupled Oscillators***

In this example, we treat a classical problem to demonstrate that the concepts of mixing scalar and vector quantities are not purely quantum mechanical. We will look at two coupled harmonic oscillators. The underlying physics is comparable to the bonding and antibonding states in a diatomic molecule. Two objects of equal mass m are connected to a fixed point with springs with a spring constant K_0, see Figure 1.6. Additionally, the masses are connected to each other with a spring with a spring constant K_1.

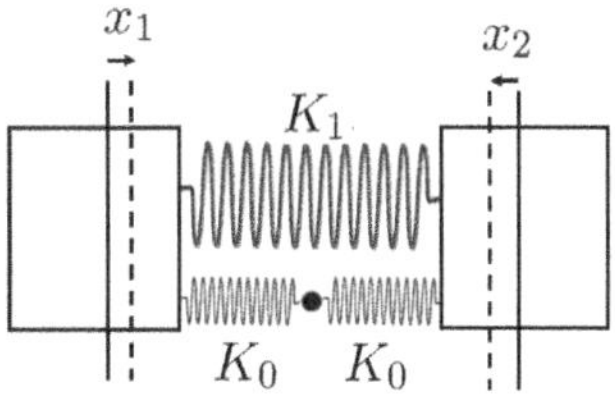

Figure 1.6 Two objects of equal mass m are coupled to a fixed position with a spring with a spring constant K_0. Additionally, they are coupled to each other with a spring with a spring constant K_1.

a) Write down Newton's second law in terms of the displacements x_i, with $i = 1, 2$, of the two masses.
b) Assume an oscillatory solution of the type $x_i = A_i e^{i\omega t}$ and derive a pair of coupled equations for the angular frequency ω.
c) Write the equations from (b) in terms of 2×2 matrices and express the result in terms of the identity matrix $\mathbb{1}_2$ and the Pauli matrices $\mathbf{e}_i$, with $i = x, y, z$.
d) Calculate the eigenfrequencies ω by calculating the norm of the vector.
e) Calculate the spinors of the system.

Solution

a) The force of a spring is given by $F = -Kx$. Newton's equation of motion for each mass gives

$$m\frac{d^2x_1}{dt^2} = -K_0x_1 - K_1(x_1 - x_2), \tag{1.157}$$

$$m\frac{d^2x_2}{dt^2} = -K_0x_2 - K_1(x_2 - x_1). \tag{1.158}$$

Note that the extension of the connecting spring depends on the displacement of both masses.

b) Inserting a solution of the type $x_i = A_i e^{i\omega t}$ gives

$$-m\omega^2 A_1 = -K_0A_1 - K_1(A_1 - A_2), \tag{1.159}$$

$$-m\omega^2 A_2 = -K_0A_2 - K_1(A_2 - A_1), \tag{1.160}$$

dividing out the complex exponential. Using $\omega_j^2 = K_j/m$, this can be rewritten as

$$(\omega_0^2 + \omega_1^2 - \omega^2)A_1 - \omega_1^2 A_2 = 0, \tag{1.161}$$

$$-\omega_1^2 A_1 + (\omega_0^2 + \omega_1^2 - \omega^2)A_2 = 0. \tag{1.162}$$

c) In terms of 2×2 unit vectors and scalars, the result can be written as

$$\left((\omega_0^2 + \omega_1^2 - \omega^2)\mathbb{1}_2 - \omega_1^2\mathbf{e}_x\right)\begin{pmatrix} A_1 \\ A_2 \end{pmatrix} = 0, \tag{1.163}$$

d) In order to solve the equation, we require

$$(\omega_0^2 + \omega_1^2 - \omega^2)\mathbb{1}_2 = \omega_1^2\mathbf{e}_x. \tag{1.164}$$

Squaring this gives

$$(\omega_0^2 + \omega_1^2 - \omega^2)^2 = \omega_1^4 \quad \Rightarrow \quad \omega_0^2 + \omega_1^2 - \omega^2 = \mp\omega_1^2. \tag{1.165}$$

The eigenfrequencies are then

$$\omega = \sqrt{\omega_0^2 + 2\omega_1^2}, \omega_0. \tag{1.166}$$

This result can also be obtained by solving the determinant.

e) Since the perturbation caused by the coupling of the two masses via a spring is in the direction $-\mathbf{e}_x$, the system aligns itself in this direction. The scalar term is irrelevant for the spinors. Now since $-\mathbf{e}_x = e^{-\boldsymbol{i}_{xz}\frac{\pi}{2}}\mathbf{e}_z$, the two spinors (the eigenstates) are given by

$$\hat{\mathbf{r}}_{\frac{1}{2}} = e^{-\boldsymbol{i}_{xz}\frac{\pi}{4}}\mathbf{e}_z = \frac{1}{\sqrt{2}}(\mathbf{e}_z - \mathbf{e}_x) \text{ and } \hat{\mathbf{r}}_{-\frac{1}{2}} = \boldsymbol{i}_{xz}\hat{\mathbf{r}}_{\frac{1}{2}} = \frac{1}{\sqrt{2}}(\mathbf{e}_z + \mathbf{e}_x). \quad (1.167)$$

The z and x axes are associated with A_1 and A_2, respectively, and the two eigenstates correspond to the situation where the springs are moving in-phase and 90° out-of-phase with each other, respectively. This explains why the in-phase motion has an eigenfrequency ω_0, since the connecting spring does not compress when the masses move in the same direction.

Conceptual summary. – In this section, it was demonstrated that the properties of the unit vectors can also be incorporated in matrix form. The Cartesian unit vectors for three dimensions can be represented by complex 2×2 matrices, known as the Pauli matrices. Just as the algebraic unit vectors introduced in the preceding sections, Pauli matrices are normalized and anticommute. A vector can now be expressed as a 2×2 matrix. There are two solutions for the norm corresponding to the two different directions of rotation around the vector $\mathbf{r}$. It is important to note that all results derived with anticommuting algebraic unit vectors are still valid when written in terms of Pauli matrices.

Although thinking of Pauli matrices in terms of unit vectors is intuitive, this language is often avoided, and one talks, for example of a Pauli matrix basis. Representative of this is the Pauli vector, which is generally used to express Eq. (1.152). The Pauli vector is a vector constructed out of Pauli matrices $\boldsymbol{\sigma} = \sigma_x \boldsymbol{e}_x + \sigma_y \boldsymbol{e}_y + \sigma_z \boldsymbol{e}_z$, where σ_i denotes the Pauli matrices, and $\boldsymbol{e}_i$ is a conventional unit vector as opposed to the anticommuting algebraic unit vector $\mathbf{e}_i$. Note that the Pauli vector looks somewhat peculiar in the notation used here: $\boldsymbol{\sigma} = \mathbf{e}_x \boldsymbol{e}_x + \mathbf{e}_y \boldsymbol{e}_y + \mathbf{e}_z \boldsymbol{e}_z$. A regular vector is given by $\mathbf{r} = x\boldsymbol{e}_x + y\boldsymbol{e}_y + z\boldsymbol{e}_z$. The Pauli vector is then used to map from the vector basis to the Pauli matrix basis $\mathbf{r} \cdot \boldsymbol{\sigma} = x\sigma_x + y\sigma_y + z\sigma_z$. However, this is equivalent to $\mathbf{r} = x\mathbf{e}_x + y\mathbf{e}_y + z\mathbf{e}_z$ in Eq. (1.152), which was obtained without invoking the conventional unit vectors $\boldsymbol{e}_i$ entirely. Note that the expression $\mathbf{r} \cdot \boldsymbol{\sigma}$ is somewhat misleading. The inner product between two vectors creates the impression that the end result is a scalar. Although it can be written as a single 2×2 matrix, see Eq. (1.152), the matrix has the properties of a vector and not a scalar. Since $\mathbf{r} \cdot \boldsymbol{\sigma}$ is somewhat cumbersome and confusing, it seems preferable to directly express $\mathbf{r}$ in terms of anticommuting unit vectors $\mathbf{e}_i$.

Problems

1 Let us revisit Problem 2 of Section 1.5 and find the product of $3\mathbf{e}_x + \mathbf{e}_y$ and $\mathbf{e}_x - 2\mathbf{e}_y$, but now in terms of Pauli matrices.

2 Show that $\mathbf{r}^2 = r^2$ in Pauli matrices.

3 Calculate the pseudoscalar $\boldsymbol{i}$ in terms of Pauli matrices. Also calculate $\mathbf{e}_z\mathbf{e}_y\mathbf{e}_x$. Show that the relation between the two quantities can also be obtained by anticommutation.

4 Calculate $\boldsymbol{i}_{zy}$, $\boldsymbol{i}_{zy}\mathbf{e}_y$, and $\boldsymbol{i}_{zy}\mathbf{e}_z$. Derive the same result using anticommutation.

5 The Pauli matrices express vectors in terms of 2×2 matrices. A general real 2×2 matrix can be written as

$$\boldsymbol{H} = \begin{pmatrix} \varepsilon_1 & V \\ V & \varepsilon_0 \end{pmatrix}. \tag{1.168}$$

a) Split the multivector $\boldsymbol{H}$ into a scalar E_0 and a vector $\mathbf{H}$.
b) Determine the norm of the multivector.
c) Find the unit vector parallel to the vector part and show that it is *not* an eigenvector of $\mathbf{H}$.
d) Determine the spinors and show that they are eigenvectors of $\mathbf{H}$.
e) Express $\mathbf{H}$ in the basis of the spinors and show that this is equivalent to $\mathbf{H}$ in the original basis.

1.8 Rotation Matrices

Introduction. – Using geometric algebra methods, we have seen that vectors can be rotated using a complex exponential of an imaginary unit consisting of a bivector of the unit vectors in the plane of rotation. In the preceding section, it was shown that the properties of the anticommuting unit vectors can also be described using 2×2 matrices. As we will see, the same can be done for bivectors and rotation.

In the previous section, we looked at the vector aspect of the Pauli matrices. The actual rotations are related to the bivectors, which are the dual space of the vectors. For geometric algebra, the bivectors can be obtained using Eq. (1.107), giving $\boldsymbol{i}_{ij} = \mathbf{e}_i\mathbf{e}_j = \underline{\mathbf{e}}_k = \boldsymbol{i}\mathbf{e}_k$ for $k = x, y, z$ and $ij = yz, zx, xy$. The rotations acting to the right are given by $e^{-\boldsymbol{i}_{ij}\theta} = e^{\boldsymbol{i}_{ji}\theta}$. For Pauli matrices, the results are similar, except that we need to replace the anticommuting algebraic unit vectors by the corresponding 2×2 matrices.

The pseudoscalar in three dimensions, see Eq. (1.105), is given by

$$\boldsymbol{i} = \mathbf{e}_x\mathbf{e}_y\mathbf{e}_z = \begin{pmatrix} i & 0 \\ 0 & i \end{pmatrix} = i\mathbb{1}_2, \tag{1.169}$$

which can be obtained by straightforward multiplication of the Pauli matrices in Eq. (1.149). The square of the pseudoscalar is $\boldsymbol{i}^2 = -\mathbb{1}_2$ showing that it is an imaginary unit. Note that $\mathbf{e}_z\mathbf{e}_y\mathbf{e}_x = -\boldsymbol{i} = -i\mathbb{1}_2$, which is the reason for choosing $\mathbf{e}_x\mathbf{e}_y\mathbf{e}_z$ as the definition for the pseudoscalar.

The bivectors can be written out as

$$\boldsymbol{i}_{zy} = -\boldsymbol{i}\mathbf{e}_x = \begin{pmatrix} 0 & -i \\ -i & 0 \end{pmatrix}, \quad \boldsymbol{i}_{xz} = -\boldsymbol{i}\mathbf{e}_y = \begin{pmatrix} 0 & -1 \\ 1 & 0 \end{pmatrix},$$
$$\boldsymbol{i}_{yx} = -\boldsymbol{i}\mathbf{e}_z = \begin{pmatrix} -i & 0 \\ 0 & i \end{pmatrix}. \tag{1.170}$$

Note that the results can also be obtained by multiplying the two unit vectors contained in the bivector. Since the Pauli matrices are written in the basis of the spinors $\mathbf{e}_z$ and $\mathbf{e}_x$ corresponding to rotations around $\mathbf{e}_z$, the rotation matrix $\boldsymbol{i}_{yx} = -\boldsymbol{i}\mathbf{e}_z$ is diagonal. Again, we see that the diagonal matrix elements are $\mp i$, as we have already seen in Eqs. (1.54), (1.55), and (1.135). The bivectors are expected to be imaginary units with a square $\boldsymbol{i}_{ij}^2 = -1$. This is indeed the case, for example

$$\boldsymbol{i}_{xz}^2 = \mathbf{e}_x\mathbf{e}_z\mathbf{e}_x\mathbf{e}_z = \begin{pmatrix} 0 & -1 \\ 1 & 0 \end{pmatrix}\begin{pmatrix} 0 & -1 \\ 1 & 0 \end{pmatrix} = \begin{pmatrix} -1 & 0 \\ 0 & -1 \end{pmatrix} = -\mathbb{1}_2. \tag{1.171}$$

This result can also be obtained by anticommutation of $\mathbf{e}_x\mathbf{e}_z\mathbf{e}_x\mathbf{e}_z = -\mathbf{e}_x\mathbf{e}_x\mathbf{e}_z\mathbf{e}_z = -1$ using $\mathbf{e}_i^2 = 1$.

Let us look at the rotation in the zx-plane or around the y-axis. These rotations are entirely equivalent to those in a two-dimensional system, and one can consider the column vector representing the z and x components of a vector in the zx-plane (or x and y components when using the notation in Section 1.2). A unit vector in the zx-plane is given by

$$|\hat{\mathbf{r}}(\theta)) = \cos\theta|\mathbf{e}_z\rangle + \sin\theta|\mathbf{e}_x\rangle = \begin{pmatrix} \cos\theta \\ \sin\theta \end{pmatrix}. \tag{1.172}$$

A rotation in the zx-plane is given by the complex exponential in Eq. (1.123),

$$e^{\boldsymbol{i}_{xz}\theta} = \cos\theta\ \mathbb{1}_2 + \sin\theta\ \boldsymbol{i}_{xz} = \begin{pmatrix} \cos\theta & -\sin\theta \\ \sin\theta & \cos\theta \end{pmatrix}. \tag{1.173}$$

This is the well-known matrix for a rotation in a two-dimensional plane. The operation can be checked by multiplication,

$$\begin{aligned} e^{\boldsymbol{i}_{xz}\theta'}|\hat{\mathbf{r}}(\theta)) &= \begin{pmatrix} \cos\theta' & -\sin\theta' \\ \sin\theta' & \cos\theta' \end{pmatrix}\begin{pmatrix} \cos\theta \\ \sin\theta \end{pmatrix} \\ &= \begin{pmatrix} \cos\theta'\cos\theta - \sin\theta'\sin\theta \\ \sin\theta'\cos\theta + \cos\theta'\sin\theta \end{pmatrix} = \begin{pmatrix} \cos(\theta+\theta') \\ \sin(\theta+\theta') \end{pmatrix} = |\hat{\mathbf{r}}(\theta+\theta')), \end{aligned} \tag{1.174}$$

where the vector is indicated by its polar coordinates in the zx-plane $\hat{\mathbf{r}}(\theta)$, and use has been made of the angle addition formulas in Eq. (1.32). Note that this corresponds to the operation in geometric-algebra format

$$e^{\boldsymbol{i}_{xz}\theta'}\hat{\mathbf{r}}(\theta) = e^{\boldsymbol{i}_{xz}\theta'}e^{\boldsymbol{i}_{xz}\theta}\mathbf{e}_z = e^{\boldsymbol{i}_{xz}(\theta'+\theta)}\mathbf{e}_z = \hat{\mathbf{r}}(\theta+\theta'), \tag{1.175}$$

which demonstrates that matrix multiplication can sometimes be somewhat heavy handed. The other difference is that the matrix approach makes a distinction between vectors in terms of matrices and column vectors. This is because the matrices transform both axes simultaneously. Performing the same operation as in the above equation gives

$$e^{\boldsymbol{i}_{xz}\theta'}e^{\boldsymbol{i}_{xz}\theta}\mathbf{e}_z = \begin{pmatrix} \cos(\theta+\theta) & -\sin(\theta+\theta) \\ \sin(\theta+\theta) & \cos(\theta+\theta) \end{pmatrix}\begin{pmatrix} 1 \\ 0 \end{pmatrix} = \begin{pmatrix} \cos(\theta+\theta') \\ \sin(\theta+\theta') \end{pmatrix},$$

as expected. The first and second columns in the matrix give the rotation of the z and x axes, respectively. If one is only interested in the rotation of a vector, only the first column is needed.

Since the methods to find eigenvectors of matrices have not been discussed yet, we use our knowledge of spinors to suggest a solution for the eigenvectors of the rotation and check via matrix multiplication. The spinors for the rotation around the y-axis should satisfy an equation of the type $\boldsymbol{i}_{xz}\hat{\mathbf{r}}_{\pm\frac{1}{2}} = \mp i\hat{\mathbf{r}}_{\pm\frac{1}{2}}$, see Eq. (1.135). The eigenvectors, *i.e.* the spinors, are related to the conserved quantity, which is the rotation axis. For a rotation in the zx-plane, the axis of rotation is in the y-direction. The y-axis can be written with respect to the z-axis (the reference axis of our coordinate system) as $\mathbf{e}_y = e^{\boldsymbol{i}_{yz}\frac{\pi}{2}}\mathbf{e}_z$. The spinors are effectively the square roots of the rotation axis that need to satisfy $\boldsymbol{e}_{y,\pm\frac{1}{2}}\boldsymbol{e}_{y,\pm\frac{1}{2}}\mathbf{e}_z = \pm\mathbf{e}_y$. This gives

$$\boldsymbol{e}_{y,\pm\frac{1}{2}} = e^{\pm\boldsymbol{i}_{yz}\frac{\pi}{4}} = \frac{1}{\sqrt{2}}(1 \pm \boldsymbol{i}_{yz}). \tag{1.176}$$

We can check that $\boldsymbol{i}_{xz}\mathbf{e}_{y,\pm\frac{1}{2}} = \mp i\boldsymbol{e}_{y,\pm\frac{1}{2}}$. In vector form, the spinor is written as

$$\mathbf{e}_{y,\pm\frac{1}{2}} = \boldsymbol{e}_{y,\pm\frac{1}{2}}\mathbf{e}_z = \frac{1}{\sqrt{2}}(\mathbf{e}_z \pm \mathbf{e}_y) = \frac{1}{\sqrt{2}}(\mathbf{e}_z \pm \mathbf{e}_x\boldsymbol{i}_{xy}) \to \frac{1}{\sqrt{2}}(\mathbf{e}_z \pm i\mathbf{e}_x), \tag{1.177}$$

where the spinor is rewritten in terms of the basis of the Pauli matrices, and the bivector $\boldsymbol{i}_{xy}$ has been replaced by the pseudoscalar i. Note that the spinor takes on a complex form comparable to Eq. (1.53). In column format, the expected spinors are

$$\boldsymbol{e}_{y,\pm\frac{1}{2}} = |y, \pm\frac{1}{2}\rangle = \frac{1}{\sqrt{2}}\begin{pmatrix} 1 \\ \pm i \end{pmatrix}. \tag{1.178}$$

It can be checked by straightforward matrix multiplication that $\boldsymbol{i}_{xz}|y,\pm\frac{1}{2}\rangle = \mp i|y,\pm\frac{1}{2}\rangle$. The spinors can also be obtained by finding the eigenvalues and eigenvectors of $\boldsymbol{i}_{xz}$ using standard methods. However, this leads to complex eigenvectors without providing much insight into their physical meaning.

The rotation around the y-axis should be diagonal in the spinors related to the y-axis. The diagonal terms in the spinor basis can be expressed in the original $1, \boldsymbol{i}_{xz}$ (or $\mathbf{e}_z, \mathbf{e}_x$) basis as

$$|y, \pm\frac{1}{2}\rangle\langle y, \pm\frac{1}{2}| = \frac{1}{2}\begin{pmatrix} 1 \\ \pm i \end{pmatrix}(1, \mp i) = \frac{1}{2}\begin{pmatrix} 1 & \mp i \\ \pm i & 1 \end{pmatrix}. \tag{1.179}$$

The imaginary unit is diagonal in the spinor basis but off-diagonal in the original basis

$$\boldsymbol{i}_{xz} = -i|y, \frac{1}{2}\rangle\langle y, \frac{1}{2}| + i|y, -\frac{1}{2}\rangle\langle y, -\frac{1}{2}| = \begin{pmatrix} 0 & -1 \\ 1 & 0 \end{pmatrix}, \tag{1.180}$$

where $\mp i$ are the eigenvalues in the spinor basis. The right-hand side shows that this is indeed equivalent to the imaginary unit in the original basis, see Eq. (1.170). Additionally, the identity matrix can also be written in terms of the spinors

$$\mathbb{1}_2 = |y, \frac{1}{2}\rangle\langle y, \frac{1}{2}| + |y, -\frac{1}{2}\rangle\langle y, -\frac{1}{2}| = \begin{pmatrix} 1 & 0 \\ 0 & 1 \end{pmatrix}. \tag{1.181}$$

This allows the entire rotation around the y-axis to be expressed in terms of the spinors $|y, \pm\frac{1}{2}\rangle$,

$$e^{\boldsymbol{i}_{xz}\theta} = \cos\theta\ \mathbb{1}_2 + \sin\theta\ \boldsymbol{i}_{xz} = e^{-i\theta}|y, \frac{1}{2}\rangle\langle y, \frac{1}{2}| + e^{i\theta}|y, -\frac{1}{2}\rangle\langle y, -\frac{1}{2}|. \tag{1.182}$$

The coefficients $e^{\mp i\theta}$ are now the eigenfunctions of the angular momentum operator. We have effectively made a unitary transformation, such that the y-axis, *i.e.* the axis of rotation, is now the reference axis of the system.

Example 1.6 ***Rotations Around the z-Axis***

a) Find the matrix for rotations around the z-axis.
b) Derive the matrix elements directly from the unit vectors.

Solution

a) The rotation around the z-axis can be expressed in terms of a complex exponential of the related bivector $\boldsymbol{i}_{yx}$,

$$e^{\boldsymbol{i}_{yx}\theta} = \cos\theta + \boldsymbol{i}_{yx}\sin\theta = \begin{pmatrix} e^{-i\theta} & 0 \\ 0 & e^{i\theta} \end{pmatrix}, \tag{1.183}$$

using Eq. (1.170) and $\mathbb{1}_2 \equiv 1$. Note that any rotation expressed in the basis of its associated spinors has this form, see, for example Eq. (1.180), where the y-axis is the reference axis. The results are much more easily obtained for rotations around the z-axis, since $\mathbf{e}_z$ is the reference axis of the Pauli matrices.

b) The spinors for the z-axis are $\boldsymbol{e}_{z,\pm\frac{1}{2}} = 1, \boldsymbol{i}_{xz}$. This gives the rotation axes $\hat{\mathbf{r}}_+ = \boldsymbol{e}_{z,\frac{1}{2}}\boldsymbol{e}_{z,\frac{1}{2}}\mathbf{e}_z = \mathbf{e}_z$ and $\hat{\mathbf{r}}_- = \boldsymbol{e}_{z,-\frac{1}{2}}\boldsymbol{e}_{z,-\frac{1}{2}}\mathbf{e}_z = \boldsymbol{i}_{xz}^2\mathbf{e}_z = -\mathbf{e}_z$, with $\pm$ denoting counterclockwise and clockwise rotations, respectively. Following Eq. (1.142), the matrix elements are then obtained via

$$\begin{aligned} \boldsymbol{i}_{yx}\boldsymbol{e}_{z,\frac{1}{2}} &= \boldsymbol{i}_{yx} \times 1 = 1 \times \boldsymbol{i}_{yx} = \boldsymbol{e}_{z,\frac{1}{2}}\mathbf{e}_z(-\boldsymbol{i}), \\ \boldsymbol{i}_{yx}\boldsymbol{e}_{z,-\frac{1}{2}} &= \boldsymbol{i}_{yx}\boldsymbol{i}_{xz} = \boldsymbol{i}_{xz}(-\boldsymbol{i}_{yx}) = \boldsymbol{e}_{z,-\frac{1}{2}}\mathbf{e}_z\boldsymbol{i}, \end{aligned} \tag{1.184}$$

with $\boldsymbol{i}_{yx} = \boldsymbol{i}_{yx}\mathbf{e}_z\mathbf{e}_z = -\mathbf{e}_x\mathbf{e}_y\mathbf{e}_z\mathbf{e}_z = -\mathbf{e}_z\boldsymbol{i}$. The total rotation is then given by

$$e^{\boldsymbol{i}_{yx}\theta}\boldsymbol{e}_{z,\pm\frac{1}{2}} = \boldsymbol{e}_{z,\pm\frac{1}{2}}(\cos\theta \pm \sin\theta\,\boldsymbol{i}_{yx}) = \boldsymbol{e}_{z,\pm\frac{1}{2}}e^{\pm\boldsymbol{i}_{yx}\theta}. \tag{1.185}$$

Conceptual summary. – In this section, it was demonstrated that rotations can be obtained using Pauli matrices in a way very similar to anticommuting unit vectors. The advantage of Pauli matrices is the readily available (numerical) tools for matrix multiplication and the determination of eigenvalues and eigenvectors. The disadvantage is a lack of transparency in the equations, making some of the results less easily understandable. For example, the rotation in the zx-plane is a real matrix. However, it is not obvious why the eigenvectors of the imaginary unit $\boldsymbol{i}_{xz}$ are $\frac{1}{\sqrt{2}}\begin{pmatrix} 1 \\ \pm i \end{pmatrix}$ with complex eigenvalues $\mp i$. The complex eigenvalues can be understood from the fact that the rotation axis is the conserved quantity. For the rotation axis, $\mathbf{e}_y = \boldsymbol{i}\boldsymbol{i}_{xz}$, the eigenvalues become real with $i(\mp i) = \pm 1$, which correspond to counterclockwise and clockwise rotations around $\mathbf{e}_y$, respectively. Additionally, by noting that the spinors are effectively the square roots of the rotation axis, the eigenvectors could be readily written down. In order to obtain the correct spinors, the bivector $\boldsymbol{i}_{xy}$ had to be replaced by the imaginary unit i, see Eq. (1.177). However, this agrees with the choice for the imaginary unit for the Pauli matrices, see Eq. (1.148).

As a result of the usual conventions, the zx-plane in three dimensions is comparable to the xy-plane in two dimensions. Making a cyclic permutation $x \to y \to z \to x$ on the indices in Eq. (1.177) gives the often-used complex spinors $(\mathbf{e}_x \pm i\mathbf{e}_y)/\sqrt{2}$ for the xy-plane. This is equal to the complex basis used in Eq. (1.53). It was already noted before that the complex vectors looked artificial. Making the same permutation for the spinor in algebraic format in Eq. (1.177) gives $(\mathbf{e}_x \pm \mathbf{e}_y\boldsymbol{i}_{yz})/\sqrt{2} = (\mathbf{e}_x \pm \mathbf{e}_z)/\sqrt{2}$. Therefore, the imaginary unit corresponds to $\boldsymbol{i}_{yz}$, which causes a rotation from the plane of rotation (xy) to the axis of rotation (z). Since the complex vectors were originally derived for a two-dimensional system, one could have naively assumed that the imaginary unit corresponded to $\boldsymbol{i}_{xy}$. Therefore, instead of being complex vectors in the plane of rotation, the spinors are lifted out of the xy-plane. The spinors are those of the z-axis with $\mathbf{e}_{z,\pm\frac{1}{2}} = \mathbf{e}_{z,\pm\frac{1}{2}}\mathbf{e}_x = e^{\pm \boldsymbol{i}_{zx}\frac{\pi}{4}}$, where $\mathbf{e}_x$ is now the reference axis. The axis of rotation is then given by the square of the spinors $\mathbf{e}_{z,\pm\frac{1}{2}}\mathbf{e}_{z,\pm\frac{1}{2}}\mathbf{e}_x = e^{\pm \boldsymbol{i}_{zx}\frac{\pi}{2}}\mathbf{e}_x = \pm\mathbf{e}_z$. Therefore, as expected, the spinors for rotation in the xy-plane correspond to rotations around the z-axis. However, note that formally the machinery of the Pauli matrices is still correct. Therefore, even if one dispenses with an interpretation of the imaginary unit or the spinors, they still contain all the transformation properties. Using geometric algebra methods, the spinors were obtained by effectively taking the square root of the rotation axis. In the following section, we demonstrate how to obtain the rotation axis from the Pauli spinors.

Problems

1 We want to consider rotations defined by the imaginary unit $\boldsymbol{i}_{zy}$.

a) Find the (complex) eigenvalues and eigenvectors without using the standard matrix techniques such as determinants. Show that the answers are indeed eigenvectors of $\boldsymbol{i}_{zy}$ in 2×2 matrix form.

b) What are the eigenvalues and eigenvectors for the rotation axis?

c) Write the identity matrix and $\boldsymbol{i}_{zy}$ in terms of kets and bras of the spinors.

d) Demonstrate the validity of Eq. (1.135) for the bivector $\boldsymbol{i}_{zy}$ when writing the spinors in terms of Pauli matrices.

e) Evaluate $\sum_i \langle x, \pm\frac{1}{2}|\mathbf{e}_i|x, \pm\frac{1}{2}\rangle \mathbf{e}_i$.

2 Rotate $\mathbf{e}_z$ around the $(1, 1, 1)$ direction over $120°$ angle. Rotate the result again over $120°$. Use only Pauli matrices. Note that the z-axis does not lie in the plane of rotation.

1.9 Projections

Introduction. – Spinors are the eigenvectors with complex eigenvalues of the rotation matrices. At the same time, the spinors are eigenvectors of the axes of rotation with real eigenvalues. This is not entirely surprising since the matrices only differ by the imaginary unit $\boldsymbol{i}_{ij} = -\boldsymbol{i}\mathbf{e}_k$. One can view the pseudoscalar $\boldsymbol{i} = \mathbf{e}_x\mathbf{e}_y\mathbf{e}_z$ as the entire space. Multiplying by a unit vector (the rotation axis) defines a plane in which the rotation occurs. The axis and plane of rotation are effectively dual spaces.

When describing the three-dimensional space using the 2×2 Pauli matrices and products thereof, the spinors are written as a column vector. In Section 1.6 on spinors using anticommuting algebraic unit vectors, it was shown that the spinors are effectively the square root of the rotation axis. Squaring the spinors recovers the rotation axis. Although this is straightforward within geometric algebra, it is less obvious using Pauli matrices and column vectors. In order to obtain the rotation axis from the spinors, we have to look more closely at the projection of vectors.

Let us start with a vector. Using the basis in Eq. (1.137), a unit vector in the zx-plane can be written as

$$|\hat{\mathbf{r}}) = \sum_{i=z,x} |\mathbf{e}_i\rangle(\mathbf{e}_i|\hat{\mathbf{r}}) = \cos\theta|\mathbf{e}_z\rangle + \sin\theta|\mathbf{e}_x\rangle = \begin{pmatrix} \cos\theta \\ \sin\theta \end{pmatrix}. \tag{1.186}$$

The inner product between two vectors is given by

$$(\hat{\mathbf{r}}'|\hat{\mathbf{r}}) = (\cos\theta', \sin\theta')\begin{pmatrix} \cos\theta \\ \sin\theta \end{pmatrix} = \cos\theta'\cos\theta + \sin\theta'\sin\theta = \cos(\theta'-\theta),$$

where $\theta' - \theta$ is the angle between the two vectors, and the angle difference formula from Eq. (1.32) has been used. This is essentially the length of the projection of $\hat{\mathbf{r}}'$ on $\hat{\mathbf{r}}$ or, equivalently, the projection of $\hat{\mathbf{r}}$ on $\hat{\mathbf{r}}'$. This result is the same as the inner product in Eq. (1.118). In particular, the projection of a vector of length one on the unit vectors is given by

$$\langle\mathbf{e}_z|\hat{\mathbf{r}}) = (1,0)\begin{pmatrix} \cos\theta \\ \sin\theta \end{pmatrix} = \cos\theta, \quad \langle\mathbf{e}_x|\hat{\mathbf{r}}) = (0,1)\begin{pmatrix} \cos\theta \\ \sin\theta \end{pmatrix} = \sin\theta, \tag{1.187}$$

which are the coefficients of the vector $\hat{\mathbf{r}}$.

Instead of an inner product between two unequal vectors, we now would like to rewrite it in terms of two equal vectors. This can be achieved by making a rotation over $-\theta/2$, since a rotation does not affect the inner product. This leads to two new vectors, see Figure 1.7. The rotation of $\hat{\mathbf{r}}$ leads to a vector at an angle of $\theta/2$ with the z-axis

$$|\hat{\mathbf{r}}_{\frac{1}{2}}) = |\hat{\mathbf{r}}_{\uparrow}) = \begin{pmatrix} \cos\dfrac{\theta}{2} \\ \sin\dfrac{\theta}{2} \end{pmatrix}. \tag{1.188}$$

Anticipating that the objects with half angles are related to spinors, the notation $\pm\frac{1}{2} = \uparrow, \downarrow$ is used. The other vector follows from the rotation over $-\theta/2$ of the z-axis, giving

$$|\hat{\mathbf{r}}^*_{\frac{1}{2}}) = |\hat{\mathbf{r}}^*_{\uparrow}) = \begin{pmatrix} \cos\dfrac{\theta}{2} \\ -\sin\dfrac{\theta}{2} \end{pmatrix}. \tag{1.189}$$

The resulting vector $|\hat{\mathbf{r}}^*_{\frac{1}{2}})$ is the conjugate of $|\hat{\mathbf{r}}_{\frac{1}{2}})$. Note that in a complex basis, the x-axis would be the imaginary axis, and taking the conjugate means changing the sign

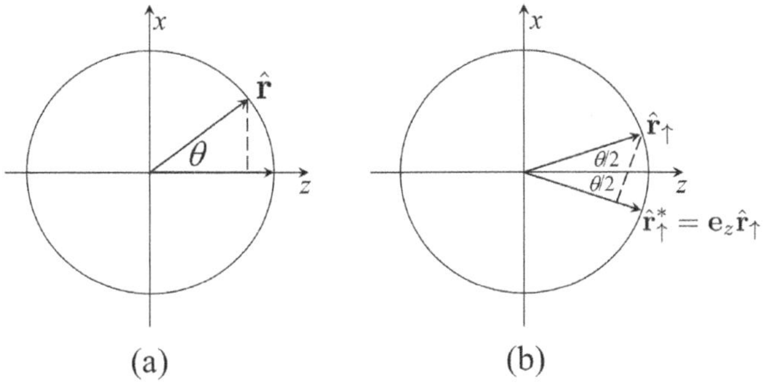

Figure 1.7 (a) The z component of the unit vector $\hat{\mathbf{r}}$ can be obtained by taking the inner product with $\mathbf{e}_z$. (b) The same value can be obtained by clockwise rotation of the vectors over $\theta/2$, giving the vectors $\hat{\mathbf{r}}_\uparrow$ and $\hat{\mathbf{r}}^*_\uparrow = \mathbf{e}_z\hat{\mathbf{r}}_\uparrow$.

of the component along the imaginary axis, $(\boldsymbol{i}_{xz})^* = -\boldsymbol{i}_{xz}$. The rotation leaves the inner product unchanged

$$(\hat{\mathbf{r}}_\uparrow|\hat{\mathbf{r}}_\uparrow^*) = \left(\cos\frac{\theta}{2}, \sin\frac{\theta}{2}\right)\begin{pmatrix} \cos\dfrac{\theta}{2} \\ -\sin\dfrac{\theta}{2} \end{pmatrix} = \cos^2\theta - \sin^2\theta = \cos\theta, \tag{1.190}$$

using Eq. (1.33). The operation of conjugation can also be achieved by the $\mathbf{e}_z$ operator in Eq. (1.149), which is essentially a reflection through the z-axis, leaving the z component unchanged and mirroring the x component,

$$|\hat{\mathbf{r}}_\uparrow^*) = \mathbf{e}_z|\hat{\mathbf{r}}_\uparrow) \quad \Rightarrow \quad (\hat{\mathbf{r}}_\uparrow|\hat{\mathbf{r}}_\uparrow^*) = (\hat{\mathbf{r}}_\uparrow|\mathbf{e}_z|\hat{\mathbf{r}}_\uparrow) = \cos\theta. \tag{1.191}$$

The z component is now expressed in terms of the expectation value between two equal objects.

For the inner product with the x-axis, the rotation of $|\hat{\mathbf{r}})$ over $-\theta/2$ leads again to $|\hat{\mathbf{r}}_\uparrow)$; the rotation of the x-axis leads to a vector at an angle $-\theta/2$ with the x-axis, see Figure 1.8. This vector can be obtained by interchanging the coefficients of $|\hat{\mathbf{r}}_\uparrow)$. This is a reflection through the line bisecting the z- and x-axis. This operation can be performed by the matrix $\mathbf{e}_x$, see Eq. (1.149). The coefficient in the x-direction can then be obtained via

$$\begin{aligned}(\hat{\mathbf{r}}_\uparrow|\mathbf{e}_x|\hat{\mathbf{r}}_\uparrow) &= \left(\cos\frac{\theta}{2}, \sin\frac{\theta}{2}\right)\begin{pmatrix} 0 & 1 \\ 1 & 0 \end{pmatrix}\begin{pmatrix} \cos\dfrac{\theta}{2} \\ \sin\dfrac{\theta}{2} \end{pmatrix} \\ &= \left(\cos\frac{\theta}{2}, \sin\frac{\theta}{2}\right)\begin{pmatrix} \sin\dfrac{\theta}{2} \\ \cos\dfrac{\theta}{2} \end{pmatrix} = 2\sin\frac{\theta}{2}\cos\frac{\theta}{2} = \sin\theta, \end{aligned} \tag{1.192}$$

using Eq. (1.33).

Therefore, the components of a vector $|\hat{\mathbf{r}})$ in the zx-plane can be expressed in terms of half angles. This can be extended for a unit vector in three dimensions. Following Eqs. (1.133) and (1.134), the spinors can be written as a column vector,

$$|\hat{\mathbf{r}}_\uparrow) = \begin{pmatrix} \cos\dfrac{\theta}{2} \\ \sin\dfrac{\theta}{2}e^{i\varphi} \end{pmatrix} \quad \text{and} \quad |\hat{\mathbf{r}}_\downarrow) = \begin{pmatrix} -\sin\dfrac{\theta}{2} \\ \cos\dfrac{\theta}{2}e^{i\varphi} \end{pmatrix}. \tag{1.193}$$

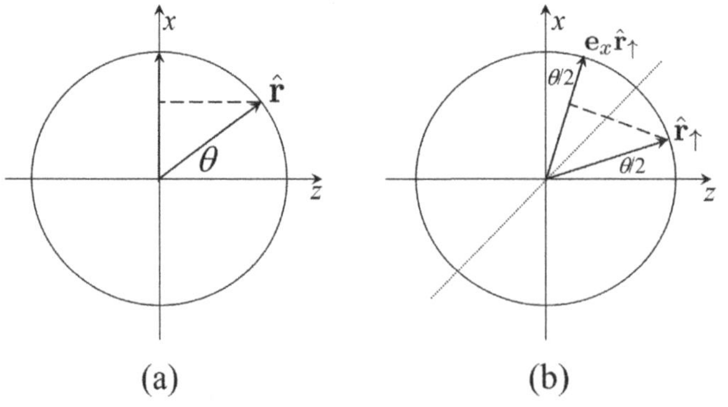

Figure 1.8 (a) The x component of the unit vector $\hat{\mathbf{r}}$ can be obtained by taking the inner product with $\mathbf{e}_x$. (b) The same value can be obtained by a clockwise rotation of the vectors over $\theta/2$, giving the vectors $\hat{\mathbf{r}}_\uparrow$ and $\mathbf{e}_x\hat{\mathbf{r}}_\uparrow$. The latter is a reflection in the dotted line at 45° with the z-axis.

The evaluation of the different components for the complete spinor is cumbersome but straightforward. For example, the y component is

$$\begin{aligned}
(\hat{\mathbf{r}}_\uparrow|\mathbf{e}_y|\hat{\mathbf{r}}_\uparrow) &= \left(\cos\frac{\theta}{2}, \sin\frac{\theta}{2}e^{-i\varphi}\right)\begin{pmatrix} 0 & -i \\ i & 0 \end{pmatrix}\begin{pmatrix} \cos\frac{\theta}{2} \\ \sin\frac{\theta}{2}e^{i\varphi} \end{pmatrix} \\
&= \left(\cos\frac{\theta}{2}, \sin\frac{\theta}{2}e^{-i\varphi}\right)\begin{pmatrix} -i\sin\frac{\theta}{2}e^{i\varphi} \\ i\cos\frac{\theta}{2} \end{pmatrix} \\
&= 2\sin\frac{\theta}{2}\cos\frac{\theta}{2}\frac{e^{i\varphi}+e^{-i\varphi}}{2i} = \sin\theta\sin\varphi,
\end{aligned} \tag{1.194}$$

giving the y coordinate in spherical polar coordinates, see Eq. (1.125). The exercise for the expectation values can be repeated for $|\hat{\mathbf{r}}_\downarrow)$, and the spherical polar coordinates are again obtained but with a negative sign. The unit vector can therefore be written as

$$\sum_{i=x,y,z} (\hat{\mathbf{r}}_{\pm\frac{1}{2}}|\mathbf{e}_i|\hat{\mathbf{r}}_{\pm\frac{1}{2}})\mathbf{e}_i = \pm\hat{\mathbf{r}} = \hat{\mathbf{r}}_\pm. \tag{1.195}$$

This is effectively the same result obtained in Eq. (1.127). The above equation is therefore the Pauli matrix equivalent of $\hat{\boldsymbol{r}}_{\pm\frac{1}{2}}\hat{\boldsymbol{r}}_{\pm\frac{1}{2}}\mathbf{e}_z = \hat{\boldsymbol{r}}_{\pm\frac{1}{2}}\mathbf{e}_z\hat{\boldsymbol{r}}^*_{\pm\frac{1}{2}} = \hat{\mathbf{r}}_\pm$.

Using geometric algebra, the spinors were designed to be the eigenvectors of the unit vector in the direction of the rotation axis, $\hat{\mathbf{r}}\hat{\boldsymbol{r}}_{\pm\frac{1}{2}} = \pm\hat{\boldsymbol{r}}_{\pm\frac{1}{2}}$, see Eq. (1.130). Using Eq. (1.149), a vector in terms of a 2×2 matrix is

$$\mathbf{r} = \sum_{i=x,y,z} r_i\mathbf{e}_i = \begin{pmatrix} z & x-iy \\ x+iy & -z \end{pmatrix} = \begin{pmatrix} r\cos\theta & re^{-i\varphi}\sin\theta \\ re^{i\varphi}\sin\theta & -r\cos\theta \end{pmatrix}. \tag{1.196}$$

The calculation takes some effort, see the Example, but the spinors are indeed eigenvectors of the vector $\mathbf{r}$ and the unit vector $\hat{\mathbf{r}} = \mathbf{r}/r$,

$$\mathbf{r}|\hat{\mathbf{r}}_{\pm\frac{1}{2}}) = \pm r|\hat{\mathbf{r}}_{\pm\frac{1}{2}}) \quad\text{and}\quad \hat{\mathbf{r}}|\hat{\mathbf{r}}_{\pm\frac{1}{2}}) = \pm|\hat{\mathbf{r}}_{\pm\frac{1}{2}}). \tag{1.197}$$

The matrix $\hat{\mathbf{r}}$ can be expressed in diagonal form in terms of the spinors as

$$\hat{\mathbf{r}} = |\hat{\mathbf{r}}_\uparrow)(\hat{\mathbf{r}}_\uparrow| - |\hat{\mathbf{r}}_\downarrow)(\hat{\mathbf{r}}_\downarrow|. \tag{1.198}$$

Expressing the right-hand side in the original basis shows that this is indeed equal to the unit vector

$$\begin{aligned}
&\begin{pmatrix} \cos^2\frac{\theta}{2} & e^{-i\varphi}\cos\frac{\theta}{2}\sin\frac{\theta}{2} \\ e^{i\varphi}\cos\frac{\theta}{2}\sin\frac{\theta}{2} & \sin^2\frac{\theta}{2} \end{pmatrix} - \begin{pmatrix} \sin^2\frac{\theta}{2} & -e^{-i\varphi}\cos\frac{\theta}{2}\sin\frac{\theta}{2} \\ -e^{i\varphi}\cos\frac{\theta}{2}\sin\frac{\theta}{2} & \cos^2\frac{\theta}{2} \end{pmatrix} \\
&= \begin{pmatrix} \cos\theta & e^{-i\varphi}\sin\theta \\ e^{i\varphi}\sin\theta & -\cos\theta \end{pmatrix} = \hat{\mathbf{r}},
\end{aligned} \tag{1.199}$$

which again shows that the square of the spinors is related to a unit vector.

Example 1.7 *Pauli Spinors*

a) Show that the spinors are eigenvectors of $\mathbf{r}$ and determine the eigenvalues.
b) The spinors can also be written in terms of Pauli matrices. Operate with a vector $\mathbf{r}$ on a spinor and express the result in terms of spinors.
c) Reproduce the answer of (b) using anticommuting vector algebra.

Solution

a) Working with $\mathbf{r}$ in spherical polar coordinates from Eq. (1.196) on the up spinor gives

$$\mathbf{r}|\hat{\mathbf{r}}_\uparrow) = \begin{pmatrix} r\cos\theta & re^{-i\varphi}\sin\theta \\ re^{i\varphi}\sin\theta & -r\cos\theta \end{pmatrix} \begin{pmatrix} \cos\frac{\theta}{2} \\ e^{i\varphi}\sin\frac{\theta}{2} \end{pmatrix}$$

$$= r\begin{pmatrix} \cos\theta\cos\frac{\theta}{2} + \sin\theta\sin\frac{\theta}{2} \\ e^{i\varphi}\left(\sin\theta\cos\frac{\theta}{2} - \cos\theta\sin\frac{\theta}{2}\right) \end{pmatrix} = r\begin{pmatrix} \cos\frac{\theta}{2} \\ e^{i\varphi}\sin\frac{\theta}{2} \end{pmatrix} = r|\hat{\mathbf{r}}_\uparrow), \tag{1.200}$$

using the angle-subtraction formula. For the down spinor, we have

$$\mathbf{r}|\hat{\mathbf{r}}_\downarrow) = \begin{pmatrix} r\cos\theta & re^{-i\varphi}\sin\theta \\ re^{i\varphi}\sin\theta & -r\cos\theta \end{pmatrix} \begin{pmatrix} -\sin\frac{\theta}{2} \\ e^{i\varphi}\cos\frac{\theta}{2} \end{pmatrix}$$

$$= r\begin{pmatrix} -\cos\theta\sin\frac{\theta}{2} + \sin\theta\cos\frac{\theta}{2} \\ e^{i\varphi}\left(-\sin\theta\sin\frac{\theta}{2} - \cos\theta\cos\frac{\theta}{2}\right) \end{pmatrix} = -r\begin{pmatrix} -\sin\frac{\theta}{2} \\ e^{i\varphi}\cos\frac{\theta}{2} \end{pmatrix} = -r|\hat{\mathbf{r}}_\downarrow). \tag{1.201}$$

b) The up spinor in terms of Pauli matrices is

$$\hat{\mathbf{r}}_\uparrow = \begin{pmatrix} \cos\frac{\theta}{2} & e^{-i\varphi}\sin\frac{\theta}{2} \\ e^{i\varphi}\sin\frac{\theta}{2} & -\cos\frac{\theta}{2} \end{pmatrix}. \tag{1.202}$$

After some work, the product of $\mathbf{r}$ and the spinor is found to be

$$\hat{\mathbf{r}}\hat{\mathbf{r}}_\uparrow = r\begin{pmatrix} \cos\frac{\theta}{2} & -e^{-i\varphi}\sin\frac{\theta}{2} \\ e^{i\varphi}\sin\frac{\theta}{2} & \cos\frac{\theta}{2} \end{pmatrix} = r\hat{\mathbf{r}}_\uparrow\mathbf{e}_z, \tag{1.203}$$

which follows after multiplying the matrices and using the angle-subtraction formulas to simplify the result. The final result is not $r\hat{\mathbf{r}}_\uparrow$ but multiplied by the reference axis $\mathbf{e}_z$. For the down spinor, we obviously find $\hat{\mathbf{r}}\hat{\mathbf{r}}_\downarrow = -r\hat{\mathbf{r}}_\downarrow\mathbf{e}_z$.

c) The answer from (b) can be reproduced using algebraic unit vectors

$$\mathbf{r}\hat{\mathbf{r}}_\uparrow = re^{\boldsymbol{i}_{x'z}\theta}\mathbf{e}_z e^{\boldsymbol{i}_{x'z}\frac{\theta}{2}}\mathbf{e}_z = re^{\boldsymbol{i}_{x'z}\theta}e^{-\boldsymbol{i}_{x'z}\frac{\theta}{2}}\mathbf{e}_z\mathbf{e}_z = re^{\boldsymbol{i}_{x'z}\frac{\theta}{2}}\mathbf{e}_z\mathbf{e}_z = r\hat{\mathbf{r}}_\uparrow\mathbf{e}_z, \tag{1.204}$$

using the anticommutation of $\boldsymbol{i}_{x'z} = e^{\boldsymbol{i}_{yx}\varphi}\boldsymbol{i}_{xz}$ and $\mathbf{e}_z$. The latter derivation only uses complex exponentials and the anticommutative property and does not require elaborate matrix multiplications and trigonometric addition/subtraction formulas. Obviously, since the Pauli matrices can be viewed as effective unit vectors that obey the exact same relations as the algebraic unit vectors, the latter result can also be viewed as a derivation in terms of Pauli matrices. In a similar fashion, we can find $\hat{\mathbf{r}}\hat{\mathbf{r}}_\downarrow = -r\hat{\mathbf{r}}_\downarrow\mathbf{e}_z$. There seems to be a discrepancy between the result $\mathbf{r}|\hat{\mathbf{r}}_\uparrow) = r|\hat{\mathbf{r}}_\uparrow)$ obtained for the column vectors and the expression $\hat{\mathbf{r}}\hat{\mathbf{r}}_\uparrow = r\hat{\mathbf{r}}_\uparrow\mathbf{e}_z$ obtained with the Pauli matrices or the anticommuting algebraic unit vectors. The reference axis is needed for the latter. When only operating on vectors, the Pauli matrices contain redundant information since there is no need to separate the vectors from the scalar quantities. Note that the column vector in Eq. (1.200) corresponds to the left column in the 2×2 matrix in Eq. (1.203). Since this part of the matrix is not affected by the multiplication with $\mathbf{e}_z$, the reference axis is absent in $\mathbf{r}|\hat{\mathbf{r}}_\uparrow) = r|\hat{\mathbf{r}}_\uparrow)$.

Conceptual summary. – In search of the eigenvectors of a vector, spinors were found, which are effectively the square root of a rotation axis $\hat{\mathbf{r}}_\pm = \pm\hat{\mathbf{r}} = \hat{\boldsymbol{r}}_{\pm\frac{1}{2}}\hat{\boldsymbol{r}}_{\pm\frac{1}{2}}\mathbf{e}_z = \hat{\boldsymbol{r}}_{\pm\frac{1}{2}}\mathbf{e}_z\hat{\boldsymbol{r}}^*_{\pm\frac{1}{2}}$, where the $\pm$ indicates the two possible rotation directions around the axis. For Pauli matrices, it is again found that two spinors are needed to obtain the rotation axis $\hat{\mathbf{r}}_\pm = \sum_{i=x,y,z}(\hat{\mathbf{r}}_{\pm\frac{1}{2}}|\mathbf{e}_i|\hat{\mathbf{r}}_{\pm\frac{1}{2}})\mathbf{e}_i$. The coefficient of the unit vector is now expressed as the expectation value $\hat{r}_i = (\hat{\mathbf{r}}_{\pm\frac{1}{2}}|\mathbf{e}_i|\hat{\mathbf{r}}_{\pm\frac{1}{2}})$ of the unit vector for a spinor.

Mathematically, the Pauli matrices become more involved. The use of matrices forces the split of the complex exponentials into trigonometric functions, which are the coefficients in different directions. Therefore, instead of simply adding angles in the exponentials, one has to resort to trigonometric addition/subtraction formulas. Also, when using matrices, different notations are often used for vectors. The vector that is changed is written as a column vector, whereas the operator (which can be a vector or a bivector) is written as a 2×2 matrix. This is a convenient short cut when only dealing with vectors. However, this operation can also be performed by only using Pauli matrices as vectors, see the example. Although, this requires some additional calculation, it has the advantage that all multivectors are treated on an equal footing and that scalars and vectors can be mixed. The same results can also be obtained using anticommuting algebraic expressions, which provides an even shorter way to obtain the same result.

In conclusion, it is important to realize that the Pauli matrices and the anticommutative algebraic vectors are equivalent. The latter is often more condensed, but it is easy to miss a sign change when anticommuting vectors and bivectors. When using matrices, the properties of matrix multiplication take care of the anticommuting properties. This is convenient when using numerical methods or analytical mathematical software. Furthermore, although spinors are somewhat strange quantities, "squaring" them leads to a vector (how to effectively square them, depends on the method one is using). The vector itself is associated with a rotation axis. This is why there are two solutions $\pm\hat{\mathbf{r}}$ corresponding to counterclockwise and clockwise rotations around the vector. The rotations effectively occur in a plane perpendicular to $\hat{\mathbf{r}}$.

Problems

1 Find the spinors for counterclockwise and clockwise rotations around an axis at a 30° angle with the z-axis toward the y-axis. Obtain the rotation axis from the spinors.

2 We want to study a rotation axis $\hat{\mathbf{r}}$ in the xy-plane at 45° with the x-axis and its related spinors. The problem illustrates the equivalence between spinors in bra-ket notation, algebraic unit vectors, or Pauli spinors.
a) Find the spinors in ket format for $\hat{\mathbf{r}}$.
b) Obtain the rotation axis from the spinors.
c) Express the spinors in bivectors (using algebraic unit vectors) and obtain the rotation axes.
d) Express the spinor in terms of Pauli matrices.
e) Obtain the rotation axis from the spinors in Pauli matrices.

1.10 Function Space in Three Dimensions

Introduction. – In the discussion of three-dimensional space, we have so far restricted ourselves to vectors. Vectors allow us to point to any position in space. However, as we saw in Section 1.4, vectors do not allow us to assign a value to each position in space, they only "store" the position. This does not allow us to describe fields, potentials, or any other type of function in space. For a unit circle in two dimensions, *i.e.* a line, it was possible to associate each position with a number by introducing a function space. For a rotating unit vector, this function space were the complex exponentials to an integer power. Complex exponentials are also used as a function space for translation. This is possible, since one can view the motion along the circumference of the circle as linear motion.

We started out with rotations, which are conceptually easier, but will become mathematically more complicated rather quickly. Linear motion is mathematically relatively easy but has some surprising conceptual issues. First, the linear motion is effectively described as a rotation, which can be somewhat confusing. Second, the direction of the motion is generally indicated with the wavevector $\mathbf{k}$ (or momentum, which is directly proportional to it since $\mathbf{p} = \hbar\mathbf{k}$, where $\hbar = h/2\pi$ is Planck's constant). In free space, finding the direction of the conserved momentum poses generally no issues, but in systems where the unit vectors are not orthonormal, the situation becomes more subtle. However, let us start with free space.

We will use the complex exponentials that followed from the rotation of a vector as a function space for three-dimensional free space. In free space, the three directions are essentially independent. Therefore, a one-dimensional system is studied first. This can then be straightforwardly extended to two and three dimensions. For a unit circle, the argument φ in the complex vector $\hat{\boldsymbol{r}} = e^{\boldsymbol{i}_{yx}\varphi}$ corresponds to the length of the arc. For a 360° rotation, $\varphi = 2\pi$, which is the circumference of the unit circle. The circumference can therefore be viewed as a one-dimensional space of length 2π with periodic boundary conditions, *i.e.* the beginning and end of the space are connected. Obviously, many physical systems are not periodic. However, periodic boundary conditions mimic well free space, since they satisfy translational symmetry.

Generally, we are more interested in one-dimensional spaces of an arbitrary length L rather than 2π. In particular, the limit $L \to \infty$, corresponding to free space, will be considered. In that case, the circle has an infinite radius, and the motion along the circumference is effectively in a straight line. Taking the circumference as the x-direction, the argument for the function space for rotations $\hat{\boldsymbol{r}}^m = e^{\boldsymbol{i}_{yx}m\varphi}$, see Eq. (1.74), can be rewritten as

$$m\varphi = m2\pi\frac{x}{L} = k_x x, \quad \text{with} \quad k_x = m\frac{2\pi}{L}, \tag{1.205}$$

where $0 \le x < L$, and k_x is known as the wavenumber. The one-dimensional function space can now be built from the following complex exponentials, known as running waves:

$$\varphi_{k_x}(x) = Ae^{ik_x x}, \tag{1.206}$$

where A is a normalization constant that needs to be determined. The imaginary unit is $\boldsymbol{i}$. Conceptually, the running waves are more complicated. For the unit circle, there is a rotating unit vector $\hat{\mathbf{r}}$. The imaginary unit times the unit vector, $\boldsymbol{i}_{yx}\hat{\mathbf{r}}$, gives a vector along the circumference. For the running wave, the motion is effectively along the circumference;

this is the x-direction that corresponds to the unit vector $\mathbf{e}_x$. The imaginary unit $\boldsymbol{i}$ connects this one-dimensional space to a dual space. This discussion, although important, is postponed till later. Often, the imaginary unit is taken to be the pseudoscalar i, and the machinery of function spaces is used without giving much thought to the implications.

To obtain the associated normalized function, we introduce, following the same procedure as in Section 1.4, $|x\rangle$ to represent a point in a one-dimensional space. In order to be able to describe a one-dimensional space, we want $|x\rangle$ to be both orthonormal and complete

$$\langle x'|x\rangle = \delta(x - x') \quad \text{and} \quad \int dx\; |x\rangle\langle x| = \mathbb{1}_\infty. \tag{1.207}$$

The orthonormality implies that each point in space can be treated independently; completeness means that the vectors $|x\rangle$ represent all points in one-dimensional space. Again, it is important to note that the point in space $|x\rangle$ is different from the vector $|x) = x\mathbf{e}_x$. The vectors $|x)$ are never orthogonal and independent in a one-dimensional space, since $(x'|x) = x'x$, whereas the points in space $|x\rangle$ are always orthogonal and independent, *i.e.* $\langle x'|x\rangle = \delta(x - x')$.

The properties of orthogonality and completeness should also apply to the function space. Orthogonality is given by

$$\langle k_x'|k_x\rangle = \langle k_x'|\mathbb{1}_\infty|k\rangle = \int dx\; \langle k_x'|x\rangle\langle x|k_x\rangle = \int dx\; \varphi^*_{k_x'}(x)\varphi_{k_x}(x), \tag{1.208}$$

with $\langle x|k_x\rangle = \varphi_{k_x}(x)$, and $\langle k_x|x\rangle = (\langle x|k_x\rangle)^*$. The ket $|k_x\rangle$ indicates the actual function. In order to obtain the value of the function at a particular position, $|k_x\rangle$ is projected onto a particular point $\langle x|$. Inserting the complex exponential from Eq. (1.206) gives

$$\langle k_x'|k_x\rangle = A^2 \int_0^L dx\; e^{i(k_x - k_x')x}. \tag{1.209}$$

When $k_x - k_x' \neq 0$, the integral over the oscillating complex exponentials is 0. For, $k_x = k_x'$, the exponential is 1, and the integral becomes $A^2 \int_0^L dx = A^2 L$. The integral becomes unity for $A = 1/\sqrt{L}$. This results in

$$\varphi_{k_x}(x) = \frac{1}{\sqrt{L}} e^{ik_x x} \quad \text{with} \quad \langle k_x'|k_x\rangle = \delta_{k_x k_x'}. \tag{1.210}$$

So far, the one-dimensional system has been assumed to have a finite length L. Due to the periodicity, the associated k_x values of $m\; 2\pi/L$ are discrete. The orthogonality is therefore expressed in terms of a Kronecker δ-function. In order to describe "free" space, the limit $L \to \infty$ is taken. The summation over k_x values is then replaced by an integral, making it more similar to the spatial variable x. The integral over the exponent in the orthogonality can be evaluated by adding a factor $e^{-\eta|x|}$ with η positive and $\eta \ll 1$. This causes the exponentials to decay slowly, thereby removing the oscillatory behavior for $x \to \pm\infty$. The integral can be split into two parts

$$\int_0^\infty dx\; e^{ik_x x - \eta x} = \frac{1}{\eta - ik_x} \quad \text{and} \quad \int_{-\infty}^0 dx\; e^{ik_x x + \eta x} = \frac{1}{\eta + ik_x}. \tag{1.211}$$

Adding the two terms gives

$$\int_{-\infty}^\infty dx\; e^{ik_x x - \eta|x|} = \frac{2\eta}{k_x^2 + \eta^2}. \tag{1.212}$$

This function is often called a Lorentzian. In the limit $\eta \to 0$, this integral is 0 for all $k_x \neq 0$. For $k_x = 0$, the value is $2/\eta$, which diverges for $\eta \to 0$. Therefore, this function has the properties of a δ-function. However, for a δ-function, the integral over x should equal 1. The integral can be evaluated as

$$\frac{2}{\eta}\int_{-\infty}^{\infty} dx\, \frac{1}{1+\left(\frac{x}{\eta}\right)^2} = \frac{2}{\eta}\left[\eta \arctan\frac{x}{\eta}\right]_{-\infty}^{\infty} = 2\pi. \tag{1.213}$$

Therefore, the δ-function is defined as

$$\delta(k_x) = \frac{1}{2\pi}\int_{-\infty}^{\infty} dx\; e^{ik_x x}. \tag{1.214}$$

For an infinite one-dimensional system, the function space is given by the exponentials

$$\varphi_{k_x}(x) = \frac{1}{\sqrt{2\pi}} e^{ik_x x}. \tag{1.215}$$

The wave functions are now normalized since

$$\langle k_x'|k_x\rangle = \int_{-\infty}^{\infty} dx\; \langle k_x'|x\rangle\langle x|k_x\rangle = \frac{1}{2\pi}\int_{-\infty}^{\infty} dk_x\; e^{i(k_x-k_x')x} = \delta(k_x - k_x'), \tag{1.216}$$

showing that $A = 1/\sqrt{2\pi}$ is the correct normalization factor in the limit $L \to \infty$.

Orthonormality in real space can now be expressed as

$$\langle x'|x\rangle = \delta(x'-x) = \frac{1}{2\pi}\int_{-\infty}^{\infty} dk_x\; e^{ik_x(x'-x)} = \int_{-\infty}^{\infty} dk_x\; \langle x'|k_x\rangle\langle k_x|x\rangle, \tag{1.217}$$

using the δ-function for real space. Comparing the left- and right-hand sides implies that

$$\int_{-\infty}^{\infty} dk_x\; |k_x\rangle\langle k_x| = \mathbb{1}_\infty, \tag{1.218}$$

which is the completeness relation for the function space. Therefore, the completeness of the running waves arises from the orthonormality of the points $|x\rangle$, and the orthogonality of the functions comes from the completeness of the functions for space. Notice that in the limit $L \to \infty$, x and k_x become effectively equivalent.

The results can be straightforwardly extended to three dimensions where we want to describe each point in space $|\mathbf{r}\rangle$ independently. Orthonormality in three dimensions is given by

$$\langle \mathbf{r}|\mathbf{r}\rangle = \prod_{i=x,y,z} \langle r_i|r_i\rangle, \tag{1.219}$$

with $r_i = x, y, z$ for $i = x, y, z$. Since, in a Cartesian coordinate system, each direction is independent, one can simply take the product of the different directions. A plane wave is then given by the product of the running waves in Eq. (1.215) in three directions

$$\varphi_{\mathbf{k}}(\mathbf{r}) = \frac{1}{(2\pi)^{\frac{3}{2}}} e^{i\mathbf{k}\cdot\mathbf{r}}. \tag{1.220}$$

This is generally known as a plane wave. The argument is now the inner product $\mathbf{k}\cdot\mathbf{r} = k_x x + k_y y + k_z z$. The basis is again complete and orthonormal

$$\langle \mathbf{k}'|\mathbf{k}\rangle = \frac{1}{(2\pi)^3}\int d\mathbf{r}\; e^{i(\mathbf{k}-\mathbf{k}')\cdot\mathbf{r}} = \delta(\mathbf{k}-\mathbf{k}') \;\text{ and }\; \int d\mathbf{k}\; |\mathbf{k}\rangle\langle\mathbf{k}| = \mathbb{1}_\infty, \tag{1.221}$$

where the shorthand $d\mathbf{k} = dk_x dk_y dk_z$ is used. The properties of the complex exponentials ensure that each point in space can be treated independently,

$$\langle \mathbf{r}'|\mathbf{r}\rangle = \int d\mathbf{k}\; \langle \mathbf{r}'|\mathbf{k}\rangle\langle\mathbf{k}|\mathbf{r}\rangle = \frac{1}{(2\pi)^3}\int d\mathbf{k}\; e^{i\mathbf{k}\cdot(\mathbf{r}'-\mathbf{r})} = \delta(\mathbf{r}-\mathbf{r}'). \tag{1.222}$$

Note the different prefactor in the δ-function compared to its one-dimensional equivalent, see Eq. (1.214).

Example 1.8 ***Matrix Element***
A very common problem is the scattering of a plane wave from a wavevector $\mathbf{k}$ to $\mathbf{k}'$ under a function $U(\mathbf{r})$.

a) Express the scattering in terms of an integral.
b) Evaluate the matrix element if the function can be expressed as a plane wave $U(\mathbf{r}) = U_{\mathbf{q}}e^{i\mathbf{q}\cdot\mathbf{r}}$.

Solution
a) The scattering is given by

$$\langle \mathbf{k}'|\, U|\mathbf{k}\rangle = \int d\mathbf{r} \int d\mathbf{r}'\ \langle \mathbf{k}'|\mathbf{r}'\rangle\langle \mathbf{r}'|\, U|\mathbf{r}\rangle\langle \mathbf{r}|\mathbf{k}\rangle, \tag{1.223}$$

where the completeness of space has been inserted twice. However, functions for a single particle are generally local, *i.e.* $\langle \mathbf{r}'|\, U|\mathbf{r}\rangle = U(\mathbf{r})\delta(\mathbf{r}' - \mathbf{r})$. This leaves only a single integral

$$\langle \mathbf{k}'|\, U|\mathbf{k}\rangle = \frac{1}{(2\pi)^3}\int d\mathbf{r}\ U(\mathbf{r})e^{i(\mathbf{k}-\mathbf{k}')\cdot\mathbf{r}}. \tag{1.224}$$

b) For a plane wave function, the matrix element becomes

$$\langle \mathbf{k}'|\, U|\mathbf{k}\rangle = \frac{1}{(2\pi)^3}\int d\mathbf{r}\ U_{\mathbf{q}}e^{i(\mathbf{k}+\mathbf{q}-\mathbf{k}')\cdot\mathbf{r}} = U_{\mathbf{q}}\delta(\mathbf{k}+\mathbf{q}-\mathbf{k}'), \tag{1.225}$$

using the δ-function in Eq. (1.222). Therefore, plane waves satisfy conservation of momentum in the same way as colliding particles conserve momentum.

Conceptual summary. – The function space for translational motion was found by adapting the function space for rotational motion. This was then extended to an infinitely large system and three dimensions. For an infinite system, the real and momentum spaces become very comparable. The function space for rotations was periodic. This causes the function space to be discrete. For an infinite translational system, the periodicity disappears, and both real and wavevector/momentum space are continuous. Both spaces are orthogonal and complete, which means that the function space can describe each point in space independently.

The reason why function spaces are important in quantum mechanics is the fact that there is no such thing as point particles. A particle is not really localized in space. The concept of point particles can be very convenient for macroscopic/classical systems but falls apart on atomic length scales.

Problems

1 Find the function space for free space in a volume $L^3 = V \to \infty$ with periodic boundary conditions and show that the functions are orthogonal and complete.

2 What is the normalized function space for one dimension if the particle is in a finite box from $0 \leq x \leq L$? Does this change the number of states compared to periodic boundary conditions in the limit $L \to \infty$?

1.11 Fourier Transform and Translation

Introduction. – In this section, we will see how to project a function onto the function space of complex exponentials or plane waves. This is generally known as a Fourier transform. This transform has the advantage that the function space contains eigenfunctions of the translation operator. The Fourier transform then allows us to understand the behavior of any function under the translation operator. We have now effectively two spaces, the real space and the momentum (wavevector) space. Properties in one space translate differently into the dual space. Of importance here is that translational symmetry in real space leads to conservation of momentum/wavevector in the dual space.

Let us start with describing an arbitrary function in space,

$$\psi(\mathbf{r}) = \langle \mathbf{r}|\psi\rangle, \tag{1.226}$$

where $|\psi\rangle$ represents an arbitrary function. The inner product $\langle \mathbf{r}|\psi\rangle$ with a point in space $\langle \mathbf{r}|$ returns the value of the function at a particular position. Since an identity matrix does not change anything, we can insert it between the bra and the ket,

$$\psi(\mathbf{r}) = \langle \mathbf{r}|\mathbb{1}_\infty|\psi\rangle = \int d\mathbf{k}\, \langle \mathbf{r}|\mathbf{k}\rangle\langle \mathbf{k}|\psi\rangle, \tag{1.227}$$

using Eq. (1.221). Now the function is first projected onto the function space $\langle \mathbf{k}|$. The projection $\langle \mathbf{k}|\psi\rangle$ is then the weight for the complex exponential $\langle \mathbf{r}|\mathbf{k}\rangle$. When written in conventional form, this reads

$$\psi(\mathbf{r}) = \frac{1}{(2\pi)^{\frac{3}{2}}} \int d\mathbf{k}\, e^{i\mathbf{k}\cdot\mathbf{r}}\psi(\mathbf{k}). \tag{1.228}$$

The function $\psi(\mathbf{r})$ is now expressed in terms of the function $\psi(\mathbf{k})$. This procedure is generally known as the inverse Fourier transform of ψ. The function $\psi(\mathbf{k})$ can be obtained from the function $\psi(\mathbf{r})$ via the relation

$$\psi(\mathbf{k}) = \langle \mathbf{k}|\psi\rangle = \int d\mathbf{r}\, \langle \mathbf{k}|\mathbf{r}\rangle\langle \mathbf{r}|\psi\rangle = \frac{1}{(2\pi)^{\frac{3}{2}}} \int d\mathbf{r}\, e^{-i\mathbf{k}\cdot\mathbf{r}}\psi(\mathbf{r}), \tag{1.229}$$

where the completeness relation for real space has been inserted. This is known as the Fourier transform.

Let us now suppose we have a function $\psi(\mathbf{r})$ that describes something in space, for example a field or a particle. Now we want to displace the function by a vector $\boldsymbol{\tau}$ to describe the motion of the object in space. Let us now introduce a translation operator $\hat{U}_{\boldsymbol{\tau}}$ that translates something by the vector $\boldsymbol{\tau}$. Operating this on a vector gives

$$\hat{U}_{\boldsymbol{\tau}}\mathbf{r} = \mathbf{r} + \boldsymbol{\tau}, \tag{1.230}$$

which simply moves the vector by $\boldsymbol{\tau}$. However, if we want to translate a function, the argument needs to be translated in the opposite direction

$$\hat{U}_{\boldsymbol{\tau}}\psi(\mathbf{r}) = \psi(\mathbf{r} - \boldsymbol{\tau}) = \psi(\hat{U}_{-\boldsymbol{\tau}}\mathbf{r}). \tag{1.231}$$

For example, if a maximum occurs at $\mathbf{r}_0$ in $\psi(\mathbf{r})$, then the same maximum occurs $\mathbf{r}_0 + \boldsymbol{\tau}$ in $\psi(\mathbf{r} - \boldsymbol{\tau})$. Therefore, the function is translated over $\boldsymbol{\tau}$.

Our goal is to find the operator $\hat{U}_{\boldsymbol{\tau}}$ for an arbitrary function $\psi(\mathbf{r})$. The displaced function in one dimension can be written in terms of a Fourier series

$$\psi(\mathbf{r} - \boldsymbol{\tau}) = \int d\mathbf{k}\, \langle \mathbf{r} - \boldsymbol{\tau}|\mathbf{k}\rangle\langle \mathbf{k}|\psi\rangle = \frac{1}{(2\pi)^{\frac{3}{2}}} \int d\mathbf{k}\, e^{i\mathbf{k}\cdot(\mathbf{r}-\boldsymbol{\tau})}\psi(\mathbf{k}). \tag{1.232}$$

In the Fourier transform, the displacement only occurs in the complex exponential. Let us first consider a small displacement $\boldsymbol{\delta}$ and consider the total displacement $\boldsymbol{\tau}$ a result of many successive small displacements $\boldsymbol{\delta}$. For a small displacement, the exponential can be expanded as

$$e^{i\mathbf{k}\cdot(\mathbf{r}-\boldsymbol{\delta})} = e^{-i\mathbf{k}\cdot\boldsymbol{\delta}}e^{i\mathbf{k}\cdot\mathbf{r}} \cong (1 - i\mathbf{k}\cdot\boldsymbol{\delta})e^{i\mathbf{k}\cdot\mathbf{r}}. \tag{1.233}$$

The factor $i\mathbf{k}$ can also be found by taking the derivative of the complex exponential

$$e^{i\mathbf{k}\cdot(\mathbf{r}-\boldsymbol{\delta})} \cong (1 - \boldsymbol{\delta}\cdot\nabla)e^{i\mathbf{k}\cdot\mathbf{r}}, \tag{1.234}$$

where the triangle is known as the nabla operator

$$\nabla = \frac{\partial}{\partial x}\mathbf{e}_x + \frac{\partial}{\partial y}\mathbf{e}_y + \frac{\partial}{\partial z}\mathbf{e}_z, \tag{1.235}$$

which turns a scalar function into a vector; $\nabla f(\mathbf{r})$ is also known as the gradient of f. The complex exponentials are eigenfunctions of the nabla operator,

$$\nabla\varphi_{\mathbf{k}}(\mathbf{r}) = i\mathbf{k}\varphi_{\mathbf{k}}(\mathbf{r}). \tag{1.236}$$

Generally, one prefers the eigenvalues to be real quantities. This can be done by defining the operator

$$\hat{\mathbf{k}} = -i\nabla, \tag{1.237}$$

where the hat indicates that this quantity is an operator. The operator is directly proportional to the quantum-mechanical momentum operator

$$\hat{\mathbf{p}} = \hbar\hat{\mathbf{k}} = -i\hbar\nabla, \tag{1.238}$$

where $\hbar$ is Planck's constant. Quantum mechanics tells us that the wavevector and momentum are the same quantities that only differ by a constant factor $\hbar$. Later, we will see that the same applies to frequency and energy, where $E = \hbar\omega$. The plane waves are eigenfunctions of $\hat{\mathbf{k}}$

$$\hat{\mathbf{k}}\varphi_{\mathbf{k}}(\mathbf{r}) = \mathbf{k}\varphi_{\mathbf{k}}(\mathbf{r}). \tag{1.239}$$

Applying nabla a second time on Eq. (1.236) gives

$$\nabla^2\varphi_{\mathbf{k}}(\mathbf{r}) + k^2\varphi_{\mathbf{k}}(\mathbf{r}) = 0. \tag{1.240}$$

This wave equation is the plane wave equivalent of Eq. (1.75) for rotations of the unit vectors to an arbitrary power. For a small displacement, we can write

$$\psi(\mathbf{r}-\boldsymbol{\delta}) \cong \frac{1}{(2\pi)^{\frac{3}{2}}}\int d\mathbf{k}\,(1 - \boldsymbol{\delta}\cdot\nabla)e^{i\mathbf{k}\cdot\mathbf{r}}\psi(\mathbf{k}) = (1 - \boldsymbol{\delta}\cdot\nabla)\psi(\mathbf{r}), \tag{1.241}$$

where the last step is possible since the derivative with respect to $\mathbf{r}$ (the ∇ operator) and the integral over $\mathbf{k}$ can be interchanged. The total displacement $\boldsymbol{\tau} = n\boldsymbol{\delta}$ can be written in terms of n small displacements. Using Eq. (1.4), this gives, in the limit $n \to \infty$,

$$\psi(\mathbf{r}-\boldsymbol{\tau}) = \hat{U}_{\boldsymbol{\tau}}\psi(\mathbf{r}) = \lim_{n\to\infty}\left(1 - \frac{\boldsymbol{\tau}}{n}\cdot\nabla\right)^n\psi(\mathbf{r}) = e^{-\boldsymbol{\tau}\cdot\nabla}\psi(\mathbf{r}). \tag{1.242}$$

The displacement operator is therefore

$$\hat{U}_{\boldsymbol{\tau}} = e^{-\boldsymbol{\tau}\cdot\nabla} = e^{-i\boldsymbol{\tau}\cdot\hat{\mathbf{k}}}. \tag{1.243}$$

Operating this on an arbitrary function gives

$$\hat{U}_{\boldsymbol{\tau}}\psi(\mathbf{r}) = \frac{e^{-i\boldsymbol{\tau}\cdot\hat{\mathbf{k}}}}{(2\pi)^{\frac{3}{2}}} \int d\mathbf{k}\; e^{-i\mathbf{k}\cdot\mathbf{r}}\psi(\mathbf{k}) = \frac{1}{(2\pi)^{\frac{3}{2}}} \int d\mathbf{k}\; e^{i\mathbf{k}\cdot(\mathbf{r}-\boldsymbol{\tau})}\psi(\mathbf{k}) = \psi(\mathbf{r}-\boldsymbol{\tau}).$$

Since the Fourier component $\psi(\mathbf{k})$ is not a function of $\mathbf{r}$, the only spatial dependence is in the complex exponential. The translation operator for a plane wave can be simply obtained by replacing the operator $\hat{\mathbf{k}}$ by its eigenvalue $\mathbf{k}$.

Example 1.9 *Translation Operator of a Point*

a) Demonstrate that $\hat{U}_{\boldsymbol{\tau}}|\mathbf{r}\rangle = |\mathbf{r}+\boldsymbol{\tau}\rangle$, but $\langle\mathbf{r}|\hat{U}_{\boldsymbol{\tau}} = \langle\mathbf{r}-\boldsymbol{\tau}|$.
b) Using this result show that translating a function with the operator $\hat{U}_{\boldsymbol{\tau}}\psi$ is equivalent to translating the argument in the opposite direction.

Solution

a) The ket $|\mathbf{r}'\rangle$ can be considered as a function describing the point $\mathbf{r}'$ in space. The function $f_{\mathbf{r}'}(\mathbf{r})$ that describes this position in space is a δ-function,

$$f_{\mathbf{r}'}(\mathbf{r}) = \langle\mathbf{r}|\mathbf{r}'\rangle = \delta(\mathbf{r}-\mathbf{r}'). \tag{1.244}$$

The function describing the point $\mathbf{r}'$ goes to infinity when $\mathbf{r}=\mathbf{r}'$ but is zero elsewhere. Note that we describe here $\mathbf{r}'$ as the fixed point in space, and $\mathbf{r}$ as the variable, but we could have done it the other way as well. Translating this function gives

$$\begin{aligned}\hat{U}_{\boldsymbol{\tau}}f_{\mathbf{r}'}(\mathbf{r}) &= \hat{U}_{\boldsymbol{\tau}}\langle\mathbf{r}|\mathbf{r}'\rangle = \hat{U}_{\boldsymbol{\tau}}\int d\mathbf{k}\langle\mathbf{r}|\mathbf{k}\rangle\langle\mathbf{k}|\mathbf{r}'\rangle = \frac{e^{-i\boldsymbol{\tau}\cdot\hat{\mathbf{k}}}}{(2\pi)^3}\int d\mathbf{k}\; e^{i\mathbf{k}\cdot(\mathbf{r}-\mathbf{r}')}\\ &= \frac{1}{(2\pi)^3}\int d\mathbf{k}\; e^{i\mathbf{k}\cdot(\mathbf{r}-\mathbf{r}'-\boldsymbol{\tau})} = \delta(\mathbf{r}-\mathbf{r}'-\boldsymbol{\tau}) = \langle\mathbf{r}|\mathbf{r}'+\boldsymbol{\tau}\rangle\\ &= \delta\left((\mathbf{r}-\boldsymbol{\tau})-\mathbf{r}'\right) = \langle\boldsymbol{\tau}-\mathbf{r}|\mathbf{r}'\rangle = \langle\mathbf{r}|\mathrm{U}_{\boldsymbol{\tau}}|\mathbf{r}'\rangle,\end{aligned} \tag{1.245}$$

where $\mathrm{U}_{\boldsymbol{\tau}}$ is no longer a derivative but a vector operation that changes the position from $\mathbf{r}'$ to $\mathbf{r}'+\boldsymbol{\tau}$, *i.e.* $\mathrm{U}_{\boldsymbol{\tau}}|\mathbf{r}'\rangle = |\mathbf{r}'+\boldsymbol{\tau}\rangle$.

b) In a similar fashion, we find for a general function $f(\mathbf{r})$

$$\hat{U}_{\boldsymbol{\tau}}f(\mathbf{r}) = \hat{U}_{\boldsymbol{\tau}}\langle\mathbf{r}|f\rangle = \langle\mathbf{r}|\mathrm{U}_{\boldsymbol{\tau}}|f\rangle = \langle\mathbf{r}-\boldsymbol{\tau}|f\rangle = f(\mathbf{r}-\boldsymbol{\tau}), \tag{1.246}$$

which is the same as was stated in Eq. (1.231).

Conceptual summary. – In this section, a general function was projected onto the function space of plane waves. This is generally known as a Fourier transform. Note that there are many different function spaces. The complex exponentials are well suited for free space, since they are the eigenfunctions of the translation operator and the momentum operator $\hat{\mathbf{p}} = \hbar\hat{\mathbf{k}}$. For rotation, the spherical harmonics form a complete function space and are eigenfunctions of the angular momentum. For harmonic oscillators, Hermite polynomials are used, and so on. Since all proper function spaces are complete, it is not necessary to use eigenfunctions of the system. For example, one can express the eigenstates of the quantum harmonic oscillator in terms of complex exponentials. Unfortunately, this is not very helpful, since the complex exponentials do not provide any insights into the problem of a harmonic oscillator. Therefore, in quantum mechanics, we spend a considerable amount of time finding the appropriate function space.

However, sometimes we are unable to find a function space that solves the entire problem that we are considering. In that case, one often uses a function space that solves part of the problem, and the eigenfunctions are then expressed in terms of those basis functions. For example, it is very natural to describe molecules and solids in terms of atomic-like basis functions. The reason for doing this is that a complete basis requires in principle an infinite number of functions. However, practical considerations generally force us to limit the number of basis functions. The more suitable the functions, the less functions we need for an acceptable description of the problem. The eigenfunctions of the system are then expressed in terms of the basis functions.

The translation operator was introduced, leading to the introduction of the momentum operator $\hat{\mathbf{p}} = \hbar\hat{\mathbf{k}} = -i\hbar\nabla$. It is important to realize that particles generally do not move in space as described by $\hat{U}_{\tau}$, as you might expect from a classical particle. This is because different Fourier components move with a different velocity. Therefore, the shape of the function changes when it propagates in time. This is known as a wave packet. An exception is when all Fourier components move with the same speed. This is the case for photons (light particles) in vacuum where all Fourier components move at the speed of light.

Problems

1 Show that the Fourier transform of $\nabla\psi(\mathbf{r})$ is $i\mathbf{k}\psi(\mathbf{k})$.

2 Derive the Fourier transform in spherical polar coordinates of a function $n(r)$ that only depends on the distance to the origin. Leave the final expression as an integral over r.

3 Show that the Fourier transform of a convolution is a product

$$A(\mathbf{r}) = \int d\mathbf{r}'\ G(\mathbf{r}-\mathbf{r}')F(\mathbf{r}') \quad \text{and} \quad A(\mathbf{k}) = G(\mathbf{k})F(\mathbf{k}). \tag{1.247}$$

Convolution is important, for example to describe the effect at a position $\mathbf{r}$ of something that happened at the positions $\mathbf{r}'$. This is called propagation. Although a unitary Fourier transform is useful when dealing with function spaces, for propagation a nonunitary transformation is more convenient

$$G(\mathbf{r}) = \frac{1}{2\pi}\int d\mathbf{k}\ e^{i\mathbf{k}\cdot\mathbf{r}}G(\mathbf{k}) \quad \text{and} \quad G(\mathbf{k}) = \int d\mathbf{r}\ e^{-i\mathbf{k}\cdot\mathbf{r}}G(\mathbf{r}). \tag{1.248}$$

Here, instead of evenly dividing the factor $1/2\pi$ between the Fourier and inverse Fourier transforms, it is now associated with the dual space. Note that, for a Fourier transform in time t, the dual is the frequency ν with the angular frequency $d\omega/2\pi = d\nu$. The exponent then becomes $e^{i\omega t} = e^{2\pi i\nu t}$. There is a change in sign in the Fourier transform with respect to time to ensure causality, *i.e.* $e^{i\mathbf{k}\cdot\mathbf{r}-i\omega t}$.

1.12 Dual Bases

Introduction. – In the preceding sections, we discussed translational motion. Although the mathematics is relatively straightforward, there are some conceptual problems that were swept under the rug. For rotation in the xy-plane, the function space was given by the unit vectors to an integer power, *i.e.* $\hat{\boldsymbol{r}}^m = e^{\boldsymbol{i}_{yx} m\varphi}$ for a two-dimensional system. Taking the derivative with respect to φ gives $d\hat{\boldsymbol{r}}^m/d\varphi = \boldsymbol{i}_{yx} m e^{\boldsymbol{i}_{yx} m\varphi}$. The factor m determines how fast the rotation is. The imaginary unit causes a 90° rotation of the position vector, creating a vector along the circumference, *i.e.* the direction where the vector is moving.

This should be compared with translational motion. Let us take motion in one particular direction. First, the motion is linear and can be compared to motion along the circumference of the circle. The function space is given by exponentials $e^{\boldsymbol{i}k_x x}$. Let us take the derivative with respect to x. This gives $\partial e^{\boldsymbol{i}k_x x}/\partial x = \boldsymbol{i}k_x e^{\boldsymbol{i}k_x x}$. The wavenumber k_x is related to the magnitude of the momentum. For rotation, the imaginary unit determined the direction of the motion. We could expect something similar for translational motion. Since the motion is in the x-direction, we obtain $\boldsymbol{i}\mathbf{e}_x = \mathbf{e}_y\mathbf{e}_z = \boldsymbol{i}_{yz}$. This defines essentially a plane perpendicular to $\mathbf{e}_x$. This can be viewed as the wavefront. However, we can represent a plane in three dimensions by a vector perpendicular to it. In this case, $\mathbf{e}_x$ is perpendicular to the yz-plane, so the wavevector or momentum is in the x-direction as expected.

The fact that a particle moving in the x-direction also has a momentum in the x-direction is so obvious that there seems no need to bring up the quantity $\boldsymbol{i}\mathbf{e}_x$. However, for more complicated systems (think, for example particles in a crystal), this intuition fails. For a coordinate system with nonorthonormal unit vectors $\mathbf{a}^i$, the quantity $\boldsymbol{i}\mathbf{a}^i$ still describes a plane corresponding to the wavefront, which is the dual space of $\mathbf{a}^i$. However, the vector normal to the wavefront is not necessarily parallel to $\mathbf{a}^i$. Therefore, spatial positions and wavevectors are no longer easily related to each other. In this section, we look more closely at the momentum in discrete and nonorthogonal systems.

The complex exponentials for three-dimensional space $e^{\mathbf{ik}\cdot\mathbf{r}}$ introduced in Section 1.10 contain an imaginary unit $\boldsymbol{i}$ whose physical meaning is not directly obvious. Multiplication of a vector by the imaginary unit $\boldsymbol{i} = \mathbf{e}_x\mathbf{e}_y\mathbf{e}_z$ forms an orthogonal space as shown in Section 1.5. Bivectors can be obtained as follows:

$$\underline{\mathbf{e}}_k = \boldsymbol{i}\mathbf{e}_k = \mathbf{e}_i\mathbf{e}_j, \tag{1.249}$$

with $ij = yz, zx, xy$ for $k = x, y, z$, respectively. The vectors $\mathbf{e}_i$ and $\mathbf{e}_j$ define a plane. The bivectors $\mathbf{e}_i\mathbf{e}_j$ form a dual basis with the vectors $\mathbf{e}_k$. Instead of expressing the dual space in bivectors, one can also define the dual space in terms of the vectors perpendicular to them. This is known as a dual or reciprocal basis. In this case, the reciprocal unit vector is $\mathbf{e}_k$ again. The real and reciprocal unit vectors are the same. This seems intuitive: when moving in a particular direction, the direction of the wavefront is in the same direction. Although this works well in a Cartesian system, it is in fact somewhat misleading. The most important aspect of the reciprocal unit vector $\mathbf{e}_k$ is not that it is parallel to the real space unit vector $\mathbf{e}_k$, but that it is perpendicular to the other unit vectors $\mathbf{e}_i$ and $\mathbf{e}_j$ with $i, j \neq k$.

Let us consider a set of basis vectors that are not orthonormal. Such a basis can occur in solids, where the basis vectors are determined by the symmetry of the crystal. A vector in real space expressed in the basis vectors $\mathbf{a}^i$ is given by

$$\mathbf{r} = (\mathbf{r}| = \sum_{i=1}^{3} (\mathbf{r}|i\rangle\langle i| = \sum_{i=1}^{3} r_i \mathbf{a}^i = r_1\mathbf{a}^1 + r_2\mathbf{a}^2 + r_3\mathbf{a}^3, \tag{1.250}$$

where the notation $(\mathbf{r}|$ is used to indicate that the norm is determined by squaring its components (as is common for a vector) and not via integration as in the point $\langle\mathbf{r}|$. In many physics applications, another change occurs. Generally, when, for example $x\mathbf{e}_x$ is used, the unit vector $\mathbf{e}_x$ is dimensionless, and the coordinate x contains the length. However, for $r_i\mathbf{a}^i$, the length is often contained in $\mathbf{a}^i$. In physical applications, $\mathbf{a}^i$ often indicates the periodicity of a system, for example a crystal lattice. The lattice vectors,

$$\mathbf{R} = n_1\mathbf{a}^1 + n_2\mathbf{a}^2 + n_3\mathbf{a}^3, \tag{1.251}$$

where n_i are integers, given the positions of the repeating unit cells. The coefficients n_i are now dimensionless integers, and the length is contained in the basis vectors $\mathbf{a}^i$. The periodicity of the crystal is not the same as the positions of the atoms. However, for many simple systems the vectors $\mathbf{R}$ can be taken equivalent to the atomic positions, which is easier to visualize.

The wavevector is expressed as

$$\mathbf{k} = |\mathbf{k}) = \sum_{i=1}^{3}|i\rangle\langle i|\mathbf{k}) = \sum_{i=1}^{3}\mathbf{a}_i k^i = \mathbf{a}_1k^1 + \mathbf{a}_2k^2 + \mathbf{a}_3k^3, \tag{1.252}$$

where $\mathbf{a}_i$ are the basis vectors of the reciprocal space. For physical applications, the unit of length generally shifts from k^i to $\mathbf{a}_i$, and the coefficients k^i are unitless. Figure 1.9 shows a schematic picture of the basis vectors $\mathbf{a}^i$, with $i = 1, 2, 3$. One reciprocal vector $\mathbf{a}_1$ is also shown. As we will see, the key feature of this vector is that it is orthogonal to both $\mathbf{a}^2$ and $\mathbf{a}^3$ not that it is parallel to $\mathbf{a}^1$. Of special importance are the reciprocal lattice vectors,

$$\mathbf{K} = 2\pi(\mathbf{a}_1m^1 + \mathbf{a}_2m^2 + \mathbf{a}_3m^3), \tag{1.253}$$

where m^i are integers. The reciprocal vectors are effectively the Fourier transform of the lattice in real space and are of special importance in many applications, in particular in solid-state physics.

Let us first look at the pseudoscalar, *i.e.* the product of all three vectors,

$$\boldsymbol{V} = \langle\mathbf{a}^1\mathbf{a}^2\mathbf{a}^3\rangle_3, \tag{1.254}$$

where $\langle\cdots\rangle_r$ indicates that of the vector product we only retain the terms of grade r. This is necessary since the basis vectors are not an orthogonal set of vectors. $\boldsymbol{V}$ is essentially the space spanned by the three basis vectors. There is only one unit multivector of grade 3 in three dimensions, namely the pseudoscalar $\boldsymbol{i} = \mathbf{e}_x\mathbf{e}_y\mathbf{e}_z$. Therefore, $\boldsymbol{V}$ must be proportional to $\boldsymbol{i}$. This means that the pseudoscalar can also be written as

$$\boldsymbol{V} = \boldsymbol{i}V \;\Rightarrow\; V = -\boldsymbol{i}\boldsymbol{V} \text{ and } \boldsymbol{i} = \frac{\boldsymbol{V}}{V} \tag{1.255}$$

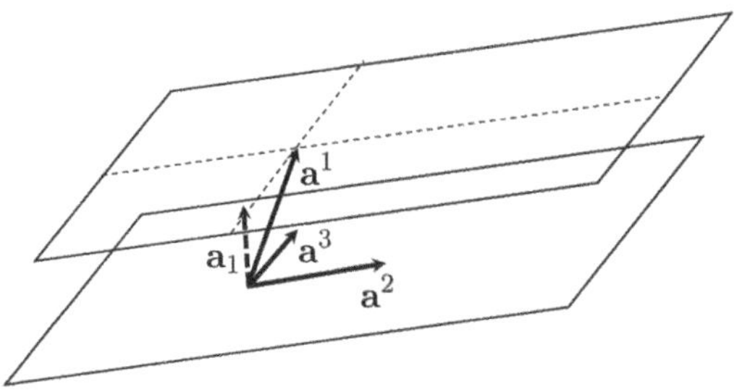

Figure 1.9 A system in three dimensions is defined by the basis vectors $\mathbf{a}^i$, with $i = 1, 2, 3$. The reciprocal vector $\mathbf{a}_1$ is perpendicular to the vectors $\mathbf{a}^2$ and $\mathbf{a}^3$, which define the wavefront (the plane where the phases of the plane waves are equal). The distance between the planes is determined by $\mathbf{a}^1$.

where V is a scalar. The product $\mathbf{a}^1\mathbf{a}^2\mathbf{a}^3$ can be expanded using Eqs. (1.110) and (1.113) as

$$\begin{aligned}\mathbf{a}^1\mathbf{a}^2\mathbf{a}^3 &= \mathbf{a}^1(\mathbf{a}^2\cdot\mathbf{a}^3 + \boldsymbol{i}\mathbf{a}^2\times\mathbf{a}^3)\\ &= \mathbf{a}^1(\mathbf{a}^2\cdot\mathbf{a}^3) + \boldsymbol{i}\mathbf{a}^1\cdot(\mathbf{a}^2\times\mathbf{a}^3) + \boldsymbol{i}\mathbf{a}^1\wedge(\mathbf{a}^2\times\mathbf{a}^3)\\ &= \mathbf{a}^1(\mathbf{a}^2\cdot\mathbf{a}^3) - \mathbf{a}^1\times(\mathbf{a}^2\times\mathbf{a}^3) + \boldsymbol{i}\mathbf{a}^1\cdot(\mathbf{a}^2\times\mathbf{a}^3).\end{aligned}\tag{1.256}$$

The first two terms are both vectors, and the term of rank 3 is therefore $\boldsymbol{V} = \langle\mathbf{a}^1\mathbf{a}^2\mathbf{a}^3\rangle_3 = \boldsymbol{i}\mathbf{a}^1\cdot(\mathbf{a}^2\times\mathbf{a}^3)$. The vector terms are zero in a Cartesian coordinate system since $\mathbf{e}_i\cdot\mathbf{e}_j = 0$ and $\mathbf{e}_i\times(\mathbf{e}_j\times\mathbf{e}_k)\sim\mathbf{e}_i\times\mathbf{e}_i = 0$ for unequal indices. If the vectors $\mathbf{a}^i$ are chosen in a right-handed fashion, the scalar $\mathbf{a}^1\cdot(\mathbf{a}^2\times\mathbf{a}^3)$ is positive and is equal to the volume of the parallelepiped spanned by the three vectors.

We want to extend the results for dual vectors in Eq. (1.249) for Cartesian unit vectors to a general basis. Multiplying a basis vector by the pseudoscalar gives

$$\underline{\mathbf{a}}_i = \boldsymbol{i}\mathbf{a}_i = \frac{\boldsymbol{V}}{V}\mathbf{a}_i = \frac{\langle\mathbf{a}^1\mathbf{a}^2\mathbf{a}^3\rangle_3}{V}\mathbf{a}_i,\tag{1.257}$$

which is the dual space of the vector $\mathbf{a}_i$, which is a bivector and therefore corresponds to a plane. The right-hand side contains products of the real and reciprocal basis. However, we want these bases to be orthonormal to each other,

$$\mathbf{a}_i\cdot\mathbf{a}^j = \delta_{ij}.\tag{1.258}$$

In fact, the orthonormality condition defines the reciprocal lattice vectors. For a Cartesian basis, this condition can be satisfied by taking $\mathbf{e}^i = \mathbf{e}_i$, with $i = x, y, z$. For a nonorthonormal basis, this condition cannot be satisfied with the basis vectors since $\mathbf{a}^i\cdot\mathbf{a}^j$ is not necessarily 0, and $\mathbf{a}^i\cdot\mathbf{a}^i$ is not necessarily 1 since $\mathbf{a}^i$ does not have to be a unit vector.

Let us take $i = 1$. Using that $\mathbf{a}_1\cdot\mathbf{a}^1 \equiv 1$, the dual basis is given by

$$\underline{\mathbf{a}}_1 = \frac{\langle\mathbf{a}^1\mathbf{a}^2\mathbf{a}^3\mathbf{a}_1\rangle_2}{V} = \frac{\langle\mathbf{a}^2\mathbf{a}^3\rangle_2}{V}.\tag{1.259}$$

This can be rewritten in terms of the wedge and outer product as

$$\underline{\mathbf{a}}_1 = \frac{\mathbf{a}^2\wedge\mathbf{a}^3}{V} = \frac{\boldsymbol{i}\mathbf{a}^2\times\mathbf{a}^3}{V}.\tag{1.260}$$

The wedge product $\mathbf{a}^2\wedge\mathbf{a}^3$ defines a plane spanned by the two vectors. The vector perpendicular to that plane (the quantity generally used in most physics applications) is

$$\mathbf{a}_1 = -\boldsymbol{i}\underline{\mathbf{a}}_1 = \frac{\mathbf{a}^2\times\mathbf{a}^3}{V}.\tag{1.261}$$

(Note that a careless application of Eqs. (1.257) and (1.261) can give that $\mathbf{a}_1$ and $\mathbf{a}^1$ are equal. However, in Eq. (1.257), we start from the pseudoscalar of rank 3 and remove a vector, giving a dual space of rank 2. Equation (1.261) then converts this dual space into a vector of rank 1. This is not the same as combining $\boldsymbol{i}$ and $-\boldsymbol{i}$ to the scalar 1 (rank 0) and then multiplying by $\mathbf{a}^1$). The reciprocal vectors defined in Eqs. (1.259) and (1.261) are indeed orthonormal to the basis vectors in real space. For example, for $\mathbf{a}^1$,

$$\mathbf{a}^1\cdot\mathbf{a}_1 = \langle\mathbf{a}^1\mathbf{a}_1\rangle_0 = -\boldsymbol{i}\langle\mathbf{a}^1\underline{\mathbf{a}}_1\rangle_3 = -\boldsymbol{i}\frac{\langle\mathbf{a}^1\mathbf{a}^2\mathbf{a}^3\rangle_3}{V} = -\boldsymbol{i}\frac{\mathbf{V}}{V} = 1,\tag{1.262}$$

using Eq. (1.255).

The direction of $\mathbf{a}_1$ is entirely determined by the vectors $\mathbf{a}^2$ and $\mathbf{a}^3$. The length of $\mathbf{a}^1$ still enters due to the division by the volume V spanned by the three vectors. The other vectors can be obtained by applying cyclic permutations $1 \to 2 \to 3 \to 1$. This gives the reciprocal vectors in three dimensions,

$$\mathbf{a}_1 = \frac{\mathbf{a}^2 \times \mathbf{a}^3}{\mathbf{a}^1 \cdot (\mathbf{a}^2 \times \mathbf{a}^3)}, \quad \mathbf{a}_2 = \frac{\mathbf{a}^3 \times \mathbf{a}^1}{\mathbf{a}^2 \cdot (\mathbf{a}^3 \times \mathbf{a}^1)}, \quad \mathbf{a}_3 = \frac{\mathbf{a}^1 \times \mathbf{a}^2}{\mathbf{a}^3 \cdot (\mathbf{a}^1 \times \mathbf{a}^2)}. \tag{1.263}$$

Note that, despite the permutations, the denominator is always equal to V. For a Cartesian coordinate system, these equations give $\mathbf{e}_i = \mathbf{e}^i$.

After all these changes, one might wonder if the plane waves are still eigenvectors of the momentum operator $\hat{\mathbf{k}} = -\boldsymbol{i}\nabla$, see Eq. (1.239). Let us first adjust the nabla operator to the general basis

$$\nabla = \sum_{i=1}^{3} \frac{\partial}{\partial r_i \mathbf{a}^i}. \tag{1.264}$$

This expression contains the basis vector $\mathbf{a}^i$ in the denominator. In contrast to vector multiplication, vector division is a concept that is generally completely avoided. However, the vector can be brought to the numerator via

$$\frac{1}{\mathbf{a}^i} = \frac{\mathbf{a}^i \cdot \mathbf{a}_i}{\mathbf{a}^i} = \frac{\langle \mathbf{a}^i \mathbf{a}_i \rangle_0}{\mathbf{a}^i} = \left\langle \frac{\mathbf{a}^i \mathbf{a}_i}{\mathbf{a}^i} \right\rangle_1 = \langle \mathbf{a}_i \rangle_1 = \mathbf{a}_i, \tag{1.265}$$

using Eq. (1.258). Therefore, division by a basis vector gives the reciprocal vector. Therefore, the nabla operator can be written as

$$\nabla = \sum_{i=1}^{3} \mathbf{a}_i \frac{\partial}{\partial r_i}. \tag{1.266}$$

Operating on the wave function

$$\varphi_{\mathbf{k}}(\mathbf{r}) = \frac{1}{(2\pi)^{\frac{3}{2}}} e^{\boldsymbol{i}\mathbf{k}\cdot\mathbf{r}} = \frac{1}{(2\pi)^{\frac{3}{2}}} e^{\boldsymbol{i}(r_1 k^1 + r_2 k^2 + r_3 k^3)} \tag{1.267}$$

gives

$$\hat{\mathbf{k}}\varphi_{\mathbf{k}}(\mathbf{r}) = -\boldsymbol{i}\nabla\varphi_{\mathbf{k}}(\mathbf{r}) = -\boldsymbol{i}\sum_{i=1}^{3} \mathbf{a}_i \frac{\partial \varphi_{\mathbf{k}}(\mathbf{r})}{\partial r_i} = -\boldsymbol{i}\sum_{i=1}^{3} \boldsymbol{i}\mathbf{a}_i k^i \varphi_{\mathbf{k}}(\mathbf{r}) = \mathbf{k}\varphi_{\mathbf{k}}(\mathbf{r}),$$

using Eq. (1.252). This gives indeed the desired result.

In summary, for linear propagation, in each dimension an effective two-dimensional space is formed of the basis vector and its dual space,

multivector	number	name
1	1	scalar
$\mathbf{a}^i, \underline{\mathbf{a}}_i$	2	vector, dual vector
$\boldsymbol{i} = \underline{\mathbf{a}}_i \mathbf{a}^i$	1	pseudoscalar

(1.268)

This is analogous to rotation in a two-dimensional system, see (1.29). However, while in a two-dimensional Cartesian system, both directions are defined by real unit vectors, we now have a situation where one direction is the basis vector $\mathbf{a}^i$ in real space, and the orthonormal direction is given by its dual space $\underline{\mathbf{a}}_i$ describing the momentum space. In both cases, the products of the two quantities form an imaginary unit. We will see something similar for the quantum harmonic oscillator, where the linear oscillation can also be viewed as a rotation in the plane of displacement and momentum.

Diffraction. – We have seen that describing space becomes more complicated when the basis vectors do not form a nice orthonormal set. Examples of these types of bases occur in crystals, when the atoms in the solid are organized in a lattice which can be described by the function

$$f(\mathbf{r}) = \sum_{\mathbf{R}} \langle \mathbf{r}|\mathbf{R}\rangle = \sum_{\mathbf{R}} \delta(\mathbf{r} - \mathbf{R}), \tag{1.269}$$

giving a system of discrete points in space. This assumes a three-dimensional crystal consisting of one type of atoms that are assumed to be point charges. (For more complicated crystals, the vectors $\mathbf{R}$ only indicate the periodicity of the crystal; the actual positions are indicated with a basis. However, the principal arguments remain the same). The lattice vectors $\mathbf{R}$, given by

$$\mathbf{R} = \sum_{i=1}^{3} n_i \mathbf{a}^i, \tag{1.270}$$

describe the periodicity of the function. We now want to describe $f(\mathbf{r})$ in terms of the complete function space $e^{i\mathbf{k}\cdot\mathbf{r}}$. If $f(\mathbf{r})$ was a random function in space, all $\mathbf{k}$ would be needed do describe it. However, since $f(\mathbf{r})$ is periodic in space, less wavevectors are needed to describe it. The Fourier transform is given by

$$\begin{aligned} f(\mathbf{k}) &= \langle \mathbf{k}|f\rangle = \int d\mathbf{r} \langle \mathbf{k}|\mathbf{r}\rangle\langle \mathbf{r}|f\rangle \\ &= \frac{1}{\sqrt{V}} \sum_{\mathbf{R}} \int d\mathbf{r}\, e^{-i\mathbf{k}\cdot\mathbf{r}} \delta(\mathbf{r} - \mathbf{R}) = \frac{1}{\sqrt{V}} \sum_{\mathbf{R}} e^{-i\mathbf{k}\cdot\mathbf{R}}. \end{aligned} \tag{1.271}$$

The summation over the exponential by $\mathbf{R}$ is again a δ-function, which is nonzero for $\mathbf{k} \cdot \mathbf{R} = 2\pi n$, with n as an integer. The wavevectors that satisfy this condition are

$$\mathbf{K} = \sum_{i=1}^{d} m_i \mathbf{b}_i = \sum_{i=1}^{d} 2\pi m_i \mathbf{a}_i, \tag{1.272}$$

using the definition of the reciprocal vectors typically employed in physics $\mathbf{b}_i = 2\pi \mathbf{a}_i$ that includes a factor 2π. The exponential is then

$$e^{-i\mathbf{k}\cdot\mathbf{R}} = \exp\left(-2\pi i \sum_{i=1}^{3} m_i n_i\right) = 1, \tag{1.273}$$

since the argument of the exponent is an integer times $2\pi i$. For all other $\mathbf{k}$, the summation of the complex exponential over $\mathbf{R}$ approaches 0. The Fourier transform of the lattice is then

$$f(\mathbf{k}) \sim \sum_{\mathbf{K}} \delta(\mathbf{k} - \mathbf{K}). \tag{1.274}$$

Therefore, the Fourier transform of the periodic lattice in real space is a discrete lattice in reciprocal space. Therefore, even though we started expressing the discrete lattice in plane waves using all $\mathbf{k}$, the reciprocal wavevectors $\mathbf{K}$ followed directly from the symmetry of the system in real space.

The wavevectors $\mathbf{K}$ are of physical importance, since they are the only wavevectors that can be transferred by the crystal without creating additional excitations. Therefore, if a plane wave of X-rays or electrons with wavevector $\mathbf{k}$ scatters elastically from a crystal lattice, the resulting wavevectors can only differ by $\mathbf{K}$, *i.e.* $\mathbf{k}' = \mathbf{k} - \mathbf{K}$. Additionally, since the scattering is elastic, one also needs to satisfy $k = k'$, that is the norm is unchanged. This gives

$$\begin{aligned} \mathbf{k}' = \mathbf{k} - \mathbf{K} \quad &\Rightarrow \quad k'^2 = (\mathbf{k} - \mathbf{K})^2 = k^2 - 2\mathbf{k}\cdot\mathbf{K} + K^2 \\ &\Rightarrow \quad \mathbf{k}\cdot\mathbf{K} = \frac{1}{2}K^2. \end{aligned} \tag{1.275}$$

These are known as the von Laue conditions for X-ray diffraction. X-rays need to be used since the wavelength of the electromagnetic radiation has to be comparable to the distance between the atoms in a solid.

Example 1.10 ***Reciprocal Space in Two Dimensions***
In solid-state physics text books, the reciprocal lattice is often defined by the vectors in Eq. (1.263). However, these expression do not work in two dimensions. Derive the expressions for the reciprocal vectors in two dimensions following the above procedure.

Solution
The pseudoscalar in two dimensions is the rank 2 component of the product of the two unit vectors

$$\boldsymbol{A} = \langle \mathbf{a}^1\mathbf{a}^2 \rangle_2. \tag{1.276}$$

The product can be straightforwardly written out by expressing the basis vectors in a Cartesian basis

$$\mathbf{a}^1 = a_{1x}\mathbf{e}_x + a_{1y}\mathbf{e}_y \quad \text{and} \quad \mathbf{a}^2 = a_{2x}\mathbf{e}_x + a_{2y}\mathbf{e}_y, \tag{1.277}$$

where $\mathbf{e}^i = \mathbf{e}_i$ are the Cartesian unit vectors. The product is then

$$\begin{aligned} \mathbf{a}^1\mathbf{a}^2 &= (a_{1x}\mathbf{e}_x + a_{1y}\mathbf{e}_y)(a_{2x}\mathbf{e}_x + a_{2y}\mathbf{e}_y) \\ &= a_{1x}a_{2x} + a_{1y}a_{2y} + (a_{1x}a_{2y} - a_{1y}a_{2x})\mathbf{e}_x\mathbf{e}_y = \mathbf{a}^1\cdot\mathbf{a}^2 + A\boldsymbol{i}_{xy}. \end{aligned} \tag{1.278}$$

where

$$\boldsymbol{A} = A\boldsymbol{i}_{xy}, \quad \text{with} \quad A = a_{1x}a_{2y} - a_{1y}a_{2x}. \tag{1.279}$$

We can show that this is equal to the area of the parallelogram spanned by the two vectors. If α is the angle between the two basis vectors, the area is given by

$$A = a^1a^2\sin\alpha = a^1a^2\cos(90^\circ - \alpha) = \mathbf{a}^1\cdot\mathbf{a}^2_\perp, \tag{1.280}$$

with the norm $a^i = |\mathbf{a}^i|$. The vector $\mathbf{a}^2_\perp = \boldsymbol{i}_{xy}\mathbf{a}^2 = a_{2y}\mathbf{e}_x - a_{2x}\mathbf{e}_y$ is $\mathbf{a}^2$ rotated by 90° clockwise. The inner product $\mathbf{a}^1\cdot\mathbf{a}^2_\perp = a_{1x}a_{2y} - a_{1y}a_{2x}$ is equal to Eq. (1.279).

The dual space is now defined as

$$\begin{aligned} \underline{\mathbf{a}}_1 &= \boldsymbol{i}_{xy}\mathbf{a}_1 = \frac{1}{A}\langle \mathbf{a}^1\mathbf{a}^2\mathbf{a}_1\rangle_1 = -\frac{\mathbf{a}^2}{A}, \\ \underline{\mathbf{a}}_2 &= \boldsymbol{i}_{xy}\mathbf{a}_2 = \frac{1}{A}\langle \mathbf{a}^1\mathbf{a}^2\mathbf{a}_2\rangle_1 = \frac{\mathbf{a}^1}{A}. \end{aligned} \tag{1.281}$$

While in three dimensions, the dual space is a plane, in two dimensions, it is a line defined by the other basis vector. The reciprocal vectors are obtained by multiplying with the imaginary unit

$$\mathbf{a}_1 = -\boldsymbol{i}_{xy}\underline{\mathbf{a}}_1 = \boldsymbol{i}_{xy}\frac{\mathbf{a}^2}{A} = \frac{a_{2y}\mathbf{e}_x - a_{2x}\mathbf{e}_y}{a_{1x}a_{2y} - a_{1y}a_{2x}},$$
$$\mathbf{a}_2 = -\boldsymbol{i}_{xy}\underline{\mathbf{a}}_1 = -\boldsymbol{i}_{xy}\frac{\mathbf{a}^1}{A} = \frac{-a_{1y}\mathbf{e}_x + a_{1x}\mathbf{e}_y}{a_{1x}a_{2y} - a_{1y}a_{2x}}, \tag{1.282}$$

with $\boldsymbol{i}_{xy}\mathbf{e}_x = -\mathbf{e}_y$ and $\boldsymbol{i}_{xy}\mathbf{e}_y = \mathbf{e}_x$.

Figure 1.10 shows two examples for a two-dimensional lattice. For a rectangular lattice, see Figure 1.10a, the basis vectors are $\mathbf{a}^1 = a_1\mathbf{e}_x$ and $\mathbf{a}^2 = a_2\mathbf{e}_y$. The reciprocal vectors are $\mathbf{a}_1 = \mathbf{e}_x/a_1$ and $\mathbf{a}_2 = \mathbf{e}_y/a_2$. Since we are very used to dealing with orthogonal basis vectors, this can create the impression that the reciprocal vectors $\mathbf{a}_i$ are entirely determined by their real space partner $\mathbf{a}^i$. However, the vectors for an oblique lattice, Figure 1.10b, demonstrate that this is not the case. The direction of $\mathbf{a}_1$ is determined by the basis vector $\mathbf{a}^2$. The basis vector $\mathbf{a}_1$ is orthogonal to $\mathbf{a}^2$ and determines the direction of the wavefront of a wave traveling in the $\mathbf{a}^1$-direction. The length of $\mathbf{a}^1$ still enters into $\mathbf{a}_1$ via the area A in the denominator.

Let us now consider the somewhat unusual case of one dimension. The real space basis vector can be defined as $\mathbf{a}^1 = a\mathbf{e}_x$. Since there is only one vector, the space spanned by all vectors is $\boldsymbol{V} = \mathbf{a}^1$ with a norm $V = a$. If we impose that the pseudoscalar is an imaginary unit, then $\mathbf{e}_x = i$, giving $\mathbf{a}^1 = ia$. The dual space is then $\underline{\mathbf{a}}_1 = \langle \mathbf{a}_1\mathbf{a}^1\rangle_0/V = 1/a$ since $\mathbf{a}_1\mathbf{a}^1 \equiv 1$. The dual space is then a scalar which is orthogonal to the complex real space. The reciprocal vector is then $\mathbf{a}_1 = -i\underline{\mathbf{a}}_1 = -i/a$, which is in the same direction as $\mathbf{a}^1$. Additionally, the inner product equals $\mathbf{a}_1 \cdot \mathbf{a}^1 = (-i/a)ia = 1$.

Conceptual summary. – The use of Cartesian coordinates has many advantages such as orthonormality and the equivalence of the real and reciprocal unit vectors. However, it can lead to the misleading idea that the reciprocal vectors $\mathbf{a}_i$ for the momentum/wavevector are parallel to the real space vectors $\mathbf{a}^i$, as it is for a Cartesian coordinate system. A better way of looking at it is to consider real space to be defined by a set of vectors $\mathbf{a}^i$. We now want to define a dual space. This can be done by defining $\boldsymbol{i} = \prod_{i=1}^{d} \mathbf{a}^i$, where d is the dimension. The pseudoscalar $\boldsymbol{i}$ can be thought of as representing the entire space.

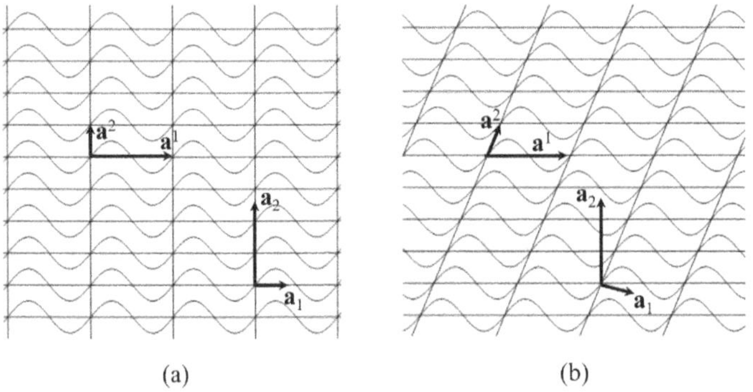

Figure 1.10 The lattice and reciprocal vectors for a two-dimensional system with a rectangular lattice in (a) and an oblique lattice in (b). The oscillations are shown for a wave traveling in the $\mathbf{a}^1$-direction.

The dual space is then given by $\underline{\mathbf{a}}_i = \boldsymbol{i}\mathbf{a}_i$, which is the space defined by all the basis vectors except $\mathbf{a}^i$. The dual space is a line and a plane in two and three dimensions, respectively. In physics, one generally prefers to represent the dual space by the vector $\mathbf{a}_i = -\boldsymbol{i}\underline{\mathbf{a}}_i$ perpendicular to it. In three dimensions, the plane is defined by the vectors $\mathbf{a}^i$ and $\mathbf{a}^j$, and the dual vector $\mathbf{a}_k$ is normal to this space with $ij = 23, 31, 12$ and $k = 1, 2, 3$. However, even though the basis and reciprocal vectors are orthonormal with $\mathbf{a}_i \cdot \mathbf{a}^j = \delta_{ij}$, this does not imply that $\mathbf{a}_i$ and $\mathbf{a}^i$ are parallel to each other. The dual space defines the direction of a wavefront of the eigenfunctions, see Figure 1.10, *i.e.* the regions in space with the same phase.

The conceptual issues encountered in this section did not really occur for rotation. The reason is that the motion for rotation is two-dimensional and given by a bivector. Note that the position vector is perpendicular to the change in position. The associated dual space is then a vector related to the axis of rotation. The angular momentum, the conserved physical quantity, is parallel to the axis of rotation. Therefore, the plane of motion and the conserved quantity are always perpendicular to each other. For translation, the motion is one-dimensional. Therefore, the associated dual space is a plane effectively related to the wavefront of the plane wave. The commonly used conserved quantity, the wavevector or momentum, is a vector perpendicular to the dual space. Since in free space and many other situations, the real and reciprocal vectors are in the same direction, it becomes natural to assume that this is always the case. That this notion is incorrect is often only first encountered in the study of crystal lattices.

Problems

1 A body-centered cubic (bcc) lattice is a simple cubic lattice with an additional atom at the center of the cube. This is a primitive lattice, *i.e.* all atomic positions can be described by a single set of vectors. One particular choice is $\mathbf{a}^1 = a\mathbf{e}_x$, $\mathbf{a}^2 = a\mathbf{e}_y$, and $\mathbf{a}^3 = \frac{a}{2}(\mathbf{e}_x + \mathbf{e}_y + \mathbf{e}_z)$, where a is the side of the cube, and $\mathbf{a}^3$ points toward the atom at the center of the cube.

a) Find the set of shortest primitive vectors that transform into each other via cyclic permutation.

b) Determine the reciprocal lattice.

1.13 Dual Basis Obtained via Matrix Inversion

Introduction. – In the preceding section, the dual space was determined by geometric considerations. We proceed here by providing an alternative approach using matrices. Matrices are a powerful tool that lends itself very well to numerical techniques and allows a straightforward generalization to arbitrary dimensions. However, as we already saw with the Pauli matrices, the approach quickly becomes abstract, and the physics can get lost.

The transformation from a basis to the reciprocal basis can also be expressed in terms of matrices. The basis vectors $\mathbf{a}^m$, with $m = 1, 2, 3$, can be written in a Cartesian basis $\mathbf{e}_i$, with $i = 1, 2, 3 = x, y, z$, by

$$\langle m| = \mathbf{a}^m = \sum_{i=1}^{3} \langle m|i\rangle\langle i| = \sum_{i=1}^{3} a_i^m \mathbf{e}^i = (a_1^m, a_2^m, a_3^m) = (a_x^m, a_y^m, a_z^m), \tag{1.283}$$

where $a_i^m = \langle m|i\rangle$ are the projections of the basis vectors onto the Cartesian coordinate system. This is a row vector since

$$\langle 1| = \mathbf{e}^1 = (1,0,0), \quad \langle 2| = \mathbf{e}^2 = (0,1,0), \quad \langle 3| = \mathbf{e}^3 = (0,0,1). \tag{1.284}$$

The dual or reciprocal vector is given by

$$|m\rangle = \mathbf{a}_m = \sum_{i=1}^{3} |i\rangle\langle i|m\rangle = \sum_{i=1}^{3} \mathbf{e}_i a_m^i = \begin{pmatrix} a_m^1 \\ a_m^2 \\ a_m^3 \end{pmatrix} = \begin{pmatrix} a_m^x \\ a_m^y \\ a_m^z \end{pmatrix}. \tag{1.285}$$

This is a column vector since the Cartesian unit vectors satisfy $\mathbf{e}_i = (\mathbf{e}^i)^T$. Note that this is not the case for the basis vectors: $\mathbf{a}_m \neq \mathbf{a}^m$.

The information of the basis vectors can also be stored in matrix form

$$\mathrm{U} = \sum_{i,m=1}^{d} |m\rangle\langle m|i\rangle\langle i| = \sum_{i,m=1}^{d} \mathbf{a}_m a_i^m \mathbf{e}^i = \sum_{m=1}^{d} \mathbf{a}_m \mathbf{a}^m. \tag{1.286}$$

Initially, this appears like a strange operator, since

$$\mathbf{a}^m \mathrm{U} = \sum_{m'=1}^{d} \mathbf{a}^m \cdot \mathbf{a}_{m'} \mathbf{a}^{m'} = \sum_{m'=1}^{d} \delta_{mm'} \mathbf{a}^{m'} = \mathbf{a}^m. \tag{1.287}$$

It therefore appears that it does not do anything. This is correct from a physical point of view, since U is a unitary transformation from one basis to another. Therefore, the vector itself has not changed at all. However, depending on U, $\mathbf{a}^m$ on the left-hand side will be expressed in a different basis than $\mathbf{a}^m$ on the far-right hand side. This can also be seen from Eq. (1.286), where $\mathbf{a}_m = |m\rangle$ is expressed in terms of the basis vectors. However, $\mathbf{a}^m = \sum_{i=1}^{3} \langle m|i\rangle\langle i| = \sum_{i=1}^{3} a_i^m \mathbf{e}^i$ is expressed in the Cartesian basis.

For simplicity, let us restrict ourselves to two dimensions. In that case, for example $|1\rangle = \mathbf{a}_1 = \begin{pmatrix} 1 \\ 0 \end{pmatrix}$ is expressed in the basis of the vectors $\mathbf{a}_m$. The dual vectors, on the other hand, are expressed in the Cartesian basis, for example $\mathbf{a}^1 = (a_x^1, a_y^1)$. The unitary transformation is given by

$$\mathrm{U} = \sum_{m=1}^{2} \mathbf{a}_m \mathbf{a}^m = \begin{pmatrix} \mathbf{a}^1 \\ \mathbf{a}^2 \end{pmatrix} = \begin{pmatrix} a_x^1 & a_y^1 \\ a_x^2 & a_y^2 \end{pmatrix}. \tag{1.288}$$

For example, the vector $\mathbf{a}^1$ can be transformed from the $\mathbf{a}^m$ basis to the Cartesian basis via

$$\mathbf{a}^1 = \mathbf{a}^1 \mathrm{U} = (1,0)\mathrm{U} = (1,0) \begin{pmatrix} a_x^1 & a_y^1 \\ a_x^2 & a_y^2 \end{pmatrix} = (a_x^1, a_y^1). \tag{1.289}$$

In the previous section, it was noted that one can think of a reciprocal vector as $\mathbf{a}_j$ as $1/\mathbf{a}^j$. We want the vectors $\mathbf{a}_j$ to be orthonormal to $\mathbf{a}^i$. Using Eq. (1.265), we can write

$$\langle j|i\rangle = \mathbf{a}^j \cdot \mathbf{a}_i = \langle \mathbf{a}^j \mathbf{a}_i \rangle_0 = \left\langle \mathbf{a}^j \frac{1}{\mathbf{a}^i} \right\rangle_0 = \left\langle \mathbf{e}^j \mathrm{U} \frac{1}{\mathbf{e}^i \mathrm{U}} \right\rangle_0. \tag{1.290}$$

Again, matrices and vectors in the denominator are generally discouraged, so we want to move them to the numerator. However, we have to be careful when dealing with products in the denominator. For example, for bivectors, we want to have the relation

$$\frac{1}{\boldsymbol{i}_{ij}} \boldsymbol{i}_{ij} = 1. \tag{1.291}$$

Now, we know that

$$\mathbf{e}_j\mathbf{e}_i\mathbf{e}_i\mathbf{e}_j = 1, \tag{1.292}$$

with $i \neq j$. This implies that

$$\frac{1}{\boldsymbol{i}_{ij}} = (\boldsymbol{i}_{ij})^{-1} = (\mathbf{e}_i\mathbf{e}_j)^{-1} = (\mathbf{e}_j)^{-1}(\mathbf{e}_i)^{-1} = \mathbf{e}_j\mathbf{e}_i, \tag{1.293}$$

since in a Cartesian coordinate system the unit vectors and their reciprocals are equal, *i.e.* $(\mathbf{e}_i)^{-1} = \mathbf{e}^i = \mathbf{e}_i$. Therefore, the inverse of the product of two unit vectors is equal to the product of the inverse in reverse order. Equation (1.290) can be written as

$$\mathbf{a}^j \cdot \mathbf{a}_i = \langle \mathbf{e}^j \mathrm{U}\mathrm{U}^{-1}\mathbf{e}_i\rangle_0 = \mathbf{e}^j \cdot \mathbf{e}_i = \mathbf{e}_j \cdot \mathbf{e}_i = \delta_{ij} \tag{1.294}$$

as long as the following condition holds:

$$\mathrm{U}\mathrm{U}^{-1} = \mathbb{1}_d, \tag{1.295}$$

where d is the dimension. The matrix U^{-1} is known as the inverse of U. The above equation can be satisfied by writing the inverse matrix in terms of the reciprocal vector,

$$\mathrm{U}^{-1} = \sum_{i,m=1}^{d} |i\rangle\langle i|m\rangle\langle m| = \sum_{i,m=1}^{d} \mathbf{e}_i a^i_m \mathbf{a}^m = (\mathbf{a}_1\ \mathbf{a}_2) = \begin{pmatrix} a_{1x} & a_{2x} \\ a_{1y} & a_{2y} \end{pmatrix}. \tag{1.296}$$

The product of U and U^{-1} gives

$$\mathrm{U}\mathrm{U}^{-1} = \begin{pmatrix} \mathbf{a}^1 \cdot \mathbf{a}_1 & \mathbf{a}^1 \cdot \mathbf{a}_2 \\ \mathbf{a}^2 \cdot \mathbf{a}_1 & \mathbf{a}^2 \cdot \mathbf{a}_2 \end{pmatrix} = \mathbb{1}_2, \tag{1.297}$$

using the orthonormality $\mathbf{a}^j \cdot \mathbf{a}_i = \delta_{ji}$ of the basis vectors and their reciprocals. We can also express U^{-1} in the components of the basis vectors as given in Eq. (1.282)

$$\mathrm{U}^{-1} = (\mathbf{a}_1\ \mathbf{a}_2) = \frac{1}{a_{1x}a_{2y} - a_{1y}a_{2x}} \begin{pmatrix} a_{2y} & -a_{1y} \\ -a_{2x} & a_{1x} \end{pmatrix}, \tag{1.298}$$

which is the inverse of U expressed in the matrix elements of U from Eq. (1.288). The prefactor is one divided by the determinant of U. This process can be generalized for higher dimensions. However, the process becomes very involved, and generally, the diagonalization of matrices is done numerically.

Example 1.11 A rotation in two dimensions can be viewed as a unitary transformation.

a) Write the rotation in matrix form and determine the inverse of the rotation.
b) Demonstrate that the Cartesian unit vectors $\mathbf{e}^i$ and $\mathbf{e}_i$ are rotated over the same angle by the rotation.

Solution

a) A rotation in two dimensions is given by

$$\mathrm{U} = \begin{pmatrix} \cos\varphi & -\sin\varphi \\ \sin\varphi & \cos\varphi \end{pmatrix}. \tag{1.299}$$

We now use Eq. (1.298) to obtain the inverse. The determinant is $\cos^2\varphi + \sin^2\varphi = 1$, which is the area spanned by two orthonormal unit vectors. The inverse is then

$$\mathrm{U}^{-1} = \begin{pmatrix} \cos\varphi & \sin\varphi \\ -\sin\varphi & \cos\varphi \end{pmatrix}. \tag{1.300}$$

Obviously, this is simply a rotation over $-\varphi$.

b) The Cartesian unit vectors can be written as

$$\mathbf{e}^x = \langle x| = (1,0) \text{ and } \mathbf{e}_x = |x\rangle = \begin{pmatrix} 1 \\ 0 \end{pmatrix}, \tag{1.301}$$

with similar expressions for the y component. Orthogonality gives

$$\langle i|j\rangle = \delta_{ij}, \tag{1.302}$$

with $i, j = x, y$. Since $\mathrm{U}^{-1}\mathrm{U} = \mathbb{1}_2$, this can be rewritten as

$$\langle i|j\rangle = \langle i|\mathrm{U}^{-1}\mathrm{U}|j\rangle = \delta_{ij}. \tag{1.303}$$

From this expression, we see that the ket is rotated counterclockwise with φ. For the bra, the inverse is used, which is a clockwise rotation over φ when working on the right. However, when operating on the left

$$\mathbf{e}^x\mathrm{U}^{-1} = (1,0)\begin{pmatrix} \cos\varphi & \sin\varphi \\ -\sin\varphi & \cos\varphi \end{pmatrix} = (\cos\varphi, \sin\varphi), \tag{1.304}$$

a counterclockwise rotation is found. Note that this is simply the transpose of the column vector $\mathrm{U}\mathbf{e}_x$. Therefore, they are rotated over the same angle. This is the same as $\mathbf{e}_i = (\mathbf{e}^i)^T$.

Example 1.12 A unitary transformation is not restricted to vectors but can also be applied to points in space $|\mathbf{r}\rangle$.

a) Give the unitary transformation from $\langle\mathbf{r}|$ to the function space $\varphi_m(\mathbf{r})$. Assume that the function space is complete, and that the functions are countable.
b) Express an arbitrary function $f(\mathbf{r})$ in terms of the function space $\varphi_m(\mathbf{r})$.

Solution

a) The unitary transformation is given by

$$\mathrm{U}^{-1} = \sum_m \int d\mathbf{r}\; |\mathbf{r}\rangle\langle\mathbf{r}|m\rangle\langle m| = \sum_m \int d\mathbf{r}\; |\mathbf{r}\rangle\varphi_m(\mathbf{r})\langle m|, \tag{1.305}$$

with $\varphi_m(\mathbf{r}) = \langle\mathbf{r}|m\rangle$.

b) Expressing $f(\mathbf{r})$ in terms of the basis functions gives

$$\begin{aligned} f(\mathbf{r}) = \langle\mathbf{r}|f\rangle = \langle\mathbf{r}|\mathrm{U}^{-1}|f\rangle &= \sum_m \int d\mathbf{r}'\; \langle\mathbf{r}|\mathbf{r}'\rangle\varphi_m(\mathbf{r}')\langle m|f\rangle \\ &= \sum_m \varphi_m(\mathbf{r})\langle m|f\rangle = \sum_m \varphi_m(\mathbf{r}) f_m, \end{aligned} \tag{1.306}$$

using $\langle\mathbf{r}|\mathbf{r}'\rangle = \delta(\mathbf{r}-\mathbf{r}')$. The coefficients f_m can be obtained from the function using the transformation

$$\mathrm{U} = \sum_m \int d\mathbf{r}\; |m\rangle\langle m|\mathbf{r}\rangle\langle\mathbf{r}| = \sum_m \int d\mathbf{r}\; |m\rangle\varphi_m^*(\mathbf{r})\langle\mathbf{r}|. \tag{1.307}$$

This gives for the coefficient

$$\begin{aligned} f_m = \langle m|f\rangle = \langle m|\mathrm{U}|f\rangle &= \sum_{m'} \int d\mathbf{r}\; \langle m|m'\rangle\varphi_{m'}^*(\mathbf{r})\langle\mathbf{r}|f\rangle \\ &= \int d\mathbf{r}\; \varphi_m^*(\mathbf{r})\langle\mathbf{r}|f\rangle = \int d\mathbf{r}\; \varphi_m^*(\mathbf{r}) f(\mathbf{r}). \end{aligned} \tag{1.308}$$

The above expressions are a generalization of the Fourier transform for an arbitrary function space. For the Fourier transform $m \to \mathbf{k}$, and $\varphi_{\mathbf{k}}(\mathbf{r}) = e^{i\mathbf{k}\cdot\mathbf{r}}/\sqrt{V}$, where V is the volume of the system.

Conceptual summary. – In this section, it was demonstrated that the dual space can also be obtained via matrix inversion, showing the close connection between matrices and vectors. Additionally, we saw that calculating the inverse is similar to dividing by a matrix or vector. There are some caveats though, since this is not always possible. A singular matrix, whose determinant is zero, has no inverse. However, this does not occur for many applications in physics. For example, suppose we have a matrix representing three basis vectors for a three-dimensional space. In order for the determinant to be zero, the vectors have to be linearly dependent, *i.e.* at least one of the vectors can be created using a linear combination of the other vectors. This also means that the three vectors can only describe a plane. Therefore, the volume spanned by the three vectors, which is equal to the determinant, is zero. However, this directly implies that the three vectors were not a proper set of basis vectors for a three-dimensional system to begin with.

Problems

1 Find the reciprocal lattice of a two-dimensional triangular lattice and describe the lattice.

2 Determine the reciprocal lattice of the body-centered cubic (bcc) lattice via matrix inversion, see Problem 1 in Section 1.12. Feel free to use mathematical software for the matrix inversion.

1.14 The Unit Sphere

Introduction. – Our interest in the coming sections lies with free rotational motion in three dimensions. We are therefore looking at problems with spherical symmetry. An important application is the electronic structure of the atom. The $1/r$ potential of the nucleus is spherically symmetric. Therefore, rotation around the nucleus is essentially free motion since the distance to the nucleus, and therefore the potential, does not change. The problem then splits into two: a one-dimensional system that describes the motion in a potential in the radial direction and a two-dimensional system (a sphere) describing free angular motion. We will deal with the radial motion later.

Our goal is to find the function space for rotation in three dimensions. As we have done for the function space of complex exponentials, we first look at a unit vector and then raise it to all positive integer powers to find the higher order vectors. The coefficients of all the higher order vectors then form the function space. This is the function space of spherical harmonics, which is also the angular part of the atomic orbitals. We want the function space and the basis vectors to reflect the conserved quantities in the system. For rotation, this is angular momentum. To achieve that, we need to start with a unit vector that has the same properties.

Equation (1.125) gives a unit vector in spherical polar coordinates. The unit vector describes a sphere, *i.e.* a two-dimensional object in three dimensions. It is given in two coordinates θ and φ but three unit vectors $\mathbf{e}_i$, with $i = x, y, z$. The latter is not satisfactory: we want a two-dimensional space described by two vector quantities. Additionally, the coefficients are not eigenfunctions of any angular momentum operator. It is not possible to find a vector where the coefficients are eigenfunctions of rotations around both the z- and y-axis, corresponding to the coordinates φ and θ, respectively. Following convention, we find eigenfunctions for rotations around the z-axis and find a set of unit vectors that is also practical for rotations around the y-axis.

Let us first look at rotations around the z-axis. Since we are dealing with rotation and are looking for two vector quantities to describe the two-dimensional space, it is natural to look into spinors. The spinors with respect to the z-axis used for constructing the Pauli spinors are 1 and $\boldsymbol{i}_{xz}$ (or, equivalently, $\mathbf{e}_z$ and $\mathbf{e}_x$). The spinors lie in the zx-plane and form therefore also a good orthogonal basis for rotation around the y-axis. However, we will pick a different pair of spinors for the following reasons. The spinors are not the same type of vector quantities since 1 is a scalar, and $\boldsymbol{i}_{xz}$ is a bivector. Additionally, the product of these spinors is an imaginary unit, and we prefer it to be a 1. This can be achieved by taking the x-axis as the reference axis for the spinors. The spinors are then each other's conjugates

$$\boldsymbol{a}^{\dagger}_{\pm\frac{1}{2}} = e^{\mp \boldsymbol{i}_{xz}\frac{\pi}{4}} = e^{\pm \boldsymbol{i}_{zx}\frac{\pi}{4}} = \frac{1}{\sqrt{2}}(1 \pm \boldsymbol{i}_{zx}). \tag{1.309}$$

The notation is changed from $\boldsymbol{r}_{\pm\frac{1}{2}}$ to $\boldsymbol{a}^{\dagger}_{\pm\frac{1}{2}}$. This agrees better with the conventional notation for the quantities that will be introduced below. The $\dagger$ indicates that this is a spinor acting on a reference vector to the right. The related vectors with respect to the x-axis are

$$\mathbf{a}^{\dagger}_{\pm\frac{1}{2}} = \boldsymbol{a}^{\dagger}_{\pm\frac{1}{2}}\mathbf{e}_x = \frac{1}{\sqrt{2}}(\mathbf{e}_x \pm \mathbf{e}_z), \tag{1.310}$$

see Figure 1.11. Note that $\boldsymbol{a}^{\dagger}_{\pm\frac{1}{2}}$ commute with each other, whereas $\mathbf{a}^{\dagger}_{\pm\frac{1}{2}}$ anticommute.

The axis of rotation is obtained by squaring the spinors with the x-direction as the reference axis,

$$\boldsymbol{a}^{\dagger}_{\pm\frac{1}{2}}\boldsymbol{a}^{\dagger}_{\pm\frac{1}{2}}\mathbf{e}_x = e^{\pm \boldsymbol{i}_{zx}\frac{\pi}{2}}\mathbf{e}_x = \pm\boldsymbol{i}_{zx}\mathbf{e}_x = \pm\mathbf{e}_z, \tag{1.311}$$

where the positive and negative signs correspond to counterclockwise and clockwise rotations around the z-axis, respectively. The spinors are eigenvectors of the rotation axis $\mathbf{e}_z$,

$$\begin{aligned}\mathbf{e}_z\boldsymbol{a}^{\dagger}_{\pm\frac{1}{2}} &= \frac{1}{\sqrt{2}}(\mathbf{e}_z \pm \mathbf{e}_z\mathbf{e}_z\mathbf{e}_x) = \frac{1}{\sqrt{2}}(\mathbf{e}_z\mathbf{e}_x \pm 1)\mathbf{e}_x \\ &= \pm\frac{1}{\sqrt{2}}(1 \pm \boldsymbol{i}_{zx})\mathbf{e}_x = \pm\boldsymbol{a}^{\dagger}_{\pm\frac{1}{2}}\mathbf{e}_x,\end{aligned} \tag{1.312}$$

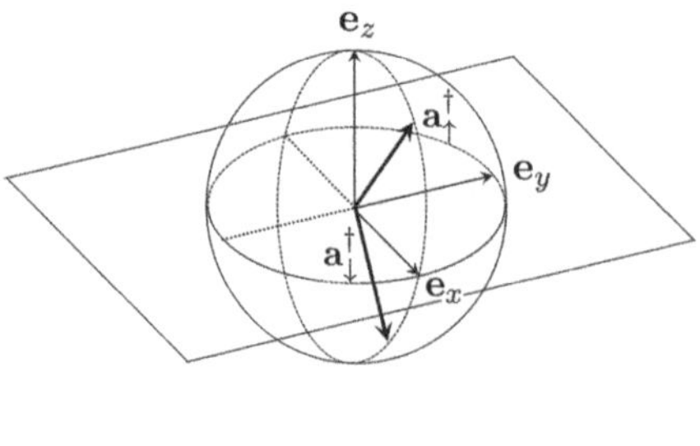

Figure 1.11 The indicated vectors $\mathbf{a}^{\dagger}_{\uparrow} = \boldsymbol{a}^{\dagger}_{\uparrow}\mathbf{e}_x$ and $\mathbf{a}^{\dagger}_{\downarrow} = \boldsymbol{a}^{\dagger}_{\downarrow}\mathbf{e}_x$ form an orthonormal right-handed set of basis vectors in the zx-plane corresponding to rotations around the y-axis. Additionally, they are spinors for the z-axis. Squaring the spinors with respect to the x-axis gives $\pm\mathbf{e}_z$ the axis for rotations around the z-axis.

where $\mathbf{e}_x$ is the reference axis for the spinors. The signs again reflect the direction of the rotation. They are then also complex eigenvectors of the imaginary unit $\boldsymbol{i}_{yx} = -\boldsymbol{i}\mathbf{e}_z$,

$$\boldsymbol{i}_{yx}\boldsymbol{a}^{\dagger}_{\pm\frac{1}{2}} = \mp\boldsymbol{i}\boldsymbol{a}^{\dagger}_{\pm\frac{1}{2}}\mathbf{e}_x. \tag{1.313}$$

This is the same expression as for the eigenvectors of a rotation in Eq. (1.135), except that the x-axis is used as the reference axis instead of the z-axis. Additionally, we can show that

$$\mathbf{e}_x\boldsymbol{a}^{\dagger}_{\pm\frac{1}{2}} = \boldsymbol{a}^{\dagger}_{\mp\frac{1}{2}}\mathbf{e}_x, \quad \mathbf{e}_y\boldsymbol{a}^{\dagger}_{\pm\frac{1}{2}} = \pm\boldsymbol{a}^{\dagger}_{\mp\frac{1}{2}}\boldsymbol{i}_{yz}\mathbf{e}_x. \tag{1.314}$$

These results can be checked using the anticommutative properties or by expressing everything in terms of Pauli matrices and using straightforward matrix multiplication. Together with Eq. (1.312), we see that the transformation properties are those of the Pauli matrices, see Eq. (1.149), except that the reference axis changes from z to x, and the imaginary unit follows the same cyclic permutation $x \to y \to z \to x$, giving $\boldsymbol{i}_{xy} \to \boldsymbol{i}_{yz}$. Therefore, the complex exponentials $\boldsymbol{a}^{\dagger}_{\pm\frac{1}{2}}$ are proper spinors that contain all the correct transformation properties.

The spinors are also relevant for the rotations around the y-axis, since they lie in the zx-plane, which is the plane of rotation. Under the imaginary unit $\boldsymbol{i}_{xz} = -\boldsymbol{i}\mathbf{e}_y$, the quantities now rotate into each other

$$\boldsymbol{i}_{xz}\boldsymbol{a}^{\dagger}_{\pm\frac{1}{2}} = \pm\boldsymbol{a}^{\dagger}_{\mp\frac{1}{2}} \quad \text{or} \quad \boldsymbol{i}_{xz}\boldsymbol{a}^{\dagger}_{\sigma} = (-1)^{\frac{1}{2}-\sigma}\boldsymbol{a}^{\dagger}_{-\sigma}. \tag{1.315}$$

Unlike spinors that are eigenvectors of an imaginary unit, these vectors rotate into each other. The above relation holds for any orthogonal set of unit vectors in the zx-plane and is the same as the rotation in two dimensions described in Section 1.2.

Just like the rotation axes, an arbitrary unit vector can also be expressed in terms of a product of spinors, see Eqs. (1.133) and (1.134),

$$\hat{\boldsymbol{r}} = e^{\boldsymbol{i}_{x'z}\frac{\theta}{2}}e^{\boldsymbol{i}_{x'z}\frac{\theta}{2}} = \left(\cos\frac{\theta}{2} + \boldsymbol{i}_{x'z}\sin\frac{\theta}{2}\right)\boldsymbol{a}^{\dagger}_{\uparrow}\boldsymbol{a}^{\dagger}_{\downarrow}\left(\cos\frac{\theta}{2} + \boldsymbol{i}_{x'z}\sin\frac{\theta}{2}\right). \tag{1.316}$$

Since we want to express the vector in terms of the spinors, the product $\boldsymbol{a}^{\dagger}_{\uparrow}\boldsymbol{a}^{\dagger}_{\downarrow} = 1$ has been inserted. The up spinor then becomes

$$\hat{\boldsymbol{r}}_{\uparrow} = \cos\frac{\theta}{2}\boldsymbol{a}^{\dagger}_{\uparrow} + \sin\frac{\theta}{2}\boldsymbol{i}_{x'z}\boldsymbol{a}^{\dagger}_{\uparrow} = \cos\frac{\theta}{2}\boldsymbol{a}^{\dagger}_{\uparrow} + e^{\boldsymbol{i}_{yx}\varphi}\sin\frac{\theta}{2}\boldsymbol{a}^{\dagger}_{\downarrow}, \tag{1.317}$$

using $\boldsymbol{i}_{x'z}\boldsymbol{a}^{\dagger}_{\uparrow} = e^{\boldsymbol{i}_{yx}\varphi}\boldsymbol{i}_{xz}e^{-\boldsymbol{i}_{xz}\frac{\pi}{4}} = e^{\boldsymbol{i}_{yx}\varphi}e^{\boldsymbol{i}_{xz}\frac{\pi}{2}}e^{-\boldsymbol{i}_{xz}\frac{\pi}{4}} = e^{\boldsymbol{i}_{yx}\varphi}e^{-\boldsymbol{i}_{zx}\frac{\pi}{4}} = e^{\boldsymbol{i}_{yx}\varphi}\boldsymbol{a}^{\dagger}_{\downarrow}$. The second term in parentheses in Eq. (1.316) can be written as

$$\hat{\boldsymbol{r}}_{\downarrow} = -\sin\frac{\theta}{2}\boldsymbol{a}^{\dagger}_{\uparrow}e^{-\boldsymbol{i}_{yx}\varphi} + \cos\frac{\theta}{2}\boldsymbol{a}^{\dagger}_{\downarrow}, \tag{1.318}$$

using that $\boldsymbol{a}^{\dagger}_{\downarrow}e^{\boldsymbol{i}_{yx}\varphi}\boldsymbol{i}_{xz} = \boldsymbol{a}^{\dagger}_{\downarrow}\boldsymbol{i}_{xz}e^{-\boldsymbol{i}_{yx}\varphi} = e^{\boldsymbol{i}_{xz}\frac{3\pi}{4}}e^{-\boldsymbol{i}_{yx}\varphi} = e^{\boldsymbol{i}_{xz}\pi}e^{\boldsymbol{i}_{zx}\frac{\pi}{4}}e^{-\boldsymbol{i}_{yx}\varphi} = -\boldsymbol{a}^{\dagger}_{\uparrow}e^{-\boldsymbol{i}_{yx}\varphi}$. The unit vector in terms of products of spinors is then

$$\hat{\boldsymbol{r}} = \hat{\boldsymbol{r}}_{\uparrow}\hat{\boldsymbol{r}}_{\downarrow} = \cos\theta\boldsymbol{a}^{\dagger}_{\uparrow}\boldsymbol{a}^{\dagger}_{\downarrow} + \frac{\sin\theta}{2}\left(-\boldsymbol{a}^{\dagger}_{\uparrow}\boldsymbol{a}^{\dagger}_{\uparrow}e^{-\boldsymbol{i}_{yx}\varphi} + e^{\boldsymbol{i}_{yx}\varphi}\boldsymbol{a}^{\dagger}_{\downarrow}\boldsymbol{a}^{\dagger}_{\downarrow}\right), \tag{1.319}$$

where the double-angle formulas for the sine and cosine have been used. This unit vector achieves the goal we set out. Although we have two different vector quantities, we are able to describe three different directions. This has been achieved using products of two spinors $\boldsymbol{a}^{\dagger}_{\sigma'}\boldsymbol{a}^{\dagger}_{\sigma}$, with $\sigma, \sigma' = \pm\frac{1}{2}$. The different directions are separated by $\sigma + \sigma'$, which is essentially the projected angular momentum with values $1, 0, -1$ for rotations around the z-axis. Additionally, the coefficients are eigenfunctions of the angular momentum operator for rotations around the z-axis, $\hat{L}_z = -\boldsymbol{i}_{yx}\partial/\partial\varphi$. Note that the angular momentum of the coefficients is opposite to that of the vector quantities. These coefficients are known as spherical coordinates, which is distinct from spherical polar coordinates.

It is useful to rewrite this result in Cartesian coordinates. Expanding the complex exponentials in Eq. (1.319) gives

$$\hat{\boldsymbol{r}} = \cos\theta \boldsymbol{a}_\uparrow^\dagger \boldsymbol{a}_\downarrow^\dagger + \sin\theta\cos\varphi\frac{1}{2}\left(-\boldsymbol{a}_\uparrow^\dagger \boldsymbol{a}_\uparrow^\dagger + \boldsymbol{a}_\downarrow^\dagger \boldsymbol{a}_\downarrow^\dagger\right)$$
$$+\sin\theta\sin\varphi\frac{1}{2}\left(\boldsymbol{a}_\uparrow^\dagger \boldsymbol{a}_\uparrow^\dagger \boldsymbol{i}_{yx} + \boldsymbol{i}_{yx}\boldsymbol{a}_\downarrow^\dagger \boldsymbol{a}_\downarrow^\dagger\right), \tag{1.320}$$

where we recognize the coefficients for the spherical polar coordinates from Eq. (1.125). This gives the two unit bivectors

$$\frac{1}{2}\left(-\boldsymbol{a}_\uparrow^\dagger \boldsymbol{a}_\uparrow^\dagger + \boldsymbol{a}_\downarrow^\dagger \boldsymbol{a}_\downarrow^\dagger\right) = \frac{1}{2}(-(-\boldsymbol{i}_{xz}) + \boldsymbol{i}_{xz}) = \boldsymbol{i}_{xz},$$
$$\frac{1}{2}\left(\boldsymbol{a}_\uparrow^\dagger \boldsymbol{a}_\uparrow^\dagger \boldsymbol{i}_{yx} + \boldsymbol{i}_{yx}\boldsymbol{a}_\downarrow^\dagger \boldsymbol{a}_\downarrow^\dagger\right) = \frac{1}{2}((-\boldsymbol{i}_{xz})\boldsymbol{i}_{yx} + \boldsymbol{i}_{yx}\boldsymbol{i}_{xz}) = \boldsymbol{i}_{yz}, \tag{1.321}$$

using that the products of the spinors can be written as

$$\boldsymbol{a}_\uparrow^\dagger \boldsymbol{a}_\downarrow^\dagger = 1, \quad \boldsymbol{a}_\uparrow^\dagger \boldsymbol{a}_\uparrow^\dagger = -\boldsymbol{i}_{xz}, \quad \boldsymbol{a}_\downarrow^\dagger \boldsymbol{a}_\downarrow^\dagger = \boldsymbol{i}_{xz}, \tag{1.322}$$

following their definition in Eq. (1.315) and using $\boldsymbol{i}_{yx}\boldsymbol{i}_{xz} = \boldsymbol{i}_{yz}$ and $\boldsymbol{i}_{xz}\boldsymbol{i}_{yx} = -\boldsymbol{i}_{yz}$. The unit vector can then be written as

$$\hat{\boldsymbol{r}} = \cos\theta + \sin\theta\cos\varphi\ \boldsymbol{i}_{xz} + \sin\theta\sin\varphi\ \boldsymbol{i}_{yz}. \tag{1.323}$$

Including the reference axis gives

$$\hat{\mathbf{r}} = \hat{\boldsymbol{r}}\mathbf{e}_z = \sin\theta\cos\varphi\ \mathbf{e}_x + \sin\theta\sin\varphi\ \mathbf{e}_y + \cos\theta\mathbf{e}_z. \tag{1.324}$$

Therefore, the unit vector reduces to the expression in spherical polar coordinates, as it should.

We are able to associate each of the coefficients in spherical coordinates with a product of spinors in Eq. (1.319). These products can be collapsed to form scalars and imaginary units. Although this can be used to correctly reproduce the unit vector in spherical polar coordinates, it does not suit our purposes. We want to use the spinors to build higher order unit vectors that we can associate with the function space in spherical symmetry. However, further multiplication of spinors does not lead to higher order unit vectors but will just keep producing scalars and imaginary units. The key to understanding the problem is that identical mathematical quantities can have different physical meanings. Take the equation $x = 2vt$, where x is the distance, v is the velocity, and t is the time. This can be interpreted in two different physical ways. We can write it as $v(2t)$, meaning that an object is traveling at a velocity v for a time $2t$. The alternative interpretation is $(2v)t$. In this case, the object is traveling at a velocity $2v$ for a time t. However, in both cases, the distance traveled is the same expression $2vt$. This issue also arises for rotations. The spinors $\boldsymbol{a}_{\pm\frac{1}{2}}^\dagger$ correspond to rotations over 45°. Applying the spinor twice gives a rotation over 90°. However, we do not want to describe the situation where we perform the same rotation twice, we want to describe the situation where the rotation goes twice as fast. Therefore, we want $(\boldsymbol{a}_{\pm\frac{1}{2}}^\dagger)^m$ to be associated with functions where the phase changes m times as fast as for $\boldsymbol{a}_{\pm\frac{1}{2}}^\dagger$. In order to associate $\boldsymbol{a}_{\pm\frac{1}{2}}^\dagger$ with the angular momentum, the spinors will be treated as independent quantities whose products can no longer be collapsed into scalars and imaginary units.

Once $\boldsymbol{a}^\dagger_{\pm\frac{1}{2}}$ are treated as independent quantities, we can no longer remove spinors via $\boldsymbol{a}^\dagger_\uparrow \boldsymbol{a}^\dagger_\downarrow = 1$. We therefore introduce an operator that cancels the independent spinors in the same way that the conjugate of the spinor cancels the spinor, $\boldsymbol{a}_{\pm\frac{1}{2}}\boldsymbol{a}^*_{\pm\frac{1}{2}} = \boldsymbol{a}_{\pm\frac{1}{2}}\boldsymbol{a}_{\mp\frac{1}{2}} = 1$. We therefore define

$$\left(\boldsymbol{a}^\dagger_{\pm\frac{1}{2}}\right)^* = e^{\pm \boldsymbol{i}_{xz}\frac{\pi}{4}} = e^{\mp \boldsymbol{i}_{zx}\frac{\pi}{4}} \quad \rightarrow \quad \boldsymbol{a}_{\pm\frac{1}{2}}, \tag{1.325}$$

where $\boldsymbol{a}_{\pm\frac{1}{2}}$ are also independent operators. Conventionally, $\boldsymbol{a}_{\pm\frac{1}{2}}$ and $\boldsymbol{a}^\dagger_{\pm\frac{1}{2}}$ are associated with the bra and the ket, respectively. Removing a single spinor can be done via

$$\boldsymbol{a}_\sigma \boldsymbol{a}^\dagger_\sigma = 1, \tag{1.326}$$

with $\sigma = \pm\frac{1}{2}$. Removal of a single $\boldsymbol{a}^\dagger_\sigma$ from a product of indistinguishable operators is given by

$$\boldsymbol{a}_\sigma(\boldsymbol{a}^\dagger_\sigma)^n = \boldsymbol{a}_\sigma(\boldsymbol{a}^\dagger_\sigma \cdots \boldsymbol{a}^\dagger_\sigma) = n(\boldsymbol{a}^\dagger_\sigma)^{n-1}. \tag{1.327}$$

Note that this is different from complex exponentials that would simply give $\boldsymbol{a}_{\pm\frac{1}{2}}(\boldsymbol{a}^*_{\pm\frac{1}{2}})^n = (\boldsymbol{a}^*_{\pm\frac{1}{2}})^{n-1}$. The machinery is similar to that of the step operators for quantum harmonic oscillators, except that there are two different operators. The above operation is comparable to the product rule in differentiation

$$\frac{d}{dx}x^n = 1 \times x^{n-1} + x \times 1 \times x^{n-2} + \cdots x^{n-1} \times 1 = nx^{n-1},$$

where the derivative works on each x separately. Note that the operation in Eq. (1.327) brings out the power of the spinor. This allows us to associate it with higher order functions. From Eq. (1.327), a number operator for a particular spin can be constructed by simply replacing the missing spinor

$$\mathrm{N}_\sigma = \boldsymbol{a}^\dagger_\sigma \boldsymbol{a}_\sigma, \tag{1.328}$$

giving

$$\mathrm{N}_\sigma(\boldsymbol{a}^\dagger_\sigma)^n = \boldsymbol{a}^\dagger_\sigma n(\boldsymbol{a}^\dagger_\sigma)^{n-1} = n(\boldsymbol{a}^\dagger_\sigma)^n. \tag{1.329}$$

Therefore, N_σ counts the number of spinors $\boldsymbol{a}_\sigma$ in a product. Again, these results are very similar to the quantum harmonic oscillator.

We now want to find an expression for the angular momentum operator in the z-direction. For rotations around the z-axis, we are looking for the equivalent of Eq. (1.312), $\mathbf{e}_z\boldsymbol{a}^\dagger_{\pm\frac{1}{2}} = \pm\boldsymbol{a}^\dagger_{\pm\frac{1}{2}}\mathbf{e}_x$. Note that the angular momentum written as a vector operation for a single spinor is the same as the unit vector in the z-direction, *i.e.* $\mathbf{L}_z = \mathbf{e}_z$. We want to add an additional modification. When describing the rotation axis, the unit vector was described in terms of single spinor. The sign ± 1 then corresponded to counterclockwise and clockwise rotations. The rotation axis was obtained by squaring the spinor. Now we are constructing a function space using the vectors in Eq. (1.319), where the spinors have already been squared. The unit vectors are now described by a product of two spinors, and a factor $\frac{1}{2}$ will be introduced to correct for that. The angular momentum is then written as

$$\mathbf{L}_z = \frac{\mathbf{e}_z}{2} = \frac{1}{2}(\boldsymbol{a}^\dagger_\uparrow \boldsymbol{a}_\uparrow - \boldsymbol{a}^\dagger_\downarrow \boldsymbol{a}_\downarrow) = \frac{1}{2}(\mathrm{N}_\uparrow - \mathrm{N}_\downarrow). \tag{1.330}$$

For those familiar with the theory of quantum angular momentum, the same factor $\frac{1}{2}$ occurs when comparing the Pauli matrices with the spin operator S. It might appear strange to have the z component of the angular momentum $\mathbf{L}_z$ written as a vector. However, we are dealing here with the unit vector part associated with the angular momentum and not with the z component of the angular momentum operator. Operating this on a single spinor gives

$$\mathbf{L}_z \boldsymbol{a}^\dagger_{\pm\frac{1}{2}} = \pm\frac{1}{2}\boldsymbol{a}^\dagger_{\pm\frac{1}{2}}. \tag{1.331}$$

Therefore, the spinors have now become half-integer objects. This definition makes sense if one considers spinors to be the square roots of a vector. Operating on a product of two equal spinors gives

$$\mathbf{L}_z \boldsymbol{a}^\dagger_{\pm\frac{1}{2}}\boldsymbol{a}^\dagger_{\pm\frac{1}{2}} = \pm\boldsymbol{a}^\dagger_{\pm\frac{1}{2}}\boldsymbol{a}^\dagger_{\pm\frac{1}{2}}. \tag{1.332}$$

The factor $\frac{1}{2}$ in the operator now cancels the 2 arising from the number operators. The factor $\frac{1}{2}$ is a matter of preference and convention. It ensures that the eigenvalues for counterclockwise and clockwise rotations are ±1, respectively. In general, we can write

$$\mathbf{L}_z \boldsymbol{a}^\dagger_{\sigma'}\boldsymbol{a}^\dagger_{\sigma} = (\sigma+\sigma')\boldsymbol{a}^\dagger_{\sigma'}\boldsymbol{a}^\dagger_{\sigma}, \tag{1.333}$$

giving eigenvalues $1, 0, -1$. In a similar fashion, we can describe the imaginary unit related to rotations around the z-axis as

$$\boldsymbol{i}_{yx} = -\frac{i}{2}(\mathrm{N}_\uparrow - \mathrm{N}_\downarrow). \quad \Rightarrow \quad \boldsymbol{i}_{yx}\boldsymbol{a}^\dagger_{\sigma'}\boldsymbol{a}^\dagger_{\sigma} = -i(\sigma+\sigma')\boldsymbol{a}^\dagger_{\sigma'}\boldsymbol{a}^\dagger_{\sigma}. \tag{1.334}$$

Since we are only dealing with rotations around the z-axis, the only remaining imaginary unit $\boldsymbol{i}_{yx}$ will be replaced by the imaginary scalar i, following convention.

Although each coefficient in the unit vector can be associated with a product of spinors, we still need to check whether the products of spinors and their conjugates are orthonormal. Two multivector quantities are considered orthonormal when the product of the multivector and its conjugate does not contain a scalar, *i.e.* the inner product is zero. Spinors of equal spin are not orthogonal $\langle \boldsymbol{a}_{\pm\frac{1}{2}}\boldsymbol{a}^\dagger_{\pm\frac{1}{2}}\rangle_0 = 1$, whereas spinors of opposite spin are $\langle \boldsymbol{a}_{\mp\frac{1}{2}}\boldsymbol{a}^\dagger_{\pm\frac{1}{2}}\rangle_0 = 0$. The latter result agrees with the original expression of the spinor in terms of complex exponentials: $\boldsymbol{a}_{\mp\frac{1}{2}}\boldsymbol{a}^*_{\pm\frac{1}{2}} = e^{\mp\boldsymbol{i}_{xz}\frac{\pi}{4}}(e^{\pm\boldsymbol{i}_{xz}\frac{\pi}{4}})^* = \mp\boldsymbol{i}_{xz}$. The product does not contain a scalar term; therefore, $\langle \boldsymbol{a}_{\mp\frac{1}{2}}\boldsymbol{a}^*_{\pm\frac{1}{2}}\rangle_0 = 0$. For the orthonormality of a product, one therefore only has to consider spinors with the same spin

$$\begin{aligned}\langle(\boldsymbol{a}_\sigma)^{n'}(\boldsymbol{a}^\dagger_\sigma)^n\rangle_0 &= n\langle(\boldsymbol{a}_\sigma)^{n'-1}(\boldsymbol{a}^\dagger_\sigma)^{n-1}\rangle_0\\ &= n(n-1)\langle(\boldsymbol{a}_\sigma)^{n'-2}(\boldsymbol{a}^\dagger_\sigma)^{n-2}\rangle_0 = \cdots = n!\delta_{nn'}.\end{aligned} \tag{1.335}$$

Note that $\langle\cdots\rangle_0$ also implies that there can be no remaining spinors between the brackets. The normalized unit vectors are

$$|n_\sigma\rangle = \frac{1}{\sqrt{n_\sigma!}}(\boldsymbol{a}^\dagger_\sigma)^{n_\sigma} \text{ and } \langle n_\sigma| = \frac{1}{\sqrt{n_\sigma!}}(\boldsymbol{a}_\sigma)^{n_\sigma}. \tag{1.336}$$

(Those familiar with quantum harmonic oscillators might be missing a "vacuum" level. However, this is not necessary here, which is related to the fact that the density function is 1 for Legendre polynomials, and e^{-x^2} for Hermite polynomials. The function space for spherical harmonics is obtained by multiplying functions and not by taking derivatives starting from a vacuum state.) Orthonormality is now given by

$$\langle n'_\sigma|n_\sigma\rangle = \frac{1}{\sqrt{n'_\sigma!n_\sigma!}}\langle(\boldsymbol{a}_\sigma)^{n'_\sigma}(\boldsymbol{a}^\dagger_\sigma)^{n_\sigma}\rangle_0 = \delta_{n'_\sigma,n_\sigma}. \tag{1.337}$$

The number of spinors can be changed using

$$\boldsymbol{a}_\sigma^\dagger|n_\sigma\rangle = \frac{1}{\sqrt{n_\sigma!}}(\boldsymbol{a}_\sigma^\dagger)^{n_\sigma+1} = \frac{\sqrt{n_\sigma+1}}{\sqrt{(n_\sigma+1)!}}(\boldsymbol{a}_\sigma^\dagger)^{n_\sigma+1} = \sqrt{n_\sigma+1}|n_\sigma+1\rangle \tag{1.338}$$

and

$$\boldsymbol{a}_\sigma|n_\sigma\rangle = \frac{n_\sigma}{\sqrt{n_\sigma!}}(\boldsymbol{a}_\sigma^\dagger)^{n_\sigma-1} = \frac{\sqrt{n_\sigma}}{\sqrt{(n_\sigma-1)!}}(\boldsymbol{a}_\sigma^\dagger)^{n_\sigma-1} = \sqrt{n_\sigma}|n_\sigma-1\rangle. \tag{1.339}$$

As a result of these properties, the operators are also known as step operators. The operators also satisfy the commutation relations

$$[\boldsymbol{a}_\sigma, \boldsymbol{a}_{\sigma'}^\dagger] = \delta_{\sigma\sigma'}, \quad [\boldsymbol{a}_\sigma^\dagger, \boldsymbol{a}_{\sigma'}^\dagger] = 0, \quad [\boldsymbol{a}_\sigma, \boldsymbol{a}_{\sigma'}] = 0. \tag{1.340}$$

For example, for the first relation with $\sigma = \sigma'$, we have

$$[\boldsymbol{a}_\sigma, \boldsymbol{a}_\sigma^\dagger]|n_\sigma\rangle = (\boldsymbol{a}_\sigma\boldsymbol{a}_\sigma^\dagger - \boldsymbol{a}_\sigma^\dagger\boldsymbol{a}_\sigma)|n_\sigma\rangle = ((n_\sigma+1) - n_\sigma)\,|n_\sigma\rangle = |n_\sigma\rangle, \tag{1.341}$$

which shows that $\boldsymbol{a}_\sigma$ and $\boldsymbol{a}_\sigma^\dagger$ do not commute with each other.

Using the normalization in Eq. (1.336), the unit vectors in the vector in Eq. (1.319) can be written as

$$\boldsymbol{e}_1 = \frac{1}{\sqrt{2}}\boldsymbol{a}_\uparrow^\dagger\boldsymbol{a}_\uparrow^\dagger, \quad \boldsymbol{e}_0 = \boldsymbol{a}_\uparrow^\dagger\boldsymbol{a}_\downarrow^\dagger, \quad \boldsymbol{e}_{-1} = \frac{1}{\sqrt{2}}\boldsymbol{a}_\downarrow^\dagger\boldsymbol{a}_\downarrow^\dagger. \tag{1.342}$$

This gives for the vector

$$\begin{aligned}\hat{\boldsymbol{r}} = |\hat{\boldsymbol{r}}) &= \sum_{m=-1}^{1}|1m\rangle\langle 1m|\hat{\boldsymbol{r}}) = \cos\theta\,\boldsymbol{e}_0 + \sin\theta\left(-\frac{e^{-i\varphi}}{\sqrt{2}}\boldsymbol{e}_1 + \frac{e^{i\varphi}}{\sqrt{2}}\boldsymbol{e}_{-1}\right)\\ &= \boldsymbol{e}_1 r^1 + \boldsymbol{e}_0 r^0 + \boldsymbol{e}_{-1} r^{-1},\end{aligned} \tag{1.343}$$

with $\boldsymbol{e}_m = |1m\rangle$, $r^m = \langle 1m|\hat{\boldsymbol{r}})$, $r^0 = \cos\theta$, and $r^{\pm 1} = \mp\sin\theta e^{\mp i\varphi}/\sqrt{2}$. In the following section, we introduce higher order unit vectors $|lm\rangle$, with $l = 0, 1, 2, 3, \ldots$. However, a vector is described only in terms of $l = 1$ unit vectors (this essentially defines a vector). The expression above corresponds to a column vector. A row vector is obtained using that

$$\boldsymbol{e}^m = \boldsymbol{e}_m^\dagger \quad \text{and} \quad r_m = (r^m)^* = -r^{-m}, \tag{1.344}$$

giving

$$(\hat{\boldsymbol{r}}| = \sum_{m=-1}^{1}(\hat{\boldsymbol{r}}|1m\rangle\langle 1m| = r_1\boldsymbol{e}^1 + r_0\boldsymbol{e}^0 + r_{-1}\boldsymbol{e}^{-1}. \tag{1.345}$$

The normalization of the vector is given by

$$\begin{aligned}(\hat{\boldsymbol{r}}|\hat{\boldsymbol{r}}) &= \sum_{m=-1}^{1}(\hat{\boldsymbol{r}}|1m\rangle\langle 1m|\hat{\boldsymbol{r}}) = \sum_{m=-1}^{1} r_m r^m = \sum_{m=-1}^{1}(r^m)^* r^m\\ &= \cos^2\theta + \sin^2\theta\left\{\left(-\frac{e^{i\varphi}}{\sqrt{2}}\right)\left(-\frac{e^{-i\varphi}}{\sqrt{2}}\right) + \frac{e^{-i\varphi}}{\sqrt{2}}\frac{e^{i\varphi}}{\sqrt{2}}\right\} = 1.\end{aligned} \tag{1.346}$$

Since we are dealing with a vector, the normalization occurs by summing over the components of the vector.

By expanding the complex exponentials, the vector can be expressed in spherical polar coordinates

$$\hat{\boldsymbol{r}} = \sin\theta\cos\varphi\;\boldsymbol{e}_x + \sin\theta\sin\varphi\;\boldsymbol{e}_y + \cos\theta\,\boldsymbol{e}_z. \tag{1.347}$$

This looks similar to Eq. (1.324). However, note in Eq. (1.324), the unit vectors are anticommuting quantities, whereas in the above equation, the unit vectors commute. The anticommuting vectors are important for performing vector operations. The commuting vectors are used to construct higher order vectors that produce a function space. The Cartesian unit vectors can then be written as

$$e_x = \frac{1}{\sqrt{2}}(-e_1 + e_{-1}), \quad e_y = \frac{i}{\sqrt{2}}(e_1 + e_{-1}), \quad e_z = e_0. \tag{1.348}$$

The unit vectors in the x and y directions are reminiscent of those in Eq. (1.52) for a two-dimensional system, see Section 1.3. Note that there is a sign change in front of e_1. This is the Condon–Shortley phase, which is an arbitrary choice. The sign could have been included in the unit vectors e_m, with $m = 1, 0, -1$ in Eq. (1.342). However, Condon and Shortley preferred not to have any sign changes in the transformations of e_m (for experts, this directly relates to the definition of the Clebsch–Gordan coefficients). The inverse can also be obtained

$$e_1 = -\frac{1}{\sqrt{2}}(e_x + ie_y), \quad e_0 = e_z, \quad e_{-1} = \frac{1}{\sqrt{2}}(e_x - ie_y). \tag{1.349}$$

When dealing with operators, the spherical unit vectors can also be expressed in terms of Pauli matrices,

$$\mathbf{e}_{\pm 1} = \mp\frac{1}{\sqrt{2}}(\mathbf{e}_x \pm i\mathbf{e}_y) = \begin{pmatrix} 0 & -\sqrt{2} \\ 0 & 0 \end{pmatrix}, \begin{pmatrix} 0 & 0 \\ \sqrt{2} & 0 \end{pmatrix}. \tag{1.350}$$

These operators effectively change the direction of the spinors.

Conceptual summary. – In order to produce a function space for free rotations in three dimensions, we have written the unit vector in an appropriate form. The goal was to have a unit vector where the coefficients are eigenfunctions of the angular momentum in the z-direction (following convention, the z-axis is taken as the reference axis for rotation). Although there are two rotation directions in spherical symmetry, we can only have conserved quantities for one rotation. At the same time, we want the unit vectors to reflect the properties of the coefficients. Since the rotations effectively form a two-dimensional system (a sphere), we want the unit vectors expressed in two independent vector quantities. Furthermore, we want to be able to build higher order unit vectors using the same quantities.

The coefficients can be written in such a way that they are eigenfunctions of the angular momentum operator in the z-direction $\hat{L}_z = -i\partial/\partial\varphi$. The φ dependence of the coefficients r^m is $e^{im\varphi}$, with $m = 1, 0, -1$. The two vector objects describing the space are the spinors $a^\dagger_{\pm\frac{1}{2}}$. In order to be able to construct higher order unit vectors, the spinors are taken as independent commuting quantities. Their behavior is comparable to the well-known step operators of the quantum harmonic oscillator. The unit vectors are expressed as a product of two spinors $a^\dagger_{\sigma'} a^\dagger_{\sigma}$, where $m = \sigma + \sigma' = 1, 0, -1$ also reflects the projected angular momentum. The value of m can be extracted with the vector operator $\mathbf{L}_z = \mathbf{e}_z/2$, where $\mathbf{e}_z$ is comparable to the z component of the Pauli matrices.

We see the appearance of a factor $\frac{1}{2}$ in the angular momentum operator. Before, vectors were expressed in terms of single spinors, which was effectively the square root of a vector.

The rotation axis could be obtained by effectively squaring the spinor. The direction of rotation is determined from a single spinor. The angular momentum operator is designed to work on unit vectors that are products of two spinors. Things become a little confusing when these angular momentum operators are used for single spinors. In that case, the angular momentum operator is indicated as $\mathbf{S}_z = \mathbf{e}_z/2$ and is usually called spin. This generally occurs when the rotation axis is fixed and therefore effectively corresponds to a planar motion. However, the physical angular momentum is still ± 1, related to complex exponentials $e^{\pm i\varphi}$. This quantity is given by $\mathbf{e}_z = g\mathbf{S}_z = 2\mathbf{S}_z$, which is the z component of the Pauli matrices. The factor $g = 2$ is also called the gyromagnetic ratio. (For the experts, for the spin arising in the nonrelativistic limit of the Dirac equation, g is not exactly equal to 2, since it also incorporates some additional higher order corrections).

In order to build higher order vectors, the complex exponentials have been replaced by indistinguishable commuting spinors. Those familiar with the quantum harmonic oscillator will have noticed that the behavior of the spinors is very similar to that of the step operators for oscillation. As we will see later, technically, it is more correct to consider one-dimensional oscillation in a harmonic potential as a rotation in the plane of displacement x vs. momentum p_x. The behavior of the unit vectors for rotation in three dimensions is very similar to a two-dimensional harmonic oscillator. The two different types of spinors correspond to counterclockwise and clockwise rotations. The number of spinors is related to the total amount of (angular) momentum that the system has. For a quantum harmonic oscillator, this is directly related to the amplitude of the oscillation.

Problems

1. Suggest an operator that connects $\boldsymbol{e}_m$ with $m = 1, 0, -1$ and find the factors.
2. Find an operator that connects $\boldsymbol{e}_x$ and $\boldsymbol{e}_y$.

1.15 Function Space for Rotation in Three Dimensions

Introduction. – Up to this point, the function spaces have been complex exponentials or related trigonometric functions. Examples are $\hat{\boldsymbol{r}}^m = e^{i_{yx} m\varphi}$ for rotation in a plane, see Section 1.4, and $e^{i\mathbf{k}\cdot\mathbf{r}}$ for translational motion in three dimensions, see Section 1.10. In this section, we look at rotations in three dimensions and find the significantly more complicated function space of spherical harmonics. The added complexity arises from the fact that it is not possible to obtain eigenfunctions for all directions of rotation. Conventionally, one picks the z-direction as the reference axis, and the spherical harmonics are eigenfunctions (again, complex exponentials) for rotations around the z-axis. The angle related to rotations around the other axis (conventionally, the y-axis) appears as the argument of trigonometric functions.

The function spaces of complex exponentials were derived by looking at an (effective) unit vector and raising it to all positive integer powers. The coefficients of the obtained higher order vectors form the function space. This procedure can also be used for spherical harmonics. In the previous section, a suitable unit vector was obtained where the coefficients are eigenvectors of the angular momentum. The basis vectors reflect the same properties and are constructed out of indistinguishable spinors.

The indistinguishability of the operators $\boldsymbol{a}^\dagger_{\pm\frac{1}{2}}$ has been introduced in the unit vectors in Eq. (1.342). Their conjugates are given by

$$\boldsymbol{e}^1 = \frac{1}{\sqrt{2}}\boldsymbol{a}_\uparrow \boldsymbol{a}_\uparrow, \quad \boldsymbol{e}^0 = \boldsymbol{a}_\uparrow \boldsymbol{a}_\downarrow, \quad \boldsymbol{e}^{-1} = \frac{1}{\sqrt{2}}\boldsymbol{a}_\downarrow \boldsymbol{a}_\downarrow. \tag{1.351}$$

The spinors $\boldsymbol{a}_{\pm\frac{1}{2}}$ related to a vector in row format are used, so that the coefficients directly correspond to the spherical harmonics. The unit vectors $\boldsymbol{e}^m$ commute with each other but do not commute with $\boldsymbol{e}^{m\dagger} = \boldsymbol{e}_m$. The unit vector in row format is

$$(\hat{\boldsymbol{r}}| = -\frac{\sin\theta}{\sqrt{2}}e^{i\varphi}\boldsymbol{e}^1 + \cos\theta\,\boldsymbol{e}^0 + \frac{\sin\theta}{\sqrt{2}}e^{-i\varphi}\,\boldsymbol{e}^{-1} = \hat{r}_1\boldsymbol{e}^1 + \hat{r}_0\boldsymbol{e}^0 + \hat{r}_{-1}\boldsymbol{e}^{-1}. \tag{1.352}$$

In the coefficients, we recognize the spherical harmonics

$$\begin{aligned}
\hat{r}_1 &= C^1_1(\hat{\boldsymbol{r}}) = \sqrt{\frac{4\pi}{3}}Y_{11}(\hat{\boldsymbol{r}}) = -\frac{\sin\theta}{\sqrt{2}}e^{i\varphi} = -\frac{x+iy}{r},\\
\hat{r}_0 &= C^1_0(\hat{\boldsymbol{r}}) = \sqrt{\frac{4\pi}{3}}Y_{10}(\hat{\boldsymbol{r}}) = \cos\theta = \frac{z}{r},\\
\hat{r}_{-1} &= C^1_{-1}(\hat{\boldsymbol{r}}) = \sqrt{\frac{4\pi}{3}}Y_{1,-1}(\hat{\boldsymbol{r}}) = \frac{\sin\theta}{\sqrt{2}}e^{-i\varphi} = \frac{x-iy}{r},
\end{aligned} \tag{1.353}$$

using the shorthand $C^1_m(\hat{\boldsymbol{r}}) \equiv C^1_m(\theta,\varphi)$. The functions $C^1_m(\hat{\boldsymbol{r}})$ are also known as renormalized spherical harmonics. The unit vector can then be written as

$$\hat{\boldsymbol{r}} = \sum_{m=-1}^{1} C^1_m(\hat{\boldsymbol{r}})\boldsymbol{e}^m = \sum_{m=-1}^{1} \boldsymbol{e}_m C^{1m}(\hat{\boldsymbol{r}}). \tag{1.354}$$

The conjugates of the coordinates are given by

$$\hat{r}^m = C^{1m}(\hat{\boldsymbol{r}}) = (C^1_m(\hat{\boldsymbol{r}}))^* = (-1)^m C^1_{-m}(\hat{\boldsymbol{r}}) = (-1)^m\hat{r}_{-m}. \tag{1.355}$$

Again, we find that the property $r^i = r_i$ is a nice aspect of Cartesian coordinates but does not hold in general. The difference between $C^1_m(\hat{\boldsymbol{r}})$ and the better-known $Y_{1m}(\hat{\boldsymbol{r}})$ lies in the normalization. The $C^1_m(\hat{\boldsymbol{r}})$ follow the normalization of a vector

$$(\hat{\boldsymbol{r}}|\hat{\boldsymbol{r}}) = \hat{\boldsymbol{r}}\cdot\hat{\boldsymbol{r}} = \sum_{m=-1}^{1} C^1_m(\hat{\boldsymbol{r}})C^{1m}(\hat{\boldsymbol{r}}) = \sum_{m=-1}^{1}(-1)^m C^1_m(\hat{\boldsymbol{r}})C^1_{-m}(\hat{\boldsymbol{r}}) = 1. \tag{1.356}$$

The $Y_{1m}(\hat{\boldsymbol{r}})$, on the other hand, are normalized by integration over the angular coordinates

$$\int d\hat{\boldsymbol{r}}\; Y^*_{1m}(\hat{\boldsymbol{r}})Y_{1m}(\hat{\boldsymbol{r}}) = \int_0^{2\pi} d\varphi \int_0^{\pi} d\theta\; \sin\theta\; Y^*_{1m}(\theta,\varphi)Y_{1m}(\theta,\varphi) = 1, \tag{1.357}$$

which is the typical normalization used in quantum mechanics.

In Section 1.4, the function space for a unit circle was obtained by taking the unit vectors to any positive integer power. For a spherical basis, we will generate the additional functions by taking the exponential of the unit vector $\hat{\boldsymbol{r}}$,

$$e^{\hat{\boldsymbol{r}}} = \sum_{l=0}^{\infty}\frac{1}{l!}\hat{\boldsymbol{r}}^l, \tag{1.358}$$

which raises $\hat{\boldsymbol{r}}$ to any arbitrary positive integer power. This is comparable to Herglotz generating function. Inserting the expression for the vector in spherical coordinates in Eq. (1.352) and using a multinomial expansion gives

$$\begin{aligned} e^{\hat{r}} &= \sum_{l=0}^{\infty} \frac{1}{l!} (\hat{r}_1 \boldsymbol{e}^1 + \hat{r}_0 \boldsymbol{e}^0 + \hat{r}_{-1} \boldsymbol{e}^{-1}.)^l \\ &= \sum_{l=0}^{\infty} \frac{1}{l!} \sum_{i+j+k=l} \frac{l!}{i!j!k!} (\hat{r}_0 \boldsymbol{e}^0)^i (\hat{r}_1 \boldsymbol{e}^1)^j (\hat{r}_{-1} \boldsymbol{e}^{-1})^k, \end{aligned} \tag{1.359}$$

where the summation goes over all the possible combinations of i, j, and k for which $i + j + k = l$. We now insert the expressions of the spherical unit vectors in terms of spinors from Eq. (1.351),

$$e^{\hat{r}} = \sum_{l=0}^{\infty} \sum_{i+j+k=l} \frac{1}{(\sqrt{2})^{j+k} i!j!k!} (\hat{r}_0)^i (\hat{r}_1)^j (\hat{r}_{-1})^k (\boldsymbol{a}_\uparrow)^{i+2j} (\boldsymbol{a}_\downarrow)^{i+2k}. \tag{1.360}$$

The higher order combinations of spinors can be rearranged as new unit vectors

$$\boldsymbol{e}^{lm} = \langle lm| = \frac{(\boldsymbol{a}_\uparrow)^{l+m} (\boldsymbol{a}_\downarrow)^{l-m}}{\sqrt{(l+m)!(l-m)!}}, \tag{1.361}$$

where $2l$ gives the total number of spinors, and $2m$ the difference between the up and down spinors. This equation is Schwinger's expression of the eigenstates of the angular momentum operator in terms of the commuting operators. The powers impose the additional condition that $j - k = m$. The length of the vector can also be included by taking $\boldsymbol{r} = r\hat{\boldsymbol{r}}$. The exponential of the vector $\boldsymbol{r}$ produces a complete set of unit vectors and their coefficients

$$e^{r} = \sum_{l=0}^{\infty} \sum_{m=-l}^{l} r^l C_m^l(\hat{\boldsymbol{r}}) \boldsymbol{e}^{lm} = \sum_{l=0}^{\infty} \sum_{m=-l}^{l} r^l (\hat{\boldsymbol{r}}|lm\rangle \langle lm|, \tag{1.362}$$

where $(\hat{\boldsymbol{r}}|lm\rangle = C_m^l(\hat{\boldsymbol{r}})$ are the renormalized spherical harmonics for arbitrary rank l and component m. The functions can be written explicitly as

$$r^l C_m^l(\hat{\boldsymbol{r}}) = \sqrt{(l+m)!(l-m)!} \sum_{\substack{i+j+k=l \\ j-k=m}} \frac{1}{i!j!k!} z^i \left(-\frac{x+iy}{2}\right)^j \left(\frac{x-iy}{2}\right)^k. \tag{1.363}$$

The renormalized spherical harmonic $C_m^l(\hat{\boldsymbol{r}})$ is related to the more often used spherical harmonic $Y_{lm}(\hat{\boldsymbol{r}})$ by a simple factor

$$(\hat{\boldsymbol{r}}|lm\rangle = C_m^l(\hat{\boldsymbol{r}}) = \sqrt{\frac{4\pi}{2l+1}} Y_{lm}(\hat{\boldsymbol{r}}) = \sqrt{\frac{4\pi}{2l+1}} \langle \hat{\boldsymbol{r}}|lm\rangle,$$

where $(\hat{\boldsymbol{r}}|$ is used for vectors that are normalized by summing over their squared components

$$\sum_{m=-l}^{l} (C_m^l(\hat{\mathbf{r}}))^* C_m^l(\hat{\mathbf{r}}) = 1. \tag{1.364}$$

The notation $\langle \hat{\boldsymbol{r}}|$ is used for vectors that are normalized by integrating over the angular components.

Example 1.13 Calculate the renormalized spherical harmonic for $l = 2$ and $m = 0$.

Solution

The expression for the spherical harmonics in Eq. (1.363) can also be written in spherical polar components

$$C^l_m(\hat{\mathbf{r}}) = \sqrt{(l+m)!(l-m)!} \sum_{\substack{i+j+k=l \\ j-k=m}} \frac{1}{i!j!k!} \cos^i\theta \left(-\frac{\sin\theta e^{i\varphi}}{2}\right)^j \left(\frac{\sin\theta e^{-i\varphi}}{2}\right)^k .$$

This seems like a rather unwieldy expression. However, only a few terms remain in the summation for particular l and m values. The restriction in the summation $j - k = m$ gives $j = k$ for $m = 0$. Additionally, $i + j + k = l$ becomes $i + 2j = 2$. Therefore, the only terms in the summation that remain are $i = 2$, $j = k = 0$ and $i = 0$, $j = k = 1$. The spherical harmonic then becomes

$$\begin{aligned} C^2_0(\hat{\mathbf{r}}) &= \sqrt{2!2!}\left(\frac{\cos^2\theta}{2!0!0!} + \frac{1}{0!1!1!}\left(-\frac{\sin\theta}{2}\right)\frac{\sin\theta}{2}\right) \\ &= \cos^2\theta - \frac{\sin^2\theta}{2} = \frac{3}{2}\cos^2\theta - \frac{1}{2}, \end{aligned} \tag{1.365}$$

using $\sin^2\theta = 1 - \cos^2\theta$. This is also equal to the Legendre polynomial $P_2(\cos\theta)$.

In the previous section, several operators describing the various properties of a vector were obtained. The function space was constructed from spinors that were eigenfunctions of the rotations around the z-axis. Therefore, the kets $|lm\rangle$ with

$$\boldsymbol{e}^l_m = |lm\rangle = \frac{(\boldsymbol{a}^\dagger_\uparrow)^{l+m}(\boldsymbol{a}^\dagger_\downarrow)^{l-m}}{\sqrt{(l+m)!(l-m)!}}, \tag{1.366}$$

following Eq. (1.361), should also be eigenstates of the z component of the angular momentum operator in Eq. (1.330)

$$\mathbf{L}_z|lm\rangle = \frac{1}{2}(N_\uparrow - N_\downarrow)|l+m; l-m\rangle = m|lm\rangle, \tag{1.367}$$

where $|n_\uparrow; n_\downarrow\rangle$ is a convenient alternative expression for the unit vectors in terms of spinor numbers, giving $|lm\rangle \equiv |l+m; l-m\rangle$. The result corresponds to half the difference $\frac{1}{2}(l + m - (l - m)) = m$ of the powers of $\boldsymbol{a}^\dagger_\uparrow$ and $\boldsymbol{a}^\dagger_\downarrow$. For the example, we see that $C^l_m(\hat{\mathbf{r}}) \sim e^{im\varphi}$. Therefore,

$$\hat{L}_z C^l_m(\hat{\mathbf{r}}) = -i\frac{\partial C^l_m(\hat{\mathbf{r}})}{\partial\varphi} = mC^l_m(\hat{\mathbf{r}}), \tag{1.368}$$

using $\hbar \equiv 1$. Therefore, $\mathbf{L}_z$ working on the unit vectors and $\hat{L}_z$ working on the functions give the same result.

The number operator can be defined by adding the number operators for each spin in Eq. (1.328),

$$\mathrm{N} = \frac{1}{2}(\boldsymbol{a}^\dagger_\uparrow\boldsymbol{a}_\uparrow + \boldsymbol{a}^\dagger_\downarrow\boldsymbol{a}_\downarrow), \quad \text{with} \quad \mathrm{N}|lm\rangle = l|lm\rangle, \tag{1.369}$$

corresponding to half the sum $\frac{1}{2}(l + m + (l - m)) = l$ of the powers of $\boldsymbol{a}^\dagger_\uparrow$ and $\boldsymbol{a}^\dagger_\downarrow$. Operating with the number operator on the kets $|lm\rangle$ from Eq. (1.361) returns the rank of the multivector. The two integers l and m are therefore representative of a particular spherical harmonic in the complete set of functions for the unit sphere. In physics, these numbers are generally known as quantum numbers.

Additionally, the spinors are also a good basis for rotations around the y-axis, *i.e.* in the zx-plane. If two orthogonal unit vectors are in the plane of rotation, then they transform into each other. For example, $\mathbf{e}_z$ and $\mathbf{e}_x$ transform as $\boldsymbol{i}_{xz}\mathbf{e}_z = \mathbf{e}_x$ and $\boldsymbol{i}_{xz}\mathbf{e}_x = -\mathbf{e}_z$. The spinors form an orthogonal set of unit vectors in the zx-plane. We therefore need operators that transform the spinors into each other. Operators that capture this are

$$\mathbf{L}_- = \boldsymbol{a}_\downarrow^\dagger \boldsymbol{a}_\uparrow \text{ and } \mathbf{L}_+ = \boldsymbol{a}_\uparrow^\dagger \boldsymbol{a}_\downarrow. \tag{1.370}$$

These are the conventional step operators for the angular momentum that use the Condon–Shortley phase convention. We return to the sign and normalization for proper spherical tensors later. In Section 1.17, we will see that the corresponding operators $\hat{L}_\pm$ that produce the same effect when working on the spherical harmonics are far from trivial.

Using Eqs. (1.338) and (1.339), the operation of these step-up operators on $|lm\rangle$ can be written as

$$\begin{aligned}\mathbf{L}_+|lm\rangle &= \boldsymbol{a}_\uparrow^\dagger \boldsymbol{a}_\downarrow |l+m; l-m\rangle = \sqrt{(l+m)+1}\sqrt{l-m}|l+m+1; l-m-1\rangle \\ &= \sqrt{(l+m+1)(l-m)}|l, m+1\rangle,\end{aligned} \tag{1.371}$$

using the alternative notation $|n_\uparrow; n_\downarrow\rangle$ introduced in Eq. (1.367). Likewise, we find

$$\mathbf{L}_-|lm\rangle = \sqrt{(l+m)(l-m+1)}|l, m-1\rangle. \tag{1.372}$$

Therefore, these operators couple unit vectors with different m but the same l to each other. Therefore, they do not change the rank of the multivector. These are also known as step operators for the angular momentum.

Conceptual summary. – In this section, the function space of the spherical harmonics was derived. This is the appropriate function space for rotations in three dimensions. The approach is different from the typical approach in many textbooks. Typically, spherical harmonics are found by writing down the Laplacian in spherical polar coordinates. The radial and angular dependencies are split by a separation of variables. The angular part is then rewritten in the differential equation for the associated Legendre polynomials that can be solved by a series solution. Although this is a perfectly valid approach, one is left with a set of functions defined by the indices l and m that are related to the angular momentum. A variety of relations between the functions can be obtained, although the underlying reasons are not directly obvious.

The approach followed here starts from the recognition that we are dealing with a higher order vector space. The unit vector in spherical coordinates is rewritten in terms of the indistinguishable spinors $\boldsymbol{a}_{\pm\frac{1}{2}}$. The entire function space is then generated by expanding the exponential in a series. This gives all the higher order unit vectors and coefficients. The coefficients are equal to the spherical harmonics. The advantage of this approach is that one has an expression for the unit vectors in terms of the spinors. The spherical harmonics are the function space for rotations. Rotations can be obtained by manipulating the unit vectors or by performing an operation on the coefficients (or the wave functions, in quantum-mechanical language). Since both methods produce the same results, the transformation properties of the functions can be obtained by studying those of the unit vectors.

The function space of spherical harmonics is complete, *i.e.* any function expressed in the angular coordinates θ and φ can be expressed in terms of spherical harmonics. This is comparable to the Fourier transform.

Problems

1 Find the spherical harmonics for $l = 2$ and $m = 1, -2$.

2 Show that $(C^l_m(\hat{\boldsymbol{r}}))^* = (-1)^m C^l_{-m}(\hat{\boldsymbol{r}})$

1.16 Higher Order Operators

Introduction. – Up to this point, we have mainly considered operations on states that are combinations of single-spinor states or unit vectors. In the previous section, we obtained unit vectors that contain products of spinors. The unit vectors correspond to the spherical harmonics. The properties under rotation of the function space can be understood by studying either the functions or the unit vectors. A comparable example was discussed in Sections 1.2 and 1.3, where we saw that a 90° counterclockwise rotation in the xy-plane transforms the unit vectors $\mathbf{e}_x \to \mathbf{e}_y$ and $\mathbf{e}_y \to -\mathbf{e}_x$ under the imaginary unit $\boldsymbol{i}_{yx}$ which is the vector operator causing a rotation. The coefficients/functions transform as $\sin\varphi \to \cos\varphi$ and $\cos\varphi \to -\sin\varphi$. The rotation operator is the derivative $d/d\varphi$. Therefore, the transformation property is that the unit vectors or coefficients are interchanged, and one obtains a minus sign. The same concepts still hold for rotations of function spaces in three dimensions, although the technical details are more complex.

In the preceding sections, we have found commuting unit vectors $\boldsymbol{e}_m$, with $m = 1, 0, -1$. The unit vectors are built up of spinors. Earlier, we have used anticommuting unit vectors $\mathbf{e}_i$, with $i = x, y, z$, also known as the Pauli matrices. These unit vectors can be viewed as operators consisting of a spinor and a conjugate spinor. The Pauli matrices transform single-spinor states. We now want to derive operators for the unit vectors $\boldsymbol{e}^l_m$, see Eq. (1.366) with an arbitrary number of spinors.

Before we can establish the relationship between these quantities, we need to go through some technical issues. Since this involves sign conventions, it is all rather painful and annoying. The crucial point in the discussion is the use of conjugates. Conjugates correspond rather well to the physics way of looking at things. The starting point is a vector quantity. We apply a vector operation, which removes the old quantity and creates a new one. This corresponds well to the bra-ket notations, where a bra removes the vector quantity, and the ket creates a new one. The bra and ket are each other conjugates. The other approach is the more mathematical way of looking at things. This basically states that all quantities involved are essentially higher order vectors. Making the creation of a "new" vector distinct from the removal of the "old" vector introduces changes in direction and additional phase factors. This makes everything extremely inelegant. For the experts, part of this discussion centers on whether to use Clebsch–Gordan coefficients or $3j$ symbols. We follow the latter approach since it leads to proper spherical tensors, but we still need something that can work like an effective conjugate.

In order to see what is going on, let us return to the spinors in terms of complex exponentials $\boldsymbol{a}^\dagger_{\pm\frac{1}{2}} = e^{\mp \boldsymbol{i}_{xz}\frac{\pi}{4}}$ from Eq. (1.309). Part of the issues are related to rotation around the y-axis and how the spinors transform into each other. From Figure 1.11, we can see that the spinors form a right-handed orthogonal set of unit vectors in the zx-plane (it is important to note that this is the zx-plane and not the xz-plane. One is generally more used to rotations in the xy-plane, with the z-axis as the rotation axis. In order to maintain a right-handed axis system, we need to make the cyclic permutation $x \leftarrow y \leftarrow z \leftarrow x$. This leads to the zx-plane, with the y-axis as rotation axis). The nice thing about spinors is that they commute and are independent of a reference axis. To visualize them as vectors, we need to pick a reference axis. Picking the z-axis gives the vectors related to the spinors $\mathbf{a}^\dagger_{\pm\frac{1}{2}} = \boldsymbol{a}^\dagger_{\pm\frac{1}{2}}\mathbf{e}_z = e^{\boldsymbol{i}_{\mp xz}\frac{\pi}{4}}\mathbf{e}_z = (\mathbf{e}_z \mp \mathbf{e}_x)/\sqrt{2}$. Taking the outer product between the spinors gives $\mathbf{a}^\dagger_\uparrow \times \mathbf{a}^\dagger_\downarrow = \mathbf{e}_y$, which is the rotation axis for the zx-plane. Note that for the reference axis $e^{\boldsymbol{i}_{xz}\frac{\pi}{4}}\mathbf{e}_z = (\mathbf{e}_z + \mathbf{e}_x)/\sqrt{2}$, one finds $\mathbf{a}^\dagger_{\pm\frac{1}{2}} = \mathbf{e}_z, \mathbf{e}_x$.

The norm between two spinors is usually obtained by taking the conjugate of one of the spinors

$$\langle \frac{1}{2}\sigma' | \frac{1}{2}\sigma \rangle = \langle \boldsymbol{a}_{\sigma'} \boldsymbol{a}^\dagger_\sigma \rangle_0 = \langle e^{\boldsymbol{i}_{xz}\sigma'\frac{\pi}{2}} e^{-\boldsymbol{i}_{xz}\sigma\frac{\pi}{2}} \rangle_0 = \langle e^{\boldsymbol{i}_{xz}(\sigma'-\sigma)\frac{\pi}{2}} \rangle_0 = \delta_{\sigma,\sigma'}, \tag{1.373}$$

where $\langle \cdots \rangle_0$ indicates that only the terms of rank 0 (scalars) are retained. The different spinors are visualized in Figure 1.12. The above result is nice, since we are left with a δ-function. This is the "physics" way of approaching things. Note that the bra effectively removes the ket for $\sigma' = \sigma$ (leaving 1). However, for $\sigma' \neq \sigma$, we have zero.

Let us see what changes for two vector objects with $\sigma' = \sigma$, but we decide that we no longer want to use conjugates

$$\boldsymbol{a}^\dagger_{\pm\frac{1}{2}} \boldsymbol{a}^\dagger_{\pm\frac{1}{2}} = e^{\mp \boldsymbol{i}_{xz}\frac{\pi}{4}} e^{\mp \boldsymbol{i}_{xz}\frac{\pi}{4}} = \mp \boldsymbol{i}_{xz}. \tag{1.374}$$

Unfortunately, the end result is an imaginary unit, which is not considered a desired quantity for a metric. The right-hand side can be made real by multiplying with $\boldsymbol{i}_{xz}$. On the left-hand side, the imaginary unit can be absorbed into one of the spinors,

$$\underline{\boldsymbol{a}}^\dagger_{\pm\frac{1}{2}} = \boldsymbol{i}_{xz} \boldsymbol{a}^\dagger_{\pm\frac{1}{2}} = \boldsymbol{i}_{xz} e^{\mp \boldsymbol{i}_{xz}\frac{\pi}{4}} = \pm e^{\pm \boldsymbol{i}_{xz}\frac{\pi}{4}} = \pm \boldsymbol{a}^\dagger_{\mp\frac{1}{2}}, \tag{1.375}$$

see Figure 1.12. This changes the metric from Eq. (1.374) into

$$\underline{\boldsymbol{a}}^\dagger_{\pm\frac{1}{2}} \boldsymbol{a}^\dagger_{\pm\frac{1}{2}} = \pm 1. \tag{1.376}$$

This is the metric used in SU(2), which is the symmetry of the Pauli matrices. SU(2) is essentially a two-dimensional space describing three dimensions. First note the similarity in the definition of the dual spinors with the comparable definition for a two-dimensional system in Eq. (1.281). Second, mathematically, it is often very hard to make a real two-dimensional system. Even when taking real rotation matrices in two dimensions, see Section 1.8, the third dimension appears as an imaginary unit.

The conjugated spinors can now be expressed in the dual spinors, *i.e.* the spinors rotated over 90° counterclockwise,

$$\underline{\boldsymbol{a}}^\dagger_{\pm\frac{1}{2}} = \pm \boldsymbol{a}_{\pm\frac{1}{2}} = \pm \boldsymbol{a}^\dagger_{\mp\frac{1}{2}}, \tag{1.377}$$

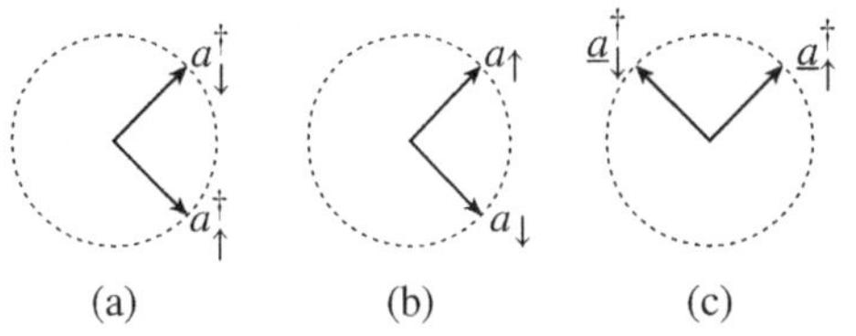

Figure 1.12 (a) The higher order vectors are described in terms of spinors $\boldsymbol{a}^\dagger_{\pm\frac{1}{2}}$, where $\pm\frac{1}{2} = \uparrow, \downarrow$. Note that the spinors are complex exponentials and therefore relative vectors. They therefore have no particular orientation in space until they are multiplied with a reference axis. (b) In physics, one often needs the conjugated spinors $\boldsymbol{a}_{\pm\frac{1}{2}}$ to remove spinors. In representation theory, one often avoids conjugation, since it creates objects with a different handedness. (c) The conjugated spinors can still be expressed in terms of the dual spinors $\underline{\boldsymbol{a}}^\dagger_{\pm\frac{1}{2}} = \pm\boldsymbol{a}_{\pm\frac{1}{2}}$, which are the spinors $\boldsymbol{a}^\dagger_{\pm\frac{1}{2}}$ rotated 90° counterclockwise.

see Figure 1.12. Dual vectors can also be obtained for the higher order vectors in Eq. (1.366) by replacing the spinors by their duals,

$$\underline{\boldsymbol{e}}^j_m = \frac{(\underline{\boldsymbol{a}}^\dagger_\uparrow)^{j+m}(\underline{\boldsymbol{a}}^\dagger_\downarrow)^{j-m}}{\sqrt{(l+m)!(l-m)!}} = (-1)^{j-m}\frac{(\boldsymbol{a}_\uparrow)^{j-m}(\boldsymbol{a}_\downarrow)^{j+m}}{\sqrt{(l+m)!(l-m)!}} = (-1)^{j-m}\boldsymbol{e}^{j,-m}, \quad (1.378)$$

where j is generally used when both half-integer and integer values are considered.

Although the conjugated spinor can be obtained from the dual spinors, see Eq. (1.377), there is a difference when performing the same action multiple times. The conjugate of the conjugate is again the original spinor, $(\boldsymbol{a}_{\pm\frac{1}{2}})^\dagger = \boldsymbol{a}^\dagger_{\pm\frac{1}{2}}$. However, the dual of the dual gives the inverse of the spinor $\boldsymbol{i}_{xz}\underline{\boldsymbol{a}}^\dagger_{\pm\frac{1}{2}} = \boldsymbol{i}_{xz}\boldsymbol{i}_{xz}\boldsymbol{a}^\dagger_{\pm\frac{1}{2}} = -\boldsymbol{a}^\dagger_{\pm\frac{1}{2}}$. This is not entirely surprising since two rotations over 90° equal a rotation over 180°. This is also observed for the general multivectors. Taking the dual of $\underline{\boldsymbol{e}}^j_m$ gives $(-1)^{2j}\boldsymbol{e}^j_m$. This gives $\boldsymbol{e}^l_m$ when l is integer. Therefore, for the unit vectors for the function space there is no issue. However, for half-integer j, this gives $-\boldsymbol{e}^j_m$. This might appear that we have to conjugate four times before returning back to the original unit vector $\boldsymbol{e}^j_m$. However, conjugation was replaced by rotating over 90°, and we indeed have to rotate four times over 90° to make a full circle.

The dual spinors allow us to find the relation between the commuting unit vectors $\boldsymbol{e}_m$, with $m = 1, 0, -1$, that are used to construct the function space and the anticommuting unit vectors $\mathbf{e}_m$ that are effective vector operations. First, we replace one of the spinors by their equivalent dual spinor, $\boldsymbol{a}^\dagger_{\pm\frac{1}{2}} \rightarrow \mp\underline{\boldsymbol{a}}^\dagger_{\mp\frac{1}{2}}$, see Figure 1.12. This can be done using the operator

$$C = -\underline{\boldsymbol{a}}^\dagger_\downarrow\boldsymbol{a}_\uparrow + \underline{\boldsymbol{a}}^\dagger_\uparrow\boldsymbol{a}_\downarrow. \quad (1.379)$$

Subsequently, the dual vector is used to represent the conjugate, *i.e.* $\underline{\boldsymbol{a}}^\dagger_{\pm\frac{1}{2}} \rightarrow \boldsymbol{a}_{\pm\frac{1}{2}}$. For the unit vector $\boldsymbol{e}_0 = \boldsymbol{a}^\dagger_\uparrow\boldsymbol{a}^\dagger_\downarrow$, see Eq. (1.342), this gives

$$C\boldsymbol{e}_0 = \boldsymbol{a}^\dagger_\uparrow\underline{\boldsymbol{a}}^\dagger_\uparrow - \underline{\boldsymbol{a}}^\dagger_\downarrow\boldsymbol{a}^\dagger_\downarrow \quad \Rightarrow \quad \mathbf{e}_0 = \boldsymbol{a}^\dagger_\uparrow\boldsymbol{a}_\uparrow - \boldsymbol{a}^\dagger_\downarrow\boldsymbol{a}_\downarrow. \quad (1.380)$$

Likewise, we obtain for $\boldsymbol{e}_1 = \boldsymbol{a}^\dagger_\uparrow\boldsymbol{a}^\dagger_\uparrow/\sqrt{2}$, see Eq. (1.342),

$$C\boldsymbol{e}_1 = -\frac{1}{\sqrt{2}}2\boldsymbol{a}^\dagger_\uparrow\underline{\boldsymbol{a}}^\dagger_\downarrow \quad \Rightarrow \quad \mathbf{e}_1 = -\sqrt{2}\boldsymbol{a}^\dagger_\uparrow\boldsymbol{a}_\downarrow. \quad (1.381)$$

The same can be done for $\boldsymbol{e}_{-1}$. The unit vector operators are now

$$\mathbf{e}_1 = -\sqrt{2}\boldsymbol{a}^\dagger_\uparrow\boldsymbol{a}_\downarrow, \quad \mathbf{e}_0 = \boldsymbol{a}^\dagger_\uparrow\boldsymbol{a}_\uparrow - \boldsymbol{a}^\dagger_\downarrow\boldsymbol{a}_\downarrow, \quad \mathbf{e}_{-1} = \sqrt{2}\boldsymbol{a}^\dagger_\downarrow\boldsymbol{a}_\uparrow. \quad (1.382)$$

If we calculate the matrices for a basis of the spinors $a_\uparrow^\dagger$ and $a_\downarrow^\dagger$, we obtain the following matrices:

$$\mathbf{e}_1 = \begin{pmatrix} 0 & -\sqrt{2} \\ 0 & 0 \end{pmatrix}, \quad \mathbf{e}_0 = \begin{pmatrix} 1 & 0 \\ 0 & -1 \end{pmatrix}, \quad \mathbf{e}_{-1} = \begin{pmatrix} 0 & 0 \\ \sqrt{2} & 0 \end{pmatrix}. \tag{1.383}$$

The matrices $\mathbf{e}_{\pm 1}$ are the same as those obtained in Eq. (1.350) for the complex unit vectors. The unit vectors $\mathbf{e}_m$ are the Pauli matrices in spherical coordinates. The expressions in terms of spinors in Eq. (1.382) are more general since they can be applied to vectors consisting of an arbitrary product of spinors.

When dealing with integer total angular momentum, one generally prefers to use

$$\mathbf{L}_0 = \mathbf{L}_z = \frac{\mathbf{e}_0}{2} = \frac{1}{2}(a_\uparrow^\dagger a_\uparrow - a_\downarrow^\dagger a_\downarrow) \tag{1.384}$$

$$\mathbf{L}_{\pm 1} = \mp \frac{1}{\sqrt{2}}(\mathbf{L}_x \pm i\mathbf{L}_y) = \frac{\mathbf{e}_{\pm 1}}{2} = \mp \frac{1}{\sqrt{2}} a_{\pm\frac{1}{2}}^\dagger a_{\mp\frac{1}{2}}. \tag{1.385}$$

Note that the step operators $\mathbf{L}_\pm = \mp\sqrt{2}\mathbf{L}_{\pm 1}$, see Eq. (1.370), used in many textbooks, are not proper spherical tensors.

Example 1.14 ***Rotations in Three Dimensions***

We now want to make a rotation of vectors in three dimensions using the vectors described by a product of two spinors. This problem is also relevant for rotations of atomic p-orbitals.

a) Determine $\mathbf{L}_0 = \mathbf{L}_z$ for the basis e_1, e_0, and e_{-1}.
b) We are interested in rotating a vector in Cartesian coordinates (or equivalently, look at the transformation properties of the p_x, p_y, and p_z orbitals under rotation). Derive the unitary transformation between the Cartesian and spherical basis.
c) Write $\mathbf{L}_z$ in a Cartesian basis.
d) Give an expression for the rotation over an angle φ around the z-axis.
e) Expand the result from (d) in trigonometric functions.

a) Using Eq. (1.384), we find for the basis e_1, e_0, and e_{-1} from Eq. (1.342)

$$\mathbf{L}_0 = \begin{pmatrix} 1 & 0 & 0 \\ 0 & 0 & 0 \\ 0 & 0 & -1 \end{pmatrix}. \tag{1.386}$$

Note that, although the $\mathbf{L}_z = \mathbf{e}_z/2$ contains a factor $\frac{1}{2}$, this does not appear in the matrix since the basis consists of two-spinor states.

b) The unitary transformation between the Cartesian and spherical basis is given by

$$\mathrm{U} = \sum_{i=x,y,z} \sum_{m=-1}^{1} |1m\rangle\langle 1m|1i\rangle\langle 1i| = \begin{pmatrix} e_x & e_y & e_z \end{pmatrix}. \tag{1.387}$$

Writing the Cartesian column unit vectors e_i in terms of the spherical unit vectors, see Eq. (1.348), gives

$$\mathrm{U} = \begin{pmatrix} -\frac{1}{\sqrt{2}} & \frac{i}{\sqrt{2}} & 0 \\ 0 & 0 & 1 \\ \frac{1}{\sqrt{2}} & \frac{i}{\sqrt{2}} & 0 \end{pmatrix}. \tag{1.388}$$

c) The angular momentum in the Cartesian basis is

$$\mathbf{L}_z = \mathrm{U}^\dagger \mathbf{L}_0 \mathrm{U} = \begin{pmatrix} 0 & -i & 0 \\ i & 0 & 0 \\ 0 & 0 & 0 \end{pmatrix}. \tag{1.389}$$

It is recommended to perform the matrix multiplications numerically. Since $\mathbf{L}_z = \mathbf{L}_0$, there is no need to make a unitary transformation of the components of the angular momentum, as we should for the x and y components.

d) In two dimensions, we had seen that a 90° counterclockwise rotation in the xy can be obtained by the imaginary unit $\boldsymbol{i}_{yx}$. This can also be expressed in terms of the rotation axis via $-\boldsymbol{i}\mathbf{e}_z = -\mathbf{e}_x\mathbf{e}_y\mathbf{e}_z\mathbf{e}_z = \mathbf{e}_y\mathbf{e}_x = \boldsymbol{i}_{yx}$. Above, we have seen that, for products of spinors, $\mathbf{L}_m$ effectively takes the role of the unit vectors $\mathbf{e}_m$. A rotation around the z-axis can therefore be described in terms of $\boldsymbol{I}_{yx} = -i\mathbf{L}_z$ (a different notation is used, since $\boldsymbol{I}_{yx}$ is not exactly the same as $\boldsymbol{i}_{yx}$ due to the fact that the basis consists of two spinors). The rotation is therefore given by

$$\mathrm{R}^z_\varphi = e^{-i\mathbf{L}_z\varphi} = e^{\boldsymbol{I}_{yx}\varphi}. \tag{1.390}$$

e) The expansion of the exponential gives

$$\mathrm{R}^z_\varphi = \sum_{n=0}^{\infty} \frac{(-i\mathbf{L}_z\varphi)^n}{n!} = \sum_{n=0}^{\infty} \frac{(\boldsymbol{I}_{yx}\varphi)^n}{n!}. \tag{1.391}$$

The imaginary unit is given by

$$\boldsymbol{I}_{yx} = -i\mathbf{L}_z = \begin{pmatrix} 0 & -1 & 0 \\ 1 & 0 & 0 \\ 0 & 0 & 0 \end{pmatrix}. \tag{1.392}$$

It is left up to the reader to demonstrate that this is equivalent to $\boldsymbol{I}_{yx} = \mathbf{L}_y\mathbf{L}_x - \mathbf{L}_x\mathbf{L}_y$. We have to be somewhat careful with this imaginary unit since it does not behave in exactly the same fashion as $\boldsymbol{i}_{yx} = \mathbf{e}_y\mathbf{e}_x$ in terms of Pauli matrices or as the scalar imaginary unit i. Multiplication with those units gives $i^2 = -1$, $i^3 = -i$, $i^4 = 1$, and so on. However, $\boldsymbol{I}_{yx}$ is an imaginary unit for the xy-plane but leaves the z-axis unchanged. For example,

$$\mathbb{1}_{xy} \equiv \boldsymbol{I}^4_{yx} = \begin{pmatrix} 1 & 0 & 0 \\ 0 & 1 & 0 \\ 0 & 0 & 0 \end{pmatrix}, \tag{1.393}$$

which is defined as the identity matrix for the x and y directions. This matrix gives an identity for the x and y components, but not the z component. Since the rotation around the z-axis does not affect the z component, the result is not equal to the identity matrix in three dimensions, *i.e.* $\mathbb{1}_{xy} \neq \mathbb{1}_3$. However, we still find $\boldsymbol{I}^n_{yx} = (-1)^{\frac{n}{2}}\mathbb{1}_{xy}$ for n is even and $n \neq 0$, and $\boldsymbol{I}^n_{yx} = (-1)^{\frac{n-1}{2}}\boldsymbol{I}_{yx}$ for n is odd. Additionally, we have $\boldsymbol{I}^0_{yx} = \mathbb{1}_3$. Therefore, $\boldsymbol{I}_{yx}$ behaves as an imaginary unit for the xy-plane, except for the $n = 0$ term which is the identity matrix for all directions. The series therefore becomes

$$e^{\boldsymbol{I}_{yx}\varphi} = \mathbb{1}_3 + \mathbb{1}_{xy}\left(-\frac{\varphi^2}{2!} + \frac{\varphi^4}{4!} - \frac{\varphi^6}{6!} + \cdots\right) + \boldsymbol{I}_{yx}\left(\varphi - \frac{\varphi^3}{3!} + \cdots\right). \tag{1.394}$$

Recognizing the Taylor series for the sine and cosine gives

$$e^{\boldsymbol{I}_{yx}\varphi} = \mathbb{1}_3 + \mathbb{1}_{xy}(\cos\varphi - 1) + \boldsymbol{I}_{yx}\sin\varphi. \tag{1.395}$$

Writing this in matrix form gives

$$\mathrm{R}^z_\varphi = e^{\boldsymbol{I}_{yx}\varphi} = \begin{pmatrix} \cos\varphi & -\sin\varphi & 0 \\ \sin\varphi & \cos\varphi & 0 \\ 0 & 0 & 1 \end{pmatrix}. \tag{1.396}$$

This is the well-known rotation around the z-axis in terms of Cartesian coordinates. The rotation occurs for the x and y coordinates, whereas the z component is constant.

Conceptual summary. – Unfortunately, representation theory can be confusing due to the abstract nature of the topic and the different phase conventions and coordinate systems. However, the essence of what we are doing is still the same as that of Sections 1.2 and 1.3. In two dimensions, we found that the rotation could be achieved in two ways: The derivative operator $\hat{\imath}_{yx} = d/d\varphi$ could be applied to the coefficients, or the unit vectors could be modified with the vector operator $\boldsymbol{i}_{yx} = \mathbf{e}_y\mathbf{e}_x$. Since the end result should be the same, the transformation properties of the functions and the unit vectors are the same.

This section is concerned with essentially the same thing. However, instead of describing the rotation of a unit vector in two dimensions, we are now looking at the rotations of higher order vectors in three dimensions. Therefore, the transformation properties of the coefficients (which have now turned into the function space of the spherical harmonics) and those of the higher order unit vectors are still the same. The vector operators working on the unit vectors are those of the angular momentum $\mathbf{L}_m = \mathbf{e}_m/2$, with $m = 1, 0, -1$. The vector operator (the angular momentum) is simply a unit vector (note that in two dimensions, the vector operator $\boldsymbol{i}_{yx}$ was a bivector related to the rotation in the plane. We are now looking at the unit vector corresponding to the axis of rotation). In the following section, we will look at the operators working on the coefficient/functions of the higher order vectors.

It might appear strange that the angular momentum is represented by a unit vector. However, "physical" quantities are often related to the symmetry of the system. Momentum and angular momentum are directly related to the translational and rotational properties of free space. For example, the splitting of the atomic levels into s, p, d, etc. orbitals is directly imposed by the spherical symmetry of the problem and is unrelated to the detailed nature of the nuclear potential. Once the symmetry properties of an interaction are known (in technical terms, the interaction can be described as a spherical tensor of a particular rank), all its transformation properties are known. Therefore, regardless of how complicated an interaction is, if it is a vector, its corresponding vector operation is described by the unit vector operators $\mathbf{e}_m$. This is an aspect of what is known as the Wigner–Eckart theorem. A vector is a tensor of rank 1 (which directly implies that it has three components). In Section 4.4, we will see that operators of higher rank also exist.

The connection between the symmetry of a problem and its physical properties has important implications. For example, the initial explorations into hadrons in terms of quarks were based on the assumption of the symmetry underlying the interactions. The symmetry then imposes a level structure that can be explored without understanding the detailed nature of the interactions.

Problems

1 Derive the rotation matrix around the z-axis in a spherical basis. Show that this indeed rotates the spherical harmonics.

2 We want to study an interaction proportional to $\mathbf{L}_x$ (this could be due to a magnetic field in the x-direction) on the spherical harmonics, with $l = 1$.

a) Find the matrix for $\mathbf{L}_x$ for $l = 1$.
b) Find the eigenvectors and eigenenergies (you can do this numerically).
c) One can also approach this problem as a reorientation of the reference axis. Find the appropriate spinors for the x-axis.
d) Find the eigenvectors in the new spinors and rewrite them in terms of the spinors with respect to the z-axis.
e) Derive the operator $\mathbf{L}_x$ in the new spinor basis.

1.17 Operator Techniques for Angular Momentum

Introduction. – In this chapter, we have treated translation and rotation in free space, *i.e.* in the absence of potentials. It was shown that extended nonrigid particles should be described by higher order vectors. This creates a function space that allows one to describe the shape of the particle in space. The motion of the particle can be achieved in two ways: by manipulation of the (higher order) unit vectors or by applying an operation on the coefficients/functions. In this section, we focus on the operators related to rotation, *i.e.* the angular momentum operators. These operators work on the function space for rotation in three dimensions, *i.e.* the spherical harmonics. In Section 1.3, a rotation in two dimensions was obtained by taking the derivative with respect to φ on the coefficients. At first, one might think that one can only derive the transformation properties if one knows the basis functions. However, we will see that many of the transformation properties are a direct result of the nature of the rotation operators.

First, we need an operator that causes rotations in three dimensions. In Section 1.3, an angular momentum operator was derived for rotations in a plane. However, rotations in many ways are always two-dimensional with a plane of rotation defined by a rotation axis. Therefore, let us generalize the earlier result from Eq. (1.43),

$$\hat{\mathbf{L}} = -\boldsymbol{i}_{x'y'}\frac{\partial}{\partial\varphi'}, \tag{1.397}$$

with $\boldsymbol{i}_{x'y'} = \boldsymbol{i}\mathbf{e}_{z'}$, taking the conventional definition of an operator working on a vector to the left. The hats indicate here that we are dealing with operators. The primed Cartesian axis system is oriented in such a way that the z' lies along the rotation axis, and the x' and y' axes are in the plane of rotation. The partial derivative is taken since there are additional coordinates in three dimensions. The derivative with respect to φ' can be rewritten in terms of the Cartesian coordinates

$$\begin{aligned}\hat{\mathbf{L}} &= -\boldsymbol{i}_{x'y'}\left(\frac{\partial x'}{\partial\varphi'}\frac{\partial}{\partial x'} + \frac{\partial y'}{\partial\varphi'}\frac{\partial}{\partial y'} + \frac{\partial z'}{\partial\varphi'}\frac{\partial}{\partial z'}\right)\\ &= -\boldsymbol{i}\left(x'\frac{\partial}{\partial y'} - y'\frac{\partial}{\partial x'}\right)\mathbf{e}_{z'} = (\mathbf{r}'\times(-\boldsymbol{i}\nabla'))_{z'}\mathbf{e}_{z'} = (\mathbf{r}'\times\hat{\mathbf{p}}')_{z'}\mathbf{e}_{z'},\end{aligned} \tag{1.398}$$

using $\hat{\mathbf{p}} = -i\nabla$, with $\hbar \equiv 1$, and that $\partial x'/\partial\varphi' = -r'\sin\theta'\sin\varphi' = -y'$, $\partial y/\partial\varphi' = r'\sin\theta'\cos\varphi' = x'$, and $\partial z'/\partial\varphi' = \partial r'\cos\theta'/\partial\varphi' = 0$ for spherical polar coordinates, see Eq. (1.125). Due to the orientation of the primed axis system, the angular momentum operator only has a z' component. However, for an arbitrary orientation of the Cartesian coordinate system, we have

$$\hat{\mathbf{L}} = -\mathbf{r} \wedge \nabla = -i\mathbf{r} \times \nabla = \mathbf{r} \times \hat{\mathbf{p}}, \tag{1.399}$$

which is the well-known expression for the angular momentum but in operator form.

The properties of a function space described in terms of unit vectors are defined by the commutation relationships. This is also the case for the operators that need to describe similar transformation properties of the functions. Before going into more detail on the angular momentum operator, let us take a look at the momentum operator. Most readers will be familiar with the commutation relation $[x, \hat{p}_x] = i\hbar$. This relation is intimately connected to Heisenberg's uncertainty principle $\Delta x \Delta p_x \geq \hbar/2$. However, it is not clear how the commutation relation affects the function space of running waves.

The properties of the function space of running waves can, in principle, also be expressed in terms of indistinguishable quantities $\boldsymbol{a}^\dagger$. However, this can only be done when separating the motion in the positive and negative directions, comparable to the split of counterclockwise and clockwise rotations. Since this is not something we would generally use, we limit the discussion to positive wavenumbers. In terms of $\boldsymbol{a}^\dagger$, the function space can be described as

$$|m\rangle = \frac{1}{\sqrt{m!}}(\boldsymbol{a}^\dagger)^m. \tag{1.400}$$

The function space is given by raising an effective unit vector to a positive integer power $(e^{ik_0x})^m = e^{imk_0x}$, with $k_0 = 2\pi/L$, where L is the size of the one-dimensional system. The conserved quantity, the wavenumber, can be extracted by the use of number operators

$$\mathrm{N} = \boldsymbol{a}^\dagger \boldsymbol{a} \;\Rightarrow\; \mathrm{k} = k_0\mathrm{N} \;\Rightarrow\; \mathrm{k}|m\rangle = mk_0|m\rangle = \frac{2\pi}{L}m|m\rangle. \tag{1.401}$$

By multiplying the result with k_0, we obtain the wavenumber/momentum. However, up to this point, there is no restriction on the value of m, it can still be a real number. That the m values are separated by 1 follows from the commutation relation

$$[\boldsymbol{a}, \boldsymbol{a}^\dagger]|m\rangle = (\boldsymbol{a}\boldsymbol{a}^\dagger - \boldsymbol{a}^\dagger\boldsymbol{a})|m\rangle = ((m+1) - m)\,|m\rangle = |m\rangle, \tag{1.402}$$

giving $[\boldsymbol{a}, \boldsymbol{a}^\dagger] = 1$, which is the well-known commutation relation. By multiplying with $\mathbf{a}^\dagger$, the commutation can also be written in terms of the number operator

$$\boldsymbol{a}^\dagger[\boldsymbol{a}, \boldsymbol{a}^\dagger] = \boldsymbol{a}^\dagger\boldsymbol{a}\boldsymbol{a}^\dagger - \boldsymbol{a}^\dagger\boldsymbol{a}^\dagger\boldsymbol{a} = \mathrm{N}\boldsymbol{a}^\dagger - \boldsymbol{a}^\dagger\mathrm{N} \;\Rightarrow\; [\mathrm{N}, \boldsymbol{a}^\dagger] = \boldsymbol{a}^\dagger. \tag{1.403}$$

Let us see how this translates into an operator form. The momentum is now obtained by taking the derivative of the function

$$\hat{p}_x = \hat{k}_x = -i\frac{d}{dx} \;\Rightarrow\; \hat{p}_x\varphi_k(x) = -i\frac{d}{dx}\left(\frac{1}{\sqrt{L}}e^{imk_0x}\right) = mk_0\varphi_k(x),$$

when taking $\hbar \equiv 1$. Note that we can also redefine the momentum operator as an effective number operator (comparable to the angular momentum operator),

$$\hat{N} = \frac{\hat{p}_x}{k_0} = -\frac{i}{k_0}\frac{d}{dx} \;\Rightarrow\; \hat{N}\varphi_m(x) = m\varphi_m(x), \tag{1.404}$$

with $\varphi_k(x) \to \varphi_m(x)$, and $m = k/k_0$.

Raising the value of m can be done through multiplication, $\hat{a}^\dagger = e^{ik_0x}$, so that $e^{ik_0x}e^{imk_0x} = e^{i(m+1)k_0x}$. While for the vector operators, the number operators can be expressed as $N = \boldsymbol{a}^\dagger\boldsymbol{a}$, this is not possible in the operator form. For the commutator $[\mathrm{N}, \boldsymbol{a}^\dagger]$, the first term $\mathrm{N}\boldsymbol{a}^\dagger$ is then effectively replaced by

$$\hat{p}_x e^{ik_0x}e^{imk_0x} = \hat{p}_x e^{i(m+1)k_0x} = (m+1)k_0 e^{i(m+1)k_0x}, \tag{1.405}$$

where the product works on a wave function e^{imk_0x}. The term $\boldsymbol{a}^\dagger\mathrm{N}$ becomes

$$e^{ik_0x}\hat{p}_x e^{imk_0x} = mk_0 e^{i(m+1)k_0x}. \tag{1.406}$$

Dividing by $e^{im_xk_0x}$ and subtracting the two equations give

$$\hat{p}_x e^{ik_0x} - e^{ik_0x}\hat{p}_x = k_0 e^{ik_0x} \quad \Rightarrow \quad [\hat{p}_x, e^{ik_0x}] = k_0 e^{ik_0x}, \tag{1.407}$$

which is the operator equivalent of Eq. (1.403). In the limit of a large system, *i.e.* $L \to \infty$, $e^{ik_0x} \cong 1 + ik_0x$.

$$\hat{p}_x(1 + ik_0x) - (1 + ik_0x)\hat{p}_x \cong k_0 \quad \Rightarrow \quad i(\hat{p}_x x - x\hat{p}_x) = 1. \tag{1.408}$$

This can also be written as

$$[x, \hat{p}_x] = i, \tag{1.409}$$

which is the commutation relation for the position and the momentum. Obviously, this relation can be straightforwardly derived in a different fashion. However, by itself, it is not directly obvious what the result implies for the function space of running/plane waves.

The commutation relations above can be extended to three dimensions, since the directions in a Cartesian coordinate system are independent from each other

$$[r_i, \hat{p}_j] = i\delta_{ij}, \tag{1.410}$$

with $i, j = x, y, z$.

The commutation relationship of the angular momentum operator is generally derived as follows:

$$\begin{aligned}[\hat{L}_x, \hat{L}_y] &= \hat{L}_x\hat{L}_y - \hat{L}_y\hat{L}_x = (y\hat{p}_z - z\hat{p}_y)(z\hat{p}_x - x\hat{p}_z) - (z\hat{p}_x - x\hat{p}_z)(y\hat{p}_z - z\hat{p}_y) \\ &= y\hat{p}_z z\hat{p}_x - y\hat{p}_z x\hat{p}_z - z\hat{p}_y z\hat{p}_x + z\hat{p}_y x\hat{p}_z \\ &\quad - z\hat{p}_x y\hat{p}_z + z\hat{p}_x z\hat{p}_y + x\hat{p}_z y\hat{p}_z - x\hat{p}_z z\hat{p}_y.\end{aligned} \tag{1.411}$$

Many of the terms on the right-hand side cancel. The terms that contain z or $\hat{p}_z$ twice cancel using the fact that momentum and position commute with themselves. The remaining terms can be rearranged as

$$[\hat{L}_x, \hat{L}_y] = x\hat{p}_y(z\hat{p}_z - \hat{p}_z z) - y\hat{p}_x(z\hat{p}_z - \hat{p}_z z) = i(x\hat{p}_y - y\hat{p}_x) = iL_z, \tag{1.412}$$

making use of the commutation relation in Eq. (1.410). Commutation relations for the other components can be obtained by cyclic permutation $x \to y \to z \to x$, giving

$$[\hat{L}_x, \hat{L}_y] = i\hat{L}_z, \quad [\hat{L}_y, \hat{L}_z] = i\hat{L}_x \quad \text{and} \quad [\hat{L}_z, \hat{L}_x] = i\hat{L}_y. \tag{1.413}$$

Again, these results are obtained with hard work, but its meaning is not altogether obvious. However, we have found before that the angular momentum behaves as generalized unit vectors, *i.e.* $\mathbf{L}_i = \mathbf{e}_i/2$. This implies that

$$[\mathbf{L}_x, \mathbf{L}_y] = \mathbf{L}_x\mathbf{L}_y - \mathbf{L}_y\mathbf{L}_x = \frac{\mathbf{e}_x}{2}\frac{\mathbf{e}_y}{2} - \frac{\mathbf{e}_y}{2}\frac{\mathbf{e}_x}{2} = \frac{\boldsymbol{i}\mathbf{e}_z}{2} = \boldsymbol{I}_{xy} = \boldsymbol{i}\mathbf{L}_z, \tag{1.414}$$

as we saw in the previous section. This shows the above result in a more intuitive fashion. Therefore, the commutation relation for angular momentum is related to the fact that the product of two vectors is a bivector, which can be expressed in terms of its dual vector. However, for higher order operators, we have to be more careful. When the space is restricted to single spinors, the anticommutation is zero, *i.e.* $\mathbf{e}_x\mathbf{e}_y + \mathbf{e}_y\mathbf{e}_x = 0$. However, this is no longer the case when products of spinors are allowed. The combinations $\mathbf{e}_x\mathbf{e}_y + \mathbf{e}_y\mathbf{e}_x$ and $\mathbf{L}_x\mathbf{L}_y + \mathbf{L}_y\mathbf{L}_x$ then correspond to higher order operators or tensors. Therefore, $\mathbf{e}_x\mathbf{e}_y \neq -\mathbf{e}_y\mathbf{e}_x$ when dealing with spaces containing multiple spinors. This will be discussed in Section 4.4.

When constructing the unit vectors $|1m\rangle$ for the unit sphere in Section 1.14, it was noted that it was not possible to find complex exponentials that were spinors for rotation around the x, y, and z axes simultaneously. In terms of functions, it should therefore not be possible to find functions that have nonzero eigenvalues for all angular momentum components $\hat{L}_i$, with $i = x, y, z$. To demonstrate this, let us assume that there was a function $\varphi_{m_x m_y m_z}$ for which $\hat{L}_i\varphi_{m_x m_y m_z} = m_i\varphi_{m_x m_y m_z}$. Using Eq. (1.413), we can write

$$\begin{aligned}\hat{L}_z\varphi_{m_x m_y m_z} &= m_z\varphi_{m_x m_y m_z} = -i(\hat{L}_x\hat{L}_y - \hat{L}_y\hat{L}_x)\varphi_{m_x m_y m_z} \\ &= -i(\hat{L}_x m_y - \hat{L}_y m_x)\varphi_{m_x m_y m_z} = -i(m_x m_y - m_y m_x)\varphi_{m_x m_y m_z} = 0. \quad (1.415)\end{aligned}$$

Since this can only be satisfied for $m_z = 0$, no eigenfunction with a finite angular momentum can be found that is also an eigenfunction of $\hat{L}_x$ and $\hat{L}_z$. Therefore, eigenfunctions can only be constructed for one rotation axis at a time. Conventionally, one chooses the z-direction as the axis of rotation

$$\hat{L}_z Y_{lm}(\hat{\mathbf{r}}) = mY_{lm}(\hat{\mathbf{r}}). \quad (1.416)$$

Since we know that the eigenfunctions are the spherical harmonics, the functions are written in the usual notation. Although m is an eigenvalue, this equation gives no insight whether m is real or integer. Let us define the step operators connecting different m values as complex combinations of the x and y components,

$$\hat{L}_\pm = \hat{L}_x \pm i\hat{L}_y. \quad (1.417)$$

Let us now consider the effect of the operator $L_z L_\pm$. Using the commutation relations in Eq. (1.413), this product can be rewritten as

$$\begin{aligned}\hat{L}_z\hat{L}_\pm &= \hat{L}_z(\hat{L}_x \pm i\hat{L}_y) = \hat{L}_x\hat{L}_z + i\hat{L}_y \pm i(\hat{L}_y\hat{L}_z - i\hat{L}_x) \\ &= (\hat{L}_x \pm i\hat{L}_y)(\hat{L}_z \pm 1) = \hat{L}_\pm(\hat{L}_z \pm 1). \quad (1.418)\end{aligned}$$

Following Eq. (1.414), these relations can also be derived for the unit vector operators $\mathbf{L}_\pm$. Operating this on an eigenfunction gives

$$\hat{L}_z\hat{L}_\pm Y_{lm} = \hat{L}_\pm(\hat{L}_z \pm 1)Y_{lm} = \hat{L}_\pm(m \pm 1)Y_{lm} = (m \pm 1)\hat{L}_\pm Y_{lm}.$$

So, $\hat{L}_\pm Y_{lm}$ are eigenfunctions of $\hat{L}_z$ with eigenvalues $(m \pm 1)$. Therefore, the application of $\hat{L}_\pm$ leads to eigenfunctions with an eigenvalue $m \pm 1$, *i.e.* $\hat{L}_\pm Y_{lm} \sim Y_{l,m\pm 1}$. So, starting from a particular eigenvalue m we can create a set of eigenfunctions with eigenvalues

$$m' = \cdots, m-2, m-1, m, m+1, m+2, \ldots, \quad (1.419)$$

by successively applying the $\hat{L}_\pm$. As a result of this property, $\hat{L}_\pm$ are also known as step operators. Note that it does not show that m is an integer, only that they are separated by an integer. It is important to realize that these properties follow from the commutation relations that the angular momentum operators need to satisfy.

Having discarded the other components of $\hat{L}$, we need an additional operator to account for the eigenvalues l. The angular momentum is a vector. Classically, the length of a vector is invariant under rotations. Let us therefore consider the operator

$$\hat{\mathbf{L}}^2 = \hat{L}_x^2 + \hat{L}_y^2 + \hat{L}_z^2. \tag{1.420}$$

If the wave functions are eigenstates of both $\hat{L}_z$ and $\hat{L}^2$, then these operators need to commute with each other:

$$\begin{aligned}[\hat{\mathbf{L}}^2, \hat{L}_z] &= \hat{\mathbf{L}}^2\hat{L}_z - \hat{L}_z\hat{\mathbf{L}}^2 = L_x^2 L_z - L_z L_x^2 + \hat{L}_y^2\hat{L}_z - \hat{L}_z\hat{L}_y^2 + \hat{L}_z^2\hat{L}_z - \hat{L}_z L_z^2 \\ &= \hat{L}_x(\hat{L}_z\hat{L}_x - i\hat{L}_y) - (\hat{L}_x\hat{L}_z + i\hat{L}_y)\hat{L}_x \\ &\quad + \hat{L}_y(\hat{L}_z\hat{L}_y + i\hat{L}_x) - (\hat{L}_y\hat{L}_z - i\hat{L}_x)\hat{L}_y = 0,\end{aligned} \tag{1.421}$$

making use of the commutation relations between the components of $\hat{\mathbf{L}}$. Therefore, the spherical harmonics are also eigenfunctions of the norm of the angular momentum

$$\hat{\mathbf{L}}^2 Y_{lm} = \lambda_l Y_{lm}, \tag{1.422}$$

where λ_l is a function of l. Rewriting the norm of the angular momentum gives

$$(\hat{L}_x^2 + \hat{L}_y^2)Y_{lm} = (\hat{\mathbf{L}}^2 - \hat{L}_z^2)Y_{lm} = (\lambda_l - m^2)Y_{lm}. \tag{1.423}$$

Since the operator $\hat{L}_x^2 + \hat{L}_y^2$ is positive, this implies that $|m| \leq \sqrt{\lambda_l}$. However, since the m values are bound through $|m| \leq \sqrt{\lambda_l}$, the range of m values in Eq. (1.419) has to be finite. Therefore, there have to be certain values of $m = l_\pm$ for which

$$\hat{L}_\pm Y_{ll_\pm} = 0. \tag{1.424}$$

Since the wave functions are now zero, stepping down/up after the final step-up/down operation should still give zero.

In order to express the above equation in terms of the eigenvalues of $\hat{\mathbf{L}}^2$ and $\hat{L}_z$, it is multiplied by the conjugate of the step over

$$\hat{L}_\pm^*\hat{L}_\pm = \hat{L}_x^2 + \hat{L}_y^2 \pm \hat{L}_z, \tag{1.425}$$

making use of the commutation relation in Eq. (1.412). Applying this to the wave functions with extremal $m = l_\pm$ values gives

$$(\hat{L}_x^2 + \hat{L}_y^2 \pm \hat{L}_z)Y_{ll_\mp} = \frac{1}{2}(\lambda_l - l_\mp^2 \pm l_\mp)Y_{ll_\mp} = 0, \tag{1.426}$$

using Eq. (1.423). This gives two equations for $l_\pm$

$$(\lambda_l - l_-^2 + l_-) = (\lambda_l - l_+^2 - l_+) = 0. \tag{1.427}$$

After removing λ_l by subtracting the two equations, we have

$$(l_+ + l_-)(l_+ - l_- + 1) = 0. \tag{1.428}$$

Since $l_+ \geq l_-$, the second term is never zero, and we find that $l_- = -l_+$ in order to make this equation zero. We are therefore left with one quantum number, and we take $l = l_+$. The allowed values of m are therefore

$$m = l, l-1, \ldots, -l+1, -l. \tag{1.429}$$

Since $-l$ and l are separated by integers, l has to be an integer, and therefore, m has to be an integer as well. However, note that another option is half integers, although these are not valid eigenvalues for the spherical harmonics.

Equation (1.427) implies that $\lambda_l = l(l+1)$ and

$$\hat{\mathbf{L}}^2 Y_{lm}(\hat{\mathbf{r}}) = l(l+1)Y_{lm}(\hat{\mathbf{r}}). \tag{1.430}$$

The above equation written in operator form using the angular part of the Laplacian ∇^2,

$$-\left(\frac{1}{\sin\theta}\frac{\partial}{\partial\theta}\left(\sin\theta\frac{\partial Y_{lm}}{\partial\theta}\right)+\frac{1}{\sin^2\theta}\frac{\partial^2 Y_{lm}}{\partial\varphi^2}\right)=l(l+1)Y_{lm}, \tag{1.431}$$

is a standard way to obtain expressions for the spherical harmonics. This is done in many textbooks and omitted here.

Equation (1.425) can also be used to obtain the factors for the step operators

$$\begin{aligned}\hat{L}_\pm^*\hat{L}_\pm Y_{lm} &= (L_x^2+L_y^2\mp L_z)Y_{lm} = (\hat{\mathbf{L}}^2 - L_z^2 \mp L_z)Y_{lm}\\ &= \left(l(l+1)-m^2\mp m\right)Y_{lm} = (l\mp m)(l\pm m+1)Y_{lm}.\end{aligned} \tag{1.432}$$

This gives the relation how the step operators $\hat{L}_\pm$ combine different spherical harmonics

$$\hat{L}_\pm Y_{lm} = \sqrt{(l\mp m)(l\pm m+1)}Y_{l,m\pm1}. \tag{1.433}$$

This is equal to the values obtained using unit vectors in Eqs (1.372) and (1.371). Although the properties of the step operators can be obtained from the commutation relations, the actual operators,

$$\hat{L}_\pm = e^{\pm i\varphi}\left(\pm\frac{\partial}{\partial\theta}+i\cot\theta\frac{\partial}{\partial\varphi}\right), \tag{1.434}$$

are not particularly insightful compared to step operators for the unit vectors, $\mathbf{L}_+ = \boldsymbol{a}_\uparrow^\dagger \boldsymbol{a}_\downarrow$ and $\mathbf{L}_- = \boldsymbol{a}_\downarrow^\dagger \boldsymbol{a}_\uparrow$. Demonstrating that these operators satisfy Eq. (1.433) when operating on the spherical harmonics is a far-from-trivial task.

The above operator for a rotation around the z-axis can also be used to rotate a function. In analogy to translation, see Eq. (1.231), a function can be rotated by rotating the argument of a function in the opposite direction. Let us restrict ourselves to a rotation around the z-axis. The application of the rotation operator $\hat{R}_\varphi^z$ on an arbitrary function gives

$$\hat{R}_\varphi^z f(\mathbf{r}) = f(\hat{R}_{-\varphi}^z\mathbf{r}). \tag{1.435}$$

First, the limit of a rotation over a small angle $d\varphi$ around the z- axis is considered. The rotated argument on the right-hand side $\hat{R}_{-d\varphi}^z\mathbf{r} = \mathbf{r} - \mathbf{r}_\perp d\varphi$, with $\mathbf{r}_\perp = -y\mathbf{e}_x + x\mathbf{e}_y$. Note that the rotation only affects the x and y components and leaves the z component unchanged. The function rotated over a small angle can then be expanded in a Taylor series

$$f(x+yd\varphi, y-xd\varphi, z) \cong f(x,y,z) + \frac{\partial f}{\partial x}yd\varphi - \frac{\partial f}{\partial y}xd\varphi. \tag{1.436}$$

The rotation operator is therefore

$$\begin{aligned}\hat{r}_{d\varphi}^z &= 1-\left(x\frac{\partial}{\partial y}-y\frac{\partial}{\partial x}\right)d\varphi = 1-(\mathbf{r}\times\nabla)_z d\varphi\\ &= 1-i(\mathbf{r}\times\hat{\mathbf{p}})_z d\varphi = 1-i\hat{L}_z d\varphi.\end{aligned} \tag{1.437}$$

A rotation over a finite angle φ can be described by N steps of $d\varphi = \varphi/N$

$$\hat{R}^z_\varphi = \lim_{N\to\infty}\left(1 - i\hat{L}_z\frac{\varphi}{N}\right)^N = e^{-i\hat{L}_z\varphi}, \tag{1.438}$$

using Eq. (1.4). Therefore, the rotation operator working on the function space is a complex exponential of the angular momentum operator. The same result was found when rotating an object using vector operators, see Eq. (1.390).

Conceptual summary. – This section shows that the transformation properties of the operators on the basis functions are equivalent to those of the angular momentum in terms of unit vectors. The transformation properties can already be obtained from the commutation properties of the angular momentum operators and do not require actually applying the operators on the function space. However, the operator approach can be less insightful than the unit vector approach, although this obviously can be a matter of opinion. The strength of the operator method is its potential in brute-force applications, where unit vectors are not easily obtained. However, it is important to note that the methods are complementary.

Problems

1 We want to look at the transformations of the spherical harmonics using the step operators $\hat{L}_{\pm 1}$.
  a) Determine the step operators $\hat{L}_{\pm 1}$.
  b) Determine the spherical harmonic $Y_{11}(\theta,\varphi)$ by applying the step-up operator $\hat{L}_1$. Normalize the spherical harmonic.
  c) Find $Y_{10}(\theta,\varphi)$ from $Y_{11}(\theta,\varphi)$ using the step-down operator.

2

Spacetime

Introduction

Chapter 1 described the motion in free space. Since free space has translational or rotational symmetry, there are conserved quantities, namely momentum and angular momentum, respectively. The vectors corresponding to these quantities form a dual space. In order to describe an extended nonrigid quantum particle, a function is needed. We saw that function spaces could be derived that are characterized by the conserved quantities in the dual space. For free space, the dual space is entirely static. Classically, this is comparable to, for example Newton's first law that states that, in the absence of a force, an object moves with a constant momentum.

Chapter 1's focus was on the motion of the real space vector in free space, and the quantities in its dual space (momentum or angular momentum) were conserved quantities. This chapter looks into how the quantities in the dual space can be modified by the presence of interactions. There are obviously different ways to approach this. Often, the starting point is the nonrelativistic limit, where the potential in the Schrödinger equation breaks the conservation of momentum. From a geometric point of view, this is somewhat less satisfactory, and we start from the relativistic limit. In the Schrödinger equation, everything is essentially reduced to a scalar. We therefore do not see how the direction of the momentum changes, only its amplitude (comparable to using Newton's equations of motion versus conservation of energy in a classical framework). Additionally, we are primarily concerned with the electromagnetic fields. Since photons move at the speed of light, they are always relativistic particles. Therefore, the relativistic results will look very elegant, and things get messy when deriving the nonrelativistic limit.

The outline of this chapter is as follows.

2.1 For spacetime, we need to introduce an additional dimension, namely time. Since the norm of the time component differs from the spatial ones, the geometry changes, leading to the Minkowski space expressed in four-vectors.

2.2 Unlike space, where the norm of the momentum can assume any value, the norm of the four-momentum is fixed and related to the mass of the particle. This leads to the Dirac equation. While space is dominated by trigonometric functions, spacetime is determined by hyperbolic functions. Rotation in spacetime leads to the well-known Lorentz transformations and Einstein's mass-energy equivalence formula.

2.3 For spacetime in the absence of an electromagnetic field, the four-momentum is a good quantum number, and the associated function space can be readily obtained. Just like free space, it is written in terms of complex exponentials.

2.4 In a fashion similar to spinors for rotation that correspond to the square root of the rotation axis, spacetime spinors can be obtained that correspond to the effective square root of the four-momentum which is the axis for hyperbolic rotations in spacetime.

2.5 In physics, the geometric algebra approach for spacetime is usually incorporated with 4×4 matrices, known as γ matrices. These approaches are effectively the same. The γ matrices are the spacetime equivalent of the Pauli matrices in space.

2.6 Using the ideas of gauge invariance, the four-potential related to the electromagnetic radiation is included. The change in the four-potential in spacetime leads to the introduction of the electromagnetic field tensor, which includes the electric and magnetic fields. The field tensor describes how a particle responds to the electromagnetic interaction. When only considering the vector aspects, this leads to the Lorentz equation.

2.7 This section looks at the creation of electromagnetic fields by charged particles. The underlying physics is similar to that of a driven harmonic oscillator. Rewriting the results in terms of electric and magnetic fields leads to Maxwell's equations.

2.8 The four-momentum has a scalar and a vector component. The nonrelativistic limit is derived by going to a reference frame that removes the vector component. The remaining scalar term leads to the Schrödinger equation.

2.9 The interaction between the particles and the electromagnetic field is described in the nonrelativistic limit. It is shown that the removal of the vector term of the four-momentum is not entirely successful, and that there are some vector terms remaining that correspond to zero point motion. These are generally interpreted as spin.

2.10 An additional remaining vector term from the four-momentum is related to the electric field. Since this term is related to the motion in the potential, it connects the vector term to the motion in real space, leading to the spin–orbit interaction. An additional potential term, the Darwin interaction, is also discussed.

2.11 The spin–orbit interaction is solved for a system with rotational symmetry. This is an example how different vector quantities can be coupled with each other. This anticipates many-particle systems where the vector aspects of different particles are coupled together.

2.12 This section looks in more detail at the Schrödinger equation and also reframes it in a different way, leading to the Heisenberg equation. While the former is more suited to derive function spaces for a particular system, the latter is more convenient when a system is excited.

2.13 The electromagnetic interaction never changes the nature of the particle, *i.e.* an electron remains an electron, and a proton remains a proton. This is no longer the case when considering different fundamental forces. Here, the mixing of different particle characters by the electroweak force is studied. So, while the electromagnetic force is essentially a scalar when it comes to the particle character, the weak interaction adds a vector component.

2.1 The Four-Vector

Introduction. – In Chapter 1, the focus was mainly on free motion, *i.e.* the space looks identical when moving through it. This could be linear motion, a particle moving in vacuum, but it can also be rotational motion. Up to this point, it has been assumed that the motion could be described by a certain physical quantity, for example the wavevector $\mathbf{k}$ for translation or the angular momentum $\mathbf{L}$ for rotational motion. These quantities are constants of motion. The functions for describing the free space could be found by solving eigenvalue equations. This gave the functions spaces of complex exponentials and spherical harmonics.

However, in physics, the values of the momentum and angular momentum are tied to the translation in another dimension, namely time. In Chapter 1, we found several dual spaces. The dual space for translation is momentum, and the dual space for rotation is angular momentum. The dual space for time is frequency/energy. The spatial and time degrees of freedom will now be tied together in spacetime. We will see that time is different from the spatial coordinates: it has a different metric. The dual quantities, momentum and energy, will be combined in a four-momentum. While the norm of the momentum of a particle can assume any real positive value, the norm of the four-momentum is constant. This is the basis of the theory of special relativity.

Let us start with the simplest example where the norm of the wavevector is directly tied to the angular frequency ω. This is applicable to the propagation of light particles or photons. The relation is given by

$$\mathbf{k} = \frac{\omega}{c} \;\Rightarrow\; \frac{\omega}{c} - \mathbf{k} = 0. \tag{2.1}$$

The proportionality constant c is the speed of light. Although Eq. (2.1) combines a vector and a scalar, it can be solved by squaring it, giving the norm of the four-momentum k'

$$\mathsf{k}'^2 = \left(\frac{\omega}{c}\right)^2 - \mathbf{k}^2 = \left(\frac{\omega}{c}\right)^2 - k^2 = 0 \;\Rightarrow\; k = \pm\frac{\omega}{c}. \tag{2.2}$$

Four-vectors, such as the four-dimensional wavevector k', are indicated by sans-serif letters. The prime indicates that we are working in the Minkowski basis, as will be explained below. Two solutions are obtained, corresponding to the two directions of motion.

Since the frequency (unit is 1/second) is the reciprocal of time (unit is seconds), we prefer to associate it with a new dimension. Therefore, instead of having a wavevector $\mathbf{k}$ with a norm given by $k^2 = (\omega/c)^2$, we now have a four-wavevector k' with a norm $\mathsf{k}'^2 = 0$, see Eq. (2.2). While the norm of k' is zero for massless photons, it is finite for particles with a mass. We now want to write down the four-wavevector. There are a variety of ways that this can be done, which can lead to confusion. Here, the notation is followed that most naturally connects to hyperbolic transformations. The four-wavevector can be written as

$$\mathsf{k}' = \frac{\omega}{c}\mathbf{e}'_0 + \mathbf{k}'. \tag{2.3}$$

The wavevector $\mathbf{k}'$ is given by

$$\mathbf{k}' = k_x\mathbf{e}'_x + k_y\mathbf{e}'_y + k_z\mathbf{e}'_z. \tag{2.4}$$

The primes are needed since the norm of $\mathbf{k}'$ appears with a negative sign in Eq. (2.2). This can be achieved by taking the norm of the unit vectors as

$$\mathbf{e}_0'^2 = 1 \quad\text{and}\quad \mathbf{e}_i'^2 = -1, \tag{2.5}$$

with $i = 1, 2, 3 = x, y, z$, as was done for Cartesian unit vectors, we assume that the unit vectors anticommute

$$\mathbf{e}'_{\mu'}\mathbf{e}'_{\mu} = -\mathbf{e}'_{\mu}\mathbf{e}'_{\mu'}, \tag{2.6}$$

for $\mu, \mu' = 0, 1, 2, 3 = 0, x, y, z$. In that case, the bivectors cancel if a four-vector is multiplied with itself. The norm in Eq. (2.2) then becomes

$$\mathbf{k}'^2 = \left(\frac{\omega}{c}\right)^2 \mathbf{e}_0'^2 + k_x^2 \mathbf{e}_x'^2 + k_y^2 \mathbf{e}_y'^2 + k_z^2 \mathbf{e}_z'^2 = \left(\frac{\omega}{c}\right)^2 - \mathbf{k}^2 = 0. \tag{2.7}$$

This equation can also be written in a four-vector format

$$\mathbf{k}' = 0. \tag{2.8}$$

The four-momentum is obtained by multiplying the four-wavevector by Planck's constant $\hbar$,

$$\mathbf{p}' = \hbar\mathbf{k}' = \frac{E}{c}\mathbf{e}_0' + \mathbf{p}'. \tag{2.9}$$

The connection between energy and momentum is given by Planck's and de Broglie's relations

$$E = \hbar\omega \quad \text{and} \quad \mathbf{p}' = \hbar\mathbf{k}'. \tag{2.10}$$

Written like this, Planck's constant is only a conversion factor. Since there is no physical difference between the four-wavevector and the four-momentum, one often takes natural units $\hbar \equiv 1$.

The time unit vector $\mathbf{e}_0'$ appears to be similar to the unit vectors in Cartesian space $\mathbf{e}_i$, with $i = x, y, z$, from Chapter 1. Let us therefore assume that the anticommutation relation for Cartesian unit vectors, see Eq. (1.101), also applies to $\mathbf{e}_0'$, *i.e.*

$$\mathbf{e}_0'\mathbf{e}_i = -\mathbf{e}_i\mathbf{e}_0'. \tag{2.11}$$

The primed Cartesian vectors can now by constructed from $\mathbf{e}_i$ and $\mathbf{e}_0'$

$$\mathbf{e}_i' = \mathbf{e}_i\mathbf{e}_0' \quad \text{with} \quad \mathbf{e}_i'^2 = \mathbf{e}_i\mathbf{e}_0'\mathbf{e}_i\mathbf{e}_0' = -\mathbf{e}_i\mathbf{e}_i\mathbf{e}_0'\mathbf{e}_0' = -1, \tag{2.12}$$

which indeed gives the desired norm of the primed unit vectors. The anticommutation relation in Eq. (2.13) still holds for the primed unit vectors

$$\mathbf{e}_0'\mathbf{e}_i' = \mathbf{e}_0'\mathbf{e}_i\mathbf{e}_0' = -\mathbf{e}_i\mathbf{e}_0'\mathbf{e}_0' = -\mathbf{e}_i'\mathbf{e}_0'. \tag{2.13}$$

Additionally, anticommutation of the spatial Minkowski vectors is also consistent with Eqs. (2.11) and (2.12),

$$\mathbf{e}_j'\mathbf{e}_i' = \mathbf{e}_j\mathbf{e}_0'\mathbf{e}_i\mathbf{e}_0' = \mathbf{e}_i\mathbf{e}_j\mathbf{e}_0'\mathbf{e}_0' = -\mathbf{e}_i\mathbf{e}_0'\mathbf{e}_j\mathbf{e}_0' = -\mathbf{e}_i'\mathbf{e}_j', \tag{2.14}$$

using $\mathbf{e}_j\mathbf{e}_i = -\mathbf{e}_i\mathbf{e}_j$.

Let us generalize the notation by writing the unit vectors as $\mathbf{e}_\mu'$, with $\mu = 0, 1, 2, 3 = 0, x, y, z$, where 0 is always taken as the index for the time component. The anticommutative properties are summarized as follows:

$$\{\mathbf{e}_\mu', \mathbf{e}_{\mu'}'\} = 2g_{\mu\mu'}, \tag{2.15}$$

using the anticommutator from Eq. (1.102); $g_{\mu\mu'}$ are elements in the Minkowski metric

$$g = \begin{pmatrix} 1 & 0 & 0 & 0 \\ 0 & -1 & 0 & 0 \\ 0 & 0 & -1 & 0 \\ 0 & 0 & 0 & -1 \end{pmatrix}. \tag{2.16}$$

Equation (2.15) defines the properties of spacetime in the same fashion as Eq. (1.101) defines the properties of space.

We take the unit vectors in time $\mathbf{e}'_0$ and the unprimed Cartesian unit vectors $\mathbf{e}_i$, with $i = x, y, z$, to be real vectors, since their norm is 1. The basis $\mathbf{e}'_\mu$ is also known as the covariant basis. The conjugate is known as the contravariant basis $\mathbf{e}'^{\mu} = \mathbf{e}'^{*}_{\mu}$. This does not affect the time component since $\mathbf{e}'^{*}_0 = \mathbf{e}'_0$ for a real vector. However, for the primed spatial unit vectors, one finds a sign change,

$$\mathbf{e}'^{i} = \mathbf{e}'^{*}_{i} = (\mathbf{e}_i \mathbf{e}'_0)^* = \mathbf{e}'_0 \mathbf{e}_i = -\mathbf{e}_i \mathbf{e}'_0 = -\mathbf{e}'_i. \tag{2.17}$$

This can be summarized with the metric tensor

$$\mathbf{e}'^{\mu} = \mathbf{e}'^{*}_{\mu} = g_{\mu\mu} \mathbf{e}'_{\mu}, \tag{2.18}$$

where $\mathbf{e}'^{\mu}$ is known as the contravariant basis. This is the dual basis in spacetime, giving the difference between real and momentum space unit vectors. It is more complicated than Cartesian coordinates, where the two spaces are equivalent. However, it is simpler than the changes observed in a nonorthogonal system, see Section 1.12.

As in three dimensions, multivectors can be created by multiplying the unit vectors. In total, there are $2^4 = 16$ multivectors for the spacetime:

multivector	number	name
1	1	scalar
$\mathbf{e}'_\mu = \mathbf{e}'_0, \mathbf{e}'_i$	4	vector
$\boldsymbol{i}'_{\mu\mu'} = \mathbf{e}'_\mu \mathbf{e}'_{\mu'} = \mathbf{e}'_i \mathbf{e}'_0, \mathbf{e}'_i \mathbf{e}'_j = \mathbf{e}_i, \mathbf{e}_j \mathbf{e}_i$	6	bivector
$\underline{\mathbf{e}}'_\mu = \boldsymbol{i} \mathbf{e}'_\mu$	4	trivector, dual vectors
$\boldsymbol{i} = \mathbf{e}'_0 \mathbf{e}'_x \mathbf{e}'_y \mathbf{e}'_z = \mathbf{e}_x \mathbf{e}_y \mathbf{e}_z = \mathbf{e}_0 \mathbf{e}_x \mathbf{e}_y \mathbf{e}_z$	1	pseudoscalar

(2.19)

with $\mu = 0, 1, 2, 3 = 0, x, y, z$ and $i = 1, 2, 3 = x, y, z$. The unprimed time unit vector is $\mathbf{e}_0 = \mathbf{e}'_0 \mathbf{e}'_0 = 1$ and therefore generally omitted. There are $4 \times 3/2 = 6$ unique bivectors. Depending on the number of unit vectors $\mathbf{e}'_\mu$ that appear, the grade goes from 0 to 4. The bivectors will again be important in transforming vectors. There are three purely spatial bivectors that we already encountered in three dimensions,

$$\mathbf{e}'_i \mathbf{e}'_j = \mathbf{e}_i \mathbf{e}'_0 \mathbf{e}_j \mathbf{e}'_0 = -\mathbf{e}_i \mathbf{e}_j \mathbf{e}'_0 \mathbf{e}'_0 = -\mathbf{e}_i \mathbf{e}_j = \mathbf{e}_j \mathbf{e}_i. \tag{2.20}$$

These bivectors were crucial for rotations. Note that the order changes for the unprimed unit vectors. Additionally, there are three spacetime bivectors

$$\mathbf{e}'_i \mathbf{e}'_0 = \mathbf{e}_i. \tag{2.21}$$

For the unprimed unit vectors, the bivector becomes a vector. As we will see, these are rotations in spacetime, also known as Lorentz boosts. In many physics applications, the six bivectors are split into two groups of three that are subsequently described in terms of vectors, such as the magnetic and electric field for rotations and Lorentz boosts, respectively, in classical electromagnetism.

A four-wavevector can be written as

$$\mathsf{k}' = \sum_{\mu=0}^{3} \mathbf{e}'_\mu k^\mu, \tag{2.22}$$

where $k^\mu = \omega/c, k_x, k_y, k_z$. A vector in the covariant basis can also be written in the contravariant basis

$$\mathbf{k}' = \sum_{\mu=0}^{3} \mathbf{e}'_\mu k^\mu = \sum_{\mu=0}^{3} g_{\mu\mu} g_{\mu\mu} \mathbf{e}'_\mu k^\mu = \sum_{\mu=0}^{3} \mathbf{e}'^\mu k_\mu, \tag{2.23}$$

using that $g_{\mu\mu} g_{\mu\mu} = 1$. The components are now

$$k_\mu = g_{\mu\mu} k^\mu = \frac{\omega}{c}, -k_x, -k_y, -k_z. \tag{2.24}$$

Using the metric the inner product between two four-vectors $\mathbf{a}'$ and $\mathbf{b}'$ is defined as

$$\begin{aligned}\mathbf{a}' \cdot \mathbf{b}' &= \sum_{\mu=0}^{3} a_\mu b^\mu = \sum_{\mu=0}^{3} g_{\mu\mu} a_\mu b_\mu \\ &= a_0 b_0 - a_x b_x - a_y b_y - a_z b_z = a_0 b_0 - \mathbf{a} \cdot \mathbf{b}.\end{aligned} \tag{2.25}$$

Thinking in terms of covariant and contravariant bases is often confusing. The metric is in the unit vectors, and the removal of the unit vectors from the definition of the inner product in Eq. (2.25) requires the use of different coefficients. Another approach to finding the inner product is to simply multiply two four-vectors, let the unit vectors take care of the metric, and then retain only the scalar terms. Following Eq. (2.7), we can write

$$\begin{aligned}\mathbf{a}' \cdot \mathbf{b}' = \langle \mathbf{a}'\mathbf{b}' \rangle_0 &= \mathbf{e}'_0 \mathbf{e}'_0 a^0 b^0 + \mathbf{e}'_x \mathbf{e}'_x a^x b^x + \mathbf{e}'_y \mathbf{e}'_y a^y b^y + \mathbf{e}'_z \mathbf{e}'_z a^z b^z \\ &= a^0 b^0 - a^x b^x - a^y b^y - a^z b^z,\end{aligned} \tag{2.26}$$

where $\langle \cdots \rangle_0$ indicates that only the terms of rank 0, *i.e.* the scalar terms arising from the products $\mathbf{e}'_\mu \mathbf{e}'_\mu$, are retained.

Conceptual summary. – In this section, we looked at the unit vectors for spacetime. The unit vectors are still anticommuting, but their norm has changed. The norm of the spatial vectors has changed from $\mathbf{e}_i^2 = 1$ to $\mathbf{e}_i'^2 = -1$, with $i = x, y, z$. The spatial unit vectors are related via $\mathbf{e}'_i = \mathbf{e}_i \mathbf{e}'_0$, where $\mathbf{e}'_0$ with $\mathbf{e}_0'^2 = 1$ is the unit vector in the time direction. Part of this is a choice. Keeping the norm of the spatial coordinates equal to 1 is equally valid as long as the norm of the unit vector in the time direction is taken negative. The opposite sign is important, since, for photons, we want the relation between the frequency and the wavenumber to be $(\omega/c)^2 = k^2$ since both ω and k are taken as real quantities. The opposite sign of the norm of the time and spatial unit vectors leads to the Minkowski metric. The reason for preferring the former convention is that it connects better to the hyperbolic functions for the coordinates.

Problems

1 What is the commutation between a unit four-vector $\mathbf{e}'_\mu$ and $\mathbf{e}_\mu$ with the pseudoscalar $\boldsymbol{i}$ in spacetime?

2 What is the product of two four-vectors $\mathbf{p}'_1 = E_1 \mathbf{e}'_0 + p_{1x} \mathbf{e}'_x + p_{1y} \mathbf{e}'_y + p_{1z} \mathbf{e}'_z$ and $\mathbf{p}'_2 = E_2 \mathbf{e}'_0 + p_{2x} \mathbf{e}'_x + p_{2y} \mathbf{e}'_y + p_{2z} \mathbf{e}'_z$ (taking $c \equiv 1$ for brevity)? What happens if $\mathbf{p}'_2 = \mathbf{p}'_1$?

3 Show that there are no multivectors of grade 5.

2.2 Four-Momentum for Particles

Introduction. – In Section 2.1, the four-wavevector was introduced. The dispersion relation for photons is given by $\mathsf{k}' = 0$, which gives $\omega = \pm ck$. The norm of the four-wavevector for photons is zero. In this section, we see that for particles the four-momentum is given by $\mathsf{p}' = mc$. Therefore, the norm of the four-momentum is related to the mass m of the particle and the speed of light c. This relatively simple looking equation is the basis of many fundamental aspects of the theory of special relativity: Einstein's famous mass-energy equivalence, relativistic mass, and Lorentz boosts and transformations.

In Section 2.1, we looked at four-wavevectors of length zero defined by the equation $\mathsf{k}' = 0$. These wavevectors are suitable for describing massless particles, for example photons. However, particles with mass m are described by four-momenta with a finite norm

$$\mathsf{p}' = mc \text{ or } \mathsf{k}' = \frac{mc}{\hbar}, \tag{2.27}$$

where c is the speed of light. The mass is effectively the norm of the four-vector. Note that unlike the norm of the momentum $\mathbf{p}$, the norm of the four-momentum p' is constant. This is a fundamental property of the theory of special relativity. The above expression is essentially Dirac's equation as we will see. Using Eq. (2.9), the four-momentum can be expressed in components to give

$$\mathsf{p}' = \frac{E}{c}\mathbf{e}_0' + \mathbf{p}' = mc \quad \Rightarrow \quad \hat{\mathsf{p}}' = \frac{E}{mc^2}\mathbf{e}_0' + \frac{\mathbf{p}'}{mc} = 1, \tag{2.28}$$

where $\hat{\mathsf{p}}'$ is the unit four-momentum. The norm of the four-momentum can be written as

$$\mathsf{p}'^2c^2 = E^2 - \mathbf{p}^2c^2 = m^2c^4, \tag{2.29}$$

which is Einstein's energy-momentum relation or the relativistic dispersion relation.

The norm of the unit four-momentum $\hat{\mathsf{p}}'$ is equal to

$$\hat{\mathsf{p}}'^2 = \left(\frac{E}{mc^2}\right)^2 - \left(\frac{\mathbf{p}'}{mc}\right)^2 = 1. \tag{2.30}$$

This equation can be written as a Pythagorean identity $a^2 - b^2 = c^2$ but for hyperbolic functions (using that $\cosh^2\alpha - \sinh^2\alpha = 1$) when we take

$$\frac{E}{c} = mc\cosh\alpha \text{ and } \mathbf{p}' = mc\sinh\alpha\,\hat{\mathbf{p}}'. \tag{2.31}$$

The hyperbolic angle α is also known as the rapidity. The four-momentum is expressed in hyperbolic functions as

$$\mathsf{p}' = mc\cosh\alpha\,\mathbf{e}_0' + mc\sinh\alpha\,\hat{\mathbf{p}}'. \tag{2.32}$$

So far, the focus was mainly on the four-momentum. The reason for starting here is that the four-momentum (or, in other words, the speed of light) is the conserved quantity in spacetime. Let us now consider the consequences in the spacetime coordinates t and $\mathbf{r}$. A vector in spacetime can be written as

$$\mathsf{r}' = ct\mathbf{e}_0' + \mathbf{r}' = \mathsf{r}\cosh\alpha\,\mathbf{e}_0' + \mathsf{r}\sinh\alpha\,\hat{\mathbf{r}}', \tag{2.33}$$

where the speed of light c ensures the proper dimension. The norm is given by

$$\mathsf{r}'^2 = (ct)^2 - r^2 = \mathsf{r}^2\cosh^2\alpha - \mathsf{r}^2\sinh^2\alpha = \mathsf{r}^2. \tag{2.34}$$

Note that the norm of the four-vector is not necessarily constant. Let us take a particle moving the x-direction with a velocity v, so that $r = x = vt$. This gives

$$\mathbf{r}' = ct\mathbf{e}_0' + x\mathbf{e}_x' = \mathsf{r}\cosh\alpha\ \mathbf{e}_0' + \mathsf{r}\sinh\alpha\ \mathbf{e}_x'. \tag{2.35}$$

For a constant velocity, the hyperbolic angle α for $\mathbf{r}'$ is the same as for $\mathbf{p}'$ in Eq. (2.32). The hyperbolic angle α can be determined from

$$\beta \equiv \tanh\alpha = \frac{x}{ct} = \frac{vt}{ct} = \frac{v}{c}. \tag{2.36}$$

The other hyperbolic functions can be expressed in terms of β. The hyperbolic cosine is

$$\gamma = \cosh\alpha = \frac{1}{\sqrt{1-\tanh^2\alpha}} = \frac{1}{\sqrt{1-\beta^2}} = \frac{1}{\sqrt{1-\left(\frac{v}{c}\right)^2}}. \tag{2.37}$$

The factor γ is also known as the Lorentz factor. The hyperbolic sine can then be written as

$$\sinh\alpha = \cosh\alpha\tanh\alpha = \gamma\beta. \tag{2.38}$$

For a constant velocity, the hyperbolic angle α is equal for the four-position and four-momentum. The energy can then be expressed as

$$\gamma = \cosh\alpha = \frac{E}{mc^2} \quad \text{or} \quad E = \gamma mc^2 = \frac{mc^2}{\sqrt{1-\left(\frac{v}{c}\right)^2}}, \tag{2.39}$$

which is Einstein's mass-energy equivalence formula. The hyperbolic tangent can also be obtained from the four-momentum in Eqs. (2.32)

$$\beta = \tanh\alpha = \frac{pc}{E} \quad \Rightarrow \quad p = \beta\frac{E}{c} = \frac{v}{c}\frac{E}{c} = \gamma mv, \tag{2.40}$$

using Eq. (2.39). While the velocity cannot exceed the speed of light, the relativistic momentum can approach infinity.

The results can also be expressed in a hyperbolic rotation

$$R_\alpha = e^{\hat{\mathbf{p}}\alpha} = \sum_{n=0}^{\infty}\frac{(\hat{\mathbf{p}}\alpha)^n}{n!}. \tag{2.41}$$

This rotation is known as a Lorentz boost. Rotations in real space were described by a bivector $\mathbf{e}_i\mathbf{e}_j$, with $i \neq j$, which is an imaginary unit. The argument in the hyperbolic rotation can also be expressed as a bivector rotating from the energy to the momentum axis (or time to spatial axis). This bivector is $\hat{\mathbf{p}}'\mathbf{e}_0'$. However, this can be rewritten as $\hat{\mathbf{p}}\mathbf{e}_0'\mathbf{e}_0' = \hat{\mathbf{p}}$, giving the unit vector in the momentum direction. Using that for even and odd powers, $\hat{\mathbf{p}}^{2m} = 1$ and $\hat{\mathbf{p}}^{2m+1} = \hat{\mathbf{p}}$, respectively, the series splits into two, giving

$$R_\alpha = \cosh\alpha + \hat{\mathbf{p}}\sinh\alpha, \tag{2.42}$$

using the series expansion for the hyperbolic functions

$$\cosh\alpha = \sum_{m=0}^{\infty}\frac{\alpha^{2m}}{(2m)!} \quad \text{and} \quad \sinh\alpha = \sum_{m=0}^{\infty}\frac{\alpha^{2m+1}}{(2m+1)!}. \tag{2.43}$$

Applying the hyperbolic rotation to the time unit vector gives

$$\begin{aligned} \mathsf{p}' &= mc\hat{\mathbf{p}}' = mcR_\alpha \mathbf{e}_0' = mce^{\alpha\hat{\mathbf{p}}}\,\mathbf{e}_0' \\ &= mc(\cosh\alpha\;\mathbf{e}_0' + \sinh\alpha\;\hat{\mathbf{p}}') = \frac{E}{c}\mathbf{e}_0' + \mathbf{p}'. \end{aligned} \tag{2.44}$$

Let us now give the particle moving in the x-direction from Eq. (2.33) an additional Lorentz boost in the same direction,

$$R_{\alpha'}\mathsf{r}' = e^{\hat{\mathbf{p}}\alpha'}(\mathsf{r}e^{\hat{\mathbf{p}}\alpha}\mathbf{e}_0') = \mathsf{r}e^{\hat{\mathbf{p}}(\alpha+\alpha')}\mathbf{e}_0' = R_{\alpha+\alpha'}\mathsf{r}\mathbf{e}_0', \tag{2.45}$$

where the original rapidity has been taken α, and $\hat{\mathbf{p}} = \mathbf{e}_x$. The final result can be expressed in terms of a hyperbolic rotation over an angle $\alpha + \alpha'$ starting from a particle at rest, *i.e.* in the time direction. The Lorentz boost is usually written in a different way. Expressing the original four-vector as $\mathsf{r}' = ct\mathbf{e}_0' + x\mathbf{e}_x'$ gives for the Lorentz boost

$$\begin{aligned} R_\alpha \mathsf{r}' &= e^{\mathbf{e}_x\alpha}\mathsf{r}' = (\cosh\alpha + \mathbf{e}_x'\mathbf{e}_0'\sinh\alpha)(ct\mathbf{e}_0' + x\mathbf{e}_x') \\ &= (\cosh\alpha\;ct + \sinh\alpha\;x)\mathbf{e}_0' + (\cosh\alpha\;x + \sinh\alpha\;ct)\;\mathbf{e}_x' = ct'\mathbf{e}_0' + x'\mathbf{e}_x'. \end{aligned} \tag{2.46}$$

Using the expressions for the hyperbolic functions in Eqs. (2.37) and (2.38) gives the conventional expressions for the coordinates x' and t' after an (inverse) Lorentz boost,

$$x' = \gamma(x + \beta ct) = \gamma(x + vt), \tag{2.47}$$

$$t' = \frac{\gamma}{c}(ct + \beta x) = \gamma\left(t + \frac{vx}{c^2}\right). \tag{2.48}$$

Unfortunately, in this notation, the idea of a hyperbolic rotation is rather lost.

We end this section by pointing out that there is an alternative way to look at the energy-momentum relation in Eq. (2.29). Rearranging the term gives

$$m^2c^4 + \mathbf{p}^2c^2 = E^2. \tag{2.49}$$

Obviously, this is entirely equivalent to the other equation. However, this corresponds more closely to a Pythagorean way of thinking of the norm of a vector, where the norm is now given by the energy E. Equation (2.28) can be multiplied from the left by $\mathbf{e}_0'$ to give

$$\frac{E}{c} - \mathbf{p} - mc\mathbf{e}_0' = 0 \;\Rightarrow\; \mathsf{p} = \frac{E}{c}, \tag{2.50}$$

using that $\mathbf{e}_0'\mathbf{e}_i' = \mathbf{e}_0'\mathbf{e}_i\mathbf{e}_0' = -\mathbf{e}_i\mathbf{e}_0'\mathbf{e}_0' = -\mathbf{e}_i$. The four-momentum p in the Dirac notation is

$$\mathsf{p} = mc\mathbf{e}_0' + \mathbf{p}, \tag{2.51}$$

which should be compared with the four-momentum in the Minkowski notation $\mathsf{p}' = (E/c)\mathbf{e}_0' + \mathbf{p}'$. Note that the unit vector associated with the energy becomes the identity after multiplying with $\mathbf{e}_0'$, since $\mathbf{e}_0 = \mathbf{e}_0'\mathbf{e}_0' = 1$. Therefore, the energy E and the rest energy mc^2 have effectively switched places. This follows more closely the approach that Dirac originally followed. In analogy with the Schrödinger equation, the energy was set as the norm. Initially, this approach fits better to a nonrelativistic limit. For small momenta, we obtain

$$E_\mathbf{p} = \sqrt{m^2c^4 + \mathbf{p}^2c^2} = mc^2\left(1 + \frac{\mathbf{p}^2}{m^2c^2}\right)^{\frac{1}{2}} \cong mc^2 + \frac{\mathbf{p}^2}{2m} + \cdots, \tag{2.52}$$

where mc^2 is the rest energy. In the nonrelativistic limit, energy becomes a constant of motion due to time translation symmetry. However, in the relativistic limit, energy is no longer a conserved quantity due to the time dilation in Eq. (2.48). Therefore, the norm of the Dirac four-momentum is not a conserved quantity. Additionally, one of the components of the four-momentum in Eq. (2.51) is a constant, which is also unsatisfactory for a vector. Therefore, the Minkowski metric is generally preferred.

Conceptual summary. – The section started with the assumption that the four-momentum is equal to the quantity mc, *i.e.* $\mathbf{p}' = mc$. Unlike the classical relation $\mathbf{p} = m\mathbf{v}$, the right-hand side is a constant. Therefore, the norm of the four-momentum is constant. Due to the different signs of the norms of the unit vectors, the coefficients of the four-momentum are not expressed in trigonometric functions but in hyperbolic functions. For a particle moving with a constant velocity v, hyperbolic functions are often removed, and the results are expressed in terms of v. This leads to many familiar results in special relativity, such as Einstein's mass-energy equivalence $E = \gamma mc^2$, where m is the rest mass. The Lorentz factor γ is often combined with the mass, giving $E = m_{\text{rel}}c^2$, where m_{rel} is the relativistic mass. Although this expression is popular, the change in energy is not a result of a change in properties of the particle but due to the geometric properties of spacetime. We therefore do not use relativistic mass in the following discussion.

Rewriting the hyperbolic functions in terms of v also produces the Lorentz transformations between different rest frames, see Eq. (2.48). Although the transformations are often introduced in this fashion, these expressions rather obscure the fact that we are dealing with a hyperbolic rotation. Note that the Lorentz boost in Eq. (2.41) can be written as an exponential of a vector. This should be contrasted with rotations in real space that are written as an exponential of a bivector, leading to trigonometric functions.

Problems

1 How do $e^{\hat{\mathbf{p}}\alpha}$ and $\mathbf{e}'_0$ commute?

2 What is the product of two unit four-vectors in the same direction $\hat{\mathbf{p}}'_1 = \cosh\alpha_1\mathbf{e}'_0 + \sinh\alpha_1\hat{\mathbf{p}}'$ and $\hat{\mathbf{p}}'_2 = \cosh\alpha_2\mathbf{e}'_0 + \sinh\alpha_2\hat{\mathbf{p}}'$.

3 How does the wavelength of light change for a moving source with velocity v? Explain the results.

2.3 Function Space for Spacetime

Introduction. – In Chapter 1, we found that the function space for systems with translational symmetry are complex exponentials. This function space is complete, *i.e.* any function can be expressed in terms of complex exponentials. This is generally known as a Fourier transform. We now want to find the function space for spacetime. Since, apart from their norm, the unit vectors for space and time behave similarly, we can expect their function spaces to be comparable as well. In space, the complex exponentials are eigenfunctions of the momentum operator. Likewise, we will see that the function space for spacetime are eigenfunctions of the four-momentum operator.

Following the arguments in Sections 1.10, we would now like to develop a function space that allows each point in spacetime to be independent from each other, *i.e.*

$$\langle \mathbf{r}_1' | \mathbf{r}_2' \rangle = \delta(\mathbf{r}_1' - \mathbf{r}_2'). \tag{2.53}$$

For massless particles, we are looking for functions that return the wavevector when being operated upon by the four-momentum operator,

$$\hat{\mathbf{p}}' \varphi_{\mathbf{k}}(\mathbf{r}') = \hbar \mathbf{k}' \varphi_{\mathbf{k}'}(\mathbf{r}') = 0 \quad \text{or} \quad \hat{\mathbf{k}}' \varphi_{\mathbf{k}'}(\mathbf{r}') = \mathbf{k}' \varphi_{\mathbf{k}'}(\mathbf{r}') = 0, \tag{2.54}$$

where Planck's constant is included in the momentum operator. The hat indicates that these quantities are operators. Note that the result is zero due to Eq. (2.8). The four-momentum operator now represents translational symmetry in both space and time. In three dimensions, the momentum operator is given by $\hat{\mathbf{p}} = -\boldsymbol{i}\hbar\nabla$, *i.e.* the derivative is taken with respect to each of the coordinates. A comparable operator is introduced for the four-momentum

$$\hat{\mathbf{p}}' \equiv \boldsymbol{i}\hbar \triangledown \quad \text{and} \quad \hat{\mathbf{k}}' \equiv \boldsymbol{i} \triangledown. \tag{2.55}$$

First, we need to check the imaginary unit which is needed to build spaces orthogonal to the unit vectors $\mathbf{e}_\mu'$. Taking the highest multivector in four dimensions gives

$$\boldsymbol{i} = \mathbf{e}_0' \mathbf{e}_x' \mathbf{e}_y' \mathbf{e}_z', \quad \text{with} \quad \boldsymbol{i}^2 = -1, \tag{2.56}$$

where the identity on the right can be checked using the relations in Eq. (2.15). This is the same imaginary unit as in three dimensions, since

$$\boldsymbol{i} = \mathbf{e}_0' \mathbf{e}_x \mathbf{e}_0' \mathbf{e}_y \mathbf{e}_0' \mathbf{e}_z \mathbf{e}_0' = \mathbf{e}_0' \mathbf{e}_0' \mathbf{e}_x \mathbf{e}_y \mathbf{e}_z \mathbf{e}_0' \mathbf{e}_0' = \mathbf{e}_x \mathbf{e}_y \mathbf{e}_z. \tag{2.57}$$

The last term on the right-hand side shows the imaginary unit in Dirac unit vectors.

The derivatives are given by $\triangledown$, which is the spacetime equivalent of the nabla operator in Eq. (1.235). Splitting the operator into its components gives

$$\begin{aligned} \triangledown &= \sum_\mu \triangledown^\mu = \sum_\mu \frac{\partial}{\mathbf{e}_\mu' \partial r^\mu} = \sum_\mu g_{\mu\mu} \mathbf{e}_\mu' \mathbf{e}_\mu' \frac{\partial}{\mathbf{e}_\mu' \partial r^\mu} = \\ &= \sum_\mu g_{\mu\mu} \mathbf{e}_\mu' \frac{\partial}{\partial r^\mu} = \sum_\mu \mathbf{e}'^\mu \frac{\partial}{\partial r^\mu} = \mathbf{e}_0' \frac{\partial}{\partial(ct)} - \nabla', \end{aligned} \tag{2.58}$$

with $g_{\mu\mu}\mathbf{e}_\mu'\mathbf{e}_\mu' = 1$, using the metric in Eq. (2.16). The nabla operator ∇' differs from that in Eq. (1.235) in that it is given in the unit vectors $\mathbf{e}_i'$. The components of the momentum operator are therefore

$$\hat{\mathbf{p}}' = \frac{\hat{E}}{c}\mathbf{e}_0' + \hat{\mathbf{p}}' = \boldsymbol{i}\hbar \frac{\partial}{\partial(ct)} \mathbf{e}_0' - \boldsymbol{i}\hbar\nabla', \tag{2.59}$$

or

$$\hat{E} = \boldsymbol{i}\hbar \frac{\partial}{\partial t} \quad \text{and} \quad \hat{\mathbf{p}}' = -\boldsymbol{i}\hbar\nabla' \quad (\text{or } \hat{\mathbf{p}} = -\boldsymbol{i}\hbar\nabla), \tag{2.60}$$

where the momentum operator is the same as in three dimensions. These are the well-known quantum-mechanical operators for energy and momentum.

The plane waves in Section 1.10 can be straightforwardly extended to include the time/energy coordinate:

$$\langle \mathbf{r}' | \mathbf{k}' \rangle = \varphi_{\mathbf{k}'}(\mathbf{r}') = \frac{1}{(2\pi)^2} e^{-\boldsymbol{i}\mathbf{k}' \cdot \mathbf{r}'}, \tag{2.61}$$

where there is a $1/\sqrt{2\pi}$ factor for each dimension. These plane waves satisfy Eq. (2.54), since

$$\hat{\mathbf{p}}'\varphi_{\mathbf{k}}(\mathbf{r}') = i\hbar\nabla\varphi_{\mathbf{k}}(\mathbf{r}') = \hbar\mathbf{k}\varphi_{\mathbf{k}}(\mathbf{r}') = 0, \tag{2.62}$$

where the norm of the four-momentum for massless particles is zero, *i.e.* $\mathbf{k} = 0$. There appears to be a sign change in the exponent in Eq. (2.64). However, this follows naturally when looking at the inner product in spacetime from Eq. (2.25)

$$\mathbf{k}' \cdot \mathbf{r}' = \omega t - \mathbf{k} \cdot \mathbf{r}, \tag{2.63}$$

so effectively the sign in front of $\mathbf{k} \cdot \mathbf{r}$ is the same as in Eq. (1.220). This gives for the plane wave from Eq. (2.61)

$$\varphi_{\mathbf{k}'}(\mathbf{r}') = \frac{1}{(2\pi)^2} e^{i(\mathbf{k}\cdot\mathbf{r}-\omega t)} = \frac{1}{(2\pi)^2} e^{\frac{i}{\hbar}(\mathbf{p}\cdot\mathbf{r}-Et)}. \tag{2.64}$$

The opposite sign between the two terms in the exponent imposes causality, *i.e.* the wave is moving in the direction of the momentum when time increases. Instead of Eq. (2.54), the plane waves can also be obtained from the norm of the four-wavevector in Eq. (2.2). The functional equivalent of $\hat{\mathbf{k}}'^2 = 0$ is given by

$$\hat{\mathbf{k}}'^2 \varphi_{\mathbf{k}'}(\mathbf{r}') = -\nabla'^2 \varphi_{\mathbf{k}'}(\mathbf{r}') = \nabla^2 \varphi_{\mathbf{k}'} - \frac{1}{c^2}\frac{\partial^2 \varphi_{\mathbf{k}'}(\mathbf{r}')}{\partial t^2} = 0, \tag{2.65}$$

which is the usual wave equation.

Example 2.1 ***Gaussian Wave Packet***

While the function space for free space in Section 1.10 allowed us to describe any function in space, we are now able to add a time component as well. Let us consider the motion of a Gaussian wave packet in one dimension. At $t = 0$,

$$\psi(x, 0) = \langle x, 0|\psi\rangle = \phi(x)e^{ik_0 x}, \quad \text{with } \phi(x) = \frac{1}{\sqrt{\sigma\sqrt{2\pi}}} e^{-\frac{x^2}{4\sigma^2}}, \tag{2.66}$$

where we separate the spatial and time variables as opposed to using four-vectors. The wave packet is defined such that $\langle x\rangle = 0$ and $\langle x^2\rangle = \sigma^2$ at $t = 0$. This gives $\Delta x = \sqrt{\langle x^2\rangle - \langle x\rangle^2} = \sigma$.

(a) Find the Fourier transform of $\psi(x, 0)$.
(b) Express the wave packet $\psi(x', t')$ at a different time t' and position x' in terms of $\psi(x, 0)$ and running waves (note the primes have nothing to do with Lorentz boosts).
(c) Evaluate the integrals over the running waves in the nonrelativistic limit to obtain $\psi(x', t')$.
(d) The wave packets move with the group velocity $v_g = \partial\omega_{k_x}/\partial k_x$. Derive v_g in the relativistic and nonrelativistic limits.

Solution

(a) The Fourier transform is

$$\psi(k_x) = \frac{1}{\sqrt{\sigma}(2\pi)^{3/4}} \int_{-\infty}^{\infty} dx\, e^{-\frac{x^2}{4\sigma^2}} e^{i(k_0-k_x)x} = \sqrt{\frac{2\sigma}{\sqrt{2\pi}}} e^{-\sigma^2(k_x-k_0)^2}, \tag{2.67}$$

where the integral over the Gaussian is straightforward. The Fourier transform has no ω coefficient, since for free particles $\hbar\omega_k = E_{k_x}$. Therefore, ω is not a free variable. The Fourier transform is a Gaussian with width $1/2\sigma$ around k_0. The expectation value of the momentum is then $\langle p_x \rangle = \hbar k_0$. So, we have $\Delta x \Delta p_x = \Delta x \, \hbar \Delta k_x = \sigma\hbar/(2\sigma) = \hbar/2$.

(b) The idea to understanding the propagation of a wave packet is that, while we lack information on the development in time of $\psi(x', t')$, we do understand how running waves behave since they are the eigenfunctions in free space. Therefore, $\psi(x', t')$ is first projected onto running waves

$$\psi(x', t') = \langle x', t' | \psi \rangle = \int dk_x \, \langle x', t' | k_x \rangle \langle k_x | \psi \rangle, \tag{2.68}$$

with the running waves

$$\varphi_{k_x}(x, t) = \langle x, t | k_x \rangle = \frac{1}{\sqrt{2\pi}} e^{i(k_x x - \omega_{k_x} t)}, \tag{2.69}$$

which are allowed to propagate freely. Note that there is no $1/\sqrt{2\pi}$ for the frequency, as in Eq. (2.64), since it is no longer a free variable. The Fourier transform $\langle k_x | \psi \rangle$ was determined in Eq. (2.67),

$$\psi(x', t') = \langle x', t' | \psi \rangle = \int dk_x \int dx \, \langle x', t' | k_x \rangle \langle k_x | x, 0 \rangle \langle x, 0 | \psi \rangle \tag{2.70}$$

$$\equiv i\hbar \int dx \, G^0(x', t'; x, 0) \psi(x, 0), \tag{2.71}$$

where $\langle x, 0 | \psi \rangle = \psi(x, 0)$ as given in Eq. (2.66). The function that connects the wave-functions at different times and position, with $t' > t$,

$$\begin{aligned} G^0(x', t'; x, t) &= -\frac{i}{\hbar} \int dk_x \, \langle x', t' | k_x \rangle \langle k_x | x, t \rangle \\ &= -\frac{i}{2\pi\hbar} \int dk_x \, e^{ik_x(x'-x) - i\omega_{k_x}(t'-t)}, \end{aligned}$$

is called a propagator or Green's function. The factor $-i/\hbar$ follows convention. Note that the propagation between different points in space and time is entirely described in terms of running waves.

The concept of a propagator is more general. Removing the eigenfunctions via completeness gives

$$G^0(x', t'; x, t) = -\frac{i}{\hbar} \langle x', t' | x, t \rangle \theta(t' - t). \tag{2.72}$$

The step function is crucial for causality, *i.e.* the particle has to arrive at a time t' later than t. The propagator essentially tells you how to get from one point in space and time $|x, t\rangle$ to another $\langle x', t'|$. Written in this way, the expression becomes independent of the system as described by the Hamiltonian. Since we are dealing with free space it becomes natural to insert the completeness of running waves $|k_x\rangle$, but for other systems, a different complete set of functions could be preferable. The superscript 0 indicates that we are dealing with eigenstates. This means at a certain point x, t we project our wave packet onto the eigenstates, these propagate independently in space and time until point x', t' is reached.

(c) Using that $\omega_{k_x} = \hbar k_x^2/(2m)$ in the nonrelativistic limit, the integral can be written in the form

$$\frac{1}{2\pi} \int_{-\infty}^{\infty} dk_x \, e^{-ak_x^2 + bk_x} \quad \text{with} \quad a = i\frac{\hbar}{2m}(t' - t) \quad \text{and} \quad b = i(x' - x).$$

The idea is to transform this into the well-known Gaussian integral $\int_{-\infty}^{\infty} d\kappa\, e^{-\kappa^2} = \sqrt{\pi}$. We can achieve this by introducing a new variable $\kappa = \sqrt{a}k_x - \frac{b}{2\sqrt{a}}$, giving $dk_x = d\kappa/\sqrt{a}$. We can then rewrite the integral as

$$\int_{-\infty}^{\infty} dk_x\, e^{-ak_x^2+bk_x} = e^{\frac{b^2}{4a}} \int_{-\infty}^{\infty} \frac{d\kappa}{\sqrt{a}} e^{-\kappa^2} = \sqrt{\frac{\pi}{a}}\, e^{\frac{b^2}{4a}}. \tag{2.73}$$

This gives a solution for the propagator

$$G^0(x', t'; x, t) = -i\sqrt{\frac{\alpha}{i\pi}} e^{i\alpha(x'-x)^2}, \tag{2.74}$$

where we define $\alpha = m/[2\hbar(t'-t)]$. Without any loss of generality, we can take $t = 0$. Inserting the propagator into Eq. (2.71) and collecting the results gives for the wavefunction

$$\psi(x', t') = \frac{1}{\sqrt{\sigma\sqrt{2\pi}}}\sqrt{\frac{\alpha}{i\pi}} \int dx\, e^{-\frac{\alpha}{i}(1+i\tau')x^2 + \frac{2\alpha}{i}(x'-v_g t')x + i\alpha x'^2}, \tag{2.75}$$

where we define the timescale $T = 2m\sigma^2/\hbar$, giving a reduced time $\tau' = t'/T$ and the constant $\alpha = m/(2\hbar t') = 1/(4\sigma^2\tau')$. The velocity is defined as $v_g = p_0/m = \hbar k_0/m$. This is known as the group velocity. We can again use Eq. (2.73) but now for the integral over x to obtain

$$\psi(x', t') = \frac{1}{\sqrt{\sigma\sqrt{2\pi}(1+i\tau')}} e^{i\alpha x'^2} e^{-\frac{i\alpha(x'-v_g t')^2}{1+i\tau'}}. \tag{2.76}$$

The probability can then be written as

$$|\psi(x', t')|^2 = \frac{1}{\sqrt{2\pi\sigma^2(1+\tau'^2)}} e^{-\frac{(x'-v_g t')^2}{2\sigma^2(1+\tau'^2)}}. \tag{2.77}$$

Comparison with the probability at $t = 0$ in Eq. (2.66) shows that the wave packet remains Gaussian. However, there are some important differences. First, the maximum of the distribution is no longer at $x = 0$, but at $x' = v_g t'$. Therefore, the particle moves as a function of time with the momentum $p_0 = \hbar k_0$. However, additionally, the wave packet broadens with $\sigma^2(t') = \sigma^2(1 + (t'/T)^2)$, with $T = 2m\sigma^2/\hbar$. Note that the broadening time decreases when the initial wave packet is narrower.

(d) The group velocity is given by

$$v_g = \frac{1}{\hbar}\frac{\partial E_{k_x}}{\partial k_x} = \frac{1}{\hbar}\frac{\partial}{\partial k_x}\left(\sqrt{m^2c^4 + \hbar^2 k_x^2 c^2}\right)$$
$$= \frac{\hbar k_x c}{\sqrt{m^2c^4 + \hbar^2 k_x^2 c^2}} c \cong \begin{cases} \frac{\hbar k_x}{m} & \text{for} \quad \hbar k_x \ll mc, \\ c & \text{for} \quad \hbar k_x \gg mc. \end{cases} \tag{2.78}$$

This reproduces the nonrelativistic result from (c) for small momenta ($\hbar k_x \ll mc$) and approaches the speed of light for $\hbar k_x \gg mc$. The group velocity should be compared with the phase velocity $v_p = \omega_{k_x}/k_x$, which is the velocity of a particular point in a running wave. For relativistic and massless particles, this gives $v_p = c$, which is equal to the group velocity. Therefore, for particles with a linear dispersion, the group velocity is always finite. However, one can still combine finite group velocities in opposite directions, creating a standing wave. For nonrelativistic particles, this gives $v_p = \hbar k_x/2m = \frac{1}{2}v_g$.

Conceptual summary. – In this section, we looked at the function space for free spacetime, *i.e.* in the absence of a potential or any other forces. The solutions are plane waves. This is comparable to what was found for free space, except that it now also includes the time component. In addition to the momentum operator $\hat{\mathbf{p}}' = -i\hbar\nabla'$, we now also find the energy operator $\hat{E} = i\hbar\partial/\partial t$. The difference in sign is related to causality, *i.e.* we want waves with positive frequency and positive wavenumber traveling forward in time. The complex exponentials are eigenfunctions of the four-momentum operator $\hat{\mathbf{p}}' = \hbar\hat{\mathbf{k}}'$. However, they are also eigenfunctions of the norm $\hat{\mathbf{k}}'^2$, which leads to the well-known wave equation.

Problems

1 There are very few wave packets that can be solved analytically. A relatively easy numerical (toy) problem is that of a block wave in reciprocal space

$$f(x,t) = \frac{\Delta k}{2} \sum_{k=-1,1,\Delta k} \cos(kx - \omega_k t), \tag{2.79}$$

where the summation over k goes from -1 to 1 in steps Δk. The dispersion relation is given by $\omega_k = ak + bk^2$. Calculate $f(x,t)$ numerically for $t = 0, 1, 2, 3, 4, 5$ using $a = 5$, $b = 0$ and $\Delta k = 0.2$ and using $(a, b) = (5, 0), (10, 0), (5, 1), (0, 1)$ and $\Delta k = 0.02$.

2.4 Spacetime Spinors

Introduction. – In Section 1.6, the concept of spinors was introduced. We were looking for vectors that describe the counterclockwise and clockwise rotations around an axis given by a unit vector. Classically, this problem is generally described as $\hat{\mathbf{r}}\hat{\mathbf{r}}_\pm = \pm 1$. Its solutions are simply $\hat{\mathbf{r}}_\pm = \pm\hat{\mathbf{r}}$, where the direction of rotation follows using the right-hand rule. This solution was deemed unsatisfactory, since it was not an eigenvalue equation. One might wonder on the insistence of eigenvectors. The reason is that we are generally dealing with multiple operations. If the application of an operation on a vector leads to a scalar, then we can find out the direction of the vector. For the rotation axis, we are effectively performing a projection of $\hat{\mathbf{r}}_\pm$ onto $\hat{\mathbf{r}}$, giving ± 1, which is then related to the direction of rotation. However, the vector itself is lost in the operation. This was solved by taking the "square root" of the rotation axis. The problem can then be written as an eigenvalue equation $\hat{\mathbf{r}}\hat{\mathbf{r}}_{\pm\frac{1}{2}} = \pm\hat{r}_{\pm\frac{1}{2}}$, where $\hat{\mathbf{r}}_{\pm\frac{1}{2}} = \hat{r}_{\pm\frac{1}{2}}\mathbf{e}_z$, taking $\mathbf{e}_z$ as our reference axis. The operation still returns the value ± 1 of the projection but also returns the same vector quantity. This allows us to perform successive operations on a particular vector quantity. In this section, we obtain the spinors for spacetime.

For hyperbolic rotations, there are also two directions. For rotation, our starting point were the vectors $\pm\hat{\mathbf{r}}$. So, it might seem natural to take something such as $\pm\hat{\mathbf{p}}$, the unit vectors parallel and antiparallel to the four-momentum, to obtain the spacetime spinors (the hat indicates the unit vector aspect). However, this does not work. For rotation, we took effectively the square root of $\pm\hat{\mathbf{r}}$. The square root of -1 can be written as a spatial bivector, which is directly related to rotation. However, the bivector related to Lorentz boosts $\hat{\mathbf{p}}'\mathbf{e}_0' = \hat{\mathbf{p}}$ is not an imaginary unit. In fact, we will see that both spinors are related to the same four-momentum.

However, it is straightforward to find one of the spinors by taking the square root of the Lorentz boost from the rest frame

$$\mathbf{p}' = mc\, e^{\hat{\mathbf{p}}\alpha}\mathbf{e}_0' = mc\, e^{\hat{\mathbf{p}}\frac{\alpha}{2}}e^{\hat{\mathbf{p}}\frac{\alpha}{2}}\mathbf{e}_0' = mc\, e^{\hat{\mathbf{p}}\frac{\alpha}{2}}\mathbf{e}_0' e^{-\hat{\mathbf{p}}\frac{\alpha}{2}}, \tag{2.80}$$

where the negative sign in the exponent comes from the anticommutation of $\mathbf{e}_0'$ and $\hat{\mathbf{p}}$. The spacetime spinor $\hat{\mathbf{p}}^+$ is defined as the exponentials arising in the above equation

$$\hat{\mathbf{p}}^+ = e^{\hat{\mathbf{p}}\frac{\alpha}{2}} = \cosh\frac{\alpha}{2} + \hat{\mathbf{p}}\sinh\frac{\alpha}{2}. \tag{2.81}$$

Working with the unit four-momentum $\hat{\mathbf{p}}' = \mathbf{p}/mc$ on the spinor gives

$$\hat{\mathbf{p}}'\hat{\mathbf{p}}^+ = e^{\hat{\mathbf{p}}\frac{\alpha}{2}}\mathbf{e}_0' e^{-\hat{\mathbf{p}}\frac{\alpha}{2}}e^{\hat{\mathbf{p}}\frac{\alpha}{2}} = e^{\hat{\mathbf{p}}\frac{\alpha}{2}}\mathbf{e}_0' = \hat{\mathbf{p}}^+\mathbf{e}_0', \tag{2.82}$$

using Eq. (2.80). Operating with the unit four-momentum $\hat{\mathbf{p}}'$ on the spinor $\hat{\mathbf{p}}^+$ returns the spinor and the reference axis $\mathbf{e}_0'$. The spinor $\hat{\mathbf{p}}^+$ is therefore an eigenvector of $\hat{\mathbf{p}}'$. For rotations, the spinor for clockwise rotations is obtained by multiplying the spinor for counterclockwise rotations with the bivector related to the rotation, see Eq. (1.134). The spacetime bivector in Minkowski unit vectors is given by $\hat{\mathbf{p}}'\mathbf{e}_0' = \hat{\mathbf{p}}\mathbf{e}_0'\mathbf{e}_0' = \hat{\mathbf{p}}$, where $\hat{\mathbf{p}}$ is a unit vector in the Dirac basis. The bivector rotates from the energy axis to the momentum axis. The other spinor can then be written as

$$\hat{\mathbf{p}}^- = \hat{\mathbf{p}}\hat{\mathbf{p}}^+ = \hat{\mathbf{p}}e^{\hat{\mathbf{p}}\frac{\alpha}{2}} = \sinh\frac{\alpha}{2} + \hat{\mathbf{p}}\cosh\frac{\alpha}{2}. \tag{2.83}$$

For those not used to working in the Minkowski metric, the spinors in Eqs. (2.81) and (2.83) might not appear orthogonal. However, multiplying the associated vectors $\hat{\mathbf{p}}^\pm\mathbf{e}_0'$ and retaining only the scalar part gives

$$\begin{aligned}\langle\hat{\mathbf{p}}^+\mathbf{e}_0'\hat{\mathbf{p}}^-\mathbf{e}_0'\rangle_0 &= \langle(\cosh\frac{\alpha}{2}\mathbf{e}_0' + \sinh\frac{\alpha}{2}\hat{\mathbf{p}}')(\sinh\frac{\alpha}{2}\mathbf{e}_0' + \cosh\frac{\alpha}{2}\hat{\mathbf{p}}')\rangle_0\\ &= \cosh\frac{\alpha}{2}\sinh\frac{\alpha}{2} - \sinh\frac{\alpha}{2}\cosh\frac{\alpha}{2} = 0,\end{aligned} \tag{2.84}$$

where $\langle\cdots\rangle_0$ indicates that only terms of rank zero are kept. Therefore, the spinors form an orthogonal basis. The norm of the vector associated with the + spinor is

$$\hat{\mathbf{p}}^+\mathbf{e}_0'\hat{\mathbf{p}}^+\mathbf{e}_0' = e^{\hat{\mathbf{p}}\frac{\alpha}{2}}\mathbf{e}_0' e^{\hat{\mathbf{p}}\frac{\alpha}{2}}\mathbf{e}_0' = e^{\hat{\mathbf{p}}\frac{\alpha}{2}}e^{-\hat{\mathbf{p}}\frac{\alpha}{2}}\mathbf{e}_0'\mathbf{e}_0' = 1, \tag{2.85}$$

using the commutation relations and norms for $\mathbf{e}_0'$ and $\hat{\mathbf{p}}$. The norm of the − spinor, on the other hand,

$$\hat{\mathbf{p}}^-\mathbf{e}_0'\hat{\mathbf{p}}^-\mathbf{e}_0' = \hat{\mathbf{p}}e^{\hat{\mathbf{p}}\frac{\alpha}{2}}\mathbf{e}_0'\hat{\mathbf{p}}e^{\hat{\mathbf{p}}\frac{\alpha}{2}}\mathbf{e}_0' = -\hat{\mathbf{p}}e^{\hat{\mathbf{p}}\frac{\alpha}{2}}\hat{\mathbf{p}}\mathbf{e}_0' e^{\hat{\mathbf{p}}\frac{\alpha}{2}}\mathbf{e}_0' = -\hat{\mathbf{p}}\hat{\mathbf{p}}e^{\hat{\mathbf{p}}\frac{\alpha}{2}}e^{-\hat{\mathbf{p}}\frac{\alpha}{2}}\mathbf{e}_0'\mathbf{e}_0' = -1,$$

is negative. Therefore, for rotation, the spinors form an orthogonal basis in the plane formed by the vectors $\mathbf{e}_i$ and $\mathbf{e}_j$, where $\mathbf{e}_i$ is a reference axis in the plane of rotation, and $\mathbf{e}_j$ is the unit vector along the rotation axis (conventionally $i = z$ and $j = x$). Both spinors are spatial vectors. For hyperbolic rotations, the spinors form an orthogonal basis in the plane defined by the vectors $\mathbf{e}_0'$ and $\hat{\mathbf{p}}$. The + spinor has a time metric, whereas the − spinor has a spatial metric.

The − spinor is also an eigenvector of the four-momentum but with an eigenvalue of −1,

$$\hat{\mathbf{p}}'\hat{\mathbf{p}}^- = e^{\hat{\mathbf{p}}\frac{\alpha}{2}}\mathbf{e}_0' e^{-\hat{\mathbf{p}}\frac{\alpha}{2}}\hat{\mathbf{p}}e^{\frac{\alpha}{2}\hat{\mathbf{p}}} = -\hat{\mathbf{p}}e^{\hat{\mathbf{p}}\frac{\alpha}{2}}\mathbf{e}_0' = -\hat{\mathbf{p}}^-\mathbf{e}_0', \tag{2.86}$$

since $\hat{\mathbf{p}}$ commutes with itself but anticommutes with $\mathbf{e}_0'$. The eigenvalue equations for the two spacetime spinors can be summarized as

$$\hat{\mathbf{p}}'\hat{\mathbf{p}}^\pm = \pm\hat{\mathbf{p}}^\pm\mathbf{e}_0' \quad \text{or} \quad \mathbf{p}'\hat{\mathbf{p}}^\pm = \pm mc\hat{\mathbf{p}}^\pm\mathbf{e}_0'. \tag{2.87}$$

This is the spacetime equivalent of the eigenvalue equation for rotations in real space $\hat{\mathbf{r}}\hat{\mathbf{r}}_{\pm\frac{1}{2}} = \pm\hat{\mathbf{r}}_{\pm\frac{1}{2}}\mathbf{e}_z$, see Eq. (1.130). Note that in spacetime, the reference axis is the unit vector for time, whereas in real space it is the unit vector in the z-direction (the latter is a matter of convention, any direction can be used as a reference axis).

The four-vector associated with this spinor is

$$mc\hat{\mathbf{p}}^-\hat{\mathbf{p}}^-\mathbf{e}_0' = mc\,\hat{\mathbf{p}}e^{\hat{\mathbf{p}}\frac{\alpha}{2}}\hat{\mathbf{p}}e^{\hat{\mathbf{p}}\frac{\alpha}{2}}\mathbf{e}_0' = mc\,\hat{\mathbf{p}}\hat{\mathbf{p}}e^{\hat{\mathbf{p}}\frac{\alpha}{2}}e^{\hat{\mathbf{p}}\frac{\alpha}{2}}\mathbf{e}_0' = mc\,e^{\hat{\mathbf{p}}\alpha}\mathbf{e}_0' = \mathbf{p}', \tag{2.88}$$

since $\hat{\mathbf{p}}$ commutes with itself. This produces the same four-momentum $\mathbf{p}'$.

However, this is not the only way to construct the opposite spinor. In Section 1.12, we saw that a dual set of vectors can be obtained by multiplying with the pseudoscalar $\boldsymbol{i}$. For spacetime, there is also a pseudoscalar, which is an imaginary unit, see Eq. (2.56). However, since we are dealing with a hyperbolic space, we prefer to define a real unit as opposed to an imaginary unit

$$\mathbf{1} = -i\boldsymbol{i} = -i\mathbf{e}_0'\mathbf{e}_x'\mathbf{e}_y'\mathbf{e}_z' \quad \text{with} \quad \mathbf{1}^2 = 1. \tag{2.89}$$

Note that $\mathbf{1}$ is not the identity matrix. The quantity $\mathbf{1}\mathbf{e}_\mu'$ creates a space perpendicular to $\mathbf{e}_\mu'$ for Minkowski space, in the same way that $\boldsymbol{i}\mathbf{e}_i$ creates a space perpendicular to $\mathbf{e}_i$ in real space. Using the results from problem 2.1 in Section 2.1, it is straightforward to show that

$$\mathbf{1}\mathbf{e}_\mu' = -\mathbf{e}_\mu'\mathbf{1} \quad \text{and} \quad \mathbf{1}\mathbf{e}_\mu = \mathbf{e}_\mu\mathbf{1}. \tag{2.90}$$

Therefore, unit vectors with the Minkowski and Dirac metric anticommute and commute with $\mathbf{1}$, respectively. Following Eq. (2.83), the other spinor can also be defined as

$$\hat{\mathbf{p}}^- = \mathbf{1}\hat{\mathbf{p}}^+ = \mathbf{1}e^{\hat{\mathbf{p}}\frac{\alpha}{2}} = \mathbf{1}\sinh\frac{\alpha}{2} + \mathbf{1}\hat{\mathbf{p}}\cosh\frac{\alpha}{2}. \tag{2.91}$$

This spinor is an eigenvector of the four-momentum since

$$\hat{\mathbf{p}}'\hat{\mathbf{p}}^- = e^{\hat{\mathbf{p}}\frac{\alpha}{2}}\mathbf{e}_0'e^{-\hat{\mathbf{p}}\frac{\alpha}{2}}\mathbf{1}e^{\frac{\alpha}{2}\hat{\mathbf{p}}} = -\,\mathbf{1}e^{\hat{\mathbf{p}}\frac{\alpha}{2}}\mathbf{e}_0' = -\hat{\mathbf{p}}^-\mathbf{e}_0', \tag{2.92}$$

where $\mathbf{1}$ commutes with $e^{\frac{\alpha}{2}\hat{\mathbf{p}}}$ but anticommutes with $\mathbf{e}_0'$. The four-vector associated with this spinor is

$$mc\hat{\mathbf{p}}^-\hat{\mathbf{p}}^-\mathbf{e}_0' = mc\,\mathbf{1}e^{\hat{\mathbf{p}}\frac{\alpha}{2}}\mathbf{1}e^{\hat{\mathbf{p}}\frac{\alpha}{2}}\mathbf{e}_0' = mc\,\mathbf{1}\mathbf{1}e^{\hat{\mathbf{p}}\frac{\alpha}{2}}e^{\hat{\mathbf{p}}\frac{\alpha}{2}}\mathbf{e}_0' = mc\,e^{\hat{\mathbf{p}}\alpha}\mathbf{e}_0' = \mathbf{p}'. \tag{2.93}$$

This reproduces the results above.

The spinors are effective unit vectors for the function space derived in Section 2.3. The function space is given by

$$\begin{aligned}|\mathbf{r}'\rangle + \langle\mathbf{r}'| &= \int d\mathbf{p}\,(|\mathbf{p}\rangle\langle\mathbf{p}|\mathbf{r}'\rangle + \langle\mathbf{r}'|\mathbf{p}\rangle\langle\mathbf{p}|) \\ &= \frac{1}{(2\pi)^{\frac{3}{2}}}\int d\mathbf{p}\left(\hat{\mathbf{p}}^+e^{-\frac{i}{\hbar}(\mathbf{p}\cdot\mathbf{r}-E_\mathbf{p}t)} + e^{\frac{i}{\hbar}(\mathbf{p}\cdot\mathbf{r}-E_\mathbf{p}t)}\hat{\mathbf{p}}^-\right).\end{aligned} \tag{2.94}$$

A factor $1/\sqrt{2\pi}$ has been removed from the normalization of the wavefunction in Eq. (2.64), since the energy has been tied to the momentum via $E_\mathbf{p} = \sqrt{m^2c^4 + \mathbf{p}^2c^2}$. Note the similarity with rotation in a plane randomly oriented in a three-dimensional space in Eq. (1.136). The spinors $\hat{\mathbf{p}}^\pm$ effectively function as unit vectors that contain information on the momentum and the direction of the hyperbolic rotation. The latter can be obtained using Eq. (2.87). As seen above, there are a variety of ways to incorporate the negative sign in $\hat{\mathbf{p}}^-$. The four-momentum $\mathbf{p}$ can be obtained by squaring the spinor. For Dirac spinors, see problem 2.1, this leads to a negative four-momentum,

$|E|\hat{\mathbf{p}}^-\hat{\mathbf{p}}^-\mathbf{e}_0' = -|E|\hat{\mathbf{p}} = -\mathbf{p}$. The signs associated with the spinors are related to the signs in the complex exponential. They should not be interpreted as having a negative energy $-|E|$ which leads to a variety of issues, which then need to be resolved by introducing a Fermi sea. In the same fashion, the square of a spinor for rotation gives a negative sign $r\hat{\mathbf{r}}_\downarrow\hat{\mathbf{r}}_\downarrow\mathbf{e}_z = -\mathbf{r}$. Again, this should not be interpreted as a negative norm $-r$, but $-\mathbf{r}$ is the vector associated with clockwise rotations. Since the spinors are the relevant unit vectors, we are more interested in satisfying Eq. (2.87) than the sign related to the square of the spinor. Therefore, Dirac and Minkowski spinors are both suitable unit vectors, although they produce different four-momenta when squaring them (for Minkowski spinors, the square $mc\hat{\mathbf{p}}^-\hat{\mathbf{p}}^-\mathbf{e}_0' = mc\hat{\mathbf{p}}' = \mathbf{p}'$ produces the four-momentum and not $-\mathbf{p}'$).

So far, spin has not been included. We can assume that the results so far have been obtained for the up spin, *i.e.* $\hat{\mathbf{p}}^\pm \to \hat{\mathbf{p}}^\pm_\uparrow$. In Section 1.7, the Pauli matrices were expressed in a basis $\mathbf{e}_z$ and $\mathbf{e}_x$, where the y-direction arose from the imaginary unit $\boldsymbol{i}_{yx}$. Transitions between the up and down spinor directions could be obtained via $-\boldsymbol{i}\mathbf{e}_y = \mathbf{e}_x\mathbf{e}_z = \boldsymbol{i}_{xz}$, which is the imaginary unit for rotations around the y-axis. Changing the direction of the spin can then be achieved by

$$\hat{\mathbf{p}}^\pm_\downarrow = \mathbf{e}_x\mathbf{e}_z\hat{\mathbf{p}}^\pm_\uparrow = \boldsymbol{i}_{xz}\hat{\mathbf{p}}^\pm_\uparrow. \tag{2.95}$$

Obtaining the four-momentum takes some care, since $\boldsymbol{i}_{xz}$ is an imaginary unit that does not commute with the spinor

$$mc(\hat{\mathbf{p}}^\pm_\downarrow)^*\hat{\mathbf{p}}^\pm_\downarrow\mathbf{e}_0' = mc\hat{\mathbf{p}}^\pm_\uparrow(-\boldsymbol{i}_{xz})\boldsymbol{i}_{xz}\hat{\mathbf{p}}^\pm_\uparrow\mathbf{e}_0' = \mathbf{p}', \tag{2.96}$$

using that $\boldsymbol{i}_{xz}^2 = -1$. This gives the same four-momentum.

Conceptual summary. – For spacetime, there are two different spinors,

$$\mathbf{p}'\hat{\mathbf{p}}^\pm = mc\hat{\mathbf{p}}'\hat{\mathbf{p}}^\pm = \pm mc\hat{\mathbf{p}}^\pm\mathbf{e}_0', \tag{2.97}$$

corresponding to the two different directions of Lorentz transformations. Although operating with the four-momentum $\hat{\mathbf{p}}'$ on the spinors $\hat{\mathbf{p}}^\pm$ produces a different sign, both spinors produce the same four-momentum

$$\mathbf{p}' = mc(\hat{\mathbf{p}}^\pm)^*\hat{\mathbf{p}}^\pm\mathbf{e}_0'. \tag{2.98}$$

This can be somewhat confusing, since for rotation we had $\hat{\boldsymbol{r}}_{\pm\frac{1}{2}}\hat{\boldsymbol{r}}_{\pm\frac{1}{2}}\mathbf{e}_z = \pm\hat{\mathbf{r}}$, see Eq. (1.127). Therefore, the square of the spinors gave different vectors, which can also be related to the rotation axis and the direction of rotation. However, for $\alpha = 0$ (*i.e.* the rest frame), $\mathbf{p}' = mc\mathbf{e}_0' = (E/c)\mathbf{e}_0'$. This gives $E = mc^2$. Therefore, for both spinors, the rest energy of the particle is positive.

Problems

1 In Eq. (2.51), we saw that there is an alternative way of writing the four-momentum.
   (a) Write the four-momentum in Dirac notation in exponential form and find the associated spinors.
   (b) Derive the response of the spinor upon applying the four-momentum.
   (c) What four-momenta are obtained by squaring the spinors?

2.5 γ Matrices

Introduction. – In Section 1.7, it was demonstrated that the properties of algebraic unit vectors in real space could be expressed in terms of Pauli matrices. A vector can then be expressed in terms of a 2×2 matrix. Since the spinors are eigenvectors of a vector, there are also eigenvectors of the Pauli matrices. A similar approach can be followed for the spacetime vectors and spinors. This initially leads to 2×2 matrices expressed on the basis of energy $\mathbf{e}'_0$ and the direction of the momentum $\hat{\mathbf{p}}'$. Expressing the direction in a three-dimensional basis leads to the 4×4 matrices known as the γ matrices.

Multiplying Eq. (2.80) from the left by $ce^{-\hat{\mathbf{p}}\frac{\alpha}{2}}$ and then from the right by $\mathbf{e}'_0$ gives

$$e^{-\hat{\mathbf{p}}\frac{\alpha}{2}} c\mathbf{p}' = mc^2 e^{\hat{\mathbf{p}}\frac{\alpha}{2}} \mathbf{e}'_0 \quad \Rightarrow \quad e^{-\hat{\mathbf{p}}\frac{\alpha}{2}} c\mathbf{p}'\mathbf{e}'_0 = mc^2 e^{\hat{\mathbf{p}}\frac{\alpha}{2}}, \tag{2.99}$$

where $c\mathbf{p}'\mathbf{e}'_0 = E + c\mathbf{p}$, using Eq. (2.28). Writing out the exponentials gives

$$\left(\cosh\frac{\alpha}{2} - \sinh\frac{\alpha}{2}\hat{\mathbf{p}}\right)(E + c\mathbf{p}) = mc^2\left(\cosh\frac{\alpha}{2} + \sinh\frac{\alpha}{2}\hat{\mathbf{p}}\right). \tag{2.100}$$

After expanding, this equation splits into a scalar and a vector term

$$\left[(E - mc^2)\cosh\frac{\alpha}{2} - c\mathbf{p}\hat{\mathbf{p}}\sinh\frac{\alpha}{2}\right] + \left[c\mathbf{p}\cosh\frac{\alpha}{2} + (-E - mc^2)\hat{\mathbf{p}}\sinh\frac{\alpha}{2}\right] = 0,$$

noting that $\mathbf{p}\hat{\mathbf{p}} = p$ is the norm of the momentum. Both the scalar and vector terms need to be zero. The two independent equations can be written in matrix form as

$$\begin{pmatrix} E - mc^2 & -c\mathbf{p} \\ c\mathbf{p} & -E - mc^2 \end{pmatrix} \begin{pmatrix} \cosh\frac{\alpha}{2} \\ \hat{\mathbf{p}}\sinh\frac{\alpha}{2} \end{pmatrix} = 0. \tag{2.101}$$

This equation is essentially Dirac's equation in antisymmetric matrix form. In analogy, to the Pauli matrices, unit vectors can be defined in the time and spatial directions

$$\mathbf{e}'_0 = \begin{pmatrix} 1 & 0 \\ 0 & -1 \end{pmatrix} \quad \text{and} \quad \mathbf{e}'_1 = \hat{\mathbf{p}}' = \begin{pmatrix} 0 & -\hat{\mathbf{p}} \\ \hat{\mathbf{p}} & 0 \end{pmatrix}, \tag{2.102}$$

where the $\hat{\mathbf{p}}$ in the matrix is an anticommuting algebraic unit vector. Using matrix multiplication, it can be shown that

$$\mathbf{e}'^2_0 = \mathbb{1}_2 \quad \text{and} \quad \mathbf{e}'^2_1 = -\mathbb{1}_2, \tag{2.103}$$

with $\hat{\mathbf{p}}^2 = 1$. Additionally, the matrices anticommute. Therefore, the matrices follow the same anticommutation relations as the unit vectors, see Eq. (2.15). The matrices can be written in 2×2 format, for motion in one particular direction, namely $\hat{\mathbf{p}}$. For a general direction, this needs to be expanded into 4×4 matrices, as we will see below. Within this basis, the norm of the four-momentum can be written as

$$c\mathbf{p}' - mc^2 = E\mathbf{e}'_0 + c\mathbf{p}' - mc^2\mathbb{1}_2 = \begin{pmatrix} E - mc^2 & -c\mathbf{p} \\ c\mathbf{p} & -E - mc^2 \end{pmatrix} = 0, \tag{2.104}$$

which is the Dirac equation.

Another important set of 2×2 matrices are those in the Dirac basis

$$\mathbf{e}_0 = \mathbf{e}'_0\mathbf{e}'_0 = \mathbb{1}_2 = \begin{pmatrix} 1 & 0 \\ 0 & 1 \end{pmatrix} \quad \text{and} \quad \hat{\mathbf{p}} = \hat{\mathbf{p}}'\mathbf{e}'_0 = \begin{pmatrix} 0 & \hat{\mathbf{p}} \\ \hat{\mathbf{p}} & 0 \end{pmatrix}. \tag{2.105}$$

The energy component is now the identity matrix and is therefore associated with the norm of the four-momentum in Dirac notation, as opposed to mc in the basis of the γ matrices. The Dirac four-momentum from Eq. (2.51) can be written as

$$\mathsf{p} = mc\mathbf{e}_0' + \mathbf{p} = \begin{pmatrix} mc & \mathbf{p} \\ \mathbf{p} & -mc \end{pmatrix}. \tag{2.106}$$

Solving the eigenvalue equation gives

$$\mathsf{p}c = E \quad \Rightarrow \quad E^2 = m^2c^4 + \mathbf{p}^2c^2, \tag{2.107}$$

which also gives the relativistic energy-momentum relation in Eq. (2.49).

Since the matrices satisfy the same relations as the noncommuting unit vectors, previous results can also be obtained via matrix multiplication. For example, Equation (2.80) expressing the four-momentum in terms of spinors can be derived using

$$\hat{\mathsf{p}}^+ = e^{\hat{\mathbf{p}}\frac{\alpha}{2}} = \cosh\frac{\alpha}{2} + \hat{\mathbf{p}}\sinh\frac{\alpha}{2} = \begin{pmatrix} \cosh\frac{\alpha}{2} & \hat{\mathbf{p}}\sinh\frac{\alpha}{2} \\ \hat{\mathbf{p}}\sinh\frac{\alpha}{2} & \cosh\frac{\alpha}{2} \end{pmatrix}, \tag{2.108}$$

which defines a hyperbolic rotation. The unit four-momentum is then in terms of spinors

$$\begin{aligned} \hat{\mathsf{p}}' = \hat{\mathsf{p}}^+\hat{\mathsf{p}}^+\mathbf{e}_0' &= \begin{pmatrix} \cosh\alpha & \hat{\mathbf{p}}\sinh\alpha \\ \hat{\mathbf{p}}\sinh\alpha & \cosh\alpha \end{pmatrix}\begin{pmatrix} 1 & 0 \\ 0 & -1 \end{pmatrix} \\ &= \begin{pmatrix} \cosh\alpha & -\hat{\mathbf{p}}\sinh\alpha \\ \hat{\mathbf{p}}\sinh\alpha & -\cosh\alpha \end{pmatrix} = \cosh\alpha\,\mathbf{e}_0' + \sinh\alpha\,\hat{\mathbf{p}}', \end{aligned} \tag{2.109}$$

which reproduces Eq. (2.80). The first matrix is now obtained after some annoying matrix multiplication and the use of double-angle formulas for the hyperbolic functions as opposed to simply using $\hat{\mathsf{p}}^+\hat{\mathsf{p}}^+ = e^{\hat{\mathbf{p}}\frac{\alpha}{2}}e^{\hat{\mathbf{p}}\frac{\alpha}{2}} = e^{\hat{\mathbf{p}}\alpha}$. However, with analytical mathematical programs, this exercise becomes easy.

Since the vectors are denoted as 2×2 matrices, there seems to be some redundancy. It is true that the same information can be contained in row and column vectors when only dealing with vectors. Although this had some advantages in the case of rotation, it leads to additional bookkeeping in Minkowski space. Unfortunately, this approach is often followed. To express the spinors in bras and kets, it is convenient to rewrite the unit four-momentum in Eq. (2.109) as $\hat{\mathsf{p}}' = \hat{\mathsf{p}}^+\hat{\mathsf{p}}^+\mathbf{e}_0' = \hat{\mathsf{p}}^+\mathbf{e}_0'\overline{\hat{\mathsf{p}}}^+$, where $\overline{\hat{\mathsf{p}}}^+ = e^{-\hat{\mathbf{p}}\alpha}$ is the transpose of $\hat{\mathsf{p}}^+ = e^{\hat{\mathbf{p}}\alpha}$. The minus sign occurs due to the anticommutation of $\mathbf{e}_0'$ and $\hat{\mathbf{p}}$. The spinors $\hat{\mathsf{p}}^+$ and $\overline{\hat{\mathsf{p}}}^+$ are associated with the bra and ket, respectively. However, conventionally, the ket is associated with $\hat{\mathsf{p}}^+$. Reversing the order gives the transpose of the four-momentum, $\overline{\hat{\mathsf{p}}}^+\mathbf{e}_0'\hat{\mathsf{p}}^+ = \overline{\hat{\mathsf{p}}}^+\overline{\hat{\mathsf{p}}}^+\mathbf{e}_0' = \overline{\hat{\mathsf{p}}}'$, with $\overline{\hat{\mathsf{p}}}' = e^{-\hat{\mathbf{p}}\alpha}\mathbf{e}_0'$. We now take the spinors in bra-ket notation as

$$|\hat{\mathsf{p}}^+\rangle = \hat{\mathsf{p}}^+ = \begin{pmatrix} \cosh\frac{\alpha}{2} \\ \hat{\mathbf{p}}\sinh\frac{\alpha}{2} \end{pmatrix}, \quad \langle\hat{\mathsf{p}}^+| = \overline{\hat{\mathsf{p}}}^+ = \left(\cosh\frac{\alpha}{2}, -\hat{\mathbf{p}}\sinh\frac{\alpha}{2}\right). \tag{2.110}$$

Note that in $\langle\hat{\mathsf{p}}^+|$, it is implicitly assumed that the transpose needs to be taken, and the bar is omitted.

We now would like to use the spinors to obtain the coefficients of the four-momentum. Following Eq. (1.195), we expect that the coefficients can be obtained by taking the expectation value of the unit vectors in the spinors. This gives

$$\hat{\mathsf{p}}_0 = \langle \hat{\mathbf{p}}^+ | \, \mathbf{e}_0' \, | \hat{\mathbf{p}}^+ \rangle = \cosh \alpha \quad \text{and} \quad \hat{\mathsf{p}}_1 = \langle \hat{\mathbf{p}}^+ | \, \mathbf{e}_1' \, | \hat{\mathbf{p}}^+ \rangle = -\sinh \alpha, \tag{2.111}$$

with $\mathbf{e}_1' = \hat{\mathbf{p}}'$. These are the components of the covariant vector. The four-momentum can be obtained by combining them with the corresponding unit vectors

$$\mathsf{p}' = mc \sum_{\mu=0,1} \langle \hat{\mathbf{p}}^+ | \, \mathbf{e}_\mu' \, | \hat{\mathbf{p}}^+ \rangle \, \mathbf{e}'^\mu = \sum_{\mu=0,1} \mathsf{p}_\mu \mathbf{e}'^\mu, \tag{2.112}$$

with $\mathsf{p}_\mu = mc\hat{\mathsf{p}}_\mu$. The four-vector is expressed in a manner similar to a vector in three dimensions, see Eq. (1.195). Note that the above equation is the matrix formulation of $\mathsf{p}' = mc\hat{\mathbf{p}}^+\hat{\mathbf{p}}^+\mathbf{e}_0'$ from Eq. (2.98).

Unfortunately, the conventional use of matrices for the unit vectors and row/column vectors for the spinors rather complicates matters, and keeping track of the covariance and contravariance of the vectors and the coefficients is somewhat painful. Note that we are only trying to obtain the four-momentum from the spinors. The same is done in Eq. (2.80) in a more convenient matter. Even Eq. (2.109), which only uses matrices and no column and row vectors, produces the same results in a more intuitive way.

While the expression of the four-momentum in terms of noncommuting unit vectors, $\mathsf{p}'c = E\mathbf{e}_0' + \mathbf{p}'$, is general, the matrices above are simplified. An effective two-dimensional space is taken of the energy and the momentum in the direction that the particle is moving. However, when discussing the unit vectors in three dimensions, it was demonstrated that they could also be expressed in terms of 2×2 matrices, known as the Pauli matrices, see Eq. (1.149). It is straightforward to extend the unit matrices in spacetime to a general axis system by writing the momentum in the Pauli matrices $\mathbf{e}_i$, giving

$$\mathbf{e}_0' = \gamma_0 = \begin{pmatrix} \mathbb{1}_2 & 0 \\ 0 & -\mathbb{1}_2 \end{pmatrix}, \quad \mathbf{e}_i' = \gamma_i = \begin{pmatrix} 0 & -\mathbf{e}_i \\ \mathbf{e}_i & 0 \end{pmatrix}, \tag{2.113}$$

for $i = 1, 2, 3 = x, y, z$, and where the unit vectors $\mathbf{e}_i$ inside the matrices are the Pauli matrices. These are the γ matrices that are often used in the Dirac equation. Here, they are indicated as unit vectors, since they satisfy the same anticommutative property, see Eq. (2.15). The unit vectors in Dirac form are given by

$$\mathbf{e}_0 = \mathbf{e}_0'\mathbf{e}_0' = \mathbb{1}_4, \quad \mathbf{e}_i = \mathbf{e}_i'\mathbf{e}_0' = \begin{pmatrix} 0 & \mathbf{e}_i \\ \mathbf{e}_i & 0 \end{pmatrix}. \tag{2.114}$$

These unit vectors have the norm $\mathbf{e}_\mu^2 = 1$. We encountered some additional matrices. In Eq. (2.89), the real unit was introduced

$$\mathbf{1} = \gamma_5 = -\boldsymbol{i}\boldsymbol{i} = -\boldsymbol{i}\mathbf{e}_0'\mathbf{e}_x'\mathbf{e}_y'\mathbf{e}_z' = \begin{pmatrix} 0 & \mathbb{1}_2 \\ \mathbb{1}_2 & 0 \end{pmatrix}. \tag{2.115}$$

This can be obtained by straightforward multiplication of the γ matrices. The real unit creates the dual vectors for Minkowski space, in the same way that the imaginary unit $\boldsymbol{i}$ creates the dual vectors for real space. Or, in terms of particle physics, it causes transitions between the particle and antiparticle states. The other dual vector of interest is, see Eq. (2.95),

$$\boldsymbol{i}_{xz} = \mathbf{e}_x\mathbf{e}_z = \begin{pmatrix} \boldsymbol{i}\mathbf{e}_y & 0 \\ 0 & \boldsymbol{i}\mathbf{e}_y \end{pmatrix}, \tag{2.116}$$

which is the imaginary unit associated with rotations around the y-axis that changes the direction of the spin.

The four-momentum in the basis of 4×4 matrices is given by

$$\mathsf{cp}' = E\mathsf{e}'_0 + c\mathbf{p}' = \sum_\mu \mathsf{e}'_\mu \mathsf{p}^\mu = \begin{pmatrix} E & -c\mathbf{p} \\ c\mathbf{p} & -E \end{pmatrix}$$
$$= \begin{pmatrix} E & 0 & -cp_z & -c(p_x - ip_y) \\ 0 & E & -c(p_x + ip_y) & cp_z \\ cp_z & c(p_x - ip_y) & -E & 0 \\ c(p_x + ip_y) & -cp_z & 0 & -E \end{pmatrix}. \tag{2.117}$$

The spinors can also be expressed as a four-component vector. For example, for the spin-up component of the + spinor, one finds

$$|\hat{\mathsf{p}}^+_\uparrow\rangle = e^{\hat{\mathsf{p}}\frac{\alpha}{2}} \begin{pmatrix} 1 \\ 0 \\ 0 \\ 0 \end{pmatrix} = \begin{pmatrix} \cosh\frac{\alpha}{2}\begin{pmatrix} 1 \\ 0 \end{pmatrix} \\ \sinh\frac{\alpha}{2}\,\hat{\mathbf{p}}\begin{pmatrix} 1 \\ 0 \end{pmatrix} \end{pmatrix} = \begin{pmatrix} \cosh\frac{\alpha}{2} \\ 0 \\ \sinh\frac{\alpha}{2}\hat{p}_z \\ \sinh\frac{\alpha}{2}(\hat{p}_x + i\hat{p}_y) \end{pmatrix}, \tag{2.118}$$

where, for the momentum, $\begin{pmatrix} 1 \\ 0 \end{pmatrix}$ is multiplied by the unit vector in the direction of the momentum $\hat{\mathbf{p}}$ expressed in terms of Pauli matrices.

There is an additional tendency to remove hyperbolic functions, which further obfuscates the results. Using the expressions for the double-argument, $\cosh\alpha = 2\cosh^2\frac{\alpha}{2} - 1 = 2\sinh^2\frac{\alpha}{2} + 1$, the following identities are obtained:

$$\cosh\frac{\alpha}{2} = \sqrt{\frac{E_\mathrm{p}}{mc^2} + \frac{1}{2}} = \sqrt{\frac{E_\mathrm{p} + mc^2}{2mc^2}} \tag{2.119}$$

and

$$\sinh\frac{\alpha}{2} = \sqrt{\frac{E_\mathrm{p} - mc^2}{2mc^2}} = \sqrt{\frac{E_\mathrm{p}^2 - m^2c^4}{2mc^2(E_\mathrm{p} + mc^2)}} = \frac{pc}{\sqrt{2mc^2(E_\mathrm{p} + mc^2)}},$$

using Eqs. (2.31) and (2.52). This leads to the following often-used expression for the spinor:

$$|\mathsf{p}^+_\uparrow\rangle = \sqrt{\frac{E_\mathrm{p} + mc^2}{2mc^2}} \begin{pmatrix} 1 \\ 0 \\ \dfrac{\hat{p}_z c}{E_\mathrm{p} + mc^2} \\ \dfrac{(\hat{p}_x + i\hat{p}_y)c}{E_\mathrm{p} + mc^2} \end{pmatrix}. \tag{2.120}$$

The problems discuss how to obtain the other spinors. Equation (2.121) can be generalized to give

$$\mathsf{p}' = mc\sum_{\mu=0}^{3} \langle \hat{\mathsf{p}}^\pm_{\pm\frac{1}{2}}|\mathsf{e}'_\mu|\hat{\mathsf{p}}^\pm_{\pm\frac{1}{2}}\rangle \mathsf{e}'^\mu = \sum_{\mu=0,1} \mathsf{p}_\mu \mathsf{e}'^\mu, \tag{2.121}$$

which produces the same four-momentum for each of the four spinors.

Conceptual summary. – In this section, the results for anticommuting algebraic unit vectors have been expressed in matrix form, leading to the γ matrices. The γ matrices obey the same anticommutation relations as the algebraic unit vectors. When expressed in terms of 4×4 matrices or four-dimensional vectors, the results can become unwieldy, in particular, when one also attempts to remove all traces of the hyperbolic functions. Although the results are reproduced here to connect to other textbooks, it should be noted that the matrix and geometric algebraic approaches contain the same information.

Problems

1 The column vector in Eq. (2.118) for a particle at rest corresponds to $\mathbf{e}'_0$. Show how the other three column vectors with a one in a particular row can be obtained.

2 Find the spinors for a particle with a rapidity α in the direction $\hat{\mathbf{p}}$ expressed in spherical polar coordinates and show that they are eigenvectors of $\hat{\mathbf{p}}$. Determine the four-momentum associated with each spinor (the use of analytical mathematical software is recommended).

3 Using the γ matrices, show that $\hat{\mathbf{p}}^{\pm}_{\uparrow}\hat{\mathbf{p}}^{\pm}_{\uparrow}\mathbf{e}'_0 = \hat{\mathbf{p}}$ (the use of analytical mathematical software is recommended). Determine $\hat{\mathbf{p}}\hat{\mathbf{p}}^{\pm}_{\uparrow}$.

2.6 Motion in an Electromagnetic Field

Introduction. – Particles rarely experience free motion and generally move in fields created by other particles. Although gravity is more important in the consideration of macroscopic objects, the most common interaction one initially encounters in quantum-mechanical problems is the electromagnetic interaction between charged particles such as protons and electrons. The interaction can be split into two parts. First, the charged particles create a field. This will be considered in Section 2.7. Second, charged particles experience the fields created by other charged particles. This will affect the momentum. We will also look at the point-particle limit, which reduces to the well-known Lorentz equation.

In order to make each point in spacetime independent, see Eq. (2.53), a function space in terms of plane waves was created. However, this choice is not unique. Using Eq. (2.61), each point in spacetime can also be expanded as

$$|\mathbf{r}'\rangle = \int d\mathbf{k}' \, |\mathbf{k}'\rangle\langle\mathbf{k}'|\mathbf{r}'\rangle = \frac{1}{(2\pi)^2}\int d\mathbf{k}' \, |\mathbf{k}'\rangle e^{-i\mathbf{k}'\cdot\mathbf{r}'+i\frac{q}{\hbar}\chi(\mathbf{r}')}, \tag{2.122}$$

with $d\mathbf{k}' = d(\omega/c)dk_x dk_y dk_z$. The prime indicates that the four-vectors are expressed in a Minkowski basis. A position-dependent phase factor has been added to the wavefunction. The introduction of $q/\hbar$ becomes apparent below. Orthonormality is then given by

$$\langle\mathbf{r}'_1|\mathbf{r}'_2\rangle = \delta(\mathbf{r}'_1 - \mathbf{r}'_2)e^{i\frac{q}{\hbar}(\chi(\mathbf{r}'_1)-\chi(\mathbf{r}'_2))} = \delta(\mathbf{r}'_1 - \mathbf{r}'_2). \tag{2.123}$$

The phase factor can be removed, since the δ-function is only nonzero for $\mathbf{r}'_1 = \mathbf{r}'_2$. This implies that, even in the presence of the phase factor $\chi(\mathbf{r}')$, each point in spacetime is still independent of each other. For orthonormality in momentum space, we have

$$\langle\mathbf{k}'_1|\mathbf{k}'_2\rangle = \frac{1}{(2\pi)^4}\int d\mathbf{r}' \, e^{-i\mathbf{k}'_2\cdot\mathbf{r}'+i\frac{q}{\hbar}\chi(\mathbf{r}')}e^{i\mathbf{k}'_1\cdot\mathbf{r}'-i\frac{q}{\hbar}\chi(\mathbf{r}')} = \delta(\mathbf{k}'_2 - \mathbf{k}'_1), \tag{2.124}$$

since the $\chi(\mathbf{r}')$ simply cancel. Therefore, the orthonormality properties are the same in both real and momentum spaces. Note that we do not get to pick a different $\chi(\mathbf{r}')$ for each plane wave. For example, we can pick $\frac{q}{\hbar}\chi(\mathbf{r}') = \mathbf{k}_1' \cdot \mathbf{r}'$. This gives

$$e^{i\mathbf{k}_1'\cdot\mathbf{r}' - i\frac{q}{\hbar}\chi(\mathbf{r}')} = 1 \quad \text{and} \quad e^{i\mathbf{k}_2'\cdot\mathbf{r}' - i\frac{q}{\hbar}\chi(\mathbf{r}')} = e^{i(\mathbf{k}_2' - \mathbf{k}_1')\cdot\mathbf{r}'}. \tag{2.125}$$

This means that one plane wave becomes a constant (a particle at rest). However, the wavevector for the other wave then becomes $\mathbf{k}_2' - \mathbf{k}_1'$. In the end, the only quantity that is relevant is the difference in momentum between the plane waves. Therefore, the additional phase does not affect the properties of the real and momentum spaces. This is often called a local symmetry. However, one can also view it as the functional equivalent of a change in reference frame, where in the above choice one of the particles is now in the rest frame.

The additional phase factor $\chi(\mathbf{r}')$ therefore seems a completely unnecessary addition. However, since the phase factor is related to space and its associated fields, they are helpful in showing us how the fields enter into the theory. Applying the momentum operator $\hat{\mathbf{p}}'$ on the plane wave $\varphi_{\mathbf{k}'}(\mathbf{r}') = \langle \mathbf{r}' | \mathbf{k}' \rangle$ including the phase factor gives

$$\hat{\mathbf{p}}'\varphi_{\mathbf{k}'}(\mathbf{r}') = i\hbar\nabla\varphi_{\mathbf{k}'}(\mathbf{r}') = \left(\mathbf{p}' - q\nabla\chi(\mathbf{r}')\right)\varphi_{\mathbf{k}'}(\mathbf{r}'). \tag{2.126}$$

Therefore, the phase factor has a direct influence on the momentum. However, that all does not matter too much. A particle has no idea whether it is moving at a particular velocity. It might as well be standing still. What is important is whether there is an acceleration taking place. Operating with the four-nabla on the phase factor gives

$$\nabla\chi(\mathbf{r}') = \frac{\partial\chi(\mathbf{r}')}{\partial(ct)}\mathbf{e}_0' - \nabla'\chi(\mathbf{r}'). \tag{2.127}$$

For someone familiar with electromagnetism, these terms might look familiar. They are in fact the gauge transformations for the electromagnetic potentials. The gauge transformations

$$V' = V - \frac{\partial\chi(\mathbf{r}')}{\partial t} \quad \text{and} \quad \mathbf{A}' = \mathbf{A} + \nabla\chi(\mathbf{r}'), \tag{2.128}$$

where V and $\mathbf{A}$ are the potential and the vector potential, respectively, have no effect on the fields experienced by the particle. Since the inclusion of $\chi(\mathbf{r}')$ does not affect the physical properties, this is known as a gauge symmetry. It provides insights on how to include a four-potential. By replacing the gauge invariance $\nabla\chi(\mathbf{r}')$ by a four-vector in Eq. (2.122),

$$\mathbf{p}' \rightarrow \mathbf{p}' - q\mathbf{A}', \tag{2.129}$$

where $\mathbf{A}'$ is known as the four-potential; an actual interaction on the particle is included. The factor q is the charge of the particle indicating the strength of the coupling to the four-potential. Since the only thing that is relevant is the total interaction $q\mathbf{A}'$, there is some arbitrariness in the definition of the interaction strength q and the magnitude of the four-potential $\mathbf{A}'$. The Dirac equation in the presence of an electromagnetic field is then

$$\mathbf{p}' - q\mathbf{A}' = mc. \tag{2.130}$$

We would now like to understand how the motion of a particle is affected by a displacement in spacetime. The components of the four-potential are conventionally split in terms of the scalar potential V and the vector potential $\mathbf{A}$:

$$\mathbf{A}' = \frac{V}{c}\mathbf{e}_0' + \mathbf{A}'. \tag{2.131}$$

At a particular point in spacetime, we can always take $\mathbf{A}' = 0$, assuming that all effects of the potential up to that point have already been included in the four-momentum. A small displacement then gives

$$\mathbf{p}' + d\mathbf{p}' - q\nabla\mathbf{A}'\, d\mathbf{r}', \tag{2.132}$$

and the change in four-momentum should simply be equal to the change in the four-potential: $d\mathbf{p}' = q\nabla\mathbf{A}'\, d\mathbf{r}'$. The four-vector is given by

$$\mathbf{r}' = ct\,\mathbf{e}_0' + \mathbf{r}', \quad \text{and} \quad d\mathbf{r}' = cdt\,\mathbf{e}_0' + d\mathbf{r}', \tag{2.133}$$

where the position vector in Minkowski space is $\mathbf{r}' = \mathbf{r}\mathbf{e}_0'$, with $\mathbf{r}'^2 = -r^2$. We now want to absorb the change in four-vector potential into the four-momentum, giving

$$d\mathbf{p}' = q\nabla\mathbf{A}'\, d\mathbf{r}' = q\left(\frac{1}{c}\frac{\partial}{\partial t}\mathbf{e}_0' - \nabla'\right)\left(\frac{V}{c}\mathbf{e}_0' + \mathbf{A}'\right) d\mathbf{r}'. \tag{2.134}$$

In order to obtain the conventional expressions for electromagnetism, we want to separate the time and real space components. Expanding the $\nabla\mathbf{A}'$ gives

$$\nabla\mathbf{A}' = \frac{1}{c^2}\frac{\partial V}{\partial t} - \frac{1}{c}\nabla' V\mathbf{e}_0' + \frac{1}{c}\frac{\partial(\mathbf{e}_0'\mathbf{A}')}{\partial t} - \nabla'\mathbf{A}'. \tag{2.135}$$

The product of the nabla operator and the vector potential is

$$-\nabla'\mathbf{A}' = \nabla\mathbf{A} = \nabla\cdot\mathbf{A} + \nabla\wedge\mathbf{A}$$

using $-\nabla'\mathbf{A}' = -\nabla\mathbf{e}_0'\mathbf{A}\mathbf{e}_0' = \nabla\mathbf{A}\mathbf{e}_0'\mathbf{e}_0' = \nabla\mathbf{A}$ and Eq. (1.110). Collecting terms, we find

$$\nabla\mathbf{A}' = \nabla\cdot\mathbf{A}' + \nabla\wedge\mathbf{A}' = \nabla\cdot\mathbf{A}' + \frac{1}{c}\left(-\nabla V - \frac{1}{c}\frac{\partial\mathbf{A}}{\partial t}\right) + \nabla\wedge\mathbf{A}, \tag{2.136}$$

where the scalar and bivector terms have been separated. The first term $\nabla\cdot\mathbf{A}'$ is the inner product between the four-nabla and the four-vector potential and contains the two scalar terms. It is equivalent to the Lorenz gauge

$$\nabla\cdot\mathbf{A}' = \frac{1}{c^2}\frac{\partial V}{\partial t} + \nabla\cdot\mathbf{A} \to 0, \tag{2.137}$$

and we can choose this to be zero by taking the appropriate gauge condition. Taking the gauge transformation in Eq. (2.128) gives

$$\nabla^2\chi - \frac{1}{c^2}\frac{\partial^2\chi}{\partial t^2} = 0. \tag{2.138}$$

The solutions to these equations are plane waves.

In the remaining vector and bivector terms, we recognize the expressions for the electric and magnetic fields in terms of the potential and the vector potential,

$$\mathbf{E} = -\nabla V - \frac{1}{c}\frac{\partial\mathbf{A}}{\partial t}, \quad \mathbf{B} = \nabla\times\mathbf{A}. \tag{2.139}$$

The wedge product of the four-vector potential $\mathbf{A}$ can then be written as

$$\mathbf{F} = \nabla\wedge\mathbf{A}' = \frac{\mathbf{E}}{c} + \boldsymbol{i}\mathbf{B} = \frac{\mathbf{E}}{c} + \underline{\mathbf{B}}, \tag{2.140}$$

where $\mathbf{F}$ is known as the electromagnetic field tensor. This looks like the combination of a vector and a bivector. However, in Minkowski unit vectors, these become $\mathbf{e}_i = \mathbf{e}_i'\mathbf{e}_0' = \boldsymbol{i}_{i0}'$ and $\boldsymbol{i}_{ij} = \mathbf{e}_i\mathbf{e}_j = \mathbf{e}_i\mathbf{e}_j\mathbf{e}_0'\mathbf{e}_0' = -\mathbf{e}_i\mathbf{e}_0'\mathbf{e}_j\mathbf{e}_0' = -\mathbf{e}_i'\mathbf{e}_j' = \mathbf{e}_j'\mathbf{e}_i' = \boldsymbol{i}_{ji}'$. The electromagnetic field tensor therefore contains six bivector terms

$$\mathbf{F} = F_{x0}\boldsymbol{i}_{x0}' + F_{y0}\boldsymbol{i}_{y0}' + F_{z0}\boldsymbol{i}_{z0}' + F_{zy}\boldsymbol{i}_{zy}' + F_{xz}\boldsymbol{i}_{xz}' + F_{yx}\boldsymbol{i}_{yx}'. \tag{2.141}$$

Conventionally, the six bivectors are described in terms of two real-space vectors, namely the electric and magnetic fields. The first three terms have a spacetime bivector $\boldsymbol{i}'_{i0}$, with $i = x, y, z$, and therefore correspond to Lorentz boosts. The amplitude is given by the electric field $F_{i0} = E_i/c$. The last three terms correspond to pure spatial rotations given by the bivector $\boldsymbol{i}'_{ji}$, with $i \neq j$. The magnitude is the magnetic field $F_{ji} = B_k$, with $k = x, y, z$, for $ji = zy, xz, yx$.

Writing out the four-vectors $\mathbf{r}' = ct\mathbf{e}'_0 + \mathbf{r}' = (ct + \mathbf{r})\mathbf{e}'_0$, and using that $d\mathbf{r} = \mathbf{v}dt$, gives $d\mathbf{r}' = (c + \mathbf{v})dt\mathbf{e}'_0$. Equation (2.134) then becomes after multiplying with $\mathbf{e}'_0$ from the right

$$\frac{d\mathbf{p}'}{dt}\mathbf{e}'_0 = \frac{d\mathbf{p}}{dt} = \frac{1}{c}\frac{dE}{dt} + \frac{d\mathbf{p}}{dt} = q\left(\frac{\mathbf{E}}{c} + \underline{\mathbf{B}}\right)(c + \mathbf{v}). \tag{2.142}$$

The vector products can be split into an inner and an outer product using Eq. (1.110), for example

$$\underline{\mathbf{B}}\mathbf{v} = \boldsymbol{i}\mathbf{B}\mathbf{v} = \boldsymbol{i}\mathbf{B} \cdot \mathbf{v} + \boldsymbol{i}\mathbf{B} \wedge \mathbf{v} = \boldsymbol{i}\mathbf{v} \cdot \mathbf{B} - \mathbf{B} \times \mathbf{v} = \boldsymbol{i}\mathbf{v} \cdot \mathbf{B} + \mathbf{v} \times \mathbf{B}. \tag{2.143}$$

Collecting the terms gives

$$\begin{aligned}\frac{1}{c}\frac{dE}{dt} + \frac{d\mathbf{p}}{dt} = q&\left(\frac{1}{c}\mathbf{v} \cdot \mathbf{E} + \mathbf{E} + \mathbf{v} \times \mathbf{B}\right)\\ &+ \boldsymbol{i}qc\left(\frac{1}{c}\mathbf{v} \cdot \mathbf{B} + \mathbf{B} - \frac{1}{c^2}\mathbf{v} \times \mathbf{E}\right).\end{aligned} \tag{2.144}$$

The scalar terms are the power equation

$$\frac{dE}{dt} = q\mathbf{E} \cdot \mathbf{v}. \tag{2.145}$$

The vector terms correspond to the motion of a particle experiencing a Lorentz force

$$\frac{d\mathbf{p}}{dt} = q\mathbf{E} + q\mathbf{v} \times \mathbf{B}. \tag{2.146}$$

The last term in Eq. (2.144) corresponds to the motion of a particle with a magnetic charge $\boldsymbol{i}q$. Magnetic charges are not found in nature. However, even if magnetic charges were present, they can be removed by a duality transformation. Taking the magnetic charge equal to zero is therefore a convention.

Example 2.2 ***Lorentz Boost of Electric Field***
An electromagnetic field tensor is given by

$$\mathsf{F} = \frac{E_x}{c}\mathbf{e}_x + \frac{E_y}{c}\mathbf{e}_y. \tag{2.147}$$

(a) A lorentz boost is given in the x-direction using $\mathsf{F}' = R\mathsf{F}R^{-1}$. Determine R.
(b) Determine F'.
(c) After the Lorentz boost, the electromagnetic field tensor F' can be written in terms of an electric and a magnetic field. If the velocity of the frame is given by $\mathbf{v} = v\mathbf{e}_x$, find the components of the electric and magnetic field in terms of v and the speed of light c.

Solution

(a) A Lorentz boost is a hyperbolic transformation over a certain hyperbolic angle α. The associated bivector for a boost in the x-direction is $\mathbf{e}'_x\mathbf{e}'_0 = \mathbf{e}_x$. The rotation R is only over half the angle

$$R = e^{\mathbf{e}_x\frac{\alpha}{2}}, \quad R^{-1} = e^{-\mathbf{e}_x\frac{\alpha}{2}}. \tag{2.148}$$

We cannot use $\mathsf{F}' = e^{\mathbf{e}_x\alpha}\mathsf{F}$ since terms that are constant under the Lorentz transformation would not transform properly.

(b) The Lorentz boost gives

$$\mathsf{F}' = R\mathsf{F}R^{-1} = e^{\mathbf{e}_x\frac{\alpha}{2}}\left(\frac{E_x}{c}\mathbf{e}_x + \frac{E_y}{c}\mathbf{e}_y\right)e^{-\mathbf{e}_x\frac{\alpha}{2}}. \tag{2.149}$$

Since $\mathbf{e}_x$ commutes with itself, but anticommutes with $\mathbf{e}_y$, the result can be rewritten as

$$\mathsf{F}' = e^{\mathbf{e}_x\frac{\alpha}{2}}e^{-\mathbf{e}_x\frac{\alpha}{2}}\frac{E_x}{c}\mathbf{e}_x + e^{\mathbf{e}_x\frac{\alpha}{2}}e^{\mathbf{e}_x\frac{\alpha}{2}}\frac{E_y}{c}\mathbf{e}_y = \frac{E_x}{c}\mathbf{e}_x + e^{\mathbf{e}_x\alpha}\frac{E_y}{c}\mathbf{e}_y. \tag{2.150}$$

Expanding the exponential gives

$$\mathsf{F}' = \frac{E_x}{c}\mathbf{e}_x + \cosh\alpha\,\frac{E_y}{c}\mathbf{e}_y + \sinh\alpha\,\frac{E_y}{c}\mathbf{e}_x\mathbf{e}_y. \tag{2.151}$$

The electric field in the x-direction is therefore constant under the Lorentz boost.

(c) After the boost, the position is given by

$$\mathsf{r}' = ct\mathbf{e}_0' + \mathbf{r}' = ct\mathbf{e}_0' + \mathbf{v}'t = \mathsf{r}\cosh\alpha\,\mathbf{e}_0' + \mathsf{r}\sinh\alpha\,\hat{\mathbf{r}}'. \tag{2.152}$$

This gives the following relations:

$$\beta \equiv \tanh\alpha = \frac{x}{ct} = \frac{vt}{ct} = \frac{v}{c}. \tag{2.153}$$

Additionally,

$$\gamma = \cosh\alpha = \frac{1}{\sqrt{1-\left(\frac{v}{c}\right)^2}}, \quad \sinh\alpha = \cosh\alpha\tanh\alpha = \gamma\beta. \tag{2.154}$$

The electromagnetic field tensor can then be written as

$$\mathsf{F}' = \frac{E_x}{c}\mathbf{e}_x + \gamma\frac{E_y}{c}\mathbf{e}_y + \frac{v\gamma}{c^2}E_y\mathbf{e}_x\mathbf{e}_y. \tag{2.155}$$

Note that the last term is a space–space rotation and therefore a magnetic field term. It can also be written as

$$\frac{v\gamma}{c^2}E_y\mathbf{e}_x\mathbf{e}_y = \frac{v\gamma}{c^2}E_y\boldsymbol{i}\mathbf{e}_z = \frac{\gamma}{c^2}vE_y\boldsymbol{i}\mathbf{e}_x \times \mathbf{e}_y = \frac{\gamma}{c^2}\boldsymbol{i}\mathbf{v} \times \mathbf{E}, \tag{2.156}$$

with $\mathbf{v} = v\mathbf{e}_x$ and $\mathbf{E} = E_x\mathbf{e}_x + E_y\mathbf{e}_y$, where the x component does not contribute to the magnetic field. The components of the electromagnetic field tensor after the Lorentz boost are therefore

$$E_x' = E_x, \quad E_y' = \gamma E_y, \quad B_z' = \frac{v\gamma}{c^2}E_y. \tag{2.157}$$

Conceptual summary. – In this section, we looked at the effect of an electromagnetic field on the motion of a particle. Gauge theory was used to understand how the electromagnetic potential enters the Dirac equation. However, the end result is relatively straightforward since the electromagnetic potential is also a four-vector, and p' is simply replaced by $\mathsf{p}' - q\mathsf{A}'$. The Dirac equation then becomes $\mathsf{p}' - q\mathsf{A}' = mc$. The factor q is the charge and represents the coupling strength to the electromagnetic field. The value of q can be positive or negative (*e.g.* for protons and electrons, respectively), so particles can respond in opposite ways to the field. It can also be zero (for example, for neutrons), and those particles are not affected by the electromagnetic field. The four-potential

$\mathsf{A}' = (V/c)\mathbf{e}'_0 + \mathbf{A}'$ contains the well-known classical electrostatic potential related to the electric field and the vector potential which affect both the electric and magnetic fields.

The results were then reduced to those of a point particle. Since this is a rigid object, we are only concerned with changes in the group velocity and omit the spatial extent of the particle. In this limit, the motion can be studied using vector operations as opposed to differential operators. Unfortunately, reductions from the elegant-looking relativistic results to the more conventional expressions are usually rather cumbersome and technical. The end result is that changes in the momentum are expressed via bivectors that change a vector into another vector. In spacetime, there are six bivectors, which leads to the electromagnetic field tensor F. The bivectors are usually split into three spacetime bivectors and three space–space bivectors. These are associated with the electric and magnetic fields, respectively, leading to Lorentz boosts (hyperbolic rotations in spacetime) and rotations in real space, respectively. When looking at the time dependence of the momentum, the results reduce to the well-known Lorentz force.

Problems

1 In Eq. (2.155), the electric fields in the x and y directions transform as $E_x\mathbf{e}_x$ and $\gamma E_y\mathbf{e}_y$, respectively, for a Lorentz boost in the x-direction. Show that this is consistent with the deformation of a large parallel-plate capacitor oriented in different directions to the velocity.

2 Determine the electromagnetic field tensor resulting from a Lorentz boost in the x-direction of the magnetic field $\mathbf{B} = B_y\mathbf{e}_y + B_z\mathbf{e}_z$.

3 Describe the motion of a particle with charge q in an electric field in the z-direction and a magnetic field in the x-direction starting at rest at the origin.

2.7 Creation of Electromagnetic Fields: Maxwell's Equations

Introduction. – In Section 2.6, the equations of motion of a charged particle in an electromagnetic field were derived. However, these fields themselves are the result of the charge of other particles and the motion of charged particles. In this section, we look at the creation of electromagnetic fields by charged particles. This is done by drawing a parallel with a driven harmonic oscillator. It will be demonstrated that the results reduce to Maxwell's equations.

The four-potential appearing in Eq. (2.129) is due to other charged particles. The generation of the four-potential is comparable to that of a driven harmonic oscillator

$$\frac{d^2x(t)}{dt^2} + \omega_0^2 x(t) = \frac{f(t)}{m}, \tag{2.158}$$

where x is the displacement, $f(t)$ is the driving force, m is the mass, and ω_0 is the frequency of the oscillator. Let us take the time dependence of the driving force to be a

δ-function, *i.e.* $f(t) = F\delta(t)$. After a Fourier transform to frequency space, one obtains

$$x(\omega) = -\frac{F}{m(\omega^2 - \omega_0^2)} = -FG(\omega) \text{ with } G(\omega) = \frac{1}{m}\frac{1}{(\omega^2 - \omega_0^2)}, \tag{2.159}$$

where the Fourier transform of the driving force is constant, *i.e.* $f(\omega) = F$. Since the time dependence is completely localized in time, the frequency dependence is completely delocalized, *i.e.* it is constant. The function $G(\omega)$ essentially describes the response in Fourier space of a spring to a $\delta(t)$ function. This is known as a Green's function or a propagator. The analogous expression for the four-vector potential is

$$\frac{d^2\mathsf{A}'(\mathsf{r}')}{dt^2} + \omega_{\mathbf{q}}^2\mathsf{A}'(\mathsf{r}') = \frac{\mathsf{J}'(\mathsf{r}')}{\varepsilon_0}, \tag{2.160}$$

where ε_0 is the vacuum permittivity, which effectively plays the role of the mass. In addition to the time dependence, there is also a spatial dependence to the driving force $\mathsf{J}'(\mathsf{r}')$, which is also written as a four-vector. Instead of a single frequency, the four-vector potential can oscillate with any frequency $\omega_{\mathbf{q}}$. For electromagnetic radiation, the frequency is directly related to the norm of the wavevector, $\omega_{\mathbf{q}}^2 = c^2\mathbf{q}^2$. The value of the momentum can be obtained from the spatial derivative of the vector potential, *i.e.* $-c^2\nabla^2\mathsf{A}'(\mathsf{r}') = c^2\mathbf{q}^2\mathsf{A}'(\mathsf{r}') = \omega_{\mathbf{q}}^2\mathsf{A}'(\mathsf{r}')$. The driven wave equation can then also be written as

$$c^2\,\nabla\!\!\!\!\bigtriangledown^2\mathsf{A}'(\mathsf{r}') = \frac{\mathsf{J}(\mathsf{r}')}{\varepsilon_0} \quad \Rightarrow \quad \nabla\!\!\!\!\bigtriangledown^2\mathsf{A}'(\mathsf{r}') = \mu_0\mathsf{J}(\mathsf{r}'), \tag{2.161}$$

where $\mu_0 = 1/(c^2\varepsilon_0)$ is the vacuum permeability. The Laplacian in spacetime is given by

$$\nabla\!\!\!\!\bigtriangledown^2 = \frac{\partial^2}{\partial(ct)^2} - \nabla^2. \tag{2.162}$$

From Eq. (2.65), we know that taking the Laplacian on the left-hand side of Eq. (2.161) effectively corresponds to the norm of the four-potential. This norm is then fixed by the driving force on the right-hand side.

The driving force is known as the four-current, which can be split into a scalar and a vector term

$$\mathsf{J}'(\mathsf{r}') = c\rho(\mathsf{r}')\mathsf{e}_0' + \mathbf{J}'(\mathsf{r}'), \tag{2.163}$$

where ρ is the charge density, and $\mathbf{J}'$ is the charge current. These quantities are expressed in macroscopic terms. However, the charge densities and currents consist of charged particles, such as electrons and protons. Since charge is a conserved quantity, the current four-vector should satisfy

$$\frac{\partial\rho}{\partial t} = -\nabla\cdot\mathbf{J} \quad \text{or} \quad \nabla\!\!\!\!\bigtriangledown\cdot\mathsf{J}' = 0. \tag{2.164}$$

Solutions for the driven wave equation of the four-potential can be found by going to Fourier space. The Fourier transform of the four-potential is given by $\mathsf{A}'(\mathsf{q}')$. This Fourier component is directly proportional to the Fourier component $\mathsf{J}'(\mathsf{q}')$ of the four-current,

$$\mathsf{A}'(\mathsf{q}') = -\frac{\mu_0\mathsf{J}'(\mathsf{q}')}{\mathsf{q}'^2}, \tag{2.165}$$

where $\mathsf{q}' = (\omega/c)\mathsf{e}_0' + \mathbf{q}'$, and $\mathsf{q}'^2 = (\omega/c)^2 - \mathbf{q}^2$. The above equation can be written as the product of two functions

$$\mathsf{A}'(\mathsf{q}') = -G^0(\mathsf{q}')\mathsf{J}'(\mathsf{q}'), \tag{2.166}$$

where, using $\mu_0 = 1/\varepsilon_0 c^2$,

$$G^0(\mathbf{q}') = \frac{\mu_0}{\mathbf{q}'^2} = \frac{1}{\varepsilon_0(\omega^2 - \omega_{\mathbf{q}}^2)} \tag{2.167}$$

is the free propagator or Green's function of the photon. It essentially tells us how photons propagate in free space. The function $G^0(\mathbf{q}')$ peaks strongly at $\omega = \omega_{\mathbf{q}} = cq$, see Eq. (2.167). This shows how the frequency and wavenumber are related to each other in free space. This is known as the dispersion relation. We can obtain the following differential equation for the free propagator:

$$-\mathbf{q}'^2 G^0(\mathbf{q}') = -\mu_0 \quad \Rightarrow \quad \nabla'^2 G^0(\mathbf{r}') = -\mu_0 \delta(\mathbf{r}'). \tag{2.168}$$

Using Eq. (2.162), this can also be written as

$$\nabla^2 G^0(\mathbf{r}') - \frac{1}{c^2}\frac{\partial^2 G^0(\mathbf{r}')}{\partial t^2} = \mu_0 \delta(\mathbf{r}'). \tag{2.169}$$

Therefore, the Green's function describes the response to a perturbance completely localized in spacetime. This information can then be used to describe charge distributions that depend on space and time.

It is also instructive to look at the time dependence of the Green's function. For a plane wave with wavevector $\mathbf{q}$, $\nabla^2 G^0_{\mathbf{q}}(\mathbf{r}') = -\mathbf{q}^2 G^0_{\mathbf{q}}(\mathbf{r}')$. This gives a differential equation

$$\frac{d^2 G^0_{\mathbf{q}}(t)}{dt^2} + \omega_{\mathbf{q}}^2 G^0_{\mathbf{q}}(t) = -\frac{\delta(t)}{\varepsilon_0}, \tag{2.170}$$

with $\omega_{\mathbf{q}}^2 = c^2\mathbf{q}^2$. We take $G^0_{\mathbf{q}}(t) = 0$ for $t < 0$, so that the δ function creates the electromagnetic wave at $t = 0$. Note that for $t > 0$, there is no longer a driving force, and the electromagnetic wave propagates freely, giving

$$G^0_{\mathbf{q}}(t) = Ae^{i\omega_{\mathbf{q}} t} + Be^{-i\omega_{\mathbf{q}} t} \tag{2.171}$$

Since the Green's function is continuous, and $G_{\mathbf{q}}(0) = 0$, $A = -B$, implying that $G^0_{\mathbf{q}}(t) = 2iA \sin \omega_{\mathbf{q}} t$. Therefore, the four-potential is a real field. The driving force only changes the derivative $dG^0_{\mathbf{q}}(t)/dt$. The change can be determined by integrating over an infinitesimally small region τ around $t = 0$,

$$\lim_{\tau \to 0} \int_{-\tau}^{\tau} dt \left(\frac{d^2 G^0_{\mathbf{q}}(t)}{dt^2} + \omega_{\mathbf{q}}^2 G^0_{\mathbf{q}}(t) \right) = \frac{dG^0_{\mathbf{q}}(0)}{dt} = -\frac{1}{\varepsilon_0}. \tag{2.172}$$

Since $G^0_{\mathbf{q}}(t)$ is continuous, its integral over an infinitesimally small region is zero. However, the derivative is discontinuous, and its integral is finite. Solving for the boundary conditions at $t = 0$ gives

$$G^0_{\mathbf{q}}(t) = -\frac{1}{\varepsilon_0 \omega_{\mathbf{q}}} \sin \omega_{\mathbf{q}} t \quad \text{for } t \geq 0. \tag{2.173}$$

Therefore, if a photon is instantaneously created at $t = 0$, it will propagate freely for $t > 0$.

The Green's function provides the solution of an instantaneous excitation in space and time of the four-potential field. A more complex driving force can be expressed in terms of instantaneous excitations at different points in space and time using the superposition principle. A product in the momentum-energy dual space is equivalent to a convolution in spacetime, see Eq. (1.247). The vector potential is then the convolution of the free-particle propagator and the four-current

$$\mathsf{A}'(\mathsf{r}') = -\int ds'G^0(\mathsf{r}' - \mathsf{s}')\mathsf{J}'(\mathsf{s}'), \tag{2.174}$$

with $d\mathsf{s}' = d(ct)dxdydz$.

Using Eq. (2.140), the differential equation in (2.161) can be written in terms of the electromagnetic field tensor

$$\nabla\mathsf{F}(\mathsf{r}) = \mu_0\mathsf{J}'(\mathsf{r}'), \tag{2.175}$$

which are Maxwell's equations in spacetime form. Using the electromagnetic field tensor, the expression can also be written in terms of the electric and magnetic fields

$$\nabla\left(\frac{\mathbf{E}}{c} + \boldsymbol{i}\mathbf{B}\right) - \mu_0\mathsf{J}'(\mathsf{r}') = 0. \tag{2.176}$$

Unfortunately, after splitting the terms into scalars, vectors, and their duals, the expressions look somewhat less elegant than Eq. (2.175). Using the expression for the electromagnetic tensor in Eq. (2.140) and the product of the four-nabla and a spatial vector $\mathbf{E}$,

$$\nabla\mathbf{E} = \left(\frac{\mathbf{e}_0'}{c}\frac{\partial}{\partial t} - \nabla'\right)\mathbf{E} = -\frac{1}{c}\frac{\partial\mathbf{E}}{\partial t}\mathbf{e}_0' - \nabla'\cdot\mathbf{E} - \nabla'\wedge\mathbf{E}. \tag{2.177}$$

Removing the primes and rearranging gives $-\nabla'\cdot\mathbf{E} = -(\nabla\,\mathbf{e}_0')\cdot\mathbf{E} = \nabla\cdot\mathbf{E}\,\mathbf{e}_0'$ and

$$-\nabla'\wedge\mathbf{E} = \nabla\wedge\mathbf{E}\,\mathbf{e}_0' = \boldsymbol{i}\nabla\times\mathbf{E}\,\mathbf{e}_0' = \nabla\times\mathbf{E}\,\boldsymbol{i}\mathbf{e}_0', \tag{2.178}$$

using $\boldsymbol{i}\mathbf{e}_i = \mathbf{e}_i\boldsymbol{i}$. Collecting the terms gives

$$\nabla\mathbf{E} = \left(\nabla\cdot\mathbf{E} - \frac{1}{c}\frac{\partial\mathbf{E}}{\partial t}\right)\mathbf{e}_0' + \nabla\times\mathbf{E}\underline{\mathbf{e}}_0', \tag{2.179}$$

with the time dual $\underline{\mathbf{e}}_0' = \boldsymbol{i}\mathbf{e}_0'$. Likewise, one finds

$$\begin{aligned}\nabla(\boldsymbol{i}\mathbf{B}) &= \frac{1}{c}\frac{\partial(\mathbf{e}_0'\boldsymbol{i}\mathbf{B})}{\partial t} - (\nabla\mathbf{e}_0')\cdot(\boldsymbol{i}\mathbf{B}) - (\nabla\mathbf{e}_0')\wedge(\boldsymbol{i}\mathbf{B})\\ &= \nabla\times\mathbf{B}\mathbf{e}_0' + \left(-\nabla\cdot\mathbf{B} + \frac{1}{c}\frac{\partial\mathbf{B}}{\partial t}\right)\underline{\mathbf{e}}_0,\end{aligned} \tag{2.180}$$

with $\mathbf{e}_0'\boldsymbol{i} = -\boldsymbol{i}\mathbf{e}_0' = \underline{\mathbf{e}}_0$. Using these expressions, Equation (2.175) can be expanded as

$$\begin{aligned}&\left[\frac{1}{c}\left(\nabla\cdot\mathbf{E} - \frac{\rho}{\varepsilon_0}\right) + \left(\nabla\times\mathbf{B} - \frac{1}{c^2}\frac{\partial\mathbf{E}}{\partial t} - \mu_0\mathbf{J}\right)\right]\mathbf{e}_0'\\ &\quad + \left[-\nabla\cdot\mathbf{B} + \frac{1}{c}\left(\nabla\times\mathbf{E} + \frac{\partial\mathbf{B}}{\partial t}\right)\right]\underline{\mathbf{e}}_0' = 0,\end{aligned} \tag{2.181}$$

grouping the same multivectors together. This results in four independent equations that we recognize as Maxwell's equations. We can also write them in a more conventional form

$$\begin{aligned}&\nabla\cdot\mathbf{E} = \frac{\rho}{\varepsilon_0}, \qquad \nabla\times\mathbf{E} + \frac{\partial\mathbf{B}}{\partial t} = 0,\\ &\nabla\cdot\mathbf{B} = 0, \qquad \nabla\times\mathbf{B} - \frac{1}{c^2}\frac{\partial\mathbf{E}}{\partial t} = \mu_0\mathbf{J}.\end{aligned} \tag{2.182}$$

The four Maxwell equations are the foundation of the field of electromagnetism. The terms containing a divergence are known as Gauss's laws for the electric and magnetic fields. The equation with the rotation of the electric field is Faraday's law.

The equation with the rotation of $\mathbf{B}$ is Ampère's law, where the time derivative of the electric field is called Maxwell's addition. Applications of these laws are treated in many electricity and magnetism textbooks and will be omitted here.

Conceptual summary. – In this section, Maxwell's equations are derived. This can be done by viewing the creation of the electromagnetic four-potential as a driven harmonic oscillator. The driving force $\mathbf{J}(\mathbf{r}')$ is also a four-vector, known as the four-current, which can depend on space and time. However, the solutions can be written in terms of sine waves created at different points in space and time as determined by the four-current. Expressing the results in terms of the electromagnetic field tensor gives $\nabla\mathbf{F}(\mathbf{r}) = \mu_0\mathbf{J}'(\mathbf{r}')$. These are Maxwell's equations in four-vector format. Rewriting these elegant results into the well-known four Maxwell equations is a somewhat painful process. Essentially, the six bivectors in the electromagnetic field tensor are split into the electric field (the spacetime bivectors) and the magnetic field (the space–space bivectors). Additionally, the four-current is separated into the scalar charge density and the current, which is a vector. These lead to four independent components, with each corresponding to a different Maxwell equation.

Problems

1 Show that in the static limit Eq. (2.166) reduces to Poisson's equation and to the Coulomb potential when $\mathbf{J}'(\mathbf{q}') = ce\mathbf{e}'_0$.

2.8 Nonrelativistic Limit of Dirac Equation

Introduction. – Although the Dirac equation is exact, and the results are elegant, it is generally complex to work with. One therefore often includes the relativistic effects in a nonrelativistic equation. In the presence of an electromagnetic field, this approximation is relatively complex. The approach taken here is known as the Foldy–Wouthuysen transformation. The essential idea behind the approximation is that the four-momentum has scalar (energy) and vector (momentum) components. In the nonrelativistic limit, one tries to bring the particle to a rest frame, meaning that the vector term is zero. This is done using a canonical transformation. The remaining scalar term is the Schrödinger equation. This differential equation is generally used to derive function spaces in the presence of a potential $U(\mathbf{r})$.

The starting point is the Dirac equation in the presence of four-potential $\mathbf{A}'$

$$\mathbf{p}' - q\mathbf{A}' = mc. \tag{2.183}$$

In components, this can be written as

$$(E - U)\mathbf{e}'_0 + c(\mathbf{p}' - q\mathbf{A}') = mc^2. \tag{2.184}$$

In the nonrelativistic limit, there is translational symmetry along the time axis, *i.e.* we are no longer considering the effects of time dilation. Therefore, energy E becomes a conserved quantity. By multiplying the above equation from the right with $\mathbf{e}_0'$, the energy E becomes associated with the identity 1,

$$E - U + c(\mathbf{p} - q\mathbf{A}) - mc^2\mathbf{e}_0' = 0. \tag{2.185}$$

Nonrelativistic dynamics are usually described within an Hamiltonian framework, which separates the energy from the four-momentum. The Hamiltonian is given by

$$\boldsymbol{H} = mc^2\mathbf{e}_0' - c(\mathbf{p} - q\mathbf{A}) + U = mc^2\mathbf{e}_0' - c\mathbf{P} + U, \tag{2.186}$$

with $\mathbf{P} = \mathbf{p} - q\mathbf{A}$. The norm is now given by

$$\boldsymbol{H} = E. \tag{2.187}$$

This leads to what is essentially the Schrödinger (or Pauli's) equation, although the energy is still relativistic. However, note that the Hamiltonian $\boldsymbol{H}$ is not a scalar but is still a multivector. The energy conservation is a direct result of translation symmetry in time. This symmetry disappears in the relativistic limit, due to effects such as time dilation. Additionally, energy is not conserved if the Hamiltonian depends on time. Therefore, all explicit time dependence in the potentials is removed.

Up to this point, we have only rearranged terms. To obtain the nonrelativistic limit, we want to remove the vector term from Eq. (2.186). In terms of spinors, the four-momentum could also be obtained via the hyperbolic rotation $\mathbf{p} = mce^{\hat{\mathbf{p}}\alpha}\mathbf{e}_0' = mce^{\hat{\mathbf{p}}\frac{\alpha}{2}}e^{\hat{\mathbf{p}}\frac{\alpha}{2}}\mathbf{e}_0' = mce^{\hat{\mathbf{p}}\frac{\alpha}{2}}\mathbf{e}_0'e^{-\hat{\mathbf{p}}\frac{\alpha}{2}}$. In order to remove the vector term $\mathbf{P}$, we want to undo this hyperbolic rotation. This can be done via

$$\boldsymbol{H}_r = e^{\frac{\boldsymbol{\alpha}'}{2}}\boldsymbol{H}e^{-\frac{\boldsymbol{\alpha}'}{2}}. \tag{2.188}$$

The vector angle $\boldsymbol{\alpha}'$ now includes the four-potential. Taking the nonrelativistic limit $P \ll mc$ gives

$$\boldsymbol{\alpha}' \cong \sinh \boldsymbol{\alpha}' = \frac{\mathbf{p}' - q\mathbf{A}'}{mc} = \frac{\mathbf{P}'}{mc}, \tag{2.189}$$

with $\mathbf{P}' = \mathbf{P}\mathbf{e}_0'$. The quantity $\mathbf{P}$ is also known as the generalized momentum. Expanding the exponentials for small $\boldsymbol{\alpha}'$ gives

$$\begin{aligned}\boldsymbol{H}_r &\cong \left(1 + \frac{\boldsymbol{\alpha}'}{2} + \frac{\boldsymbol{\alpha}'^2}{8}\right)\boldsymbol{H}\left(1 - \frac{\boldsymbol{\alpha}'}{2} + \frac{\boldsymbol{\alpha}'^2}{8}\right)\\ &= \boldsymbol{H} + \frac{1}{2}(\boldsymbol{\alpha}'\boldsymbol{H} - \boldsymbol{H}\boldsymbol{\alpha}') + \frac{1}{8}(\boldsymbol{\alpha}'^2\boldsymbol{H} + \boldsymbol{H}\boldsymbol{\alpha}'^2 - 2\boldsymbol{\alpha}'\boldsymbol{H}\boldsymbol{\alpha}').\end{aligned} \tag{2.190}$$

This can be shortened using the following commutation relations:

$$\begin{aligned}[\boldsymbol{\alpha}', \boldsymbol{H}] &= \boldsymbol{\alpha}'\boldsymbol{H} - \boldsymbol{H}\boldsymbol{\alpha}',\\ [\boldsymbol{\alpha}', [\boldsymbol{\alpha}', \boldsymbol{H}]] &= \boldsymbol{\alpha}'(\boldsymbol{\alpha}'\boldsymbol{H} - \boldsymbol{H}\boldsymbol{\alpha}') - (\boldsymbol{\alpha}'\boldsymbol{H} - \boldsymbol{H}\boldsymbol{\alpha}')\boldsymbol{\alpha}'.\end{aligned} \tag{2.191}$$

This gives for the rotated Hamiltonian

$$\begin{aligned}\boldsymbol{H}_r \cong{}& \boldsymbol{H} + \frac{1}{2}[\boldsymbol{\alpha}', \boldsymbol{H}] + \frac{1}{2^2 2!}[\boldsymbol{\alpha}', [\boldsymbol{\alpha}', \boldsymbol{H}]]\\ &+ \frac{1}{3^2 3!}[\boldsymbol{\alpha}', [\boldsymbol{\alpha}', [\boldsymbol{\alpha}', \boldsymbol{H}]]] + \frac{1}{4^2 4!}[\boldsymbol{\alpha}', [\boldsymbol{\alpha}', [\boldsymbol{\alpha}', [\boldsymbol{\alpha}', \boldsymbol{H}]]]],\end{aligned} \tag{2.192}$$

where some additional terms in the expansion have been added.

Let us now evaluate the commutation for different terms in the Hamiltonian $\boldsymbol{H}$. The rest energy term gives in the lowest order

$$[\boldsymbol{\alpha}', mc^2\mathbf{e}_0'] = \frac{\mathbf{P}'}{mc}mc^2\mathbf{e}_0' - mc^2\mathbf{e}_0'\frac{\mathbf{P}'}{mc}mc^2 = 2c\mathbf{P}, \tag{2.193}$$

using $\mathbf{P}'\mathbf{e}_0' = \mathbf{P}\mathbf{e}_0'\mathbf{e}_0' = \mathbf{P}$ and $\mathbf{e}_0'\mathbf{P}' = \mathbf{e}_0'\mathbf{P}\mathbf{e}_0' = -\mathbf{P}$. Adding this to the original Hamiltonian, see Eq. (2.190), gives

$$\boldsymbol{H} + \frac{1}{2}[\boldsymbol{\alpha}', mc^2\mathbf{e}_0'] = mc^2\mathbf{e}_0' - c\mathbf{P} + U + \frac{1}{2}(2c\mathbf{P}) = mc^2\mathbf{e}_0' + U. \tag{2.194}$$

We have therefore succeeded in removing the vector term $-c\mathbf{P}$ from the original Hamiltonian $\boldsymbol{H}$ in Eq. (2.186). This is not entirely surprising since the canonical transformation was set up to do this. Unfortunately, as we will see below, although the dominant vector term has been removed, smaller vector terms still remain in the Hamiltonian. However, let us first look for additional scalar terms.

The lowest order commutation with the momentum gives

$$[\boldsymbol{\alpha}', -c\mathbf{P}] = \frac{\mathbf{P}'}{mc}(-c\mathbf{P}) - (-c\mathbf{P})\frac{\mathbf{P}'}{mc} = \frac{2\mathbf{P}^2}{m}\mathbf{e}_0', \tag{2.195}$$

using $\mathbf{P}'\mathbf{P} = \mathbf{P}\mathbf{e}_0'\mathbf{P} = -\mathbf{P}^2\mathbf{e}_0'$ and $\mathbf{P}\mathbf{P}' = \mathbf{P}\mathbf{P}\mathbf{e}_0' = \mathbf{P}^2\mathbf{e}_0'$. However, higher order commutations can produce similar terms. The only comparable term to Eq. (2.195) follows from

$$[\boldsymbol{\alpha}', [\boldsymbol{\alpha}', mc^2\mathbf{e}_0']] = [\boldsymbol{\alpha}', 2c\mathbf{P}] = -\frac{4\mathbf{P}^2}{m}\mathbf{e}_0', \tag{2.196}$$

using the results from Eqs. (2.193) and (2.195). Adding those terms gives

$$\frac{1}{2}[\boldsymbol{\alpha}', -c\mathbf{P}] + \frac{1}{8}[\boldsymbol{\alpha}', [\boldsymbol{\alpha}', mc^2\mathbf{e}_0']] = \frac{1}{2}\frac{2\mathbf{P}^2}{m}\mathbf{e}_0' - \frac{1}{8}\frac{4\mathbf{P}^2}{m}\mathbf{e}_0' = \frac{\mathbf{P}^2}{2m}\mathbf{e}_0'. \tag{2.197}$$

This is the well-known kinetic energy in the classical limit.

Higher order terms in the momentum are obtained from the third-order commutator,

$$\frac{1}{3^2 3!}[\boldsymbol{\alpha}', [\boldsymbol{\alpha}', [\boldsymbol{\alpha}', -c\mathbf{P}]]] = \frac{1}{3^2 3!}\frac{1}{(mc)^3}(-c)[\mathbf{P}', [\mathbf{P}', [\mathbf{P}', \mathbf{P}]]] = -\frac{\mathbf{P}^4}{6m^3c^2}\mathbf{e}_0',$$

using that

$$[\mathbf{P}', [\mathbf{P}', [\mathbf{P}', \mathbf{P}]]] = 8\mathbf{P}^4\mathbf{e}_0', \tag{2.198}$$

and from the fourth-order commutator,

$$\frac{1}{4^2 4!}[\boldsymbol{\alpha}', [\boldsymbol{\alpha}', [\boldsymbol{\alpha}', [\boldsymbol{\alpha}', mc^2\mathbf{e}_0']]]] = \frac{1}{4^2 4!}\frac{1}{(mc)^3}2c[\mathbf{P}', [\mathbf{P}', [\mathbf{P}', \mathbf{P}]]] = \frac{\mathbf{P}^4}{24m^3c^2}\mathbf{e}_0',$$

using the results from Eqs. (2.193) and (2.198). Adding these gives $-\mathbf{P}^4/8m^3c^2$. The total transformed Hamiltonian is $\boldsymbol{H}_r = \boldsymbol{H}_0 + \boldsymbol{H}_1$, where $\boldsymbol{H}_1$ contains all the higher order corrections, some of which will be described in Sections 2.9 and 2.10. The terms in $\boldsymbol{H}_0$ are

$$\boldsymbol{H}_0 = \left(mc^2 + \frac{\mathbf{P}^2}{2m} - \frac{\mathbf{P}^4}{8m^3c^2}\right)\mathbf{e}_0' + U(\mathbf{r}). \tag{2.199}$$

The lowest order Hamiltonian is still a multivector. Using the Dirac basis, the energy was taken as the norm, *i.e.* the scalar term. The potential appears as a local correction to the norm. In Eq. (2.194), the energy term of the four-momentum initially only contained the rest energy mc^2. However, additional energy terms appeared from higher order corrections. Although the terms $\mathbf{P}^2$ and $\mathbf{P}^4$ contain the generalized momentum to an even power, they are not scalar terms but contain vector components as well, as we will see in Section 2.9. Note that the kinetic energy terms can also be obtained by expanding the relativistic energy from Eq. (2.49),

$$E_{\mathbf{P}} = \sqrt{m^2c^4 + c^2\mathbf{P}^2} = mc^2 + \frac{\mathbf{P}^2}{2m} - \frac{\mathbf{P}^4}{8m^3c^2} + \cdots \tag{2.200}$$

In the absence of the vector potential $\mathbf{A}$, the eigenenergies E follow from the norm of the Hamiltonian $\boldsymbol{H}_0 = E$,

$$\boldsymbol{H}_0 = \left(mc^2 + \frac{\mathbf{p}^2}{2m}\right)\mathbf{e}'_0 + U(\mathbf{r}) = E \quad \Rightarrow \quad mc^2 + \frac{\mathbf{p}^2}{2m} = \pm(E - U(\mathbf{r})),$$

neglecting the $\mathbf{P}^4$ term. There are positive and negative energy solutions. There are different ways to interpret this. We can assume the existence of negative energy solutions. However, in that case all these states need to be filled with particles, otherwise a particle with positive energy can decay into one of these negative energy states, which is not observed. This is known as the Dirac sea. The other option is to view these states as traveling "backward in time." This can be seen by looking at the time dependence of the function space in Eq. (2.64). For negative energy states with $E = -\hbar\omega$ and $\omega > 0$, the complex exponential can be rewritten as $e^{-i(-\omega)t} = e^{-i\omega(-t)} = e^{i\omega t}$. So, the solution can also be viewed as a wave with a negative time argument. This might also sound weird. However, notice that for real trigonometric functions, both exponentials $e^{\pm i\omega t}$ are needed. This occurs, for example for the photon in Eqs. (2.171) and (2.173) that is created at a particular point in time. To describe a sine, both clockwise and counterclockwise rotations in time are needed.

The nonrelativistic energy is taken relative to the rest mass, *i.e.* $E_{\text{n.r.}} = E - mc^2$. When we only consider the positive energy value, the problem can be rewritten in terms of a purely scalar Hamiltonian H_S,

$$H_S = \frac{\mathbf{p}^2}{2m} + U(\mathbf{r}) = E_{\text{n.r.}}, \tag{2.201}$$

which is the classical conservation of energy. In order to find the function space associated with the space described by H_S, the momentum and energy need to be replaced by their respective operators, giving

$$\hat{H}_S\psi(\mathbf{r},t) = \frac{\hat{\mathbf{p}}^2}{2m}\psi(\mathbf{r},t) + U(\mathbf{r})\psi(\mathbf{r},t) = \hat{E}_{\text{n.r.}}\psi(\mathbf{r},t), \tag{2.202}$$

using Eq. (2.60). This gives the time-dependent Schrödinger equation

$$-\frac{\hbar^2}{2m}\nabla^2\psi(\mathbf{r},t) + U(\mathbf{r})\psi(\mathbf{r},t) = i\hbar\frac{\partial\psi(\mathbf{r},t)}{\partial t}. \tag{2.203}$$

This differential equation allows us to determine function spaces in the presence of a potential. This will be done for several potentials in Chapter 3. While in Chapter 1, the function space could be associated with higher-order vectors, this is not possible in general. The strength of the Schrödinger equation is that it provides a brute-force way to derive the function space.

Conceptual summary. – In this section, we have studied the nonrelativistic limit of the Dirac equation. This was done by a canonical transformation that brings the particle to a rest frame. As a result, the vector part of the four-momentum is removed, leaving only the energy terms. In Section 2.9, we will see that this operation is not entirely successful. Since the momentum is an operator, it does not commute with the potential U and the vector potential $\mathbf{A}$. Therefore, some vector terms remain in the rest frame.

The lowest terms of the Foldy–Wouthuysen transformation simply lead to the classical conservation of energy. In operator form this leads to the Schrödinger equation. So far, the considerations in this chapter have mainly been in terms of vectors. This is fine when dealing with rigid objects or point particles. However, in quantum mechanics, we are generally interested in describing deformable particles with a finite extent, meaning that the shape of the particle adjusts to the local potentials. We are therefore interested in obtaining the best function space to describe these particles. In Chapter 1 on space, these functions were obtained by finding the eigenfunctions of a conserved quantity, such as the momentum and angular momentum. However, in the presence of an electromagnetic field, these quantities are not conserved, and the Schrödinger equation is a better option. Several examples of potentials are treated in Chapter 3.

Problems

1 In Chapter 3, we look in detail at more complex potentials. Here, we consider several problems with discrete potentials, where the value of the potential is constant, but suddenly changes value. A typical one-dimensional problem is a potential with $U(x) = 0, U$ (with $U > 0$) for $x \leq 0$ and $x > 0$, respectively. Find the probability of a running wave traveling in the positive x-direction to be reflected and transmitted for $E > U$ and $E < U$.

2 As in the preceding problem, calculate the reflection and transmission, but now for a potential $U < 0$ in the region $0 \leq x \leq L$ and 0 elsewhere. Look for solutions with $E > 0$ and $E < 0$.

3 Repeat the preceding problem, but now for $U > 0$.

4 Calculate the bound state ($E < 0$) for a potential $U(x) = -U_0\delta(x)$, with $U > 0$.

2.9 Interactions Between Particles and Electromagnetic Field

Introduction. – In this section, we have a look at how the electromagnetic field is incorporated in the nonrelativistic framework. The scalar term, the electric potential $U(\mathbf{r})$, is already nicely separated in the Schrödinger equation in Eq. (2.203). The vector potential is trickier since it appears in the square $(\hat{\mathbf{p}} - q\mathbf{A})^2$ with the momentum operator. Although the goal of the Foldy–Wouthuysen transformation was to remove all the vector terms from the four-momentum, we see that there is a remaining vector component which is generally identified as the spin.

In Section 2.8, it was demonstrated how to expand the Hamiltonian in the nonrelativistic limit. Some additional terms in the expansion will be obtained in Section 2.10, but let us initially look at the lowest-order terms containing interaction between a particle and the field. The most important term is the potential energy $U(\mathbf{r})$ which directly shifts the energy when the particle is moving in space. This term is already separated from the momentum operator in the Schrödinger equation, see Eq. (2.203). The next term is contained in the expansion of the kinetic energy with the generalized momentum. The leading term is

$$\frac{\hat{\mathbf{P}}^2}{2m} = \frac{(\hat{\mathbf{p}} - q\mathbf{A})^2}{2m} = \frac{\hat{\mathbf{p}}^2}{2m} - \frac{q}{2m}(\hat{\mathbf{p}}\mathbf{A} + \mathbf{A}\hat{\mathbf{p}}) + \frac{\mathbf{A}^2}{2m}, \tag{2.204}$$

where the momentum $\hat{\mathbf{p}} = -i\hbar\nabla$ is expressed as an operator. This is necessary to account for the changes of the vector potential in space. The first term on the right-hand side is the kinetic energy of the particle. The last term only contains the vector potential. Since it is squared, it describes the scattering of electromagnetic radiation. The interaction between the particle and the field is contained in the remaining terms, which can be written as

$$\boldsymbol{H}_Z = -\frac{q}{m}\mathbf{A} \cdot \mathbf{p} - \frac{q}{2m}[\hat{\mathbf{p}}\mathbf{A}], \tag{2.205}$$

where the brackets on the right-hand side indicate that the momentum operates only on the vector potential and not on any other functions to the right. Below, the discussion will focus on electrons with $q = -e$, where these effects are most commonly studied. The first term on the right-hand side describes the effect of the vector potential on the momentum. Note that by summing the different products of $\hat{\mathbf{p}}$ and $\mathbf{A}$ the wedge product cancels, see Eqs. (1.115) and (1.116), and only the inner product remains. The operator sign has been removed from the momentum. It still operates on the function space, but we have the freedom to express it as a derivative or a vector operation. In the second term on the right-hand side, the momentum only operates on the vector potential and not on the function space.

The term $\mathbf{A} \cdot \mathbf{p}$ describes the effect of the vector potential on the momentum at a particular point in space. Since it is a scalar, it becomes a potential energy term. We want to reexpress this in terms of the magnetic field, which is the bivector that changes the direction of the momentum. Given a magnetic field $\mathbf{B}$ at a particular point in space, the vector potential can be expressed in $\mathbf{B}$ via

$$\mathbf{A} = \frac{1}{2}\mathbf{B} \times \mathbf{r}. \tag{2.206}$$

This can be checked using vector calculus identities taking $\mathbf{B}$ to be constant

$$\mathbf{B} = \nabla \times \mathbf{A} = \frac{1}{2}\nabla \times (\mathbf{B} \times \mathbf{r}) = \frac{1}{2}(\mathbf{B}(\nabla \cdot \mathbf{r}) - (\mathbf{B} \cdot \nabla)\mathbf{r}) = \frac{1}{2}(3\mathbf{B} - \mathbf{B}) = \mathbf{B}.$$

Inserting this in the first term gives

$$-\frac{q}{2m}(\mathbf{B} \times \mathbf{r}) \cdot \mathbf{p} = -\frac{q}{2m}\mathbf{B} \cdot (\mathbf{r} \times \mathbf{p}) = -\frac{q}{2m}\mathbf{B} \cdot \mathbf{L}, \tag{2.207}$$

using that the angular momentum is $\mathbf{L} = \mathbf{r} \times \mathbf{p}$. This can be classically interpreted as the interaction between the magnetic moment of the particle and the magnetic field. The classical magnetic moment is $\mu = IA$, where I is the current, and A is the area. Suppose a charged particle is moving in a circle, then the current is $I = q/T$, where T is the period. In terms of the velocity v, the period is $2\pi r/v$. The magnetic moment is then $\mu = IA = (qv/2\pi r)(\pi r^2) = qvr/2 = qpr/2m$. For circular motion, $\mathbf{r}$ and $\mathbf{p}$ are perpendicular, so the magnitude of the angular momentum is $L = rp$. The magnetic moment is therefore $\mu = qL/2m$. The vector $\boldsymbol{\mu}$ is perpendicular to the plane of rotation and therefore parallel to the angular momentum. The potential of a magnetic moment in an external magnetic field is $U = -\mathbf{B} \cdot \boldsymbol{\mu} = -(q/2m)\mathbf{B} \cdot \mathbf{L}$, which is exactly the term above. This term needs aligns the magnetic moment (or, equivalently, the angular momentum) of the charged particle to the magnetic field.

For the second term, we have

$$-\frac{q}{2m}[\hat{\mathbf{p}}\mathbf{A}] = \boldsymbol{i}\frac{q\hbar}{2m}\nabla\mathbf{A} = \boldsymbol{i}\frac{q\hbar}{2m}(\nabla \cdot \mathbf{A} + \nabla \wedge \mathbf{A}). \tag{2.208}$$

Using gauge invariance, we can take $\nabla \cdot \mathbf{A} = 0$. The wedge term gives

$$\frac{q\hbar}{2m}\boldsymbol{i}\nabla \wedge \mathbf{A} = -\frac{q\hbar}{2m}\nabla \times \mathbf{A} = -\frac{q\hbar}{2m}\mathbf{B}. \tag{2.209}$$

The total interaction is then

$$\boldsymbol{H}_Z = -\frac{q}{2m}(\mathbf{B} \cdot \mathbf{L} + \hbar\mathbf{B}). \tag{2.210}$$

Note that the Hamiltonian contains a scalar and a vector term. Although the leading vector term $\mathbf{p} - q\mathbf{A}$ was removed by the Foldy–Wouthuysen transformation, vector terms due to the noncommutativity of the operator $\hat{\mathbf{p}}$ and the vector potential $\mathbf{A}$ still remain.

Let us find the eigenfunctions of $\boldsymbol{H}_Z$. Let us look at an electron in an atom with $q = -e$ where the transformation properties of the spherical harmonics can be expressed in terms of the unit vectors $|lm\rangle$ in Eq. (1.361) and the dimensionless angular momentum operators $\mathbf{L}_i$ from Eqs. (1.330) and (1.370). The change in energy due to $\boldsymbol{H}_Z$ for the state can be obtained from the norm of $\boldsymbol{H}_Z$

$$\boldsymbol{H}_Z = \mu_B(\mathbf{B} \cdot \mathbf{L} + \mathbf{B}) = E \;\Rightarrow\; E - \mu_B\mathbf{B} \cdot \mathbf{L} = \mu_B\mathbf{B}, \tag{2.211}$$

where the scalar and vector terms have been separated. The constant

$$\mu_B = \frac{e\hbar}{2m} \tag{2.212}$$

is known as the Bohr magneton. The angular momentum is given in dimensionless units. Squaring gives

$$(E - \mu_B\mathbf{B} \cdot \mathbf{L})^2 = (\mu_B\mathbf{B})^2 \;\Rightarrow\; E = \mu_B(\mathbf{B} \cdot \mathbf{L} \pm B). \tag{2.213}$$

This result is actually still in vector format since it corresponds to $\boldsymbol{H}_Z = \mu_B(\mathbf{B} \cdot \mathbf{L} + B\mathbf{e}_z)$. However, since we have aligned the spin parallel or antiparallel with the magnetic field, the vector term now appears as two possible scalar changes. When the magnetic field is taken along the z-axis, the inner product becomes $\mathbf{B} \cdot \mathbf{L} = B\mathbf{L}_z$, where the angular momentum is given as a vector operator. Working with the interaction on $|lm\rangle$ gives

$$\mu_B B(\mathbf{L}_z \pm 1)|lm\rangle = \mu_B B(m \pm 1)|lm\rangle \;\Rightarrow\; E = \mu_B B(m \pm 1). \tag{2.214}$$

An aspect that can be confusing is that the angular momentum term is a scalar for the Hamiltonian. However, $\mathbf{L}_i$ is a vector operator when working in the function space $|lm\rangle$ of spherical harmonics.

This result is known as the Zeeman splitting. The magnetic field splits the energies of the different values of the projected angular momentum m. Therefore, the magnetic field lowers the symmetry of the atom from spherical to cylindrical. The lowest state has $m = -l$. The electron's angular momentum is then opposite to the magnetic field. However, since the electron is negatively charged, its magnetic moment aligns parallel to the field. The vector term seems to behave like a rotation with eigenvalues $\pm\hbar$. This is comparable to rotations in a plane, where the rotation axis is defined by the magnetic field $\mathbf{B}$. This is generally known as the spin of the particle. We return to this below.

Note that the same result can be obtained using the Pauli matrices in Eq. (1.149) as the unit vectors. For an arbitrary direction of the magnetization,

$$(E - \mu_B \mathbf{B}\cdot\mathbf{L})\mathbb{1}_2 = \mu_B \mathbf{B} = \mu_B(B_x \mathbf{e}_x + B_y \mathbf{e}_y + B_z \mathbf{e}_z). \tag{2.215}$$

The eigenvalues are obtained by solving the determinant (which calculates the norm of a vector)

$$\begin{vmatrix} \mu_B \mathbf{B}\cdot\mathbf{L} - E + \mu_B B_z & \mu_B(B_x - iB_y) \\ \mu_B(B_x + iB_y) & \mu_B \mathbf{B}\cdot\mathbf{L} - E - \mu_B B_z \end{vmatrix}$$
$$= (\mu_B \mathbf{B}\cdot\mathbf{L} - E)^2 - \mu_B^2 B_z^2 - \mu_B^2(B_x^2 + B_y^2) = 0. \tag{2.216}$$

This gives the same equation as Eq. (2.213).

Using the results from Section 1.2, rotational motion in a plane is described by complex exponentials

$$\hat{\boldsymbol{s}}_{\pm\frac{1}{2}}(\varphi) = e^{\pm \boldsymbol{i}_B \varphi}. \tag{2.217}$$

The bivector is given by $\boldsymbol{i}_B = -\boldsymbol{\imath}\mathbf{e}_B$, where $\mathbf{e}_B = \mathbf{B}/B$ is the unit vector in the direction of the magnetic field. The subscripts are half-integer since the rotation is around a "fixed" axis (the magnetic field) and therefore effectively two-dimensional. The only freedom is whether to perform a counterclockwise or a clockwise rotation. Additionally, it is directly related to the magnetic field vector. For a single particle, there are therefore no higher order vector terms, such as $e^{\pm m_s \boldsymbol{i}_B \varphi}$. The relative unit vector is indicated as $\hat{\boldsymbol{s}}$. This is still a unit vector in real space in a plane perpendicular to the magnetic field. However, it is a vector that is independent of the coordinates $\boldsymbol{r}$. While $\hat{\boldsymbol{s}}$ is a purely rotational motion intimately related to the direction of the magnetic field, $\boldsymbol{r}$ is connected to the motion of the particle, which often has little to do with the direction of the magnetic field due to the presence of much stronger electric fields. However, in some cases, for example for the Zeeman effect in atoms, the rotation of $\hat{\boldsymbol{s}}$ and $\boldsymbol{r}$ is around the same axis. Note that $\hat{\boldsymbol{s}}_{\pm\frac{1}{2}}$ defines the motion in a plane perpendicular to the magnetic field. The associated angular momentum $\mathbf{S}$, generally called the spin, is parallel to the magnetic field. When the electron is in an eigenstate, the angular momentum related to the spin is $\pm\hbar$. When the electron is in a mixed state, it can assume values between $-\hbar$ and $\hbar$. In Section 2.10, we will see the presence of additional vector terms that are coupled to the motion of the particle in space.

The complex exponentials are eigenvectors of the angular momentum operator for the spin with respect to the bivector $\boldsymbol{i}_B$, see Eq. (1.43),

$$2\hat{S}_z = \hat{\mathbf{e}}_B = -\boldsymbol{i}_B \frac{\partial}{\partial \varphi} \quad \Rightarrow \quad \hat{\mathbf{e}}_B \hat{\boldsymbol{s}}_{\pm\frac{1}{2}} = \pm \hat{\boldsymbol{s}}_{\pm\frac{1}{2}}. \tag{2.218}$$

As for angular momentum, we want to associate a unit vector with the two functions $\hat{\mathbf{s}}_{\pm\frac{1}{2}}$. Following Section 1.14, the rotation axis (the direction of the magnetic field) can be associated with spinors. This gives two spinors

$$|\frac{1}{2}, \pm\frac{1}{2}\rangle = \boldsymbol{b}^\dagger_{\pm\frac{1}{2}}, \tag{2.219}$$

where we switch directly to the notation of indistinguishable spinors. The operator working on the unit vector can then be written as

$$\mathbf{e}_B = \boldsymbol{b}^\dagger_\uparrow \boldsymbol{b}_\uparrow - \boldsymbol{b}^\dagger_\downarrow \boldsymbol{b}_\downarrow \quad \Rightarrow \quad \mathbf{e}_B |\frac{1}{2}, \pm\frac{1}{2}\rangle = \pm|\frac{1}{2}, \pm\frac{1}{2}\rangle. \tag{2.220}$$

Note that the operation is equivalent to the z component of the Pauli spin matrices, see Eq. (1.149), which is expressed in terms of a single spinor basis.

Although all eigenvalues are related to integers (± 1), the quantities are often denoted by half-integers. Although a rotation in a plane or a rotation around a fixed axis can be indicated by single spinors, the extension to a three-dimensional space required the use of products of two spinors, see Section 1.14. This added a factor $\frac{1}{2}$ in the angular momentum operator, see Eq. (1.330). Defining the spin operator in the same fashion gives

$$\mathbf{S}_B = \frac{1}{2}\mathbf{e}_B = \frac{1}{2}(\boldsymbol{b}^\dagger_\uparrow \boldsymbol{b}_\uparrow - \boldsymbol{b}^\dagger_\downarrow \boldsymbol{b}_\downarrow) \quad \Rightarrow \quad \mathbf{S}_B|\frac{1}{2}, \pm\frac{1}{2}\rangle = \pm\frac{1}{2}|\frac{1}{2}, \pm\frac{1}{2}\rangle. \tag{2.221}$$

The operators on the unit vectors are taken dimensionless. We follow here the conventional notation of $\mathbf{S}$ for the angular momentum of the spin.

A complete basis for a unit sphere for the combined orbital and spin degrees of freedom is

$$\begin{aligned}\sum_{\sigma=\pm\frac{1}{2}} \boldsymbol{b}^\dagger_\sigma \hat{\boldsymbol{s}}^\sigma e^{\hat{\boldsymbol{r}}} &= \sum_{\sigma=\pm\frac{1}{2}} \sum_{l=0}^{\infty} \sum_{m=-l}^{l} \boldsymbol{b}^\dagger_\sigma \, e^l_m e^{-2\sigma \boldsymbol{i}_B \varphi'} C^{lm}(\hat{\boldsymbol{r}}) \\ &= \sum_{\sigma=\pm\frac{1}{2}} \sum_{l=0}^{\infty} \sum_{m=-l}^{l} |lm\rangle |\frac{1}{2}\sigma\rangle \langle \frac{1}{2}\sigma|\hat{\boldsymbol{s}}\rangle \langle lm|\hat{\boldsymbol{r}}\rangle, \end{aligned} \tag{2.222}$$

using Eq. (1.362) and with $\hat{\mathbf{s}}^\sigma = (\hat{\mathbf{s}}_\sigma)^*$. This fixes a problem that was not discussed before. The first term in the expansion of $e^{\hat{\boldsymbol{r}}}$ is simply 1. Additionally, $|00\rangle = 1$. However, applying 1 to something does not change anything. However, now the lowest term is $\boldsymbol{b}^\dagger_{\pm\frac{1}{2}} \, e^{\mp \boldsymbol{i}_B \varphi}$, *i.e.* there is a rotational zero-point motion. Therefore, atomic s orbitals with $l = m = 0$ rotate around the direction of the magnetic field. In the absence of a magnetic field, the direction of rotation is random. Note that, unlike, for example for the quantum harmonic oscillator, the zero-point motion is not needed to construct the higher order wavefunctions.

The vector term is generally written differently. To separate the zero-point motion from the magnetic field, the Pauli vector is introduced,

$$\mathbf{S} = \frac{\hbar \boldsymbol{\sigma}}{2} = \frac{\hbar}{2} \sum_{i=x,y,z} \mathbf{e}_i \boldsymbol{e}_i, \quad \text{with } \mathbf{S}^2 = \frac{\hbar^2}{4} \sum_{i=x,y,z} \mathbf{e}_i^2 = \frac{3}{4}\hbar^2, \tag{2.223}$$

where the unit $\hbar$ has been included, $\boldsymbol{\sigma} = \sum_{i=x,y,z} \mathbf{e}_i \boldsymbol{e}_i$ is a vector of unit vectors, and $\mathbf{e}_i^2 = 1$. Conventionally, the Pauli matrices $\mathbf{e}_i$ are indicated as σ_i. In our notation, this looks somewhat strange, since it expresses anticommuting unit vectors $\mathbf{e}_i$ in a basis of commuting unit vectors $\boldsymbol{e}_i$. The magnetic field can then also be expressed in the commuting basis $\boldsymbol{B} = B_x \boldsymbol{e}_x + B_y \boldsymbol{e}_y + B_z \boldsymbol{e}_z$. The interaction with an external magnetic field then becomes

$$\boldsymbol{H}_Z = -\frac{q}{2m} \boldsymbol{B} \cdot (\mathbf{L} + \hbar \boldsymbol{\sigma}) = -\frac{q}{2m} \boldsymbol{B} \cdot (\mathbf{L} + g\mathbf{S}). \tag{2.224}$$

However, note that $\boldsymbol{B} \cdot \boldsymbol{\sigma} = \mathbf{B} = B_x \mathbf{e}_x + B_y \mathbf{e}_y + B_z \mathbf{e}_z$, which is the magnetic field in the anticommuting basis. Therefore, the Pauli vector makes the vector term appear as a scalar. However, it does bring out the similarities between orbital and spin momentum. The factor $g = 2$ in Eq. (2.224) is known as the gyromagnetic ratio. Higher order corrections make this factor slightly different from 2.

Example 2.3 ***Change of Magnetic Field Direction***

Let us take an initial magnetic field in an arbitrary direction $\hat{\mathbf{r}}_1$. The initial motion is in the counterclockwise direction. The final magnetic field is in the $\hat{\mathbf{r}}_2$-direction. This is essentially the Stern–Gerlach experiment.

(a) Calculate the probabilities of ending up in counterclockwise and clockwise rotations around the $\hat{\mathbf{r}}_2$-axis using spinors.
(b) Obtain the same result using the φ-dependent function in Eq. (2.217).
(c) Give a physical interpretation of the results.

Solution

(a) Essentially, the experiment changes the axis for a two-dimensional rotation described in terms of oriented complex exponentials, see Eq. (1.136). The probabilities can be obtained from the unit vectors (the spinors) or the oriented complex exponentials. Let us start with the spinors. In column notation, the spinors for rotation around an axis in an arbitrary direction $\hat{\mathbf{r}}_\alpha$ are

$$|\hat{\mathbf{r}}_{\alpha\uparrow}\rangle = \begin{pmatrix} \cos\dfrac{\theta_\alpha}{2} \\ e^{i_{yx}\varphi_\alpha} \sin\dfrac{\theta_\alpha}{2} \end{pmatrix}, \quad |\hat{\mathbf{r}}_{\alpha\downarrow}\rangle = \begin{pmatrix} -\sin\dfrac{\theta_\alpha}{2} \\ e^{i_{yx}\varphi_\alpha} \cos\dfrac{\theta_\alpha}{2} \end{pmatrix}, \tag{2.225}$$

where the basis of the column vectors is $\mathbf{e}_z$ and $\mathbf{e}_x$. The unit vectors $\hat{\mathbf{r}}_\alpha$, with $\alpha = 1, 2$, indicate the initial and final directions of the magnetic field, respectively. The counterclockwise and clockwise rotations around the fixed axis $\hat{\mathbf{r}}$ are given by $\sum_i \langle \hat{\mathbf{r}}_{\pm\frac{1}{2}} | \mathbf{e}_i | \hat{\mathbf{r}}_{\pm\frac{1}{2}} \rangle \mathbf{e}_i = \pm\hat{\mathbf{r}} = \hat{\mathbf{r}}_\pm$, where $\mathbf{e}_i$, with $i = x, y, z$, are the Pauli spinors. This expression is the Pauli matrix equivalent of Eq. (1.127). The probabilities for different final states are $|\langle \hat{\mathbf{r}}_{2,\pm\frac{1}{2}} | \hat{\mathbf{r}}_{1\uparrow} \rangle|^2$.

(b) Let us consider the transition between two up spins as an integral over complex exponentials oriented in space

$$\int_0^{2\pi} d\varphi \, \chi^*_{\hat{\mathbf{r}}_2\uparrow}(\varphi) \chi_{\hat{\mathbf{r}}_1\uparrow}(\varphi) = \frac{1}{2\pi} \int_0^{2\pi} d\varphi \, e^{\hat{\imath}\hat{\mathbf{r}}_2 \varphi} e^{-\hat{\imath}\hat{\mathbf{r}}_1 \varphi}, \tag{2.226}$$

where φ is related to the argument of the complex exponential, which is unrelated to the φ_i indicating the direction of the axes of rotation. The oriented complex exponentials can be expanded to give

$$\frac{1}{2\pi}\int_0^{2\pi} d\varphi(\cos^2\varphi + \hat{\mathbf{r}}_1\hat{\mathbf{r}}_2\sin^2\varphi - \frac{\boldsymbol{i}}{2}(\hat{\mathbf{r}}_1 - \hat{\mathbf{r}}_2)\sin 2\varphi)$$
$$= \frac{1}{2}(1+\hat{\mathbf{r}}_1\hat{\mathbf{r}}_2) = \frac{1}{2}\left(1+e^{\boldsymbol{i}_{x_1'z}\theta_1 - \boldsymbol{i}_{x_2'z}\theta_2}\right), \qquad (2.227)$$

where $\boldsymbol{i}_{x'_\alpha z} = e^{\boldsymbol{i}_{yx}\varphi_\alpha}\boldsymbol{i}_{xz}$ is the imaginary unit related to a rotation in the zx'_α-plane, which is a plane rotated by an angle φ_α with the x-axis. The integrals over φ are straightforward, and the term containing $\sin 2\varphi$ cancels. Taking out a complex exponential over half the angle and expanding the remaining complex exponentials gives

$$e^{\boldsymbol{i}_{x_1'z}\frac{\theta_1}{2} - \boldsymbol{i}_{x_2'z}\frac{\theta_2}{2}}\left(\cos\frac{\theta_1}{2}\cos\frac{\theta_2}{2} + e^{\boldsymbol{i}_{yx}(\varphi_1-\varphi_2)}\sin\frac{\theta_1}{2}\sin\frac{\theta_2}{2}\right)$$
$$= e^{\boldsymbol{i}_{x_1'z}\frac{\theta_1}{2} - \boldsymbol{i}_{x_2'z}\frac{\theta_2}{2}}\langle\hat{\mathbf{r}}_{2\uparrow}|\hat{\mathbf{r}}_{1\uparrow}\rangle. \qquad (2.228)$$

The result for the down spinor can be obtained by changing the direction of the rotation axis in the final state by taking $\theta_2 \to \theta_2 + \pi$, giving $\cos\frac{\theta_2+\pi}{2} = -\sin\frac{\theta_2}{2}$ and $\sin\frac{\theta_2+\pi}{2} = \cos\frac{\theta_2}{2}$. This shows that, apart from a phase factor, the transformation between the two oriented complex exponentials is expressed in terms of the inner product $\langle\hat{\mathbf{r}}_{2,\pm\frac{1}{2}}|\hat{\mathbf{r}}_{1\uparrow}\rangle$ between two spinors, see Eq. (2.225). If one is only interested in the probability, one can square the expression in Eq. (2.227) or the matrix elements between the spinors given by Eqs. (2.225) and (2.228)

$$|\langle\hat{\mathbf{r}}_{2,\pm\frac{1}{2}}|\hat{\mathbf{r}}_{1\uparrow}\rangle|^2 = \left(\frac{1}{2}\pm\frac{1}{2}\hat{\mathbf{r}}_1\cdot\hat{\mathbf{r}}_2\right)^2 + \left(\frac{1}{2}|\hat{\mathbf{r}}_1\times\hat{\mathbf{r}}_2|\right)^2$$
$$= \frac{1}{2}(1\pm\cos\theta_1\cos\theta_2 \pm \cos(\varphi_1-\varphi_2)\sin\theta_1\sin\theta_2). \qquad (2.229)$$

Note that the calculation using Eq. (2.227) contains no half-angles. This result reduces to the well-known result $\cos^2\frac{\theta_1-\theta_2}{2}$, $\sin^2\frac{\theta_1-\theta_2}{2}$ for $\varphi_1 = \varphi_2$, a typical setup of the Stern–Gerlach experiment.

(c) The spin is a remaining vector component of the four-momentum that arises because the momentum and the vector potential do not commute. It can be viewed as a rest motion, essentially a two-dimensional rotation around a fixed axis. The direction of the axis is usually determined by the magnetic field and should correspond to the square of the spinors, see Eq. (1.195). Although the axis of rotation can be randomly oriented in space, there are only two eigenstates $e^{\pm\boldsymbol{i}_B\varphi}$, corresponding to counterclockwise and clockwise rotations around the magnetic field direction $\mathbf{e}_B$, with $\boldsymbol{i}_B = -\boldsymbol{i}\mathbf{e}_B$. When the magnetic field is switched off, the rotation continues in the same direction. When a magnetic field is switched on in a different direction, there are again only two states: counterclockwise and clockwise rotations around the new magnetization direction. Equation (2.229) calculates the probability of ending up in each direction.

Unfortunately, confusion often arises in the description of the Stern–Gerlach experiment, where this projection onto a new axis of rotation is described as measuring the S_z component. The initial state is interpreted as having "classical" expectation values $\langle S_z\rangle = \frac{\hbar}{2}\cos\theta$ and $\langle S_x\rangle = \frac{\hbar}{2}\sin\theta$ for a spin in the zx-plane. Therefore, naively, one would expect to measure $\langle S_z \rangle$. Instead, one observes $\pm\frac{\hbar}{2}$. However, if one insists on a classical analogy, it is better to visualize a top spinning in free space at a constant angular velocity (the angular velocity is constant, since we are looking at the rest rotation). Taking $\theta = 90^\circ$, the top is initially spinning counterclockwise around the x-axis. Now, we suddenly change the axis of rotation to $\mathbf{e}_z$. However, there are only two states: counterclockwise and clockwise rotations around the z-axis with the same constant angular velocity. Since, there is no particular preference for either direction, the probabilities are equal, as expected from Eq. (2.229).

Conceptual summary. – In this section, we looked at the interaction between the particle and the vector potential $\mathbf{A}$, which is associated with the magnetic field. Classically, one expects that the magnetic field causes a particle to move in a circle. A scalar term is obtained that corresponds to this motion. The result can be expressed in terms of the inner product $\mathbf{B} \cdot \mathbf{L}$ of the magnetic field and the angular momentum, in agreement with the classical expectation. There are additional effects related to the term $\mathbf{A} \cdot \mathbf{p}$. Since the vector potential only occurs once, this interaction also describes the absorption and emission of electromagnetic radiation by particles. We return to this in Section 3.6.

However, additionally, a vector term is obtained due to the fact that the momentum and the vector potential do not commute. Therefore, our attempt to move to the rest frame of the particle and to remove the vector part from the four-momentum is not entirely successful. The remaining vector part is generally known as the spin. Since it is due to the noncommutativity of the momentum and the electromagnetic field, this can be viewed as a zero point motion. In Section 2.10, we will see that there are additional vector terms that remain in the Foldy–Wouthuysen transformation. Ignoring those additional terms for the moment, we see that the direction of the remaining vector term is entirely determined by the magnetic field, and the two solutions correspond to clockwise and counterclockwise rotations around $\mathbf{B}$. There is so far nothing else in the Hamiltonian that affects the vector part. This is not the case for the term associated with the angular momentum $\mathbf{L}$. Although the vector potential wants the particle to move in a circular fashion, the potential energy (which often dominates the physics) can drastically change the motion. This is not the case for the spin which makes it often appear as an independent degree of freedom.

Those familiar with the zero-point motion of the quantum harmonic oscillator can consider the following analogy. The zero-point motion for the harmonic oscillator arises because the momentum does not commute with the potential. In one dimension, there is no directional aspect, so let us look in two dimensions. For an isotropic two-dimensional oscillator, there is no preferred direction. However, let us take the oscillation frequencies $\omega_x \ll \omega_y$. This means that the amplitude of the zero-point motion is always predominantly in the x-direction. Even if we give the system a very large amplitude in the y-direction, this zero-point motion in the x-direction always remains. The direction of the zero-point motion is entirely determined by the potential landscape and independent of the motion of the particle. The spin occurs because there is a vector term arising from the fact that the momentum does not commute with the vector potential. This is again a zero-point motion, however, now corresponding to a rotation. Regardless of the motion of the particle, this zero-point rotation always points in a direction solely determined by the electromagnetic interaction, in this case the magnetic field (ignoring additional vector terms in the Hamiltonian).

Since the spin describes a rotation around a fixed axis, there are only two possible directions given by $e^{\pm i_B \varphi}$ (since it is the zero-point rotation, there are no higher order terms $e^{\pm i_B m \varphi}$). Therefore, it appears as a planar motion which can be described by a single spinor. However, when we look at the consequences on the energy, we have to multiply the spin $\mathbf{S}$ by $g = 2$ to account for the fact that it is really a vector. This is because $\mathbf{S}$ is written as the angular momentum operator designed for a three-dimensional space, where a vector is described by products of two spinors. This problem does not occur when using Pauli spinors $\mathbf{e}_i$ which are designed for single spinors.

Problems

1 Determine how the atomic levels split for $l = 1$ for a magnetic field in the z-direction. Calculate the magnitude of the splitting.

2.10 Spin–Orbit and Darwin Interactions

Introduction. – In Sections 2.8 and 2.9, the nonrelativistic limit of the Dirac equation was obtained by moving to a reference frame where the momentum part of the four-momentum was zero. The remaining energy part leads to energy conservation if the potential is not explicitly time dependent. This becomes the Schrödinger equation when written in operator form. However, the vector part cannot be entirely removed since the momentum operator and the vector potential $\mathbf{A}$ do not commute with each other. In Section 2.9, this remaining vector part of the four-momentum was identified as the spin. Generally, the momentum operator also does not commute with the electrostatic potential $U(\mathbf{r})$. However, U already enters into the Schrödinger equation in (2.203) in lowest order, where the commutation plays no role. However, when including higher order terms in the Foldy–Wouthuysen transformation, the noncommutativity of the momentum and the electrostatic potential becomes important.

Since electric fields are generally stronger than magnetic fields, the vector potential in the transformation is ignored, giving $\boldsymbol{\alpha}' \cong \hat{\mathbf{p}}'/mc$. We now look for additional relativistic corrections to the Hamiltonian in the nonrelativistic limit. The focus is for electrons in an atom. The next term is $\frac{1}{2}[\boldsymbol{\alpha}', U]$. This produces a vector term related to the electric field that can be removed by choosing a better $\boldsymbol{\alpha}'$ for the Foldy–Wouthuysen transformation. The next term in Eq. (2.192) is

$$\boldsymbol{H}'' = \frac{1}{8}[\frac{\hat{\mathbf{p}}'}{mc}, [\frac{\hat{\mathbf{p}}'}{mc}, U]] = \frac{1}{8m^2c^2}[\hat{\mathbf{p}}', [\hat{\mathbf{p}}', U]]. \tag{2.230}$$

The commutator of the momentum and the potential energy gives

$$[\hat{\mathbf{p}}', U] = \hat{\mathbf{p}}'U - U\hat{\mathbf{p}}' = [\hat{\mathbf{p}}'U] = -i q\hbar \nabla V \mathbf{e}_0' = i q\hbar \mathbf{E}', \tag{2.231}$$

where $[\hat{\mathbf{p}}'U]$ indicates that $\hat{\mathbf{p}}'$ only operates on U and not on any other functions to the right. The potential is $V = U/q$, and the electric field is $\mathbf{E} = -\nabla V$. The additional commutation with the momentum gives

$$\frac{iq\hbar}{8m^2c^2}[\hat{\mathbf{p}}', \mathbf{E}'] = \frac{iq\hbar}{8m^2c^2}([\hat{\mathbf{p}}'\mathbf{E}'] - 2\mathbf{E}' \wedge \mathbf{p}'), \tag{2.232}$$

where $[\hat{\mathbf{p}}'\mathbf{E}']$ implies that the momentum only operates on the electric field. The first term is

$$\frac{iq\hbar}{8m^2c^2}[\hat{\mathbf{p}}'\mathbf{E}'] = -\frac{iq\hbar}{8m^2c^2}[\hat{\mathbf{p}}\mathbf{E}] = -\frac{q\hbar^2}{8m^2c^2}\nabla\mathbf{E} = -\frac{q\hbar^2}{8m^2c^2}\nabla \cdot \mathbf{E}. \tag{2.233}$$

The wedge product of $\hat{\mathbf{p}}'\mathbf{E}'$ is zero since $\nabla \wedge \mathbf{E} = i\nabla \times \mathbf{E} = -i\partial\mathbf{B}/\partial t = 0$ using Maxwell's equations, see Eq. (2.182). This term is zero since we have assumed that none of the fields are time dependent to ensure energy conservation. The divergence is, according to Maxwell's equations, related to the charge density. For an atom, the assumption is that the charge density is dominant at the nucleus at the origin, giving $\rho(\mathbf{r}) = Z_{\text{eff}}e\delta(\mathbf{r})$, where e is the elementary charge, and Z_{eff} indicates that the nuclear charge Z can be screened by strongly bound electrons. Since the radial extent of the nucleus is small compared to that of the electron, it is treated as a point charge. We end up with the scalar term

$$H''_{\text{Darwin}} = \frac{Z_{\text{eff}}e^2\hbar^2}{8\varepsilon_0 m^2c^2}\delta(\mathbf{r}), \tag{2.234}$$

taking $q = -e$ for the electron. This is known as the Darwin term.

The second term in Eq. (2.232) is

$$\frac{iq\hbar}{8m^2c^2}2\mathbf{E}' \wedge \mathbf{p}' = -\frac{q\hbar}{4m^2c^2}\mathbf{E} \times \mathbf{p}. \tag{2.235}$$

For a central potential, $\mathbf{E} = (-dV/dr)\hat{\mathbf{r}}$. This leads to the vector term

$$\mathbf{H}''_{\text{SOI}} = \frac{q\hbar}{4m^2c^2}\frac{dV}{dr}\hat{\mathbf{r}} \times \mathbf{p} = -\frac{\mu_B}{2mc^2}\frac{dV}{rdr}\mathbf{L} = \mu_B\mathbf{B}_{\text{n}}, \tag{2.236}$$

with $q = -e$ and using the Bohr magneton from Eq. (2.212). This term is known as the spin–orbit interaction. It is an additional vector term in the Hamiltonian. In analogy to Eq. (2.210), it is expressed as an effective magnetic field

$$\mathbf{B}_{\text{n}} = -\frac{1}{2mc^2}\frac{dV}{rdr}\mathbf{L}, \tag{2.237}$$

due to the potential of the nucleus.

An expression for the effective field $\mathbf{B}_{\text{n}}$ can also be found differently. Instead of having the electron move around the nucleus, one can also view the nucleus as moving around the electron. This creates a current I which leads to a magnetic field on the electron. The magnitude of the magnetic field can be obtained using Biot–Savart's law, which is a solution of Ampère's law $\int \mathbf{B} \cdot d\mathbf{l} = \mu_0 I$. For a circular orbit and a nucleus with charge $Z_{\text{eff}}e$, the current is $I = Z_{\text{eff}}e/T$ with the oscillation period $T = 2\pi r/v$. The effective magnetic field is then

$$\mathbf{B}_{\text{n}} = \frac{\mu_0}{4\pi}\int\frac{Id\mathbf{l} \times \mathbf{r}}{r^3} = \frac{\mu_0}{4\pi}\frac{Z_{\text{eff}}ev}{2\pi}\int\frac{d\mathbf{l} \times \mathbf{r}}{r^4} = \frac{\mu_0}{4\pi}\frac{Z_{\text{eff}}e}{2\pi}\frac{\mathbf{v} \times \mathbf{r}}{r^4}\int dl, \tag{2.238}$$

where $\mathbf{v}$ is in the same direction as the path $\mathbf{l}$. For a circular orbit, the integral simply gives $\int dl = 2\pi r$. This leaves us with

$$\mathbf{B}_{\text{n}} = -\frac{Z_{\text{eff}}e}{4\pi\varepsilon_0 mc^2}\frac{\mathbf{r} \times (m\mathbf{v})}{r^3} = -\frac{1}{mc^2}\frac{d}{rdr}\left(-\frac{Z_{\text{eff}}e}{4\pi\varepsilon_0 r}\right)\mathbf{L}, \tag{2.239}$$

where $\mu_0 = 1/c^2\varepsilon_0$, and the angular momentum is $\mathbf{L} = \mathbf{r} \times (m\mathbf{v})$. The term inside the parentheses $V = -Z_{\text{eff}}e/4\pi\varepsilon_0 r$ is the potential due to the nucleus. The end result is twice as large as $\mathbf{B}_{\text{n}}$ in Eq. (2.236). The additional factor $\frac{1}{2}$ comes from converting the classical angular momentum vector written in terms of commuting unit vectors into a vector operator, see Eqs. (1.384) and (1.385).

Using the expression for the angular momentum and the Pauli vector from Eq. (2.223), the interaction in Eq. (2.236) can then also be written as

$$\mathbf{H}''_{\text{SOI}} = \frac{g\mu_B}{\hbar}\mathbf{B}_{\text{n}} \cdot \mathbf{S} = \zeta(r)\mathbf{L} \cdot \mathbf{S}, \tag{2.240}$$

with

$$\zeta(r) \cong \frac{1}{2m^2c^2}\frac{dU}{rdr}, \tag{2.241}$$

with $U = -qV = eV$ and taking $g = 2$. Writing the interaction in this fashion explains why it is generally called the spin–orbit interaction.

Conceptual summary. – Since the momentum operator and the electrostatic potential do not commute, additional vector terms are found when looking at the higher order terms in the Foldy–Wouthuysen expansion. While the lowest order vector term due to the vector potential was the magnetic field, the lowest order vector term due to the electrostatic potential is related to the angular momentum. However, the latter term can again be reinterpreted as an effective magnetic field. Viewed from the perspective of a particle rotating around a charge, the charge giving rise to the electrostatic potential can also be viewed as rotating around the particle. This moving charge is an effective current producing a magnetic field on the particle. It is not entirely surprising that the vector terms can be thought as effective magnetic field terms. The higher-order corrections that can be thought of as related to an effective electric field can all be written in terms of a potential. They are therefore scalar terms. Since the effective magnetic field is due to the circular motion of a particle through an electric field, it is directly related to the angular momentum of the particle. This interaction is generally called the spin–orbit interaction, where the spin is identified with the vector term in the four-momentum in the rest frame.

2.11 Spin–Orbit Coupling

Introduction. – In Sections 2.9 and 2.10, we have found two interactions in the nonrelativistic limit of the Dirac equation that contain vector terms, the Zeeman and the spin–orbit interactions. Both arise from the fact that the momentum operator does not commute with the electromagnetic potentials. The spin–orbit interaction plays an important role in atoms, particularly when the atomic number increases. The orbitals are defined by the angular momentum. In spherical symmetry, the angular momentum has no preferred direction. The presence of a magnetic field leads to the alignment of **L**, the angular momentum of the orbital motion, and **S**, the angular momentum of the zero point motion arising from the noncommutativity of the momentum and the electromagnetic potentials. Although the vector part of the Hamiltonian often needs to be viewed as a quantity that behaves independently of the motion of the particle in space, it makes often sense for atoms to combine **L** and **S** into a single quantum number. This is known as spin–orbit coupling.

The expansion of $e^{\hat{r}}$ in Eq. (2.222) led to spherical harmonics that are eigenfunctions of the angular momentum operator $\hat{L}_z$. We now want to add the spin function $e^{\pm i_B \varphi}$ and consider the situation where the angular momentum and spin rotate around the same axis, *i.e.* $\boldsymbol{i}_B \to \boldsymbol{i} \equiv \boldsymbol{i}_{yx}$, see Eq. (2.217). The usual convention is followed that the rotation is around the z-axis, and the subscript of the imaginary unit is dropped. The product of the spherical harmonics and the spin function $|lm\rangle|\frac{1}{2}\sigma\rangle$ is still an eigenfunction of $\hat{L}_z + \hat{S}_z$. The eigenvalue is $m + \sigma$ in units of $\hbar$. However, there are now different combinations with the same eigenvalue, for example $m + \frac{1}{2}$ and $(m+1) - \frac{1}{2}$. Therefore, while the product functions are eigenfunctions of the z components of the angular momentum and spin operators, they are not necessarily eigenfunctions of the spin–orbit interaction $\mathbf{L} \cdot \mathbf{S}$. Therefore, angular momentum and spin are no longer conserved quantities by themselves. A new quantity **J** needs to be introduced to describe the conservation of the combined orbital and spin angular momentum. This is known as the total angular momentum.

Since the angular momentum is an integer, and the spin is half-integer, the coupled operator for a single particle must be half-integer. The total angular momentum **J** is given by

$$\mathbf{J} = \mathbf{L} + \mathbf{S}. \tag{2.242}$$

The norm of **J** is given by

$$\mathbf{J}^2 = (\mathbf{L} + \mathbf{S})^2 = \mathbf{L}^2 + \mathbf{S}^2 + 2\mathbf{L} \cdot \mathbf{S}. \tag{2.243}$$

For spherical symmetry, the effect of the angular momentum operators is known, and we can use the dimensionless number and step operators on the unit vectors. The eigenvalues for the spin–orbit interaction $\mathbf{L} \cdot \mathbf{S}$ are then

$$\begin{aligned} \mathbf{L} \cdot \mathbf{S}|jm_j\rangle &= \frac{1}{2}(\mathbf{J}^2 - \mathbf{L}^2 - \mathbf{S}^2)|jm_j\rangle \\ &= \frac{1}{2}\left(j(j+1) - l(l+1) - \frac{3}{4}\right)|jm_j\rangle, \end{aligned} \tag{2.244}$$

using Eq. (1.430); $|jm_j\rangle$ indicates the coupled wavefunction that is an eigenvalue of $\mathbf{L} \cdot \mathbf{S}$. Since j can only take the values $l \pm \frac{1}{2}$, we can also write the eigenvalues as

$$\langle jm_j|\mathbf{L} \cdot \mathbf{S}|jm_j\rangle = \begin{cases} \dfrac{1}{2}l, & j = l + \dfrac{1}{2}, \\ -\dfrac{1}{2}(l+1), & j = l - \dfrac{1}{2}. \end{cases} \tag{2.245}$$

Let us consider some examples. For $l = 0$, the spherical harmonic $C_{00} = 1$ has no angular dependence. The wavefunction is then

$$|\frac{1}{2}\sigma\rangle = |00\rangle|\frac{1}{2}\sigma\rangle = \boldsymbol{b}_\sigma^\dagger \text{ and } \langle\hat{\boldsymbol{r}}|\frac{1}{2}, \pm\frac{1}{2}\rangle = e^{\pm i\varphi}. \tag{2.246}$$

Therefore, there are only counterclockwise and clockwise rotations around the axis. This is the zero-point motion of an s orbital. A more interesting example is $l = 1$. The allowed

j values are $j = \frac{3}{2}, \frac{1}{2}$. The $m_j = \frac{3}{2}$ value can only be obtained with $m = 1$ and $\sigma = \frac{1}{2}$. The corresponding wavefunction is

$$|\frac{3}{2}\frac{3}{2}\rangle = |11\rangle|\frac{1}{2}\frac{1}{2}\rangle = \frac{1}{\sqrt{2}}\boldsymbol{a}_\uparrow^\dagger\boldsymbol{a}_\uparrow^\dagger\boldsymbol{b}_\uparrow^\dagger \text{ and } \langle\hat{\boldsymbol{r}}|\frac{3}{2},\pm\frac{3}{2}\rangle = \mp\frac{e^{\pm 2i\varphi}}{\sqrt{2}}\sin\theta. \tag{2.247}$$

The situation becomes more complicated for $m_j = m + \sigma = \frac{1}{2}$, where there are two possibilities

$$\begin{aligned}|11\rangle|\frac{1}{2},-\frac{1}{2}\rangle &= \frac{1}{\sqrt{2}}\boldsymbol{a}_\uparrow^\dagger\boldsymbol{a}_\uparrow^\dagger\boldsymbol{b}_\downarrow^\dagger \quad\text{and}\quad \langle\hat{\boldsymbol{r}}|11\rangle\langle\hat{\boldsymbol{r}}|\frac{1}{2},-\frac{1}{2}\rangle = -\frac{1}{\sqrt{2}}\sin\theta,\\ |10\rangle|\frac{1}{2}\frac{1}{2}\rangle &= \boldsymbol{a}_\uparrow^\dagger\boldsymbol{a}_\downarrow^\dagger\boldsymbol{b}_\uparrow^\dagger \quad\text{and}\quad \langle\hat{\boldsymbol{r}}|10\rangle\langle\hat{\boldsymbol{r}}|\frac{1}{2}\frac{1}{2}\rangle = e^{i\varphi}\cos\theta.\end{aligned} \tag{2.248}$$

We want to find out what combinations are eigenstates of j. We know that $|\frac{3}{2}\frac{3}{2}\rangle$ is an eigenstate, since $j = \frac{1}{2}$ cannot have the same m_j value. Let us write this unit vector in terms of spinors $\boldsymbol{j}_\sigma^\dagger$,

$$|\frac{3}{2}\frac{3}{2}\rangle = |3;0\rangle = \frac{1}{\sqrt{6}}\boldsymbol{j}_\uparrow^\dagger\boldsymbol{j}_\uparrow^\dagger\boldsymbol{j}_\uparrow^\dagger. \tag{2.249}$$

The notation $|jm\rangle = |n_\uparrow; n_\downarrow\rangle = |j+m; j-m\rangle$, where n_σ is the number of creation operators for a particular σ, is convenient when using step operators $\boldsymbol{j}_\downarrow^\dagger\boldsymbol{j}_\uparrow$. This gives

$$\boldsymbol{j}_\downarrow^\dagger\boldsymbol{j}_\uparrow|n_\uparrow; n_\downarrow\rangle = \sqrt{n_\uparrow(n_\downarrow+1)}|n_\uparrow - 1; n_\downarrow + 1\rangle. \tag{2.250}$$

The unit vector for $j = \frac{3}{2}$ and $m_j = \frac{1}{2}$ can be found by stepping down

$$\boldsymbol{J}_-|\frac{3}{2}\frac{3}{2}\rangle = \boldsymbol{j}_\downarrow^\dagger\boldsymbol{j}_\uparrow|\frac{3}{2}\frac{3}{2}\rangle = \boldsymbol{j}_\downarrow^\dagger\boldsymbol{j}_\uparrow|3;0\rangle = \sqrt{3\times 1}|2;1\rangle = \sqrt{3}|\frac{3}{2}\frac{1}{2}\rangle. \tag{2.251}$$

However, since $\mathbf{J} = \mathbf{L} + \mathbf{S}$, the operation can also be applied to the right-hand side of Eq. (2.247).

$$\begin{aligned}(\mathbf{L}_- + \mathbf{S}_-)|00\rangle|\frac{1}{2}\sigma\rangle &= (\boldsymbol{a}_\downarrow^\dagger\boldsymbol{a}_\uparrow + \boldsymbol{b}_\downarrow^\dagger\boldsymbol{b}_\uparrow)|2;0\rangle|1;0\rangle\\ &= \sqrt{2\times 1}|1;1\rangle|1;0\rangle + \sqrt{1\times 1}|2;0\rangle|0;1\rangle\\ &= \sqrt{2}|10\rangle|\frac{1}{2}\frac{1}{2}\rangle + |11\rangle|\frac{1}{2},-\frac{1}{2}\rangle,\end{aligned} \tag{2.252}$$

where the first and the second ket correspond to the orbital and spin, respectively. Since this has to be equal to Eq. (2.251), we find

$$|\frac{3}{2}\frac{1}{2}\rangle = \frac{1}{\sqrt{2}}\boldsymbol{j}_\uparrow^\dagger\boldsymbol{j}_\uparrow^\dagger\boldsymbol{j}_\downarrow^\dagger = \sqrt{\frac{2}{3}}|10\rangle|\frac{1}{2}\frac{1}{2}\rangle + \frac{1}{\sqrt{3}}|11\rangle|\frac{1}{2},-\frac{1}{2}\rangle. \tag{2.253}$$

The orthogonal combination then has to belong to the $j = \frac{1}{2}$ manifold

$$|\frac{1}{2}\frac{1}{2}\rangle = \boldsymbol{j}_\uparrow^\dagger = \frac{1}{\sqrt{3}}|10\rangle|\frac{1}{2}\frac{1}{2}\rangle - \sqrt{\frac{2}{3}}|11\rangle|\frac{1}{2},-\frac{1}{2}\rangle. \tag{2.254}$$

Trying to step up from the $m_j = \frac{1}{2}$ leads to

$$\begin{aligned}&(\boldsymbol{a}_\uparrow^\dagger\boldsymbol{a}_\downarrow + \boldsymbol{b}_\uparrow^\dagger\boldsymbol{b}_\downarrow)\left(\frac{1}{\sqrt{3}}|1;1\rangle|1;0\rangle - \sqrt{\frac{2}{3}}|2;0\rangle|0;1\rangle\right)\\ &= \frac{1}{\sqrt{3}}\left(\sqrt{2\times 1}|2;0\rangle|1;0\rangle + 0 - \sqrt{2}(0 + \sqrt{1\times 1}|2;0\rangle|1;0\rangle)\right) = 0,\end{aligned}$$

which agrees with trying to step up the m_j values $\boldsymbol{J}_+\boldsymbol{j}_\uparrow^\dagger = \boldsymbol{j}_\uparrow^\dagger\boldsymbol{j}_\downarrow\boldsymbol{j}_\uparrow^\dagger = 0$.

The expressions for the functions in Eqs. (2.247) and (2.249) contain exponentials that describe the motion in a magnetic field. However, the argument in the exponential is not given by m_j but by $m + 2\sigma$. This gives the operator

$$(\hat{L}_z + 2\hat{S}_z) = -\boldsymbol{i}\frac{\partial}{\partial\varphi}, \tag{2.255}$$

which is the operator that appears in the Zeeman interaction, see Eq. (2.224). Operating on the basis functions gives

$$(\hat{L}_z + 2\hat{S}_z)(\hat{\boldsymbol{r}}|lm\rangle(\hat{\boldsymbol{r}}|\frac{1}{2}\sigma\rangle = (m + 2\sigma)(\hat{\boldsymbol{r}}|lm\rangle(\hat{\boldsymbol{r}}|\frac{1}{2}\sigma\rangle. \tag{2.256}$$

The expectation values of the operator in the on the eigenfunctions of the spin-orbit interaction gives

$$\int d\boldsymbol{r}\langle jm_j|\hat{\boldsymbol{r}}\rangle(\hat{L}_z + 2\hat{S}_z)(\hat{\boldsymbol{r}}|jm_j\rangle = \begin{cases} 2, \frac{2}{3}, -\frac{2}{3}, -2, & m_j = \frac{3}{2}, \frac{1}{2}, -\frac{1}{2}, -\frac{3}{2} \quad \left(j = \frac{3}{2}\right), \\ \frac{1}{3}, -\frac{1}{3}, & m_j = \frac{1}{2}, -\frac{1}{2} \quad \left(j = \frac{1}{2}\right). \end{cases} \tag{2.257}$$

This gives the relative splitting of the levels under a magnetic field. Although the splitting is proportional to m_j, it is not exactly equal. This is related to a more fundamental result called the Wigner–Eckart theorem. In general, the result can be written as

$$(\mathbf{L}_z + 2\mathbf{S}_z)|jm_j\rangle = g_j m_j, \tag{2.258}$$

where g_j is the Landé g-factor. An expression for the factor is easily obtained for $j = l + \frac{1}{2}$. Since the $m_j = j$ function can always be written as $|ll\rangle|\frac{1}{2}\frac{1}{2}\rangle$, the value $(\mathbf{L}_z + 2\mathbf{S}_z)|jj\rangle = l + 1$. This has to be equal to $g_{\frac{3}{2}}j = g_{\frac{3}{2}}(l + \frac{1}{2})$, giving a Landé factor of $g_{\frac{3}{2}} = (l + 1)/(l + \frac{1}{2})$. Generalizing the procedure above for an arbitrary value of l gives for the eigenfunction for the maximum m_j value for $j = l - \frac{1}{2}$,

$$|j - \frac{1}{2}, j - \frac{1}{2}\rangle = \frac{1}{\sqrt{2l+1}}|l, l-1\rangle|\frac{1}{2}\frac{1}{2}\rangle - \sqrt{\frac{2l}{2l+1}}|ll\rangle|\frac{1}{2}, -\frac{1}{2}\rangle. \tag{2.259}$$

The eigenvalue of the operator $(\mathbf{L}_z + 2\mathbf{S}_z)$ is then

$$\frac{1}{2l+1}((l-1)+1) + \frac{2l}{2l+1}(l-1) = \frac{l(2l-1)}{2l+1}. \tag{2.260}$$

This is indeed $\frac{1}{3}$ for $l = 1$, see Eq. (2.257). Dividing by $j = l - \frac{1}{2}$ gives the Landé factor $g_{\frac{1}{2}} = l/(2l + \frac{1}{2})$.

Conceptual summary. – In this section, we looked at the eigenstates of the spin-orbit interaction. In general, one could expect the relativistic corrections to be small. For light atoms, the coupling strength of the spin-orbit interaction is of the order of meV. However, for heavier atoms this can increase to 100s of meV for valence electrons. When looking at strongly bound states in heavier atoms (core levels), the interaction strength can increase to several tens to hundreds of electron volts, which are not small perturbations.

In solving the problem of the spin-orbit coupling, the vectors for angular momentum and spin are coupled together to form a new vector, known as the total angular momentum. This coupling foreshadows what we will do in Chapter 4 when the angular momentum of different particles will be coupled together to form many-body states with different angular momentum. These concepts are also important in the formation of particles from quarks.

Problems

1 Write down the matrix for the spin–orbit interaction in the basis $|m\sigma\rangle$ for $l = 1$. Solve for the eigenenergies and eigenvectors.

2.12 Schrödinger/Heisenberg Equations and Propagators

Introduction. – In Sections 2.8–2.10, the Schrödinger equation was derived including higher order vector corrections to the scalar equation. The Schrödinger equation is the most commonly used way to derive the function space in the presence of a potential. The function space is then used to describe the behavior of particles. In principle, the entire function space needs to be included. However, in practice, one can often limit the function space. For example, many atomic problems only require a few atomic orbitals.

However, once the function space has been derived, one often wants to investigate it. Although the time dependence of the wavefunctions can be calculated, one is often more interested in the energetics of the function space for the following reasons. First, it is often difficult to probe the wavefunction or even the probability of particles. Second, even if one is able to probe the wavefunction in space, the result might not be very insightful. For example, properties such as optical, thermal, and electrical conductivity of solids have very little relation with the actual positions of the electrons in the solid but are primarily due to the structure of the energy levels. Therefore, when studying physical properties, it is often advantageous to look at the propagation of certain excitations in function space. This leads to the Heisenberg equation and Green's function techniques.

Going back to the nonrelativistic limit led to the Schrödinger equation,

$$\hat{H}\varphi_n(\mathbf{r},t) = \hat{E}\varphi_n(\mathbf{r},t) = i\hbar\frac{\partial\varphi_n(\mathbf{r},t)}{\partial t}, \tag{2.261}$$

using the pseudoscalar i for the imaginary unit. The function $\varphi_n(\mathbf{r},t)$ is the eigenfunction of the equation identified by the energy E_n. The quantum number n can in principle also contain a spin component. For complicated systems, energy is often the only conserved quantity. In that case, n is simply an index labeling the eigenenergies and has no deeper physical meaning. The time-dependent part of $\varphi_n(\mathbf{r},t)$ can be easily obtained

$$\varphi_n(\mathbf{r},t) = e^{-\frac{i}{\hbar}E_n t}\varphi_n(\mathbf{r}) \quad \Rightarrow \quad i\hbar\frac{\partial\varphi_n(\mathbf{r},t)}{\partial t} = E_n\varphi_n(\mathbf{r},t). \tag{2.262}$$

By dividing by the complex exponential, the time-dependent operator can be removed from Eq. (2.261), giving the time-independent Schrödinger equation,

$$\hat{H}\varphi_n(\mathbf{r}) = E_n\varphi_n(\mathbf{r}). \tag{2.263}$$

It is convenient to view the solutions of the Schrödinger equation as basis functions, in the sense that they are used to construct other functions. For example, any arbitrary wave packet can be constructed using these functions. Additionally, the solutions of a particular Schrödinger equation often do not represent the entire problem. For example, there can be additional interactions that have not yet been included in the Hamiltonian. The real eigenfunctions of the system are then expressed in terms of the basis functions. Another common situation is that the Schrödinger equation provides single-particle solutions to the problem. However, the real system contains many particles. The many-body wavefunctions are then expressed in terms of the basis functions.

In the first chapter, the function space could be viewed as higher order vectors. Each function could be associated with a unit vector. By cleverly constructing the unit vectors, all the information on the transformation of the function could be contained in the unit vectors. For system with a complicated potential, it is still possible to associate the basis functions with a vector space $|n\rangle$. There is generally no clever construction in terms of spinors of this vector space. The only quantum number that is left is the energy. This information can be extracted by

$$\boldsymbol{H} = \sum_n E_n |n\rangle\langle n| \quad \Rightarrow \quad \boldsymbol{H}|n\rangle = E_n|n\rangle. \tag{2.264}$$

For complicated systems, the only way to obtain the values of $E_n = \hbar\omega_n$ is by solving the Schrödinger equation numerically.

Since the Schrödinger equation is a second-order linear ordinary differential equation, it follows all the properties associated with the Sturm–Liouville theory. For our purposes, this means that every Schrödinger equation leads to a complete function space, *i.e.*

$$\sum_n |n\rangle\langle n| = \mathbb{1}_\infty, \tag{2.265}$$

with an infinite number of orthogonal functions, *i.e.*

$$\langle n'|n\rangle = \int d\mathbf{r}\ \langle n'|\mathbf{r}\rangle\langle \mathbf{r}|n\rangle = \int d\mathbf{r}\ \varphi^*_{n'}(\mathbf{r})\varphi_n(\mathbf{r}) = \delta_{nn'}, \tag{2.266}$$

where the completeness of space has been inserted. These properties have already been observed for the function spaces of complex exponentials and spherical harmonics.

When a system is in its lowest energy state, all the particles occupy eigenfunctions. However, one is often more interested in the response of a system to, for example electromagnetic radiation. Due to the presence of an additional interaction, particles are excited. For a single particle, the excited wavefunction can always be expressed in terms of the basis functions

$$\psi(\mathbf{r}) = \langle \mathbf{r}|\psi\rangle = \sum_n \langle \mathbf{r}|n\rangle\langle n|\psi\rangle = \sum_n \varphi_n(\mathbf{r})\langle n|\psi\rangle, \tag{2.267}$$

where $\langle n|\psi\rangle$ is a coefficient describing the amount of the eigenfunction $\varphi_n(\mathbf{r})$ in the function $\psi(\mathbf{r})$. This is the generalization of the Fourier transform for an arbitrary function space. Since $|n\rangle$ are eigenfunctions, this directly allows us to calculate the time dependence,

$$|\psi(t)\rangle = \sum_n e^{-i\omega_n t}|n\rangle\langle n|\psi\rangle, \tag{2.268}$$

where the spatial dependence of the wavefunction has been removed. Often, we are interested in the time dependence of the expectation value of a particular operator. This can be done by calculating the expectation of this operator in the time-dependent wavefunctions

$$\langle \boldsymbol{O}(t)\rangle = \langle \psi(t)|\boldsymbol{O}|\psi(t)\rangle, \tag{2.269}$$

where any operator can also be expressed in matrix form

$$\boldsymbol{O} = \sum_{nn'} |n'\rangle\langle n'|\boldsymbol{O}|n\rangle\langle n|, \tag{2.270}$$

where the matrix elements are given by

$$\langle n'|\boldsymbol{O}|n\rangle = \int d\mathbf{r}' \int d\mathbf{r}\, \langle n'|\mathbf{r}'\rangle\langle \mathbf{r}'|\boldsymbol{O}|\mathbf{r}\rangle\langle \mathbf{r}|n\rangle = \int d\mathbf{r}\, \varphi^*_{n'}(\mathbf{r})\hat{O}(\mathbf{r})\varphi_n(\mathbf{r}),$$

where $\langle \mathbf{r}'|\boldsymbol{O}|\mathbf{r}\rangle = \hat{O}(\mathbf{r})\delta(\mathbf{r}-\mathbf{r}')$, since physical operators are local in space. Again, for complex systems, the only way to obtain the values of the matrix elements is by numerical evaluation. A special case is when the particle is in an eigenstate $|\psi\rangle = |n\rangle$. The time-dependent expectation value is then

$$\langle n(t)|\boldsymbol{O}|n(t)\rangle = \langle n|e^{i\omega_n t}\boldsymbol{O}e^{-i\omega_n t}|n\rangle = \langle n|\boldsymbol{O}|n\rangle. \tag{2.271}$$

Therefore, the expectation value of any operator in an eigenstate is independent of time. Additionally, the probability of an eigenfunction is also independent of time. This is the reason why eigenfunctions are also called stationary states.

An important operator is the Hamiltonian itself. Calculating the time-dependent expectation value gives

$$E = \langle \psi(t)|\boldsymbol{H}|\psi(t)\rangle = \sum_{nn'} \langle \psi(t)|n'\rangle\langle n'|\boldsymbol{H}|n\rangle\langle n|\psi(t)\rangle = \sum_n E_n |\langle n|\psi\rangle|^2,$$

using the fact that the Hamiltonian is a diagonal operator in this basis, see Eq. (2.264). The energy of $|\psi(t)\rangle$ is therefore a conserved quantity. However, note that this does not imply that $|\psi(t)\rangle$ is an eigenfunction of the Hamiltonian,

$$\boldsymbol{H}|\psi\rangle = \sum_n \boldsymbol{H}|n\rangle\langle n|\psi\rangle = \sum_n E_n|n\rangle\langle n|\psi\rangle \neq E|\psi\rangle. \tag{2.272}$$

So far, the time dependence of a physical property $\hat{O}$ has been connected to the state $|\psi(t)\rangle$. Alternatively, one can think of this problem as creating a state $|\psi\rangle$, and the physical property $\hat{O}(t)$ varies as a function of time in this state. Let us first separate the time dependence from the wavefunction,

$$\begin{aligned} |\psi(t)\rangle &= \sum_n e^{-i\omega_n t}|n\rangle\langle n|\psi\rangle = \sum_n e^{-\frac{i}{\hbar}\boldsymbol{H}t}|n\rangle\langle n|\psi\rangle \\ &= e^{-\frac{i}{\hbar}\boldsymbol{H}t}\sum_n |n\rangle\langle n|\psi\rangle = e^{-\frac{i}{\hbar}\boldsymbol{H}t}|\psi\rangle. \end{aligned} \tag{2.273}$$

The time-dependent expectation value can then be written as

$$\langle \boldsymbol{O}(t)\rangle = \langle \psi(t)|\boldsymbol{O}|\psi(t)\rangle = \langle \psi|e^{\frac{i}{\hbar}\boldsymbol{H}t}\boldsymbol{O}e^{-\frac{i}{\hbar}\boldsymbol{H}t}|\psi\rangle = \langle \psi|\boldsymbol{O}(t)|\psi\rangle. \tag{2.274}$$

The time dependence has now shifted from the wavefunction to the operator

$$\boldsymbol{O}(t) = e^{\frac{i}{\hbar}\boldsymbol{H}t}\boldsymbol{O}e^{-\frac{i}{\hbar}\boldsymbol{H}t}. \tag{2.275}$$

The time dependence can now also be studied by looking at the properties of $\boldsymbol{O}(t)$,

$$i\hbar\frac{d\boldsymbol{O}(t)}{dt} = e^{\frac{i}{\hbar}\boldsymbol{H}t}(-\boldsymbol{H})\boldsymbol{O}e^{-\frac{i}{\hbar}\boldsymbol{H}t} + e^{\frac{i}{\hbar}\boldsymbol{H}t}\boldsymbol{O}e^{-\frac{i}{\hbar}\boldsymbol{H}t}\boldsymbol{H}. \tag{2.276}$$

The order is important. Although $\boldsymbol{H}$ commutes with itself, it does not necessarily commute with $\boldsymbol{O}$. The above equation can also be written as a commutator

$$i\hbar\frac{d\boldsymbol{O}(t)}{dt} = [\boldsymbol{O}(t), \boldsymbol{H}]. \tag{2.277}$$

This equation is known as the Heisenberg equation.

In addition to the time-dependent calculation of an operator, one can also calculate correlation functions, such as

$$G(t) = -\frac{i}{\hbar}\langle 0|\boldsymbol{O}(t)\boldsymbol{O}^{\dagger}(0)|0\rangle\theta(t), \tag{2.278}$$

where the prefactor is introduced to obtain more nicely defined quantities when taking the Fourier transform. The function $G(t)$ is called a Green's function. An excitation is created at time $t = 0$ and removed again at a later time t. The excitation is created in the state $|0\rangle$. This can be the vacuum state. However, it can also denote any initial state of a system. The times can also be taken as t and t', but usually only the time difference is relevant. There are many different types of perturbations: addition/removal of an electron, an optical excitation, a fluctuation in charge density, etc. The function $\theta(t)$ is a step function, which is 0 and 1 for $t < 0$ and $t \geq 0$, respectively. This ensures causality, meaning that the excitation has to be created before it is removed.

Let us consider the Fourier transform in time of the Green's function

$$G(E) = \langle E|G\rangle = \int_{-\infty}^{\infty} dt\ \langle E|t\rangle\langle t|G\rangle = \int_{-\infty}^{\infty} dt\ e^{i\omega t}G(t), \tag{2.279}$$

with $E = \hbar\omega$. For Fourier transforms in real space, the normalization is often equally divided between real and momentum spaces, see Eq. (1.215). This is called a unitary Fourier transform. In order to avoid square roots of 2π in places we do not want them, Fourier transforms in time are often taken nonunitary. The factor $1/2\pi$ now only appears in the inverse Fourier transform, where the integration is over the frequency $d\nu = d\omega/2\pi$. This is called a nonunitary Fourier transform and is the reason that there is no $1/\sqrt{2\pi}$ in $\langle t|E\rangle = e^{-i\omega t}$. Setting the energy of the initial state to zero gives $\boldsymbol{H}|0\rangle = 0$. The Fourier transform then becomes

$$G(E) = -\frac{i}{\hbar}\int_0^{\infty} dt\ \langle 0|\boldsymbol{O}e^{\frac{i}{\hbar}(E-\boldsymbol{H}+i\eta)t}\boldsymbol{O}^{\dagger}|0\rangle = -\left[\langle 0|\boldsymbol{O}\frac{e^{\frac{i}{\hbar}(E-\boldsymbol{H}+i\eta)t}}{E-\boldsymbol{H}+i\eta}\boldsymbol{O}^{\dagger}|0\rangle\right]_0^{\infty}.$$

An infinitesimally small complex number $i\eta$ has been added to the energy, so that the integral converges in the limit $t \to \infty$. Note that $\boldsymbol{H}$ in the denominator is still a matrix. The Green's function in the energy domain is then

$$G(E) = \langle 0|\boldsymbol{O}\frac{1}{E-\boldsymbol{H}+i\eta}\boldsymbol{O}^{\dagger}|0\rangle. \tag{2.280}$$

The matrix $\boldsymbol{H}$ can be removed by diagonalizing the Hamiltonian. This gives

$$G(E) = \sum_n \langle 0|\boldsymbol{O}|n\rangle\frac{1}{E-E_n+i\eta}\langle n|\boldsymbol{O}^{\dagger}|0\rangle. \tag{2.281}$$

This can be split into a real and an imaginary part

$$G(E) = \sum_n |\langle n|\boldsymbol{O}^{\dagger}|0\rangle|^2\frac{E-E_n-i\eta}{(E-E_n)^2+\eta^2}. \tag{2.282}$$

In the imaginary part, we recognize the Lorentzian, see Eq. (1.212), which, in the limit $\eta \to 0$, can be related to a δ function, see Eq. (1.214). The imaginary part of the Green's function is therefore

$$I(E) = -\frac{1}{\pi}\lim_{\eta\to 0} \operatorname{Im} G(E) = \sum_n |\langle n|\boldsymbol{O}^\dagger|0\rangle|^2 \delta(E - E_n). \tag{2.283}$$

This is known as Fermi's golden rule. Starting from the initial state $|0\rangle$, an excitation is created by $\boldsymbol{O}^\dagger$, and the final states $|n\rangle$ are reached. The δ function imposes energy conservation. This is a spectral representation, where $I(E)$ gives the intensity of the excitations created by $\boldsymbol{O}^\dagger$ at a particular energy E. This is a fundamental equation in the study of many different types of spectroscopy.

Another set of useful operators are the creation and annihilation operators that put particles in and remove particles from certain states, respectively,

$$\mathbf{c}_k^\dagger|0\rangle = |k\rangle \quad \text{and} \quad \mathbf{c}_k|k\rangle = |0\rangle, \tag{2.284}$$

where $|k\rangle$ does not have to be an eigenfunction. The state $|0\rangle$ is taken as the vacuum. In principle, since electrons are fermions, these operators anticommute with each other. However, when dealing with single-electron problems, we generally do not have to consider this, and this discussion is postponed to Chapter 4. Let us consider a simple system of free particles with momentum $\mathbf{k}$ and energy $\varepsilon_{\mathbf{k}} = \hbar^2\mathbf{k}^2/2m$. Taking the operator $\boldsymbol{O}^\dagger = \mathbf{c}_{\mathbf{k}}^\dagger$ and the vacuum state $|0\rangle$ for the expectation of the correlation function in Eq. (2.278) gives

$$G_{\mathbf{k}}(t) = -\frac{i}{\hbar}\langle 0|\mathbf{c}_{\mathbf{k}} e^{-\frac{i}{\hbar}\varepsilon_{\mathbf{k}}t}\mathbf{c}_{\mathbf{k}}^\dagger|0\rangle\theta(t) = -\frac{i}{\hbar}e^{-\frac{i}{\hbar}\varepsilon_{\mathbf{k}}t}\theta(t), \tag{2.285}$$

with the expectation value $\langle 0|\mathbf{c}_{\mathbf{k}}\mathbf{c}_{\mathbf{k}}^\dagger|0\rangle = 1$, since the electron is first added to the vacuum and then removed. Therefore, the matrix elements in Eq. (2.283) are $\langle \mathbf{k}|\mathbf{c}_{\mathbf{k}}^\dagger|0\rangle = 1$. In general, this is not the case, and matrix elements can often include very specific selection rules. The result is known as a single-particle Green's function. It is also called a propagator, since it describes the propagation of a particle created at a particular time. The propagation is very simple here, since the electron stays in the free-electron eigenstate with momentum $\mathbf{k}$. It becomes more complicated when, for example scattering to different states is included. The Fourier transform of the single-particle Green's function is

$$G_{\mathbf{k}}(E) = \frac{1}{E - \varepsilon_{\mathbf{k}} + i\eta} \quad \text{and} \quad G(E) = \sum_{\mathbf{k}} G_{\mathbf{k}}(E). \tag{2.286}$$

The related spectral function is

$$\rho(E) = I(E) = -\frac{1}{\pi}\operatorname{Im} G(E) = \sum_{\mathbf{k}} \delta(E - \varepsilon_{\mathbf{k}}). \tag{2.287}$$

This is known as the density of states $\rho(E)$, since it associates each state with a δ function centered at its energy $\varepsilon_{\mathbf{k}}$.

Conceptual summary. – Although the Schrödinger equation is the basis for many quantum-mechanical problems, additional methods are often used. There are several reasons for that. The wavefunction itself is often difficult to probe, and many physical properties and experiments are more susceptible to the energies of the eigenfunctions than to their spatial aspects. Also, when looking at problems with many particles, the wavefunction often only enters on a single-particle level. It is then used to calculate matrix elements, which form the basis for calculating the many-particle states. For model systems, the matrix elements are simply taken as free parameters. Additionally, for many complicated systems the wavefunction in real space is often not very insightful.

The Green's functions are very useful to describe problems where the system is excited from the state with the lowest energy (also known as the ground state). This is important when studying, for example the optical and electrical conductivity of materials or for any type of spectroscopic experiment. Green's functions are also very well suited for perturbative approaches. This means that one is able to solve a system for a particular Hamiltonian. This solution is then used to solve the problem with an additional interaction (the perturbation).

Problems

1 The expectation value of an operator is given by

$$\langle \hat{M} \rangle = \int d\mathbf{r}\ \psi^*(\mathbf{r},t)\hat{M}\psi(\mathbf{r},t). \tag{2.288}$$

a. Calculate the time dependence of the expectation value (note that the operator can be time dependent as well).
b. Calculate the time dependence of the expectation value of the position operator $\hat{\mathbf{r}}$.
c. Calculate the time dependence of the expectation value of the momentum operator $\hat{\mathbf{p}}$.
d. Show that these results satisfy the classical equation of motion.

2 The Hamiltonian for a spin $\mathbf{s}$ of a particle with charge e in an applied magnetic field $\mathbf{B}$ is given by

$$\hat{H} = -\frac{ge}{2m}\mathbf{S}\cdot\mathbf{B}, \tag{2.289}$$

where g is the gyromagnetic ratio.
a. Calculate $d\mathbf{S}/dt$.
b. Describe the motion if the magnetic field is in the y-direction. Express the results in terms of the initial spin components.

2.13 Electroweak Interaction

Introduction. – In the Dirac equation in the presence of an electromagnetic field, the particles experience the vector potential but remain unchanged. When entering the realm of particle physics, particles can change character. One interaction that can do this is the electroweak interaction, which combines the electromagnetic and the weak interaction. Although the electroweak interaction can couple many different particles, it is most likely between certain pairs of particles. An important example of the weak interaction is β decay. A typical β^- decay process is

$$n(udd) \rightarrow p(udu) + e^- + \overline{\nu}_e. \tag{2.290}$$

This does not appear directly as the interaction between two particles, so let us look at this more closely. Let us first look at the changes in the nucleus, where the weak interaction changes a neutron (n) into a proton (p). Note that the charge of the nucleon changes from 0 to $+e$. Therefore, the force carrier that mediates the interaction carries a charge. This is already rather different from the electromagnetic interaction where the force carrier, the photon, is neutral. Although β decay was initially explained at this level, later theories discovered that neutrons and protons are composite particles consisting of three quarks. The neutron has one up and two down quarks, and the proton has two up and one down quarks. For our discussion, we do not need to know in detail what quarks are. However, we can also view the interaction as a change of a down into an up quark, which also changes the charge by $+e$.

Since the charge of the nucleus changes by $+e$, it seems natural to assume that an electron with a charge $-e$ escapes in order to obey charge conservation. However, something is still missing. Since both protons and neutrons are fermions, the spin of the nucleus has not changed. However, the electron carries a spin. Therefore, either spin is not a conserved quantity, or there is an additional fermion involved. Furthermore, the observed kinetic energy distribution of the electron was very broad, in violation of conservation of energy, unless there is an additional particle involved. Pauli and Fermi suggested as a possible solution that β decay is a four-fermion interaction, which became eventually the weak interaction. This additional fermion is called the electron neutrino. Note that in Eq. (2.290), an antineutrino appears on the right side of the decay. Therefore, the β-decay creates an electron–antineutrino pair. However, one can also view the interaction as a change of a neutrino into an electron.

The goal of this section is not to discuss the details of β decay or the electroweak interaction, which is beyond the scope of this text book. The idea is to view the particle character in geometric terms. The electromagnetic interaction $q\mathbf{A}'$ is essentially a one-dimensional system when it comes to particle character: an electron remains an electron, and a proton remains a proton. When restricting ourselves to pairs of particles, the electroweak interaction is a two-dimensional system where each axis denotes a certain particle character. Therefore, a change from neutrino to electron or from up to down quark is essentially a rotation in a plane.

The interaction between the particles is given by a four-vector, just as the interaction of the four-potential $\mathbf{A}'$ on a charged particle. However, now the interaction can also scatter between the two particles. This can be expressed in terms of a 2×2 matrix containing a scalar and a vector term. Any hermitian 2×2 matrix can always be expressed in terms of the identity $\mathbb{1}_2$ and the three Pauli matrices $\mathbf{e}_i$,

$$\boldsymbol{H}_{\mathrm{EW}} = gY\mathbf{B}'\mathbb{1}_2 + g' \sum_{i=x,y,z} \mathbf{W}'_i\mathbf{e}_i, \tag{2.291}$$

where g and g' are the strengths of the interactions. The factor Y is a symmetry term introduced to reflect the differences between an electron/neutrino and an up/down quark pair. This quantity is known as the hypercharge.

The vector character in $\boldsymbol{H}_{\mathrm{EW}}$ expresses the particle nature of the interaction. The interaction itself is still given in terms of four-vectors. Note that the interaction $\mathbf{B}'$ related to the identity matrix is not equal to the four-potential $\mathbf{A}'$, since there are also diagonal matrix elements in the $\mathbf{W}'_z$ term. The second term is essentially a vector of four-vectors $\mathbf{W}'_i$. The particles experience an electroweak vector-potential $\mathbf{W}'_i$; the unit vectors $\mathbf{e}_i$

indicate whether the vector-potential changes the nature of the particle. Expressed in a 2×2 matrix, the electroweak interaction is

$$\boldsymbol{H}_{\text{EW}} = \begin{pmatrix} gY\mathbf{B}' + g'\mathbf{W}'_z & g'(\mathbf{W}'_x - i\mathbf{W}'_y) \\ g'(\mathbf{W}'_x + i\mathbf{W}'_y) & gY\mathbf{B}' - g'\mathbf{W}'_z \end{pmatrix} \quad \begin{matrix} \nu_e & u \\ e & d, \end{matrix} \tag{2.292}$$

using Eq. (1.149). Note that the diagonal matrix elements contain both $\mathbf{B}'$ and $\mathbf{W}'_z$. These vector potentials do not change the nature of the particle. The off-diagonal terms containing $\mathbf{W}'_x$ and $\mathbf{W}'_y$ switch between the two components of the particle pair under consideration. These interaction terms carry a charge. On the right, two possible bases are given: neutrino/electron (ν_e/e) and up and down quarks (u/d). Again, the 2×2 matrix is an approximation. The electroweak interaction can couple to many different particles. We are simply focusing on the strongest couplings.

In Eq. (2.291), both components of the basis are equivalent, just as the spin-up and spin-down components of an electron are equivalent. However, in the electroweak interaction this symmetry breaks, and the interaction will act differently on the different components. We are then left with two interactions. The electromagnetic interaction, which is diagonal but acts differently on the two components. The remaining terms form the weak interaction. Later, it will be shown that the force carriers (also known as exchange particles or gauge bosons) are also dramatically different. The electromagnetic interaction is mediated by photons that have zero mass. The force carrier for the weak interaction is a vector boson almost 80 times as massive as a proton, making this interaction extremely short-ranged.

In order to bring out the way the electroweak interaction works in nature, let us rewrite the interaction constants as

$$\tan\theta = \frac{g}{g'}, \quad G = \sqrt{g^2 + g'^2} \quad \Rightarrow \quad g = G\sin\theta, \quad g' = G\cos\theta. \tag{2.293}$$

The change into the observed interactions essentially takes the form of a rotation of the diagonal matrix elements

$$\mathbf{W}'_z = \cos\theta\ \mathbf{Z}' + \sin\theta\ \mathbf{A}', \quad \mathbf{B}' = -\sin\theta\ \mathbf{Z}' + \cos\theta\ \mathbf{A}', \tag{2.294}$$

where θ is known as the Weinberg angle. The $\mathbf{A}'$ in this expression is the four-potential of the electromagnetic field. For the neutrino/electron pair, it turns out that the hypercharge $Y = -1$, and the matrix in Eq. (2.292) can be written as

$$\boldsymbol{H}_{\text{EW}} = \begin{pmatrix} G\mathbf{Z}' & \sqrt{2}g'\mathbf{W}'_- \\ \sqrt{2}g'\mathbf{W}'_+ & -G\sin 2\theta\ \mathbf{A}' - G\cos 2\theta\ \mathbf{Z}' \end{pmatrix} \quad \begin{matrix} \nu_e \\ e, \end{matrix} \tag{2.295}$$

with $\mathbf{W}'_\pm = (\mathbf{W}'_x \pm i\mathbf{W}'_y)/\sqrt{2}$. The four-potential $\mathbf{A}'$, associated with the electromagnetic interaction, only appears in the electron part. The neutrinos are therefore neutral and only have a diagonal interaction with the $\mathbf{Z}'$ four-vector. The electrons feel both the electromagnetic and weak interactions. The off-diagonal matrix elements are unchanged. The coupling strength of the electron to the electromagnetic four-potential $\mathbf{A}'$ is conventionally denoted as the elementary charge

$$e = G\sin 2\theta = \frac{G}{2}\sin\theta\cos\theta = \frac{gg'}{2G} = \frac{gg'}{2\sqrt{g^2 + g'^2}}. \tag{2.296}$$

For the up/down quark pair, the hypercharge turns out to be $Y = \frac{1}{3}$, and the electroweak interaction can be written as

$$\boldsymbol{H}_{\mathrm{EW}} = \begin{pmatrix} \frac{2}{3}e\mathbf{A}' + \frac{G}{3}(1 + 2\cos 2\theta)\mathbf{Z}' & \sqrt{2}g'\mathbf{W}'_- \\ \sqrt{2}g'\mathbf{W}'_+ & -\frac{1}{3}e\mathbf{A}' - \frac{G}{3}(2 + \cos 2\theta)\mathbf{Z}' \end{pmatrix} \begin{matrix} u. \\ d \end{matrix} \tag{2.297}$$

Both quarks interact with the electromagnetic field with charges $\frac{2}{3}e$ and $-\frac{1}{3}e$ for the up and down quarks, respectively. Note that the difference in charge is also e. Baryons are particles that consist of three quarks. Most important are the proton and the neutron that form the nucleus of everyday matter. The quark combinations are *uud* and *udd*, giving charges $+e$ and 0 for the proton and neutron, respectively. Note, however, that it is not the weak interaction that keeps the quarks together. This is mainly a result of the strong interaction as will be discussed later.

In general, the strength of the interaction with the vector-potential $\mathbf{A}'$ is directly proportional to the charge of the particle given by the diagonal 2×2 matrix

$$Q = I_3 + \frac{1}{2}Y\mathbb{1}_2 = \frac{1}{2}\begin{pmatrix} 1+Y & 0 \\ 0 & -1+Y \end{pmatrix} \begin{matrix} \nu_e & u \\ e & d, \end{matrix} \tag{2.298}$$

where the isospin is given by $I_3 = \frac{1}{2}\mathbf{e}_z$; $Y = -1, \frac{1}{3}$ for the neutrino/electron and up quark/down quark pairs, respectively.

It is important to note that without symmetry breaking, *i.e.* when the force carriers of the electromagnetic and weak interactions acquire different masses, the only thing that has been done so far is rewriting the diagonal components in Eq. (2.292). The matrices are exactly the same.

Conceptual summary. – In this section, we looked at the electroweak interaction. The strongest coupling is between certain pairs of particles, such as the electron and the electron neutrino and between the up and down quarks (or, effectively, the proton and the neutron). This interaction can be expressed with a scalar and a vector term. Note the similarities between the classical examples of two coupled oscillators, where one also finds a scalar and a vector term, see the Example in Section 1.7. The springs related to the scalar term only affect one object, whereas the vector term is related to a spring connecting the two objects. For the electroweak interaction, there is again a scalar term that does not change the nature of the particle and a vector term that causes particles to change their nature. For the coupled oscillators, a one-dimensional problem, the interaction itself is a scalar quantity. For the electroweak interaction, the matrix elements are four-vectors since the interaction is in spacetime. The interactions are often described in terms of symmetry. Since the electromagnetic field does not change the particle nature, it is described as U(1), which is effectively a one-dimensional space. The weak interaction couples most strongly between pairs of particles and is described in an effective two-dimensional space, known as SU(2). This stands for special unitary group of dimension 2. The Pauli matrices form a representation of this group. Note that this is an approximate symmetry, since there are additional couplings to other particles.

The interaction was rewritten in a different form. At this point, this does not do anything, since the matrices are effectively the same. However, in Section 5.4, we see that the nature of the force carriers changes. As it turns out, the force carrier of the electromagnetic force, the photon, remains massless. Electromagnetic interactions are therefore long range. The force carriers of the weak interaction, the Z and W bosons, are massive and have a very short range and are therefore most important in the nucleus. In terms of the classical coupled spring example, this would be a change in the nature (not the strength) of the springs. The massive and massless force carriers correspond to damped and undamped springs, respectively.

It is okay to miss some details of this section. The most important aspect is to see how one can obtain interactions that do not conserve the nature of the particles by introducing a vector component in the particle nature. The matrix elements are four-vectors, which is related to the fact that the interaction occurs in spacetime.

3

Single-Particle Problems

Introduction

In Chapter 2, the electromagnetic field was incorporated by going to spacetime. The inclusion of the four-potential to describe the changes in the four-momentum was relatively elegant. However, many applications are done using the Schrödinger equation, which is obtained by going to a rest frame where the vector component of the four-momentum is minimized. The remaining vector component of the generalized four-momentum becomes the spin of a particle. The advantage of the Schrödinger equation is that it is, apart from the spin, a scalar differential equation, and energy is the conserved quantity. It is especially convenient to study problems where the scalar part of the electromagnetic interaction, *i.e.* the electrostatic potential $U(\mathbf{r})$, dominates the physics. Since the related electric fields are often much larger than magnetic fields, this is true for a wide variety of problems.

In this chapter, we solve the Schrödinger equation for several different electrostatic potentials. Although the number of applications is legion, the focus here is on two different classes of problems. In selecting the topics, the preference was for function spaces that can be constructed in ways comparable to the function spaces of complex exponentials and spherical harmonics for free space. These problems provide the most insight into the geometric aspects. We will also look at perturbative problems. Here, the starting point is an already-solved system, and an additional interaction is added. Many introductory textbooks introduce perturbation theory, where the additional interaction is small. However, we look at a number of problems where the perturbation can be included exactly, *i.e.* to infinite order.

The chapter is divided as follows.

3.1 The well-known problem of the quantum harmonic oscillator is often the first problem where step operators are introduced. An oscillator is comparable to a rotation in two dimensions except that the rotation occurs in the plane of displacement versus momentum.

3.2 This section treats two exactly solvable perturbations to the harmonic oscillator, namely linear and quadratic in the displacement. The former is known as a displaced harmonic oscillator. The second is an effective change in oscillation frequency. These approaches occur in a much more complicated form for superconductivity and superfluidity.

3.3 The two-dimensional quantum harmonic oscillator is solved directly from the Schrödinger equation. This is an example of a series solution. This is not necessarily the most convenient approach to this problem, but the differential equation is encountered again for the radial equation of the hydrogen atom.

3.4 The two-dimensional quantum harmonic oscillator is revisited and solved with step operators. The step operators are written in polar coordinates, which corresponds better to the symmetry of the problem.

3.5 The rotational motion of an electron in the potential of a nucleus is free and described in terms of spherical harmonics, which we obtained in the first chapter. Here, the radial motion of the electron is considered, which is the motion in the presence of a $1/r$ potential. The differential equation can be rewritten in the same form as the two-dimensional harmonic oscillator.

3.6 The transitions between different atomic orbitals are described. It is shown that these transitions can again be described in terms of step operators.

3.7 Selected atomic orbitals are used to build extended systems. Although the approach is restricted to one-dimensional systems, this is sufficient to understand the concepts of molecular orbital and band formation.

3.8 The problem of solids is approached from the opposite limit. Instead of combining atoms to a solid, the starting point is free electrons, and a periodic potential due to the nuclei is introduced. The periodicity of the lattice introduces its own dual space.

3.9 The effect of a localized potential on a continuum of states is studied. The potential leads to a completely symmetric state separated from the other states. This problem uses perturbative methods summed to infinite order, which corresponds to an exact solution.

3.10 In this section, the coupling of a local state to a continuum of states is considered. When the continuum is effectively removed from the problem, the coupling translates into an effective lifetime of the state.

3.1 Quantum Harmonic Oscillator

Introduction. – A well-known one-dimensional problem in quantum mechanics is a particle in a parabolic potential. This is known as a quantum harmonic oscillator. This corresponds to the classical problem of an object attached to a spring, giving a force $-Kx$ for a spring constant $K = m\omega_0^2$, with ω_0 as the resonance frequency. This leads to a quadratic potential $U(x) = \frac{1}{2}m\omega_0^2 x^2$. A parabola is often a good approximation for a local minimum in a (one-dimensional) potential landscape. This is generally treated extensively in most introductory textbooks. The problem can be approached by solving the Schrödinger equation directly using a series solution, leading to Hermite polynomials. Alternatively, we can use operator techniques, leading to the well-known step-up and step-down operators. Here, the step operators have already been introduced in the first chapter when discussing rotations. We see that this is not a coincidence, since the harmonic oscillator can be viewed as a rotation in the plane consisting of the displacement and the momentum.

The time-independent Schrödinger equation in the case of a harmonic oscillator is

$$-\frac{\hbar^2}{2m}\frac{d^2\varphi_n(x)}{dx^2} + \frac{1}{2}m\omega_0^2 x^2 \varphi_n(x) = E_n \varphi_n(x). \tag{3.1}$$

The Schrödinger equation can be brought into a more compact form by introducing the dimensionless coordinate

$$\xi = \sqrt{\frac{m\omega_0}{\hbar}}x, \tag{3.2}$$

giving

$$\frac{\hbar\omega_0}{2}\left(-\frac{d^2}{d\xi^2}+\xi^2\right)\varphi_n(\xi) = \frac{\hbar\omega_0}{2}\left(\hat{p}_\xi^2+\xi^2\right)\varphi_n(\xi) = E_n\varphi_n(\xi), \tag{3.3}$$

where $\hat{p}_\xi = -id/d\xi$ is the momentum in the dimensionless coordinates. The Hamiltonian can be written in a separable form using the operators

$$\hat{a} = \frac{1}{\sqrt{2}}\left(\xi + \frac{d}{d\xi}\right) \quad \text{and} \quad \hat{a}^\dagger = \frac{1}{\sqrt{2}}\left(\xi - \frac{d}{d\xi}\right). \tag{3.4}$$

We will see below that $\hat{a}^\dagger$ and $\hat{a}$ work as step-up and step-down operators. There is a strong similarity between the operators defined here and the spinors discussed in Sections 1.6 and 1.14. The spinors were defined for rotations in a two-dimensional plane and are at a 45° angle in the plane of rotation. Here, the step operators are at a 45° angle in the ξ-p_ξ-plane. One can effectively view this as a spinor $\hat{a}^\dagger = (\xi - ip_\xi)/\sqrt{2}$. The other spinor is obtained by taking the conjugate, giving a right-handed coordinate system. Expressing the spinor as a vector gives $\hat{a}^\dagger\mathbf{e}_x = (\xi\mathbf{e}_x - p_\xi\underline{\mathbf{e}}_x)/\sqrt{2}$, where $\underline{\mathbf{e}}_x = i\mathbf{e}_x$ effectively defines a direction orthogonal to $\mathbf{e}_x$, see also Section 1.12. The approach is very similar to a vector in the complex plane $\boldsymbol{r} = x + iy$. When the vector rotates, x and y change continuously, but the norm of the vector $r^2 = \boldsymbol{r}^*\boldsymbol{r} = x^2 + y^2$ remains constant. For an oscillation, ξ and p_ξ change continuously, but the norm $\xi^2 + p_\xi^2$, which is related to the energy, remains constant. Switching to the spinors allows us to focus on the norm, *i.e.* the energy or amplitude of the oscillation. There are some differences: ξ and p_ξ do not commute with each other. Additionally, for rotation, we need two spinors to indicate the different directions. For linear motion, an oscillation in the opposite direction is the same solution but with a phase shift. Therefore, we only need one spinor.

Using the expressions for the step operators, the Hamiltonian can be rewritten as

$$\hat{H}\varphi_n(\xi) = \hbar\omega_0\left(\hat{a}^\dagger\hat{a} + \frac{1}{2}\right)\varphi_n(\xi) = E_n\varphi_n(\xi). \tag{3.5}$$

The factor $\frac{1}{2}$ arises from the fact that ξ and $\hat{p}_\xi$ do not commute. Since the $\varphi_n(x)$ are eigenfunctions, this implies that $\hat{a}^\dagger\hat{a}$ is a diagonal operator. Therefore, we can write it as a number operator

$$\hat{N} = \hat{a}^\dagger\hat{a}. \tag{3.6}$$

This gives for the Hamiltonian

$$\hat{H} = \hbar\omega_0\left(\hat{N} + \frac{1}{2}\right) \quad \text{and} \quad E_n = (n + \frac{1}{2})\hbar\omega_0. \tag{3.7}$$

One of the nice aspects of the function space of the quantum harmonic oscillator is that the number operator can be expressed in terms of the step operators. For complex exponentials, this can be done in vector form, but not in terms of operators, see Eqs. (1.401) and (1.404).

Up to this point, n can in principle be any real number. However, we will see that n can only be a positive integer. It is straightforward to show that the step operators commute as

$$[\hat{a}, \hat{a}^\dagger] = 1. \tag{3.8}$$

Note that the step operators commute with themselves, $[\hat{a}^\dagger, \hat{a}^\dagger] = 0$ and $[\hat{a}, \hat{a}] = 0$. Working with the number operator on the function $\hat{a}^\dagger \varphi_n(x)$ gives

$$\hat{a}^\dagger \hat{a} \hat{a}^\dagger \varphi_n(x) = \hat{a}^\dagger (\hat{a}^\dagger\ \hat{a} + 1)\varphi_n(x) = \hat{a}^\dagger (n+1)\varphi_n(x) = (n+1)\hat{a}^\dagger \varphi_n(x), \tag{3.9}$$

using Eq. (3.8). Likewise, we can show that $\hat{N}\hat{a}\varphi_n(x) = (n-1)\hat{a}\varphi_n(x)$. Therefore, $\hat{a}^\dagger\varphi_n(x)$ and $\hat{a}\varphi_n(x)$ are eigenfunctions with quantum numbers $n+1$ and $n-1$, respectively. Therefore, the values of n differ by 1 but can still be real number. However, since the Hamiltonian is the sum of two squares, $E_n \geq 0$. To avoid negative values, there must be a value of n for which $\hat{a}\varphi_{n_{\text{low}}}(x) = 0$, so that no eigenfunctions with negative energies are produced. Multiplying with $\hat{a}^\dagger$ gives $\hat{a}^\dagger \hat{a}\varphi_{n_{\text{low}}}(x) = n_{\text{low}}\varphi_{n_{\text{low}}}(x) = 0$. This gives $n_{\text{low}} = 0$. The possible values of n are then $n = 0, 1, 2, 3, \ldots$

Multiplying with $\hat{a}^\dagger$ produces eigenfunctions with higher values of n. For the spherical harmonics, the entire function space was generated by expanding the exponent of a unit vector $e^{\hat{\boldsymbol{r}}}$, see Eq. (1.358). We want to follow the same procedure for the quantum harmonic oscillator, *i.e.*

$$|\xi\rangle = e^{\hat{\boldsymbol{a}}^\dagger}\varphi_0(\xi) \quad \text{with} \quad \hat{\boldsymbol{a}}^\dagger = \hat{a}^\dagger \boldsymbol{a}^\dagger, \tag{3.10}$$

where $\hat{\boldsymbol{a}}^\dagger$ splits into the operator $\hat{a}^\dagger$ and the unit vector $\boldsymbol{a}^\dagger$. The unit vectors satisfy the same commutation relation as the operator in Eq. (3.8), *i.e.* $[\boldsymbol{a}, \boldsymbol{a}^\dagger] = 1$. An important difference with the function spaces for free space is the presence of a vacuum function $\varphi_0(\xi)$. In the absence of a potential, the function spaces are products of the coefficients of a vector. The vacuum state is simply $\varphi_0 = 1$ and can therefore be omitted. For example, for running waves in one dimension, working with the momentum operator, *i.e.* the effective number operator, on the vacuum state gives $\hat{k}_x\varphi_0 = -i(d/dx)1 = 0$. Note that $\varphi_0 = 1$ is, apart from the normalization, also the state for $k_x = 0$. In vector format, this is written as $\boldsymbol{a}^\dagger \boldsymbol{a}|0\rangle = 0$, see Eq. (1.401). Therefore, removing something from the vacuum state produces zero. The free-particle vacuum state $\varphi_0 = 1$ does not satisfy this criterium for the quantum harmonic oscillator, since $\hat{a}\varphi_0 = \hat{a}\ 1 = \xi/\sqrt{2}$ due to the additional term related to the potential energy. It is straightforward to demonstrate that $e^{-\frac{\xi^2}{2}}$ satisfies the criterium for a vacuum

$$\hat{a}e^{-\frac{\xi^2}{2}} = \frac{1}{\sqrt{2}}\left(\xi + \frac{d}{d\xi}\right)e^{-\frac{\xi^2}{2}} = \frac{1}{\sqrt{2}}e^{-\frac{\xi^2}{2}}(\xi + (-\xi)) = 0. \tag{3.11}$$

In quantum mechanics, functions are generally normalized such that the squared integral over the coordinates is 1. For the vacuum state $\varphi_0(\xi) = Ae^{-\frac{\xi^2}{2}}$, this gives

$$\int_{-\infty}^{\infty} dx\ (\varphi_0(\xi))^2 = \int_{-\infty}^{\infty} dx\ A^2 e^{-\xi^2} = A^2\sqrt{\pi}, \tag{3.12}$$

giving a normalization constant $A = 1/\pi^{1/4}$. The vacuum function is then given by

$$\varphi_0(\xi) = \frac{1}{\pi^{\frac{1}{4}}}e^{-\frac{\xi^2}{2}}. \tag{3.13}$$

The exponent can be expanded as

$$|\xi\rangle = e^{\hat{\boldsymbol{a}}^\dagger}\varphi_0(\xi) = \sum_{n=0}^{\infty}\frac{(\hat{a}^\dagger \boldsymbol{a}^\dagger)^n}{n!}\varphi_0(\xi). \tag{3.14}$$

Inserting the expression for the operator $\hat{a}^\dagger$ from Eq. (3.4) gives

$$|\xi\rangle = \sum_{n=0}^{\infty} \frac{1}{\sqrt{n!}} (\boldsymbol{a}^\dagger)^n \frac{1}{\sqrt{n!}} \left[\frac{1}{\sqrt{2}} \left(\xi - \frac{d}{d\xi} \right) \right]^n \frac{1}{\pi^{\frac{1}{4}}} e^{-\frac{\xi^2}{2}}.$$

The Hermite polynomial is defined as

$$H_n(\xi) = e^{\frac{\xi^2}{2}} \left(\xi - \frac{d}{d\xi} \right)^n e^{-\frac{\xi^2}{2}}. \tag{3.15}$$

The functions that comprise the function space are then

$$\varphi_n(\xi) = \langle n|\xi\rangle = \frac{1}{\sqrt{n!}} (\hat{a}^\dagger)^n \varphi_0(\xi) = \frac{1}{\sqrt{2^n n! \sqrt{\pi}}} e^{-\frac{\xi^2}{2}} H_n(\xi). \tag{3.16}$$

The basis vector is defined as

$$|n\rangle = \frac{1}{\sqrt{n!}} (\boldsymbol{a}^\dagger)^n |0\rangle, \tag{3.17}$$

where, in analogy to φ_0, a vacuum state has been introduced. These unit vectors follow the same properties as those defined for the angular momentum in Section 1.15. The position vector therefore splits into

$$|\xi\rangle = e^{\hat{a}^\dagger} \varphi_0(\xi) = \sum_{n=0}^{\infty} |n\rangle\langle n|\xi\rangle. \tag{3.18}$$

The wavefunction in Eq. (3.16) and the unit vector in Eq. (3.17) are constructed in the same fashion. Additionally, both the operator $\hat{a}^{(\dagger)}$ and the unit vectors $\boldsymbol{a}^{(\dagger)}$ follow the same commutation relations, see Eqs. (3.8) and (1.340), respectively. The combination of the two operators in the Hamiltonian in Eq. (3.5) is then equal to the number operator in Eq. (1.328).

Since both the momentum and position operator appear in quadratic form in the Hamiltonian in Eq. (3.3), the eigenfunctions of the quantum harmonic oscillator have the interesting property that they are also eigenfunctions in Fourier space. The Fourier transform of the lowest eigenfunction is given by

$$\varphi_0(\kappa) = \langle \kappa|0\rangle = \int d\xi \, \langle \kappa|\xi\rangle\langle \xi|0\rangle = \frac{1}{\sqrt{2\pi}} \int d\xi \, e^{-i\kappa\xi} \frac{1}{\pi^{\frac{1}{4}}} e^{-\frac{\xi^2}{2}}, \tag{3.19}$$

where κ is the dual coordinate of the dimensionless coordinate ξ. The argument of the exponentials can be written in quadratic from using $(\xi + i\kappa)^2 - \kappa^2 = \xi^2 - 2i\xi\kappa$,

$$\varphi_0(\kappa) = \frac{1}{\pi^{\frac{1}{4}}} e^{-\frac{\kappa^2}{2}} \frac{1}{\sqrt{2\pi}} \int_{-\infty}^{\infty} d\xi \, e^{-\frac{(\xi+i\kappa)^2}{2}} = \frac{1}{\pi^{\frac{1}{4}}} e^{-\frac{\kappa^2}{2}}. \tag{3.20}$$

By introducing the new variable $\xi' = \xi + i\kappa$ and noting that $d\xi' = d(\xi + i\kappa) = d\xi$, a standard integral with value $\sqrt{2\pi}$ is obtained. The end result is the same function in momentum space as in real space. Let us now look at the operators. The Fourier transform of the derivative operator is

$$\frac{d\varphi_n(\xi)}{d\xi} = \frac{1}{\sqrt{2\pi}} \frac{d}{d\xi} \int d\xi \, e^{i\kappa\xi} \varphi_n(\kappa) = \frac{1}{\sqrt{2\pi}} \int d\xi \, e^{i\kappa\xi} i\kappa\varphi_n(\kappa). \tag{3.21}$$

Therefore, the Fourier transform of the derivative is $i\kappa\varphi_n(\kappa)$. Vice versa, the derivative in Fourier space with respect to κ is $-i\xi$. Therefore, multiplication with ξ in real space corresponds to $id/d\kappa$ in momentum space. This means that the Fourier transform of the differential equation is

$$\frac{\hbar\omega_0}{2}\left(-(i\kappa)^2 + i^2\frac{d^2}{d\kappa^2}\right)\varphi_n(\kappa) = \frac{\hbar\omega_0}{2}\left(-\frac{d^2}{d\kappa^2} + \kappa^2\right)\varphi_n(\kappa) = E_n\varphi_n(\kappa).$$

Therefore, the differential equation in Fourier space is equivalent to that in real space. The step operator transforms to

$$\hat{a}^\dagger(\kappa) = \frac{1}{\sqrt{2}}\left(i\frac{d}{d\kappa} - i\kappa\right) = (-i)\frac{1}{\sqrt{2}}\left(\kappa - \frac{d}{d\kappa}\right). \tag{3.22}$$

The eigenfunctions in momentum space are then

$$\varphi_n(\kappa) = \frac{(-i)^n}{\sqrt{2^n n!\sqrt{\pi}}} e^{-\frac{\kappa^2}{2}} H_n(\kappa), \tag{3.23}$$

showing that, apart from a factor, the basis functions of the Harmonic oscillator are also eigenfunctions of the Fourier transform.

Conceptual summary. – This section briefly recapitulates the results for the quantum harmonic oscillator. Again, the entire function space can be generated using an exponential expansion of the lowest order term. The major difference with the function spaces for translation and rotation is the presence of a vacuum state. The vacuum state differs due to the presence of a potential. The lowest energy state is therefore a bound state described by a Gaussian as opposed to a completely structureless delocalized state in free space (the lowest states for translation and rotation are just a constant).

Since the quantum harmonic oscillator can be viewed as a system rotating in the ξ-p_ξ-plane, it has the unique property that the coordinates in real and momentum space are effectively equivalent. Therefore, the functional form of the Fourier transform of the wavefunction is the same as the wavefunction. Furthermore, the Hamiltonian expressed in term of momentum is the same as the Hamiltonian in spatial coordinates.

Problems

1 Find $\varphi_1(\xi)$ and $\varphi_2(\xi)$ from $\varphi_0(\xi)$ using the step operators.

2 The zero-point solution gives the correct asymptotic behavior of the eigenfunctions. This allows us to separate the wavefunction into $\varphi_n(\xi) = N_n H_n(\xi)e^{-\xi^2/2}$, where N_n is a normalization constant. What is the differential equation for $H_n(\xi)$? Find the solutions for $n = 0, 1, 2$ and their respective energies by suggesting a solution of the type $H_n(\xi) = \sum_{k=0}^{n} c_k\xi^k$.

3 Write the eigenfunction $|3\rangle$ in terms of $\boldsymbol{a}^\dagger$ and show that it is normalized. Find the results of working with $\boldsymbol{a}^\dagger$, $\boldsymbol{a}$, and $\boldsymbol{a}^\dagger\boldsymbol{a}$ on the state $|3\rangle$. What are the expectation values of $\boldsymbol{\xi} = (\boldsymbol{a} + \boldsymbol{a}^\dagger)/\sqrt{2}$ and $\boldsymbol{\xi}^2$ in the state $|3\rangle$?

3.2 Perturbed Harmonic Oscillator

Introduction. – In this section, we want to consider the harmonic oscillator where the potential is perturbed by terms proportional to x and x^2, where x is the displacement from the equilibrium position. Classically, this corresponds to a change in the equilibrium position of the spring and a change in spring constant, respectively. Despite their simplicity when considered classically, these concepts are important for a wide variety of important subjects, such as superconductivity and superfluidity. For the more complex problems, one is generally dealing with many different types of oscillators, and the perturbations are an approximation for many-body interactions, but the underlying mathematics is the same as for a single oscillator.

Linear perturbation: displaced harmonic oscillator. – We start with a change in the potential proportional to x. This problem is also known as the displaced harmonic oscillator. The most direct application occurs in molecules. In these systems, local ions often interact with the surrounding ligands (a neighboring ion that binds to the central ion). A change in the charge or the electronic configuration on the ion could repel or attract the ligands. The simplest situation is that all the surrounding ligands respond in the same way. This is known as a breathing mode which corresponds to a change in the ion–ligand distance. There can also be more complicated modes, such as Jahn–Teller distortions, where the response of the ligands depends on the direction. An understanding of the displaced harmonic oscillator also helps in the understanding of the electron–phonon interaction in superconductors, where the lattice in a solid responds to changes in the electron density.

Let us first consider the oscillation of an harmonic oscillator classically. There are two, effectively equivalent, ways to make an object attached to a spring oscillate. First, we can place the object attached to the spring out of the equilibrium position. Subsequently, we release the object. Second, one can switch on a constant force at a particular time. This effectively moves the equilibrium position, and the object starts to oscillate around the new equilibrium position. The differential equation for a displaced harmonic oscillator around an equilibrium position x_0 is given by

$$m\frac{d^2x}{dt^2} = -m\omega_0^2(x - x_0). \tag{3.24}$$

Note that this can be interpreted as adding a constant force $m\omega_0^2 x_0$ to the object. At rest, the spring has a displacement $x = x_0$. We can also solve the motion of the spring in the presence of the constant force

$$x(t) = A\cos\omega_0 t + x_0, \tag{3.25}$$

where A is the amplitude of the oscillation. If the object was at rest before adding the constant force at $t = 0$, then $A = -x_0$, so that $x(0) = 0$. The oscillation simply occurs around the new equilibrium position x_0 in the presence of the constant force. The Hamiltonian is given by

$$H = \frac{p_x^2}{2m} + \frac{1}{2}m\omega_0^2 x^2 - m\omega_0^2 x_0 x. \tag{3.26}$$

The momentum,

$$p_x(t) = -m\omega_0 A \sin \omega_0 t, \tag{3.27}$$

is independent of the constant force, that is the oscillation frequency around the new equilibrium position is exactly the same. Inserting $x(t)$ and $p_x(t)$ into the Hamiltonian gives

$$\begin{aligned} E &= \frac{m^2 A^2 \omega_0^2}{2m} \sin^2 \omega_0 t + \frac{1}{2} m\omega_0^2 (A\cos\omega_0 t + x_0)^2 - m\omega_0^2 x_0 (A\cos\omega_0 t + x_0) \\ &= \frac{1}{2} m\omega_0^2 A^2 - \frac{1}{2} m\omega_0^2 x_0^2. \end{aligned} \tag{3.28}$$

By defining the shift in energy as $\varepsilon_p = \frac{1}{2} m\omega_0^2 x_0^2$, we can write the displacement as

$$x_0 = \sqrt{\frac{2\varepsilon_p}{m\omega_0^2}}. \tag{3.29}$$

The displacement term in the Hamiltonian can then be written as

$$H' = -m\omega_0^2 x x_0 = -\sqrt{2m\omega_0^2 \varepsilon_p} x = -\sqrt{2\hbar\omega_0 \varepsilon_p} \xi. \tag{3.30}$$

Let us now cast the problem in quantum-mechanical terms. Using Eqs. (3.2) and (3.4), we can write the total Hamiltonian for the quantum harmonic oscillator in the presence of a constant force as

$$\hat{H} = \hbar\omega_0 \hat{a}^\dagger \hat{a} - \lambda(\hat{a} + \hat{a}^\dagger), \tag{3.31}$$

with $\lambda = \sqrt{\hbar\omega_0 \varepsilon_p}$. The sign in front of λ determines the direction of the displacement. The above Hamiltonian is still that of a harmonic oscillator but with a different equilibrium position. The Hamiltonian can be diagonalized by introducing displaced step operators $\hat{a}^\dagger = \hat{a}'^\dagger + \Delta$. The new Hamiltonian is

$$\hat{H} = \hbar\omega_0 (\hat{a}'^\dagger + \Delta)(\hat{a}' + \Delta) - \lambda(\hat{a}' + \Delta + \hat{a}'^\dagger + \Delta) \tag{3.32}$$

$$= \hbar\omega_0 \hat{a}'^\dagger \hat{a}' - (\lambda - \hbar\omega_0 \Delta)(\hat{a}' + \hat{a}'^\dagger) + \hbar\omega_0 \Delta^2 - 2\lambda\Delta. \tag{3.33}$$

We want the interaction to vanish, implying that $\Delta = \lambda/\hbar\omega_0$. This gives the Hamiltonian

$$\hat{H} = \hbar\omega_0 \hat{a}'^\dagger \hat{a}' - \varepsilon_p, \tag{3.34}$$

with $\varepsilon_p = \lambda^2/\hbar\omega_0$. We therefore see that the difference between the eigenstates is still $\hbar\omega_0$ (the shape of the harmonic potential has not changed), but the energy is shifted by $-\varepsilon_p$. The corresponding unit vectors for the displaced harmonic oscillator are the same as those for an harmonic oscillator but in terms of the displaced step operators

$$|n'\rangle = \frac{1}{\sqrt{n'!}} (\boldsymbol{a}'^\dagger)^{n'} |0\rangle \quad \text{with} \quad E_{n'} = n'\hbar\omega_0 - \varepsilon_p. \tag{3.35}$$

Let us now look at the situation where we take the displaced harmonic oscillator in its ground state ($n' = 0$) and release it, *i.e.* we remove the constant force by taking $\lambda = 0$. Classically, we expect the spring to oscillate with an amplitude x_0. The energy of the oscillator is equal to $\frac{1}{2} m\omega_0^2 x_0^2 = \varepsilon_p = g\hbar\omega_0$ with the dimensionless factor

$$g = \frac{\varepsilon_p}{\hbar\omega_0} = \frac{m\omega_0 x_0^2}{2\hbar} = \left(\frac{\lambda}{\hbar\omega_0}\right)^2 = \Delta^2. \tag{3.36}$$

In order to study the motion of the state $|0'\rangle$, we want to express it in the unperturbed states $|n\rangle$

$$\langle n|0'\rangle = \frac{1}{\sqrt{n!}}\langle 0|\boldsymbol{a}^n|0'\rangle = \frac{1}{\sqrt{n!}}\langle 0|(\boldsymbol{a}' + \sqrt{g})^n|0'\rangle = \frac{g^{\frac{n}{2}}}{\sqrt{n!}}\langle 0|0'\rangle, \tag{3.37}$$

using $\boldsymbol{a}'|0'\rangle = 0$. Note that the overlap between the vacuum states $\langle 0|0'\rangle$ is not 1, since they are vacuum states for different equilibrium positions. This overlap can be obtained by noting that the total integrated weight has to equal 1 (the displacement is essentially a unitary transformation between two different bases). This gives

$$1 = \sum_{n=0}^{\infty}|\langle n|0'\rangle|^2 = \sum_{n=0}^{\infty}\frac{g^n}{n!}|\langle 0|0'\rangle|^2 = e^g|\langle 0|0'\rangle|^2, \tag{3.38}$$

using the series of an exponential. This gives $\langle 0|0'\rangle = e^{-g/2}$. Note that this overlap becomes exponentially small for $\varepsilon_p/\hbar\omega_0 \gg 1$. The lowest displaced state $|0'\rangle$ can therefore be expressed in terms of the unperturbed harmonic oscillator states $|n\rangle$ as

$$|0'\rangle = e^{-\frac{g}{2}}\sum_{n=0}^{\infty}\frac{g^{\frac{n}{2}}}{\sqrt{n!}}|n\rangle \quad \Rightarrow \quad |\langle n|0'\rangle|^2 = \frac{e^{-g}g^n}{n!}. \tag{3.39}$$

The weights on the right side correspond to a Poisson distribution. This is directly related to the Franck–Condon principle in molecular spectroscopy. For example, the removal of an electron by, say, photoemission causes a sudden change in the interaction with the local vibrational modes. If the change is sudden, the nuclei cannot follow, and many vibrational states at energies $n\hbar\omega_0$ are excited. The weights are given by the Poisson distribution.

When written in the original diagonal basis, the time dependence of $|0'\rangle$ is easily determined

$$|0'(t)\rangle = e^{-\frac{g}{2}}\sum_{n=0}^{\infty}\frac{g^{\frac{n}{2}}}{\sqrt{n!}}e^{-in\omega_0 t}|n\rangle. \tag{3.40}$$

We are interested in the expectation value of the position for this wavepacket. Using Eqs. (3.2) and (3.4), the x coordinate can be expressed as

$$\boldsymbol{x} = \sqrt{\frac{\hbar}{2m\omega_0}}(\boldsymbol{a} + \boldsymbol{a}^\dagger). \tag{3.41}$$

Let us first evaluate $\boldsymbol{a}^\dagger$. Noting that $\langle n+1|\boldsymbol{a}^\dagger|n\rangle = \sqrt{n+1}$, the only nonzero terms are

$$\langle 0'(t)|\boldsymbol{a}^\dagger|0'(t)\rangle = \sum_{n=0}^{\infty}\frac{e^{-\frac{g}{2}}g^{\frac{n+1}{2}}e^{i(n+1)\omega_0 t}}{\sqrt{(n+1)!}}\frac{e^{-\frac{g}{2}}g^{\frac{n}{2}}e^{-in\omega_0 t}}{\sqrt{n!}}\sqrt{n+1}.$$

Cleaning up the expression gives

$$\langle 0'(t)|\boldsymbol{a}^\dagger|0'(t)\rangle = e^{-g}g^{\frac{1}{2}}e^{i\omega_0 t}\sum_{n=0}^{\infty}\frac{g^n}{n!} = e^{-g}\sqrt{g}e^{i\omega_0 t}e^g = \sqrt{g}e^{i\omega_0 t}, \tag{3.42}$$

using the series expression of the exponential. The matrix element for $\boldsymbol{a}$ is its conjugate. This gives for the expectation value of $\boldsymbol{x}$

$$\langle x(t)\rangle = \langle 0'(t)|\boldsymbol{x}|0'(t)\rangle = \sqrt{\frac{\hbar}{2m\omega_0}}\sqrt{g}(e^{i\omega_0 t} + e^{-i\omega_0 t}) = \sqrt{\frac{2\hbar}{m\omega_0}\frac{m\omega_0 x_0^2}{2\hbar}}\cos\omega_0 t,$$

using Eq. (3.36). The final result is then

$$\langle x(t)\rangle = x_0 \cos\omega_0 t. \tag{3.43}$$

Therefore, the motion of this wavepacket is the same as the classical motion. It is therefore a completely coherent wavepacket.

Quadratic perturbation: change of spring constant. – Within a classical framework, a perturbation proportional to the displacement squared can be written as

$$H = \frac{p_x^2}{2m} + \frac{1}{2}m\omega_0^2 x^2 + m\omega_0 v x^2 = \frac{p_x^2}{2m} + \frac{1}{2}m(\omega_0^2 + 2\omega_0 v)x^2. \tag{3.44}$$

The reason for the notation becomes apparent below. This gives an effective frequency

$$\Omega = \sqrt{\omega_0^2 + 2\omega_0 v}. \tag{3.45}$$

Obviously, we have effectively displaced the spring by one with a different spring constant.

In quantum mechanics, it is not directly obvious that a perturbation proportional to the squared displacement can be diagonalized,

$$\boldsymbol{H} = \hbar\omega_0 \boldsymbol{a}^\dagger \boldsymbol{a} + \hbar v \boldsymbol{\xi}^2 = \hbar(\omega_0 + v)\boldsymbol{a}^\dagger \boldsymbol{a} + \frac{\hbar v}{2}(\boldsymbol{a}^\dagger \boldsymbol{a}^\dagger + \boldsymbol{a}\boldsymbol{a} + 1), \tag{3.46}$$

using the step operators in vector notation; $\hbar v$ is the strength of the perturbation, and $\boldsymbol{\xi} = (\boldsymbol{a} + \boldsymbol{a}^\dagger)/\sqrt{2}$ is the unitless displacement defined in Eq. (3.2), which can be expressed in terms of the step operators using Eq. (3.4). This is generally solved using a Bogoliubov transformation,

$$\boldsymbol{b}^\dagger = \cosh\frac{\alpha}{2}\boldsymbol{a}^\dagger + \sinh\frac{\alpha}{2}\boldsymbol{a}. \tag{3.47}$$

Note the similarity between the Bogoliubov transformation and the spacetime spinors in Eq. (2.81). These operator still satisfy the commutation relations

$$\begin{aligned}[\boldsymbol{b}, \boldsymbol{b}^\dagger] &= [\cosh\frac{\alpha}{2}\boldsymbol{a} + \sinh\frac{\alpha}{2}\boldsymbol{a}^\dagger, \cosh\frac{\alpha}{2}\boldsymbol{a}^\dagger + \sinh\frac{\alpha}{2}\boldsymbol{a}] \\ &= \cosh^2\frac{\alpha}{2}[\boldsymbol{a}, \boldsymbol{a}^\dagger] + \sinh^2\frac{\alpha}{2}[\boldsymbol{a}^\dagger, \boldsymbol{a}] = \cosh^2\frac{\alpha}{2} - \sinh^2\frac{\alpha}{2} = 1,\end{aligned} \tag{3.48}$$

using the properties of the commutation relation in Eq. (1.340). Let us consider the diagonal Hamiltonian in the new basis

$$\boldsymbol{H}' = \hbar\Omega \boldsymbol{b}^\dagger \boldsymbol{b} = \hbar\Omega(\cosh^2\frac{\alpha}{2}\boldsymbol{a}^\dagger \boldsymbol{a} + \sinh^2\frac{\alpha}{2}\boldsymbol{a}\boldsymbol{a}^\dagger + \cosh\frac{\alpha}{2}\cosh\frac{\alpha}{2}(\boldsymbol{a}^\dagger \boldsymbol{a}^\dagger + \boldsymbol{a}\boldsymbol{a})).$$

Using the double-angle formulas for the hyperbolic functions, $\cosh^2\frac{\alpha}{2} + \sinh^2\frac{\alpha}{2} = \cosh\alpha$ and $2\sinh\frac{\alpha}{2}\cosh\frac{\alpha}{2} = \sinh\alpha$, this can be rewritten as

$$\boldsymbol{H}' = \hbar\Omega\left(\cosh\alpha\ \boldsymbol{a}^\dagger \boldsymbol{a} + \frac{\sinh\alpha}{2}(\boldsymbol{a}^\dagger \boldsymbol{a}^\dagger + \boldsymbol{a}\boldsymbol{a}) + \sinh^2\frac{\alpha}{2}\right). \tag{3.49}$$

$\boldsymbol{H}'$ has the same form as the original Hamiltonian in Eq. (3.46). We ignore the shifts in energy. The hyperbolic functions need to satisfy

$$\cosh\alpha = \frac{\omega_0 + v}{\Omega}, \quad \sinh\alpha = \frac{v}{\Omega}, \quad \tanh\alpha = \frac{v}{\omega_0 + v}. \tag{3.50}$$

Using the norm of the hyperbolic functions, the energy can be found

$$\cosh^2\alpha - \sinh^2\alpha = \left(\frac{\omega_0 + v}{\Omega}\right)^2 - \left(\frac{v}{\Omega}\right)^2 = 1, \tag{3.51}$$

which gives

$$\Omega^2 = (\omega_0 + v)^2 - v^2 = \omega_0^2 + 2\omega_0 v, \tag{3.52}$$

reproducing the result in Eq. (3.45).

Conceptual summary. – Understanding the perturbed harmonic oscillator is important for a wide variety of problems. The displaced harmonic oscillator is important on its own in the study of excitations in molecules and solids. Excitations with visible light can excite a system from one potential well into a one with a displaced equilibrium position. The excitation creates a wavepacket, and the system starts to oscillate. In many problems, the complexity increases, but the underlying ideas remain the same. An example is the derivation of the effective attractive electron–electron interaction mediated by lattice vibrations in conventional superconductors. Here, the lattice vibrations are the oscillators and are momentum dependent. The linear perturbation now depends on momentum and the electron density. However, the underlying physics is still that of a displaced harmonic oscillator. The perturbation quadratic in the displacement becomes relevant for superfluidity when a similar Hamiltonian arises when simplifying the interaction between the moving He atoms and the condensate of helium atoms.

Problems

1 Let us consider two quantum harmonic oscillators with the same mass m and the same frequency ω_0 coupled via the interaction $U = \frac{1}{2}m\omega_1^2(x_1 - x_2)^2$, where x_i with $i = 1, 2$ are the x coordinates of the two oscillators, and K is a constant. Find the energy levels and the wavefunctions.

2 A function can be displaced using the translation operator, see, *e.g.* Eq.(1.242). Derive the displacement operator in terms of $\boldsymbol{a}$ and $\boldsymbol{a}^\dagger$. Use the Baker–Campbell–Hausdorff formula $e^{X+Y} = e^{-\frac{1}{2}[X,Y]}e^Y e^X$ for noncommuting quantities X and Y to derive the displaced eigenstate in terms of the eigenstates $|n\rangle$.

3.3 Two-Dimensional Harmonic Oscillator via Differential Equation

Introduction. – In this section, we treat an example following the approach often taken in elementary textbooks, namely the brute-force solution of the Schrödinger equation via a series solution. This method has been avoided so far, since it does not always provide insights into the structure of function spaces. However, straightforward solving of the differential equation is often the only practical method for complicated potentials, although this is often done numerically. Here, we look at the two-dimensional harmonic oscillator. Series solutions for spherical harmonics and Hermite polynomials are easily found in many other textbooks, and we have already obtained these function spaces in a different way.

There are two approaches to the Schrödinger equation for the two-dimensional quantum harmonic oscillator: the easy and the hard way. The easy way is to treat the two directions as independent and use the solution of the harmonic oscillator in one dimension from Section 3.1. The hard way is to view the system as a problem with a harmonic potential in the radial direction. However, the latter problem itself is instructive, since one can view it as a two-dimensional "atom," and one would expect quantum numbers associated with the rotational and radial motion. In fact, the same differential equation can be obtained when rewriting the radial part of the Schrödinger equation for a hydrogen-like atomic potential. Additionally, polar coordinates correspond more closely to the symmetry of the problem, and the eigenfunctions are more naturally organized. These concepts are important when looking at the structure of the nucleus which can be viewed, in its simplest form, as protons and neutrons in a three-dimensional harmonic oscillator well.

The Hamiltonian for a two-dimensional oscillator is given by

$$\left(-\frac{\hbar^2}{2m}\left(\frac{\partial^2}{\partial x^2}+\frac{\partial^2}{\partial y^2}\right)+\frac{1}{2}m\omega_0^2(x^2+y^2)\right)\varphi_{n_x n_y}(x,y)=E\varphi_{n_x n_y}(x,y). \tag{3.53}$$

Since the problem is two dimensional, we should be able to identify the eigenstates with two quantum numbers that we choose to be n_x and n_y. The solution naturally splits into two separate problems for the x and y coordinates, and the solution can be written as a product of the solutions in one dimension given in Eq. (3.23)

$$\varphi_{n_x n_y}(x,y)=\varphi_{n_x}(x)\varphi_{n_y}(y). \tag{3.54}$$

The energy is simply given by the sum of the two independent oscillators

$$E=(n_x+n_x+1)\hbar\omega_0. \tag{3.55}$$

This is the easy solution to the two-dimensional harmonic oscillator.

This function space can also be generated by an exponential function. Following Eq. (3.14),

$$\begin{aligned} e^{\hat{a}_x^\dagger+\hat{a}_y^\dagger}\varphi_0(x,y) &= \sum_{n=0}^{\infty}\frac{1}{n!}(\hat{\boldsymbol{a}}_x^\dagger+\hat{\boldsymbol{a}}_y^\dagger)^n\varphi_0(x,y) \\ &= \sum_{n=0}^{\infty}\sum_{n_x=0}^{n}\frac{1}{n!}\binom{n}{n_x}(\hat{\boldsymbol{a}}_x^\dagger)^{n_x}(\hat{\boldsymbol{a}}_y^\dagger)^{n-n_x}\varphi_0(x,y), \end{aligned} \tag{3.56}$$

with the vacuum state given by $\varphi_0(x,y)=\varphi_0(x)\varphi_0(y)$, see Eq. (3.13). Using that $\hat{\boldsymbol{a}}_i^\dagger = \hat{a}_i^\dagger \boldsymbol{a}_i^\dagger$, with $i=x,y$, gives

$$\begin{aligned} e^{\hat{a}_x^\dagger+\hat{a}_y^\dagger}\varphi_0(x,y) &= \sum_{n=0}^{\infty}\sum_{n_x+n_y=n}\frac{(\boldsymbol{a}_x^\dagger)^{n_x}(\boldsymbol{a}_y^\dagger)^{n_y}}{\sqrt{n_x!n_y!}}\frac{(\hat{a}_x^\dagger)^{n_x}}{\sqrt{n_x!}}\varphi_0(x)\frac{(\hat{a}_y^\dagger)^{n_y}}{\sqrt{n_y!}}\varphi_0(y) \\ &= \sum_{n=0}^{\infty}\sum_{n_x+n_y=n}|n_x;n_y\rangle\langle n_x;n_y|x;y\rangle, \end{aligned} \tag{3.57}$$

where the different bra-kets are identified by the presence of a semicolon. This defines the unit vectors and the eigenfunctions

$$\begin{aligned} |n_x;n_y\rangle &= \frac{(\boldsymbol{a}_x^\dagger)^{n_x}(\boldsymbol{a}_y^\dagger)^{n_y}}{\sqrt{n_x!n_y!}}, \\ \langle n_x;n_y|x;y\rangle &= \langle n_x|x\rangle\langle n_y|y\rangle = \varphi_{n_x}(x)\varphi_{n_y}(y), \end{aligned} \tag{3.58}$$

using Eq. (3.16).

The hard way is to solve the same problem in polar coordinates. Fortunately, this will produce additional insights. The Schrödinger equation is given by

$$-\frac{\hbar^2}{2m}\nabla^2\varphi_{jm_j}(r,\varphi)+\frac{1}{2}m\omega_0^2 r^2\varphi_{jm_j}(r,\varphi)=(2j+1)\hbar\omega_0\varphi_{jm_j}(r,\varphi), \quad (3.59)$$

where $r=\sqrt{x^2+y^2}$. The angular momentum quantum numbers jm_j now replace n_x and n_y. Since this is a two-dimensional problem, half-integer values are expected for the angular momentum. The energy term has been replaced by

$$E=(2j+1)\hbar\omega_0. \quad (3.60)$$

Here, we are anticipating the notation from the solution in Eq. (3.55). For a fixed number of $n=n_x+n_y$, giving an energy $E=(n+1)\hbar\omega_0$, one finds the following wavefunctions in products of Hermite polynomials:

$n=n_x+n_y$	wavefunctions	j	$2j+1$
0	$\vert 0;0\rangle$	0	1
1	$\vert 1;0\rangle, \vert 0;1\rangle$	$\frac{1}{2}$	2
2	$\vert 2;0\rangle, \vert 1;1\rangle, \vert 0;2\rangle$	1	3
3	$\vert 3;0\rangle, \vert 2;1\rangle, \vert 1;2\rangle, \vert 0;3\rangle$	$\frac{3}{2}$	4
4	$\vert 4;0\rangle, \vert 3;1\rangle, \vert 2;2\rangle, \vert 1;3\rangle, \vert 0;4\rangle$	2	5

(3.61)

The wavefunctions are written in bra-ket notation as $|n_x;n_y\rangle=|n_x\rangle|n_y\rangle$, where $\langle\xi|n\rangle=\varphi_n(\xi)$ is an eigenfunction of the Hamiltonian of the one-dimensional quantum harmonic oscillator. Since the eigenfunctions in polar coordinates can only be a unitary transformation of the above eigenfunctions, the number of states must be conserved. The value of $n+1=2j+1$ gives the number of eigenstates and is directly related to the energy, $E/\hbar\omega_0=2j+1$. We now want to find the functions that can be associated with $m_j=j,j-1,\ldots,-j$.

The Laplacian in polar coordinates is given by

$$\nabla^2=\frac{1}{r}\frac{\partial}{\partial r}\left(r\frac{\partial}{\partial r}\right)+\frac{1}{r^2}\frac{\partial^2}{\partial\varphi^2}. \quad (3.62)$$

Since the potential does not depend on φ, the angular part can be solved easily using a complex exponential. The unit vector in a two-dimensional plane is $\hat{\boldsymbol{r}}=e^{\boldsymbol{i}\varphi}$. Since the rotation is around the z-axis, the imaginary unit corresponds to $\boldsymbol{i}\equiv\boldsymbol{i}_{yx}$. The angular solutions are given by the unit vector to an integer power

$$\varphi_{jm_j}(\mathbf{r})=R_{jm_j}(r)\,e^{2\boldsymbol{i}m_j\varphi}. \quad (3.63)$$

The factor 2 in the exponential occurs because of our choice to describe the angular momentum around a fixed axis in terms of half-integers. The angular motion is determined by the operator

$$\hat{\mathbf{e}}_z=\frac{2}{\hbar}\hat{J}_z=-\boldsymbol{i}\frac{\partial}{\partial\varphi}, \quad (3.64)$$

giving

$$\hat{\mathbf{e}}_z\varphi_{jm_j}=2m_j\varphi_{jm_j} \quad\text{and}\quad \hat{J}_z\varphi_{jm_j}=m_j\hbar\varphi_{jm_j}, \quad (3.65)$$

where $\hat{\mathbf{e}}_z$ is the operator related to the rotation axis. The φ-dependent part in the kinetic energy then becomes

$$-\frac{\hbar^2}{2m}\frac{1}{r^2}\frac{\partial^2}{\partial\varphi^2}e^{2\boldsymbol{i}m_j\varphi}=\frac{(2m_j\hbar)^2}{2mr^2}e^{2\boldsymbol{i}m_j\varphi}=\frac{(2J_z)^2}{2mr^2}e^{2\boldsymbol{i}m_j\varphi}. \quad (3.66)$$

This is known as the centrifugal barrier. This term prevents the particle from getting too close to the origin, because this causes the angular momentum to diverge. Note that, although m_j can be half-integer, the actual "physical" angular momentum is related to $2J_z$, comparable to the gyromagnetic ratio we encountered for the Zeeman splitting.

Inserting the Laplacian in polar coordinates, taking the derivative with respect to φ, and multiplying the equation by $-2mr^2/\hbar^2$ gives

$$r\frac{d}{dr}\left(r\frac{dR_{jm_j}}{dr}\right) - 4m_j^2 R_{jm_j} - \frac{m^2\omega_0^2}{\hbar^2} r^4 R_{jm_j} = -2(2j+1)\frac{m\omega_0}{\hbar} r^2 R_{jm_j},$$

where the complex exponential has been removed from the equation.

The next steps have little physical meaning. The goal is to bring the differential equation into a form that allows us to use known results from mathematics. The desired shape is $d^2\overline{R}_{jm_j}/d\rho^2 + \cdots = 0$. This means all the "physical" aspects need to be removed, leading to dimensionless coordinates. Additionally, all the terms related to r in the second derivative need to be eliminated.

Let us first write the equation into a dimensionless form, so that it can be better compared with standard differential equations. Following Eq. (3.2), the dimensionless squared radius $\rho = r^2\, m\omega_0/\hbar$ is defined. The differential equation then becomes

$$4\rho\frac{d}{d\rho}\left(\rho\frac{dR_{jm_j}}{d\rho}\right) + \left(-4m_j^2 - \rho^2 + 2(2j+1)\rho\right) R_{jm_j} = 0, \tag{3.67}$$

using that

$$r\frac{d}{dr} = \sqrt{\rho}\frac{d}{d\sqrt{\rho}} = \sqrt{\rho}\frac{d\rho}{d\sqrt{\rho}}\frac{d}{d\rho} = 2\rho\frac{d}{d\rho}. \tag{3.68}$$

There are still ρ''s in the second derivative. They can be removed by introducing $R_{jm_j} = \overline{R}_{jm_j}/\sqrt{\rho}$. The first term in Eq. (3.67) then becomes

$$4\rho\frac{d}{d\rho}\left(\rho\frac{d}{d\rho}\frac{\overline{R}_{jm_j}}{\sqrt{\rho}}\right) = \frac{4\rho^2}{\sqrt{\rho}}\frac{d^2\overline{R}_{jm_j}}{d\rho^2} + \frac{1}{\sqrt{\rho}}\overline{R}_{jm_j}. \tag{3.69}$$

Inserting this into Eq. (3.67) and dividing by $4\rho^2/\sqrt{\rho}$

$$\frac{d^2\overline{R}_{jm_j}}{d\rho^2} + \left(-\frac{m_j^2 - \frac{1}{4}}{\rho^2} - \frac{1}{4} + \frac{2j+1}{2\rho}\right)\overline{R}_{jm_j} = 0. \tag{3.70}$$

The terms between parentheses arise from the centrifugal barrier, the harmonic potential, and the energy, respectively. A very similar differential equation is found for the radial equation of the hydrogen atom.

The next step is to separate the asymptotic behavior from the oscillatory behavior. For a harmonic oscillator, the asymptotic behavior was given by a decaying exponential, which keeps the particle inside the potential well. The oscillatory character was described in terms of polynomials. In order to find a differential equation for the polynomials, the asymptotic behavior needs to be removed.

First, we look for a solution in the limit $\rho \to \infty$. The terms between parentheses proportional to $1/\rho^2$ and $1/\rho$ in Eq. (3.70) go to zero, and only the $-\frac{1}{4}$ term remains. This term arose from the harmonic potential, which dominates the $\rho \to \infty$ behavior. The differential equation then becomes

$$\frac{d^2\overline{R}_{jm_j}}{d\rho^2} - \frac{1}{4}\overline{R}_{jm_j} \cong 0. \tag{3.71}$$

This term ensures that the particle remains localized inside the potential well. The solutions in the limit are

$$\overline{R}_{jm_j} \sim e^{-\frac{\rho}{2}} = e^{-\frac{m\omega_0 r^2}{2\hbar}}. \tag{3.72}$$

This is similar to the zero-point motion term for the harmonic oscillator in one dimension, see Eq. (3.13), except that r replaces x as the distance to the equilibrium position.

The potential related to the centrifugal barrier restricts the motion in the limit $\rho \to 0$, where this term dominates. The differential equation in this limit is

$$\frac{d^2\overline{R}_{jm_j}}{d\rho^2} - \frac{m_j^2 - \frac{1}{4}}{\rho^2}\overline{R}_{jm_j} \cong 0. \tag{3.73}$$

The solutions in this limit are

$$\overline{R}_{jm_j} \sim \rho^{n_j+\frac{1}{2}} \text{ with } n_j = |m_j|. \tag{3.74}$$

The differential equation can also be satisfied with a diverging power, but this does not lead to a physical wavefunction. The integer n_j gives the magnitude of the angular motion.

The remaining terms in the solution can be expressed in terms of a polynomial

$$\overline{R}_{jm_j} = e^{-\frac{\rho}{2}}\rho^{n_j+\frac{1}{2}}L_{j-n_j}^{2n_j}(\rho) \text{ or } R_{jm_j} = e^{-\frac{\rho}{2}}\rho^{n_j}L_{j-n_j}^{2n_j}(\rho). \tag{3.75}$$

The indices in the polynomial $L_{j-n_j}^{2n_j}(\rho)$ are already given in the conventional notation to avoid relabeling later. Reinserting this into Eq. (3.73), performing some laborious differentiations, and dividing by $e^{-\rho/2}\rho^{n_j-\frac{1}{2}}$ give

$$\rho\frac{d^2L_{j-n_j}^{2n_j}}{d\rho^2} + (2n_j + 1 - \rho)\frac{dL_{j-n_j}^{2n_j}}{d\rho} + (j - n_j)L_{j-n_j}^{2n_j} = 0. \tag{3.76}$$

Note that there are no longer any diverging terms in the differential equation. Let us lighten the notation a bit and bring the differential equation into a standard form. Taking the indices of the Laguerre polynomial as

$$n_r = j - n_j \text{ and } \nu = 2n_j \tag{3.77}$$

allows us to write $L_{n_r}^{\nu}$. The integer n_r is called the degree, *i.e.* the highest power of the polynomial. This also indicates the number of times the Laguerre polynomial crosses zero. The order ν affects the coefficients in the polynomial but not its degree. The differential equation then becomes

$$\rho\frac{d^2L_{n_r}^{\nu}}{d\rho^2} + (\nu + 1 - \rho)\frac{dL_{n_r}^{\nu}}{d\rho} + n_r L_{n_r}^{\nu} = 0. \tag{3.78}$$

This is known as the associated or generalized Laguerre equation. The Laguerre equation is obtained for $\nu = 0$. The energy term in the new coefficients is

$$\frac{E}{\hbar\omega_0} = 2j + 1 = 2n_r + \nu + 1 = 2(n_r + n_j) + 1 \tag{3.79}$$

Just as n_x and n_y indicate how the energy, $E = (n_x + n_y + 1)\hbar\omega_0$, is distributed between the motion in the x and y directions, respectively, n_r and n_j show how the energy is split between the radial and angular directions, respectively. Although we know from the solution in Cartesian coordinates that the energy is quantized, there is nothing in the Schrödinger equation so far that indicates that this is the case.

The typical approach to solving the Laguerre differential equation is by a series solution. The assumption is that the solution can be written as a polynomial

$$L^{\nu}_{n_r}(\rho) = \sum_{m=0}^{\infty} c_m \rho^m, \tag{3.80}$$

where m is simply a summation index. The next step is to find an expression for the derivatives in terms of a series. The first derivative is given by

$$\frac{dL^{\nu}_{n_r}}{d\rho} = \sum_{m=0}^{\infty} m c_m \rho^{m-1} = \sum_{m'=0}^{\infty} (m'+1) c_{m'+1} \rho^{m'}, \tag{3.81}$$

where in the second step the index has been shifted to $m' = m - 1$ or $m = m' + 1$ (the prime will be dropped in the following). The second derivative is then

$$\begin{aligned} \rho \frac{d^2 L^{\nu}_{n_r}}{d\rho^2} &= \rho \sum_{m=0}^{\infty} m(m-1) c_m \rho^{m-2} = \sum_{m=0}^{\infty} m(m-1) c_m \rho^{m-1} \\ &= \sum_{m'=0}^{\infty} (m'+1) m' c_{m'+1} \rho^{m'}, \end{aligned} \tag{3.82}$$

where again we used $m' = m - 1$. Collecting the terms gives

$$\sum_{m=0}^{\infty} \left[(m+1)(m+\nu+1) c_{m+1} - (m - n_r) c_m \right] \rho^m = 0. \tag{3.83}$$

Since the ρ^m are independent functions, the only way to make this zero is to have each coefficient in front of the ρ^m equal to zero. This gives the recursion relation

$$c_{m+1} = \frac{m - n_r}{(1+m)(\nu+1+m)} c_m. \tag{3.84}$$

In the limit $m \gg n_r, \nu$, this becomes

$$c_{m+1} \cong \frac{c_m}{m} = \frac{1}{m} \frac{1}{m-1} \cdots \frac{1}{2} \frac{1}{1} c_0 = \frac{c_0}{m!}. \tag{3.85}$$

For $c_0 = 1$, the solutions are given by

$$L^{\nu}_{n_r}(\rho) = \sum_{m=0}^{\infty} \frac{\rho^m}{m!} = e^{\rho}, \tag{3.86}$$

leading to a radial eigenfunction $\rho^{\nu/2} e^{\rho/2}$. However, this solution diverges for $\rho \to \infty$ and is therefore not localized inside the potential well. To avoid this, the series has to be cut off at a particular power, leaving us with a polynomial. This can be achieved by taking

$$m_{\max} = n_r. \tag{3.87}$$

Limiting the series leads to the quantization of the energy levels in Eq. (3.79).

Limiting the series to a particular value n_r allows us to derive the Laguerre polynomials. For $n_r = 0$, only the constant remains, and

$$L^{\nu}_{0}(\rho) = 1. \tag{3.88}$$

We follow here the conventional normalization of the Laguerre polynomials, which imposes

$$L^{\nu}_{n_r}(0) = \frac{(n_r + \nu)!}{n_r! \nu!}. \tag{3.89}$$

For $n_r = 1$, the recursion relation gives

$$c_1 = -\frac{1}{\nu+1}c_0. \tag{3.90}$$

The normalization condition implies that $c_0 = L_1^\nu(0) = 1+\nu$, and the Laguerre polynomial is therefore

$$L_1^\nu(\rho) = 1+\nu-\rho. \tag{3.91}$$

For $n_r = 2$, we can write, using the recursion relation,

$$L_2^\nu(\rho) = \left(\frac{-1}{2(\nu+2)}\frac{-2}{\nu+1}\rho^2 + \frac{-2}{\nu+1}\rho + 1\right)\frac{(2+\nu)!}{2!\nu!} \tag{3.92}$$

$$= \frac{1}{2}\rho^2 - (\nu+2)\rho + \frac{1}{2}(\nu+1)(\nu+2). \tag{3.93}$$

In quantum mechanics, the preference is to normalize wavefunctions by integration over the coordinates. In this case, this leads to

$$\int_0^\infty d\rho R_{n_r\nu'}(\rho)R_{n_r\nu}(\rho) = \int_0^\infty d\rho\, e^{-\rho}\rho^\nu L_{n_r}^{\nu'}(\rho)L_{n_r}^\nu(\rho) = \frac{(n_r+\nu)!}{n_r!}\delta_{\nu,\nu'}, \tag{3.94}$$

changing the indices to $n_r\nu$ from jm_j in Eq. (3.75). The total normalized solution for the radial part of the wavefunction can then be written as

$$R_{n_r\nu}(\rho) = \frac{\overline{R}_{n_r\nu}(\rho)}{\sqrt{\rho}} = \sqrt{\frac{n_r!}{(n_r+\nu)!}}e^{-\frac{\rho}{2}}\rho^{\frac{\nu}{2}}L_{n_r}^\nu(\rho). \tag{3.95}$$

Combining this result with the φ-dependent part in Eq. (3.63) and reinserting the original coordinate $\rho = \alpha^2 r^2$ with $\alpha = \sqrt{m\omega_0/\hbar}$ lead to the total solution of the two-dimensional harmonic oscillator in terms of n_r and ν,

$$\psi_{n_r\nu}^\pm(r,\varphi) = \sqrt{\frac{\alpha^2 n_r!}{\pi(n_r+\nu)!}}e^{-\frac{\alpha^2 r^2}{2}}(\alpha r)^\nu L_{n_r}^\nu(\alpha^2 r^2)e^{\pm i\nu\varphi}. \tag{3.96}$$

Note that there is no limitation that $\nu \le n_r$. The π appears due to the integration in polar coordinates.

Conceptual summary. – In this section, the differential equation for the two-dimensional oscillator was solved in two different ways. The easy approach is to solve it in Cartesian coordinates and take the x and y coordinates as separable. This gives solutions in terms of products of Hermite polynomials in both directions. Alternatively, one can view the two-dimensional oscillator as a problem with rotational symmetry around the origin. The problem can then be expressed in terms of complex exponentials representing rotations around the origin and Laguerre polynomials representing oscillations in the radial direction. Mathematically, this is more complicated than solving it in Cartesian coordinates. However, the same differential equation is found for the radial wavefunction of the hydrogen atom.

Additionally, note that in a series solution, quantization follows from finding a bound solution. This requires that the polynomial in the Laguerre polynomial cuts off at a finite power. When constructing the function space using step operators, the starting point is the bound state lowest in energy. The amplitude of oscillation is then increased in discrete steps.

Problems

1 An alternative way to derive Laguerre polynomials is Rodrigues formula

$$L_{n_r}^{\nu}(\rho) = \frac{\rho^{-\nu}}{n_r!}\left(\frac{d}{d\rho} - 1\right)^{n_r} \rho^{n_r+\nu}. \tag{3.97}$$

Derive the Laguerre polynomials for $n_r = 2$.

3.4 Two-Dimensional Harmonic Oscillator via Unit Vectors

Introduction. – We have now obtained two different solutions for the two-dimensional harmonic oscillator. In Cartesian coordinates, the solution is written in terms of products of Hermite polynomials. In polar coordinates, the function space is expressed in Laguerre polynomials and complex exponentials. Obviously, these solutions should be consistent with each other. An interesting thing to note while studying this chapter is that solutions are obtained with half-integer quantum numbers that are expressed in terms of single spinors. This is related to the two-dimensional nature of the problem, and positive and negative values correspond to counterclockwise and clockwise rotations in the plane.

The solution for the two-dimensional harmonic oscillator in Eq. (3.96) can be written in terms of j and m_j as

$$\psi'_{jm_j}(\mathbf{r}) = \frac{\psi_{jm_j}(\mathbf{r})}{\psi_{00}(\mathbf{r})} = (-1)^{j-n_j}\sqrt{\frac{(j-n_j)!}{(j+n_j)!}}\,(\boldsymbol{r}_{\sigma_{m_j}})^{2n_j} L_{j-n_j}^{2n_j}(\mathbf{r}^2), \tag{3.98}$$

with $n_j = |m_j|$ and $\sigma_{m_j} = \text{sign}(m_j) = \pm 1$ corresponding to counterclockwise and clockwise rotations. To limit the number of constants, reduced coordinates are used, defined as

$$\mathbf{r}^2 = \xi^2 + \eta^2, \quad \mathbf{r} = \xi \mathbf{e}_x + \eta \mathbf{e}_y, \quad \text{and} \quad \boldsymbol{r}_{\pm 1} = \mathbf{r}_{\pm}\mathbf{e}_x = \xi \pm \boldsymbol{i}\eta, \tag{3.99}$$

where $\xi = \alpha x$ and $\eta = \alpha y$, with $\alpha = \sqrt{m\omega_0/\hbar}$, and $\boldsymbol{i} = \mathbf{e}_y\mathbf{e}_x$ is the imaginary unit in two dimensions. The vector in the plane is then $\boldsymbol{r}_\sigma = re^{\boldsymbol{i}\sigma_{m_j}\varphi}$ or $\boldsymbol{r}_{\pm 1} = re^{\pm \boldsymbol{i}\varphi}$. The additional phase vector $(-1)^{j-n_j}$ is needed to agree with some phase conventions. Without loss of generality, the vacuum state,

$$\psi_{00}(\mathbf{r}) = \frac{1}{\sqrt{\pi}}e^{-\frac{\mathbf{r}^2}{2}}, \tag{3.100}$$

has been removed from the function. We want to compare the results of Section 3.3 with Eq. (3.54) where the solution in two dimensions is written as a product of the one-dimensional solutions in the x and y directions

$$\varphi'_{n_x;n_y}(\mathbf{r}) = \varphi'_{n_x}(\xi)\varphi'_{n_y}(\eta). \tag{3.101}$$

Following Eq. (3.16), the solutions for the harmonic oscillator in one dimension can be written in reduced coordinates as

$$\varphi'_n(\xi) = \frac{\varphi_n(\xi)}{\varphi_0(\xi)} = \frac{1}{\sqrt{2^n n!}}H_n(\xi), \tag{3.102}$$

again dividing out the vacuum state $\varphi_0(\xi)$ in Eq. (3.13).

The first step is to write the step operators in polar format

$$\boldsymbol{a}^\dagger_{\pm\frac{1}{2}} = \frac{1}{\sqrt{2}}(\boldsymbol{a}^\dagger_x \pm \boldsymbol{i}\boldsymbol{a}^\dagger_y) = \frac{1}{\sqrt{2}}(|1;0\rangle + \boldsymbol{i}|0;1\rangle). \tag{3.103}$$

The last step expresses the step operators in terms of the product functions $|n_x;n_y\rangle = |n_x\rangle|n_y\rangle$ in bra-ket notation, with $\langle\xi|n\rangle = \varphi'_n(\xi)$. Since we are only looking at the polynomials, $\langle\xi|0\rangle = 1$. Following Eq. (1.361), a general unit vector $|jm\rangle$ can be written as

$$|jm_j\rangle = |j+m_j;j-m_j\rangle = \frac{(\boldsymbol{a}^\dagger_\uparrow)^{j+m_j}(\boldsymbol{a}^\dagger_\downarrow)^{j-m_j}}{\sqrt{(j+m_j)!(j-m_j)!}}, \tag{3.104}$$

with $\langle\mathbf{r}|jm_j\rangle = \psi'_{jm_j}(\mathbf{r})$. The notation $|j+m_j;j-m_j\rangle$ is the state $|jm_j\rangle$ but expressed in terms of the number of step operators $|n_\uparrow;n_\downarrow\rangle$, which is a convenient form for performing operations. The reason for the use of half-integer quantum numbers is related to the dimension of the system and the resulting degeneracy when raising a unit vector to a particular integer power. There are three terms corresponding to three directions in the multinomial expansion in Eq. (1.359). For a positive integer power l, the number of terms with a distinct m value in the expansion goes as $2l+1$, with $m = l, l-1, \dots, -l$. For a two-dimensional system there are only two directions. When raising a unit vector with two components to a positive integer power $2j$, one only finds $2j+1$ distinct m_j values, with $m_j = j, j-1, \dots, -j$, where j is half-integer.

The lowest polynomial for $j=0$ is simply a constant, $|00\rangle = |0;0\rangle$ and $\psi'_{00}(\mathbf{r}) = \psi'_{0;0}(\mathbf{r}) = 1$. The next term with $j=\frac{1}{2}$ is somewhat more interesting. Let us first obtain the eigenfunctions using Hermite polynomials. In terms of polar step operators from Eq. (3.103), we obtain

$$|\frac{1}{2},\pm\frac{1}{2}\rangle = \boldsymbol{a}^\dagger_{\pm\frac{1}{2}} = \frac{1}{\sqrt{2}}(|1;0\rangle \pm \boldsymbol{i}|0;1\rangle) \text{ and } \psi'_{\frac{1}{2},\pm\frac{1}{2}}(\mathbf{r}) = \xi \pm \boldsymbol{i}\eta, \tag{3.105}$$

with $\varphi'_1(\xi) = \langle\xi|1\rangle = H_1(\xi)/\sqrt{2} = \sqrt{2}\xi$ using Eq. (3.102) and the expressions for the Hermite polynomials $H_1(\xi) = 2\xi$. Since we are looking at a two-dimensional system, we find solutions in terms of single-step operators, since the axis of rotation is fixed. This differs from the function space of the spherical harmonics that were expressed in products of an even number of spinors since the space is three-dimensional. The eigenfunctions in terms of Laguerre polynomials in Eq. (3.98) should obviously produce the same results,

$$\psi'_{\frac{1}{2},\pm\frac{1}{2}}(\mathbf{r}) = \langle\mathbf{r}|\frac{1}{2},\pm\frac{1}{2}\rangle = (\xi \pm \boldsymbol{i}\eta)L^1_0(\rho) = \xi \pm \boldsymbol{i}\eta, \tag{3.106}$$

with $\rho = \xi^2 + \eta^2$ and $L^1_0(\rho) = 1$.

The polar unit vectors $|jm_j\rangle$ can always be written as a sum of different $|n_x;n_y\rangle$. The unit vectors become increasingly complex but are straightforward to obtain. For $j=1$ and $m_j = \pm 1$, the wavefunction obtained via Hermite polynomials is

$$|1,\pm1\rangle = \frac{1}{\sqrt{2}}(\boldsymbol{a}^\dagger_{\pm\frac{1}{2}})^2 = \frac{1}{2\sqrt{2}}(\boldsymbol{a}^\dagger_x \pm \boldsymbol{i}\boldsymbol{a}^\dagger_y)^2 = \frac{1}{2}(|2;0\rangle - |0;2\rangle) \pm \frac{\boldsymbol{i}}{\sqrt{2}}|1;1\rangle,$$

with the corresponding functions

$$\psi'_{1,\pm1}(\mathbf{r}) = \frac{1}{2}(\varphi'_2(\xi) - \varphi'_2(\eta)) \pm \frac{\boldsymbol{i}}{\sqrt{2}}\varphi'_1(\xi)\varphi'_1(\eta) = \frac{1}{\sqrt{2}}(\xi^2 - \eta^2) \pm \boldsymbol{i}\sqrt{2}\xi\eta,$$

using $\varphi_2'(\xi) = (2\xi^2 - 1)/\sqrt{2}$ and similar for η. In terms of Laguerre polynomials, the same function is given by

$$\psi'_{1,\pm 1}(\mathbf{r}) = \frac{1}{\sqrt{2}}(\xi \pm i\eta)^2 L_0^2(\rho) = \frac{1}{\sqrt{2}}(\xi^2 - \eta^2) \pm i\sqrt{2}\xi\eta,$$

with $L_0^2(\rho) = 1$.

The $j = 1$ and $m_j = 0$ component is orthogonal to this

$$\begin{aligned} |10\rangle = |1;1\rangle = \boldsymbol{a}^\dagger_{\frac{1}{2}}\boldsymbol{a}^\dagger_{-\frac{1}{2}} &= \frac{1}{2}(\boldsymbol{a}^\dagger_x + i\boldsymbol{a}^\dagger_y)(\boldsymbol{a}^\dagger_x - i\boldsymbol{a}^\dagger_y) \\ &= \frac{1}{2}(\boldsymbol{a}^\dagger_x\boldsymbol{a}^\dagger_x + \boldsymbol{a}^\dagger_y\boldsymbol{a}^\dagger_y) = \frac{1}{\sqrt{2}}(|2;0\rangle + |0;2\rangle). \end{aligned} \tag{3.107}$$

In terms of Hermite polynomials, this can be written as

$$\psi'_{10}(\mathbf{r}) = \langle \mathbf{r}|10\rangle = \frac{1}{\sqrt{2}}(\varphi_2'(\xi) + \varphi_2'(\eta)) = \xi^2 + \eta^2 - 1. \tag{3.108}$$

So far, the polynomials in polar coordinates were due to the φ dependence, *i.e.* the Laguerre polynomial was $L_0^{2n_j}(\rho) = 1$. However, for $m_j = 0$, the polynomial is due to the motion in the radial direction,

$$\psi'_{10}(\mathbf{r}) = -L_1^0(\rho) = -(1 - \rho) = \xi^2 + \eta^2 - 1. \tag{3.109}$$

Note that it does not make a difference whether these functions are obtained through the Laguerre polynomials or the Hermite polynomials, as it should.

Finishing up with the $j = \frac{3}{2}$, where the unit vectors are

$$\begin{aligned} |\frac{3}{2}, \pm\frac{3}{2}\rangle &= \frac{1}{2\sqrt{2}}(|3;0\rangle \mp i|0;3\rangle) + \frac{\sqrt{3}}{2\sqrt{2}}(\pm i|2;1\rangle - |1;2\rangle), \\ |\frac{3}{2}, \pm\frac{1}{2}\rangle &= \frac{\sqrt{3}}{2\sqrt{2}}(|3;0\rangle \pm i|0;3\rangle) + \frac{1}{2\sqrt{2}}(\pm i|2;1\rangle + |1;2\rangle), \end{aligned} \tag{3.110}$$

with the corresponding functions

$$\begin{aligned} \langle \mathbf{r}|\frac{3}{2}, \pm\frac{3}{2}\rangle &= \frac{1}{\sqrt{6}}\xi^3 \pm i\sqrt{\frac{3}{2}}\xi^2\eta - \sqrt{\frac{3}{2}}\xi\eta^2 \mp \frac{i}{\sqrt{6}}\eta^3, \\ \langle \mathbf{r}|\frac{3}{2}, \pm\frac{1}{2}\rangle &= \frac{1}{\sqrt{2}}\xi^3 - \sqrt{2}\xi \pm \frac{i}{\sqrt{2}}\xi^2\eta + \frac{1}{\sqrt{2}}\xi\eta^2 \mp i\sqrt{2}\eta \pm \frac{i}{\sqrt{2}}\eta^3. \end{aligned} \tag{3.111}$$

These functions can also be obtained from the Laguerre polynomials.

Conceptual summary. – The two-dimensional oscillator for a particular energy E is described in two quantum numbers. In Cartesian coordinates, the oscillations split into two different directions with associated quantum numbers n_x and n_y. The energy is related to the sum of the two quantum numbers $n = n_x + n_y$, giving $(n + 1)\hbar\omega_0$. Therefore, the energy imposes a condition on the quantum numbers. In polar coordinates, the quantum numbers are j and m_j. The energy is $E = (2j + 1)\hbar\omega_0$. Since polar coordinates correspond better to the symmetry of the system, the energy is now expressed in a single quantum number, which is directly associated with all the eigenstates with the same energy. The multiplicity of this energy level is also given by $2j + 1$.

Although the function space might be more easily obtained in Cartesian coordinates, the polar coordinates offer additional insights. We saw that, in terms of step operators, the problem could be treated in a manner very similar to the function space of spherical harmonics. The quantum numbers for the latter are integers, which is directly related to the three-dimensional nature of the space and the expansion of three directions in a multinomial expansion. However, the quantum numbers for the two-dimensional quantum harmonic oscillator are half-integer and integer (although, one should view the latter as half-integers too). Additionally, the function space is expressed in terms of single spinors and not products of two spinors, as is the case for the spherical harmonics. Half-integer angular momentum is often met with a considerable amount of confusion, since it is often first encountered for spin, a leftover vector term from the relativistic four-momentum. However, we see that half-integer angular momentum follows naturally for rotation in a two-dimensional system for a nonrelativistic Hamiltonian. Note that we can add a three-dimensional aspect to this problem by orienting the plane of rotation in a three-dimensional space. However, this does not affect the two-dimensional nature of the motion. The levels are labeled by $m_j = j, j-1, \ldots, -j$. This degeneracy can be lifted by an additional interaction, say, a magnetic field.

The concepts discussed in this section are also relevant for three-dimensional systems. In Section 4.8, we look at the nuclear shell model, where the nucleus containing the protons and neutrons is considered as an effective three-dimensional harmonic oscillator. This can be considered in Cartesian or spherical coordinates. However, the latter has the advantage that it is significantly easier to include spin–orbit interaction.

Problems

1 Derive the wavefunctions for $j = \frac{3}{2}$ and $m_j = \pm\frac{3}{2}, \pm\frac{1}{2}$ using Laguerre polynomials.

3.5 Radial Equation for Hydrogen Atom

Introduction. – The treatment of the energy-level structure of the hydrogen atom is one of the greatest achievements of quantum mechanics and enabled a deeper understanding of atoms, molecules, and solids. The hydrogen atom is a problem with spherical symmetry, since the potential $U(\mathbf{r}) \sim 1/r$ only depends on the distance to the nucleus r and not on the angular coefficients θ and φ. The rotation around the nucleus can therefore be considered free motion, and the angular part of the function space is given by the spherical harmonics as described in Section 1.15. In this section, we solve the radial part of the equation and demonstrate its similarities with the two-dimensional harmonic oscillator.

To obtain the radial part, let us consider the Schrödinger equation for the hydrogen-like atom

$$-\frac{\hbar^2}{2m}\nabla^2\psi_{nlm}(\mathbf{r}) - \frac{1}{4\pi\varepsilon_0}\frac{Ze^2}{r}\psi_{nlm}(\mathbf{r}) = E\psi_{nlm}(\mathbf{r}). \tag{3.112}$$

The first term is the kinetic energy, and the second term is the potential energy due to the attractive Coulomb potential of the nucleus with the effective charge Ze. For actual hydrogen, $Z = 1$. The functions $\psi_{nlm}(\mathbf{r})$ can be split into a radial and an angular part

$$\psi_{nlm}(\mathbf{r}) = R_{nl}(r) Y_{lm}(\hat{\mathbf{r}}), \tag{3.113}$$

where the conventional quantum numbers have already been attached to the functions.

In order to obtain the differential equation for $R_{nl}(r)$, the Laplacian ∇^2 needs to be split into radial and angular parts. This can be done by adding a unit vector $\hat{\mathbf{r}}$ to the momentum. For the kinetic energy term, this is achieved via $\hat{\mathbf{p}}^2 = \hat{\mathbf{p}}\hat{\mathbf{r}}^2\hat{\mathbf{p}} = \hat{\mathbf{p}}\hat{\mathbf{r}}\hat{\mathbf{r}}\hat{\mathbf{p}}$ using that $\hat{\mathbf{r}}^2 = 1$. Note that the hat on the momentum $\hat{\mathbf{p}}$ and the vector $\hat{\mathbf{r}}$ indicates an operator and a unit vector, respectively. Let us first evaluate $\hat{\mathbf{p}}\hat{\mathbf{r}}$,

$$\begin{aligned}\hat{\mathbf{p}}\hat{\mathbf{r}} &= \frac{1}{2}(\hat{\mathbf{p}}\hat{\mathbf{r}} + \hat{\mathbf{r}}\hat{\mathbf{p}}) + \frac{1}{2}(\hat{\mathbf{p}}\hat{\mathbf{r}} - \hat{\mathbf{r}}\hat{\mathbf{p}}) = \hat{\mathbf{r}} \cdot \hat{\mathbf{p}} + \frac{1}{2}[\hat{\mathbf{p}}\hat{\mathbf{r}}] - \hat{\mathbf{r}} \wedge \hat{\mathbf{p}} \\ &= -i\hbar\left(\hat{\mathbf{r}} \cdot \nabla + \frac{1}{2}\nabla \cdot \hat{\mathbf{r}}\right) - i\hat{\mathbf{r}} \times \hat{\mathbf{p}},\end{aligned} \tag{3.114}$$

using Eqs. (1.115) and (1.116). The product does not simply split into an inner and wedge product, since $\hat{\mathbf{p}}$ and $\hat{\mathbf{r}}$ do not commute. The square brackets indicate that $\hat{\mathbf{p}}$ is only working on $\hat{\mathbf{r}}$. The second term is the divergence of the vector $\hat{\mathbf{r}}$. The radial component of a unit vector is simply 1, and the divergence in spherical coordinates becomes

$$\nabla \cdot \hat{\mathbf{r}} = \frac{1}{r^2}\frac{\partial(r^2 \times 1)}{\partial r} = \frac{2}{r}. \tag{3.115}$$

The term $\hat{\mathbf{r}} \cdot \nabla = \partial/\partial r$ is the radial part of the gradient. The first two terms then form the radial part of the momentum

$$\hat{p}_r = -i\hbar\left(\frac{\partial}{\partial r} + \frac{1}{r}\right) = -i\hbar\frac{1}{r}\frac{\partial}{\partial r}r \quad \Rightarrow \quad \hat{p}_r^2 = -\frac{\hbar^2}{r}\frac{\partial}{\partial r}r\frac{1}{r}\frac{\partial}{\partial r}r = -\frac{\hbar^2}{r}\frac{\partial^2}{\partial r^2}r.$$

Using $\hat{\mathbf{L}}/r = \mathbf{r} \times \hat{\mathbf{p}}/r = \hat{\mathbf{r}} \times \hat{\mathbf{p}}$, the kinetic energy becomes

$$\frac{\hat{\mathbf{p}}^2}{2m} = \frac{\hat{\mathbf{p}}\hat{\mathbf{r}}\hat{\mathbf{r}}\hat{\mathbf{p}}}{2m} = \frac{1}{2m}\left|\hat{p}_r - i\frac{\hat{\mathbf{L}}}{r}\right|^2 = \frac{\hat{p}_r^2}{2m} + \frac{\hat{\mathbf{L}}^2}{2mr^2}. \tag{3.116}$$

This separates the kinetic energy into a radial and an angular part. Equation (1.430) shows that $\hat{\mathbf{L}}^2 Y_{lm}(\hat{\mathbf{r}}) = l(l+1)\hbar^2 Y_{lm}(\hat{\mathbf{r}})$. We have already generated the function space of the spherical harmonics. This gives a radial equation

$$-\frac{\hbar^2}{2m}\frac{d^2(rR_{nl})}{dr^2} + \left(\frac{\hbar^2 l(l+1)}{2mr^2} - \frac{1}{4\pi\varepsilon_0}\frac{Ze^2}{r} + |E|\right) rR_{nl} = 0, \tag{3.117}$$

where the energy for the bound states, $E = -|E|$, is negative. The radial equation can be brought into the same form as the two-dimensional harmonic oscillator equation in polar coordinates in Eq. (3.70).

Introducing the radial function $\overline{R}_{nl} = rR_{nl}$ and dividing Eq. (3.117) by $-4|E|$ gives

$$\frac{d^2\overline{R}_{nl}}{d\rho^2} + \left(-\frac{l(l+1)}{\rho^2} + \frac{n}{\rho} - \frac{1}{4}\right)\overline{R}_{nl} = 0, \tag{3.118}$$

where the following dimensionless coordinate has been introduced:

$$\rho = \alpha r \quad \text{with} \quad \alpha = \sqrt{\frac{8m|E|}{\hbar^2}}. \tag{3.119}$$

This takes over the role of the dimensionless coordinate $\rho = \alpha^2 r^2$, with $\alpha = \sqrt{m\omega_0/\hbar}$, for the two-dimensional harmonic oscillator. Note that the effective length ρ now depends on the energy. Comparison with Eq. (3.70) shows that the differential equation is comparable to that of the two-dimensional harmonic oscillator. The $1/\rho^2$ term still arises from the centrifugal barrier. However, the roles of the potential and the energy term have reversed. The $1/\rho$ term now arises from the potential energy, and the constant $-\frac{1}{4}$ comes from the energy.

The factor n is given by

$$n = \frac{Ze^2}{4\pi\varepsilon_0}\frac{\alpha}{4|E|} \quad \Rightarrow \quad n^2 = \frac{Z^2e^4}{(4\pi\varepsilon_0)^2}\frac{1}{16|E|^2}\frac{8m|E|}{\hbar^2} = \frac{Z^2me^4}{2(4\pi\varepsilon_0)^2\hbar^2|E|}, \tag{3.120}$$

using Eq. (3.119). Since n is dimensionless, the following energy can be defined:

$$R = \frac{me^4}{2(4\pi\varepsilon_0)^2\hbar^2} = 13.6 \text{ eV}, \tag{3.121}$$

which is known as the Rydberg constant. Since Eq. (3.70), and hence Eq. (3.118), leads to the Laguerre equation, the values of n are integers, and the eigenenergies are given by

$$E_n = -\frac{Z^2R}{n^2} = -\frac{1}{2}\frac{Z^2e^2}{4\pi\varepsilon_0 n^2 a_0}. \tag{3.122}$$

This defines the energies of the hydrogen-like atom. On the right, the energy is expressed as $-\frac{1}{2}U(n^2a_0)$ (the expected result from the virial theorem), where $a_0 = 4\pi\varepsilon_0\hbar^2/e^2m$ is the Bohr radius.

The quantum numbers enter in a different fashion into Eqs. (3.70) and (3.118). First, the quantum number associated with rotational motion n_j is

$$n_j^2 - \frac{1}{4} = l(l+1) \quad \Rightarrow \quad n_j^2 = l^2 + l + \frac{1}{4} = \left(l + \frac{1}{2}\right)^2 \quad \Rightarrow \quad n_j = l + \frac{1}{2}. \tag{3.123}$$

Second, we have the quantum number related to the total motion n,

$$\frac{2j+1}{2} = n \quad \Rightarrow \quad j = n - \frac{1}{2}. \tag{3.124}$$

The indices for the Laguerre polynomials are then, using Eq. (3.77),

$$n_r = j - n_j = n - l - 1 \quad \text{and} \quad \nu = 2n_j = 2l + 1, \tag{3.125}$$

where n_r is the quantum number associated with radial motion.

It is interesting to see what Laguerre polynomials $L_{n_r}^{\nu}(\rho)$ occur for the two-dimensional quantum harmonic oscillator and the radial part of the wavefunction for the hydrogen atom,

jn_j	nl	$n_r\nu$	n_rn_j
00	–	00	00
$\frac{1}{2}\frac{1}{2}$	10 (1s)	01	$0\frac{1}{2}$
11, 10	–	02, 10	01, 10
$\frac{3}{2}\frac{3}{2}, \frac{3}{2}\frac{1}{2}$	21 (2p), 20 (2s)	03, 11	$0\frac{3}{2}, 1\frac{1}{2}$
22, 21, 20	–	04, 12, 20	02, 11, 20
$\frac{5}{2}\frac{5}{2}, \frac{5}{2}\frac{3}{2}, \frac{5}{2}\frac{1}{2}$	32 (3d), 31 (3p), 30 (3s)	05, 13, 21	$0\frac{5}{2}, 1\frac{3}{2}, 2\frac{1}{2}$

(3.126)

For the two-dimensional quantum harmonic oscillator, the quantum numbers n_x and n_y directly indicate how the energy is divided between the x and y directions. The corresponding quantities for radial and angular motion are n_r and n_j, which are half-integers. The sum of the two is given by $j = n_r + n_j$. The energy is then given by $E = (2j+1)\hbar\omega_0 = 2(n_r + n_j + \frac{1}{2})\hbar\omega_0$.

For the hydrogen atom, the motion can also be split into radial and angular contributions, with $n_r = n - l - 1$ and $n_j = l + \frac{1}{2}$, respectively, see Eq. (3.125). The quantum number for the total motion is given by $n_r + n_j + \frac{1}{2} = n$, where the $\frac{1}{2}$ is related to the zero-point motion in the radial direction. The energy is related to n via $E_n = -Z^2R/n^2$, see Eq. (3.122). Note that for the two-dimensional harmonic oscillator, the angular motion is twofold degenerate with $m_j = \pm n_j$ and described by the complex exponential $e^{2im_j\varphi}$. This gives the counterclockwise and clockwise rotations. For the hydrogen-like atom, the angular motion $n_j = l + \frac{1}{2}$ is related to the spherical harmonics, $Y_{lm}(\hat{\boldsymbol{r}})$, where l determines the magnitude of the angular motion, and m corresponds to the direction of the rotation. However, there is also a zero-point angular motion due to the spin, so that for s orbitals ($l = 0$), we still have $n_j = \frac{1}{2}$.

Also note that the Laguerre polynomials for the hydrogen-like atoms equal those of the two-dimensional harmonic oscillator for half-integer j values, see Eq. (3.126). For example, the Laguerre polynomials of the 1s orbital (with no orbital moment) equal those of the $j = \frac{1}{2}$ states for the two-dimensional harmonic oscillator. It seems natural to associate the zero-point motion with the spin given by $e^{\pm \boldsymbol{i}_B\varphi}$ in Eq. (2.217). Therefore, each nl state is $2(2l+1)$-fold degenerate.

The partial function space of the hydrogen-like atom for a particular nl value is given by

$$\hat{\boldsymbol{r}}^l R_{nl}(\rho) = \frac{\hat{\boldsymbol{r}}^l \overline{R}_{nl}(\rho)}{\rho} = \sqrt{\frac{(n-l-1)!}{(n+l)!}} e^{-\frac{\rho}{2}} \rho^l \hat{\boldsymbol{r}}^l L_{n-l-1}^{2l+1}(\rho). \tag{3.127}$$

The unit vector to the power l produces all the spherical harmonics $Y_{lm}(\hat{\boldsymbol{r}})$ with the appropriate unit vectors, see Section 1.15. Note the similarity with solution for a two-dimensional oscillator, see Eq. (3.98). In the latter, the vector $(\boldsymbol{r}_{\sigma_{m_j}})^{2n_j} = e^{2im_j\varphi}$, where n_j and m_j are half-integers, contains a directional component $\sigma_{m_j} = \pm 1$ that corresponds to counterclockwise and clockwise rotations, respectively. Since the two directions of rotation are equivalent from the point of view of rotational symmetry, the value of σ_{m_j} does not affect the energy. Obviously, for three dimensions, the directional part is more complicated. All the possible directions of rotation for a particular l are contained in the term $\hat{\boldsymbol{r}}^l$, see Section 1.15. They are indexed by the quantum number $m = l, l-1, l-2, \ldots, -l$, which again does not affect the total energy due to the spherical symmetry of the atom. When the symmetry is lowered by, say, a magnetic field, the energy becomes dependent on m. Normalizing the wavefunction over r instead of ρ gives the conventional expression of the atomic orbitals

$$\varphi_{nlm}(\mathbf{r}) = \sqrt{\left(\frac{2}{na_0}\right)^3 \frac{(n-l-1)!}{2n(n+l)!}} e^{-\frac{\rho}{2}} \rho^l L_{n-l-1}^{2l+1}(\rho) Y_{lm}(\hat{\mathbf{r}}), \tag{3.128}$$

with $\rho = 2r/na_0$, where $a_0 = 4\pi\varepsilon_0\hbar^2/me^2$ is the Bohr radius.

Example 3.1 ***The Bohr Model***
Explain why the Bohr model for the hydrogen atom works.

Solution
Using Eq. (3.120), we can write

$$n = \frac{Ze^2}{4\pi\varepsilon_0}\frac{1}{\hbar}\sqrt{\frac{m}{2|E|}} \quad \Rightarrow \quad \sqrt{\frac{2|E|}{m}}n\hbar = \frac{Ze^2}{4\pi\varepsilon_0}, \tag{3.129}$$

$\alpha = \sqrt{8m|E|}/\hbar$. Bohr's starting point is a classical particle moving in a circular orbit. Newton's second law, $F = ma$, where a is the centripetal acceleration, gives

$$\frac{Ze^2}{4\pi\varepsilon_0}\frac{1}{r^2} = m\frac{v^2}{r} \quad \Rightarrow \quad mv^2r = \frac{Ze^2}{4\pi\varepsilon_0}. \tag{3.130}$$

The term on the left-hand side can be written as $mv^2r = (mvr)v$. The virial theorem tells us that $E_{\text{kin}} = -\frac{1}{2}E_{\text{pot}}$. The total energy is then $E = E_{\text{kin}} + E_{\text{pot}} = \frac{1}{2}E_{\text{pot}} = -E_{\text{kin}}$. Since $E < 0$, $\frac{1}{2}mv^2 = |E|$, giving $v = \sqrt{2|E|/m}$. The other term can be identified as the angular momentum for a circular orbit, which Bohr suggested to quantize as

$$L = mvr = n\hbar \quad \text{with} \quad n = 1, 2, 3, \ldots \tag{3.131}$$

This term requires some additional consideration. First, the quantum number n does not only include angular momentum, but also radial motion. However, the Bohr model focuses on the energy, and since all orbitals nlm with the same n have the same energy, we only have to look at one. Therefore, we can consider the orbitals that have no radial oscillations, *i.e.* $n_r = 0$, see Eq. (3.126). Secondly, the quantum number associated with angular momentum is $l = 0, 1, 2, 3, \ldots$ This would cause the energy of the s orbitals to diverge. However, the Bohr model is saved by the presence of zero-point motion, giving $n = (l + \frac{1}{2}) + \frac{1}{2} = l + 1$. The zero-point motions in the radial and angular (the spin) directions each contribute a $\frac{1}{2}$. This gives a term $mv^2r = \sqrt{2|E|/m}\, n\hbar$, which is equal to the left-hand side of Eq. (3.129).

Conceptual summary. – The radial wave equation for the hydrogen atom can be viewed as an effective two-dimensional oscillator. The two oscillators correspond to the radial and angular directions. For the 1s orbital, there is only zero point motion in the radial and angular directions. Even though the Hamiltonian for the hydrogen atom is treated nonrelativistically, a zero-point angular motion with $n_j = \frac{1}{2}$, *i.e.* the spin, is still needed. For $n = 2$, there is motion in addition to the zero-point motion that can be put into radial direction (the 2s orbital) or into rotation (the 2p orbital). For $n = 3$, there are two additional units of oscillatory energy beyond the zero-point motion. These can all be put into radial motion (3s orbital), equally divided among radial and rotational motion (3p) or be entirely used for rotational motion (3d).

For the two-dimensional quantum harmonic oscillator, the angular motion has an additional directional component, which is twofold degenerate since rotation can occur counterclockwise and clockwise. For an atom, the oscillation in the radial direction is one-dimensional comparable (though not equivalent due to the different potential) to a quantum harmonic oscillator. The rotational motion, however, also has a more complex directional component with a degeneracy of $2l + 1$ (or $2(2l + 1)$ when also including the directional part of the zero-point motion, the spin). This directional component corresponds to the function space of spherical harmonics.

Problems

1 After finding the radial part of the atomic wavefunctions, we can look at the effect of several relativistic corrections (take $Z = 1$).

a. Derive the higher order correction to the kinetic energy in Eq. (2.199). Rewrite the term into the total and potential energy and use the expectation values $\langle 1/r \rangle = 1/n^2 a_0$ and $\langle 1/r^2 \rangle = 1/(l + \frac{1}{2})n^3 a_0^2$.

b. Calculate the correction to the energy for the Darwin term in Eq. (2.234).

c. Obtain the correction for the spin–orbit interaction from Eqs. (2.240) and (2.241), with $g = 2$ and $\langle 1/r^3 \rangle = 1/(na_0)^3 l(l + \frac{1}{2})(l + 1)$.

d. Combine the relativistic correction with the binding energy and calculate the splitting for $n = 2$.

3.6 Transitions on Atoms

Introduction. – In Section 1.15, the function space for rotations in three dimensions was constructed. This created higher order vectors $|lm\rangle$, whose coefficients were given by the spherical harmonics, $Y_{lm}(\hat{\mathbf{r}}) = \langle \hat{\mathbf{r}} | lm \rangle$. The vectors could be expressed in terms of commuting spinors $\boldsymbol{a}^\dagger_{\pm\frac{1}{2}}$. For the quantum harmonic oscillator, we considered operators that affected the amplitude of the oscillation, for example $x \sim (\boldsymbol{a} + \boldsymbol{a}^\dagger)$, *i.e.* the number of step operators changes. For spherical harmonics, we have only considered operators that conserved the number of spinors, see Section 1.16. The rotations were expressed in terms of the angular momentum operators $\mathbf{L}_q$ that kept the number of the spinors $\boldsymbol{a}^\dagger_{\pm\frac{1}{2}}$ constant. Or, in terms of orbitals, rotation of an atom does not change the n and l quantum numbers, only the projected angular momentum m. In other words, when thinking of spherical harmonics as the coefficients of a vector of a certain rank l, a rotation does not change the rank of the vector.

In this section, we consider operations that do change the quantum numbers n and l. These transitions often occur in the interaction with an external field. For example, electromagnetic radiation can cause a transition between different orbitals. It can also occur in molecules and solids. When the local surroundings of an atom break inversion symmetry, one can, for example see mixing of s and p or p and d orbitals. In this section, we focus on transitions that do not conserve the number of spinors.

Let us first consider operators for atomic transitions from $l \to l + 1$. In Eq. (1.342), the unit vectors are expressed in terms of two equal spinors. We define the following operators:

$$\boldsymbol{T}^1_1 = \boldsymbol{e}_1 = \frac{1}{\sqrt{2}} \boldsymbol{a}^\dagger_\uparrow \boldsymbol{a}^\dagger_\uparrow, \quad \boldsymbol{T}^1_0 = \boldsymbol{e}_0 = \boldsymbol{a}^\dagger_\uparrow \boldsymbol{a}^\dagger_\downarrow, \quad \boldsymbol{T}^1_{-1} = \boldsymbol{e}_{-1} = \frac{1}{\sqrt{2}} \boldsymbol{a}^\dagger_\downarrow \boldsymbol{a}^\dagger_\downarrow. \tag{3.132}$$

The rank k and component q are indicated in the superscript and subscript in $\boldsymbol{T}^k_q$, respectively. For example, operating with $\boldsymbol{T}^1_q$ on the vacuum (rank 0) creates a vector $\boldsymbol{e}_q$ (rank 1). Therefore, the rank changes by 1. Let us now consider some examples of how these transition work out in practice. Important for spectroscopy are dipole transitions that occur due to the absorption or emission of electromagnetic radiation. Let us restrict the discussion to absorption. The interaction of an atom with a plane wave is given by Eq. (2.205),

$$\boldsymbol{D} = \frac{e}{m}\mathbf{A}\cdot\hat{\mathbf{p}} \sim \frac{e}{m}e^{i\mathbf{q}\cdot\mathbf{r}}\boldsymbol{\epsilon}\cdot\hat{\mathbf{p}} \cong \frac{e}{m}\boldsymbol{\epsilon}\cdot\hat{\mathbf{p}}, \tag{3.133}$$

where $\boldsymbol{\epsilon}$ is the polarization vector of the light. Within the dipole approximation, the plane wave is approximated to $e^{i\mathbf{q}\cdot\mathbf{r}} \cong 1$, *i.e.* the variation of the electromagnetic wave across the atom is neglected. The spectral intensities for exciting an electron between two orbitals are calculated using Fermi's golden rule in Eq. (2.283),

$$I(\omega) = \sum_{n'l'm'} |\langle n'l'm'|\boldsymbol{D}|nlm\rangle|^2 \delta(E_{n'l'm'} - E_{nlm} - \hbar\omega), \tag{3.134}$$

for the transition $nlm \rightarrow n'l'm'$. The bras $\langle n'l'm'|$ are the absorption final states. For the initial state in Eq. (2.283), the replacement $E \rightarrow E_{nlm} + \hbar\omega$ has been made, where $\hbar\omega$ is the energy of the incoming photon, and E_{nlm} is the energy of the initial orbital. For generality, we assume that the energy can depend on m due to, say, the presence of a magnetic field. $I(\omega)$ therefore has peaks at absorption energies $\hbar\omega = E_{n'l'm'} - E_{nlm}$. Generally, the spectral lines are broadened by lifetime broadening and experimental resolution. The weights of the peaks are given by the squared matrix elements $|\langle n'l'm'|\boldsymbol{D}|nlm\rangle|^2$.

The calculation of the matrix elements is generally not a trivial task due to the complexity of the wavefunctions. Additionally, we have to calculate the matrix element for the operator $\hat{\mathbf{p}}$. However, there is an important result that simplifies the calculations. The matrix elements for any spherical tensor $\boldsymbol{H}_q^k$ can always be written as

$$\langle n'l'm'|\boldsymbol{H}_q^k|nlm\rangle = (n'l'||\boldsymbol{H}^k||nl)(-1)^{l'-m'}\begin{pmatrix} l' & k & l \\ -m' & q & m \end{pmatrix}. \tag{3.135}$$

This is known as the Wigner–Eckart theorem. Let us try to understand what this means. No matter what matrix element we are looking at, the nl and $n'l'$ quantum numbers are always the same. Therefore, although the evaluation might be very complicated, it only results in an overall scaling of the matrix element. This is contained in the reduced matrix element $(n'l'||\boldsymbol{H}^k||nl)$. The only real variables are therefore m and m'. The total number of matrix elements is therefore $(2l'+1)(2l+1)$ that can be expressed in terms of a $(2l'+1)\times(2l+1)$ matrix. The matrix elements are given by the $3j$ symbols, which are the terms on the far right in Eq. (3.135). The $3j$ symbol is an analytical function with six variables usually written in this form. Apart from a factor, it is equivalent to the Clebsch–Gordan coefficients. The spherical symmetry of the problem imposes the condition

$$m' = m + q. \tag{3.136}$$

In spectroscopy, this is generally called a selection rule.

With modern analytical mathematical codes, the evaluation of the $3j$ symbols is straightforward. However, to gain a better understanding, we want to express the $3j$ symbols as matrix elements of the operators $\boldsymbol{T}_q^k$ written in terms of spinors. The relation between the $3j$ symbol and the matrix element is

$$(-1)^{l'-m'}\begin{pmatrix} l' & k & l \\ -m' & q & m \end{pmatrix} \sim \langle l'm'|\boldsymbol{T}_q^k|lm\rangle. \tag{3.137}$$

Note that the changes for the quantum numbers $l'm'$ are related to conjugation, see Eq. (1.378), since we need the bra $\langle l'm'|$ for the matrix element. The two sides are not exactly equal since the $3j$ symbols have been designed to satisfy certain normalization properties. However, this is not important for our understanding. The above equation only works if the operators $\boldsymbol{T}^k_q$ are proper spherical tensors. The majority of operators that we dealt with initially were vectors, *i.e.* tensors of rank 1 with $2 \times 1 + 1 = 3$ components. The most important example is the angular momentum operator $\mathbf{L}_q$. Note that the angular momentum in Cartesian coordinates $\mathbf{L}_i$ with $i = x, y, z$ is not a spherical tensor, although it can be expressed in spherical tensors, as is done in Eqs. (1.384) and (1.385). Additionally, the spherical harmonics $C^k_q(\hat{\mathbf{r}})$ are spherical tensors of rank k and component q. The integral over three spherical harmonics is proportional to a $3j$ symbol. Therefore, the $3j$ symbol describes how different higher order spherical vectors are coupled to each other. However, these transformation properties are also contained in the operators $\boldsymbol{T}^k_q$.

The important result is that the action of any operator $\boldsymbol{H}^k_q$ between two orbitals can always be expressed in terms of $\boldsymbol{T}^k_q$ written in spinors, as long as $\boldsymbol{H}^k_q$ can be considered a spherical tensor of rank k and component q. Therefore, a vector always transforms as a vector, regardless of how complex it is. The operator is then

$$\boldsymbol{H}^k_q = \sum_{mm'} |n'l'm'\rangle\langle n'l'm'|\boldsymbol{H}^k_q|nlm\rangle\langle nlm| = \alpha^k_{n'l'nl}\boldsymbol{T}^k_q. \tag{3.138}$$

To avoid cluttering the expressions with indices, it is implicitly assumed that $\boldsymbol{H}^k_q$ and $\boldsymbol{T}^k_q$ are between the orbitals with quantum numbers nl and $n'l'$. The factor $\alpha^k_{n'l'nl}$ is an absolute scaling that is often not calculated in, for example spectroscopy. It is proportional to the reduced matrix element $(n'l'||\boldsymbol{H}^k||nl)$, although not entirely equivalent since differences in the normalization of the $3j$ symbol and $\boldsymbol{T}^k_q$ have been absorbed into $\alpha^k_{n'l'nl}$. All the symmetry properties related to the transformation of vectors are determined by $\boldsymbol{T}^k_q$.

For $n'l' = nl$ and a tensor of rank 1, the operator is equivalent to the angular momentum, *i.e.* $\boldsymbol{T}^1_q \equiv \mathbf{L}_q$, see Eqs. (1.384) and (1.385). In that case, the number of spinors is conserved. Therefore, for a particular orbital, any spherical tensor of rank 1 (a vector) looks, apart from a scaling, the same as the angular momentum operator. However, for the spherical tensor $r_q = rC^1_q(\boldsymbol{r})$, *i.e.* the position vector in spherical coordinates, this is in principle correct, but the reduced matrix element is 0, *i.e.* $(nl||\boldsymbol{r}||nl) = 0$. This is because the operator is an odd function, whereas the square of an atomic orbital is an even function. The matrix element of an odd function (odd times even) is 0. Therefore, r_q cannot cause a transition between orbitals with the same nl. The same argument holds for the momentum operator for dipolar transitions in Eq. (3.133). In fact, the Wigner–Eckart theorem basically states that the position and momentum operator between two orbitals given by nl and $n'l'$ are the same apart from a scaling factor. This might seem strange since we tend to view $\mathbf{r}$ and $\hat{\mathbf{p}}$ as entirely different quantities. However, the quantum numbers m and m' are directional quantities that only depend on the spherical symmetry of the problem. Therefore, they only care about the symmetry aspect of the operator, and both $\mathbf{r}$ and $\hat{\mathbf{p}}$ are vectors (They are tensors of rank 1. However, make sure to write them in terms of a spherical basis in order for the Wigner–Eckart theorem to work.) All the "physical" aspects of the operators are in the reduced matrix elements.

Although there are no transitions between two orbitals with the same l for the operators $\mathbf{r}$ and $\hat{\mathbf{p}}$, there are nonzero matrix elements in the case $l' = l \pm 1$. Since vectors are tensors of rank 1, l' cannot differ from l by more than 1. The transformation properties for $l' = l + 1$ are given by the operators $\boldsymbol{T}_q^1$ in Eq. (3.132). Let us consider some examples.

Example 3.2 *Dipolar Transitions from $l \to l + 1$*

a) Give the general expressions of the dipole operator and show their effect on an atomic orbital $|lm\rangle$.
b) Derive the dipole matrices between a p and a d orbital.

Solution

a) The dipole operator can be expressed as $\boldsymbol{D} = \frac{e}{m}\boldsymbol{\epsilon} \cdot \mathbf{p}$. The polarization operator is such that $\mathbf{p}$ becomes a proper spherical tensor, *i.e.* the light is left-circularly, z, and right-circularly polarized for $q = 1, 0, -1$. Spectra for different polarizations, say, x and y polarized light, can be obtained by making the appropriate combinations of the spectra calculated with the spherical tensors. The dipole operator is proportional to the momentum operator $\mathbf{p}$, which is, in principle, a complicated operator. However, unless one is really interested in the absolute intensity, the only relevant aspect is that it is a tensor of rank 1. The dipole operator between nl and $n'l'$ can then be written as

$$\boldsymbol{D}_q = \alpha^1_{n'l',nl} \boldsymbol{T}^1_q, \tag{3.139}$$

where $l' = l + 1$ and $\boldsymbol{T}_q^1$ are given in Eq. (3.132). The unit vectors $\mathbf{e}^l_m = |lm\rangle$ for the orbitals with angular momentum l and projection m are given in Eq. (1.361). It is convenient to express unit vectors in terms of the number n_σ of up and down spinors $|lm\rangle = |n_\uparrow; n_\downarrow\rangle = |l + m; l - m\rangle$, with $l = (n_\uparrow + n_\downarrow)/2$ and $m = (n_\uparrow - n_\downarrow)/2$. For polarization $q = 0$, this gives

$$\begin{aligned} \boldsymbol{T}^1_0|lm\rangle &= \boldsymbol{a}^\dagger_\uparrow \boldsymbol{a}^\dagger_\downarrow |l + m; l - m\rangle \\ &= \sqrt{(l + m + 1)(l - m + 1)}|l + m + 1; l - m + 1\rangle \\ &= \sqrt{(l + m + 1)(l - m + 1)}|l + 1, m\rangle. \end{aligned} \tag{3.140}$$

The number of spinors in the unit vector has now increased by two, giving $l + 1$, although the projected angular momentum is still m. Apart from a constant factor the factors can also be obtained using the $3j$ symbols. The $3j$ symbol in Eq. (3.135) also imposes conservation of the projected angular momentum, *i.e.* $-m' + q + m = 0$ or $m' = m$ for $q = 0$ as found above. For $q = 1$, the matrix elements are determined by

$$\begin{aligned} \boldsymbol{T}^1_1|lm\rangle &= \frac{1}{\sqrt{2}}\boldsymbol{a}^\dagger_\uparrow \boldsymbol{a}^\dagger_\uparrow |l + m; l - 1\rangle = \frac{1}{\sqrt{2}}\boldsymbol{a}^\dagger_\uparrow \sqrt{l + m + 1}|l + m + 1; l - m\rangle \\ &= \frac{1}{\sqrt{2}}\sqrt{(l + m + 2)(l + m + 1)}|l + m + 2; l - m\rangle \qquad (3.141) \\ &= \frac{1}{\sqrt{2}}\sqrt{(l + m + 2)(l + m + 1)}|l + 1, m + 1\rangle. \qquad (3.142) \end{aligned}$$

Therefore, the projected angular momentum $m' = m + 1$ increases by 1. A similar evaluation can be made for $q = -1$.

b) Expressed in matrix form, the dipole matrix elements between $l = 1$ and $l' = 2$ are proportional to

$$\boldsymbol{T}_1^1 = \begin{pmatrix} \sqrt{6} & 0 & 0 \\ 0 & \sqrt{3} & 0 \\ 0 & 0 & 1 \\ 0 & 0 & 0 \\ 0 & 0 & 0 \end{pmatrix}, \boldsymbol{T}_0^1 = \begin{pmatrix} 0 & 0 & 0 \\ \sqrt{3} & 0 & 0 \\ 0 & 2 & 0 \\ 0 & 0 & \sqrt{3} \\ 0 & 0 & 0 \end{pmatrix}, \boldsymbol{T}_{-1}^1 = \begin{pmatrix} 0 & 0 & 0 \\ 0 & 0 & 0 \\ 1 & 0 & 0 \\ 0 & \sqrt{3} & 0 \\ 0 & 0 & \sqrt{6} \end{pmatrix}.$$

The dimension of the matrices is $(2l' + 1) \times (2l + 1) = 5 \times 3$. The values of m are given in descending order, *i.e.* $m = l, l - 1, \ldots, -l$. These matrices can also be evaluated using $3j$ symbols. In that case, they are a factor $\sqrt{30}$ smaller due to the different normalization.

Example 3.3 ***Stark Effect***
The Stark effect is the perturbation due to a constant electric field.

a) Write the interaction in terms of the tensor operators $\boldsymbol{T}_q^k$ for an electric field in the z-direction.
b) Calculate the effect between s and p orbitals with the same n.
c) Do the same for an electric field in the x-direction.

Solution

a) The Stark effect is the splitting of levels due to an external electric field. In this case, the field is constant in the z-direction. The perturbance is then given by

$$H' = eEz = eEr_0. \tag{3.143}$$

Since H' is proportional to the position vector, it is zero within one set of orbitals with the same l. Since r_0 is a tensor of rank 1, it will couple l and $l \pm 1$. Let us look at the situation where $l = 0$ and $l' = 1$ and take the same n, so that the binding energies are similar. This is the coupling between an ns and an np orbital. Since the z-direction corresponds to the $q = 0$ component, the effective operator can be written as

$$\boldsymbol{H}' = \alpha_{n1,n0}^1 \boldsymbol{T}_0^1. \tag{3.144}$$

b) Since the projected angular momentum is conserved, the s orbital only couples to the p_z orbital with $\boldsymbol{T}_0^1|00\rangle = \boldsymbol{a}_\uparrow^\dagger \boldsymbol{a}_\downarrow^\dagger|0;0\rangle = \sqrt{1 \times 1}|1;1\rangle = |10\rangle$. We additionally introduce a binding energy difference $\boldsymbol{H}_0 = \frac{\Delta}{2}(\boldsymbol{a}_\uparrow^\dagger \boldsymbol{a}_\uparrow + \boldsymbol{a}_\downarrow^\dagger \boldsymbol{a}_\downarrow)$, which is $0, \Delta$ in the case of $0, 2$ spinors for s and p orbitals, respectively. This gives the following 2×2 matrix:

$$\boldsymbol{H}_0 + \boldsymbol{H}' = \begin{pmatrix} 0 & \alpha_{n1,n0}^1 \\ \alpha_{n1,n0}^1 & \Delta \end{pmatrix} \begin{matrix} |n00\rangle \\ |n10\rangle \end{matrix}. \tag{3.145}$$

In the case that $\Delta = 0$, the energies are $\pm\alpha_{n1,n0}^1$. The eigenfunctions are $(|n00\rangle \pm |n10\rangle)/\sqrt{2}$. Therefore, hybrids of s and p orbitals are formed: One with a large lobe in the positive z-direction and a small lobe of opposite sign in the negative z-direction, and the other hybrid is the same but with the large lobe pointing in the negative z-direction. The application of an electric field therefore lowers the symmetry from spherical to cylindrical and, additionally, breaks inversion symmetry.

c) Let us now take the electric field in the x-direction, giving $H' = eEx$. Although x is not a proper spherical tensor it can be expressed in them. Following Eq. (1.348), we obtain

$$\boldsymbol{H}' = \frac{\alpha^1_{n1,n0}}{\sqrt{2}}(-\boldsymbol{T}^1_1 + \boldsymbol{T}^1_{-1}). \tag{3.146}$$

Using that $\boldsymbol{T}^1_1|00\rangle = \frac{1}{\sqrt{2}}\boldsymbol{a}^\dagger_\uparrow \boldsymbol{a}^\dagger_\uparrow|0;0\rangle = \frac{\sqrt{2\times 1}}{\sqrt{2}}|2;0\rangle = |11\rangle$, and similarly $\boldsymbol{T}^1_{-1}|0;0\rangle = |1,-1\rangle$, the matrix can be written as

$$\boldsymbol{H}_0 + \boldsymbol{H}' = \begin{pmatrix} 0 & -\alpha & \alpha \\ -\alpha & \Delta & 0 \\ \alpha & 0 & \Delta \end{pmatrix} \begin{matrix} |n00\rangle \\ |n11\rangle \\ |n1,-1\rangle \end{matrix}, \tag{3.147}$$

with $\alpha = \alpha^1_{n1,n0}/\sqrt{2}$. The eigenvalues are $\pm\sqrt{2}\alpha = \pm\alpha^1_{n1,n0}$. The eigenfunctions are $(|n00\rangle \pm |n1x\rangle)/\sqrt{2}$, where $|n1x\rangle = (-|n11\rangle + |n1,-1\rangle)/\sqrt{2}$ is the p_x orbital. Therefore, as we could have anticipated, the end result is the same but rotated by 90° around the y-axis. Therefore, although the math appears different, the only thing we effectively did was solve the same problem with a rotated axis system. It is important to note that the physics of a problem should not depend on our choice of the coordinate system.

Conceptual summary. – The expression of the function space of spherical harmonics in terms of indistinguishable spinors allows the evaluation of matrix elements in a fashion similar to the step operators for the quantum harmonic oscillator. In the first chapter we saw this for operators that conserve the number of spinors. In this section this was extended to operators that change the number of spinors. These operators are important for the study of the interaction of an atom with electromagnetic radiation, leading to transitions between different orbitals.

Problems

1 The X-ray absorption spectrum for a particular polarization q is given by

$$I_q(\omega) = \sum_f |\langle f|\boldsymbol{T}^1_q|i\rangle|^2 \delta(\hbar\omega + E_i - E_f). \tag{3.148}$$

The indices i and f denote the initial and final states, respectively. These can, in principle, be very complicated states. Show that the difference between the integrated intensities of the spectra for $q = 1$ and -1 is proportional to the expectation value of the angular momentum in the initial state. Take the situation for absorption from filled p orbitals into partially occupied d orbitals.

3.7 Molecules and Solids

Introduction. – Molecules and solids are complex systems consisting of many atoms. Understanding these materials is an important field in chemistry and physics. The Sections 3.7 and 3.8 only distill the essence of these materials, namely the competition between the attractive potentials of the nuclei and the kinetic energy in determining the eigenstates of the electrons. We limit ourselves to one-dimensional systems to avoid the complexities of molecular structures and crystal lattices. The problem is approached from two opposite sides. First, we start with bound electrons and introduce the overlap between orbitals on the neighboring sites. In Section 3.8, the starting point is free electrons, and the periodic potential is introduced. Both approaches obviously simplify the problem but give a good idea of the physics underlying the formation of molecules and solids.

Solids also change the geometry of space, which directly affects the eigenstates and the conserved quantities in the system. As we have seen before, periodicity in one space leads to discreteness in the dual space. For solids, both aspects are found in real space and its dual space, known as the momentum or reciprocal space. The periodicity of the potential of the nuclei leads to discreteness in the Fourier transform of the potential. The relation between the real space vectors and their duals is described in Section 1.12.

The problem of a single electron and a nucleus is exactly solvable under the assumption that the nucleus can be represented by an effective central potential. Any problem where more than a single particle needs to be considered becomes significantly more complex. This is true for atoms and, obviously, molecules and solids, where the number of particles becomes significantly larger. A general Hamiltonian for a molecule or a solid is given by

$$\hat{H} = \hat{H}_{\text{el}} = \sum_i \frac{\hat{\mathbf{p}}_i^2}{2m} - \frac{1}{4\pi\varepsilon_0}\sum_{ij}\frac{Z_j e^2}{|\mathbf{r}_i - \mathbf{R}_j|} + \frac{1}{2}\frac{1}{4\pi\varepsilon_0}\sum_{i\neq i'}\frac{e^2}{|\mathbf{r}_i - \mathbf{r}_{i'}|}, \tag{3.149}$$

where the indices i, i' and j go over the electrons and the nuclei in the system, respectively. The Hamiltonian already contains the approximation that the nuclei are fixed. This is often a reasonable approximation due to the large difference in speeds of the electron and the nuclei. The focus is therefore on the electrons. The first term is the kinetic energy; the second term is the potential due to the nuclei with charge $+Z_j e$; and the last term is the Coulomb interaction between the electrons (the factor $\frac{1}{2}$ corrects for the double counting in the summation). This is still an unsolvable problem, due to the large number of electrons. The goal in the treatment of the Hamiltonian for molecules and solids is to turn it into an effective single-particle model, where an electron moves in an effective potential of the nuclei and all the other electrons. It is important to realize that this approach is not always successful. For example, for solids, an independent-particle approach can predict a material to be metallic, whereas, experimentally, it is a large gap insulator. For these materials, one has to use methods that treat all the electrons explicitly. These are called many-body problems and significantly increase the complexity of the problem. Here, we limit ourselves to effective independent-particle systems. Also, we only treat some rather simple systems that nevertheless bring out some important aspects. As a first step, treating the electron–electron interaction as an effective potential could be done by replacing Z_j by an effective Z_j^{eff}. A large portion of the electrons is strongly bound to the nucleus with binding energies from several tens to thousands of electron volts. These electrons do not play a significant role in the low-energy properties of the material, such as bonding, conductivity, and the response to visible light. They can therefore be considered part of the nucleus.

We are then left with two terms in the Hamiltonian, the kinetic energy and an effective potential energy, describing both the nuclei and the other electrons. There are now two starting points. The focus could be on the kinetic energy using free electrons as a basis and then adding the potential. This approach is followed in Section 3.8. Alternatively, one can start with electrons bound in atomic orbitals and add hopping between different atoms. This seems like a logical step for molecules. Let us consider the simplest system, the H_2 molecule, and only consider the $1s$ orbitals. There are two orbitals $\langle \mathbf{r}|\mathbf{R}_1\rangle = \varphi(\mathbf{r}-\mathbf{R}_1)$ and $\langle \mathbf{r}|\mathbf{R}_2\rangle = \varphi(\mathbf{r}-\mathbf{R}_2)$, where $\varphi(\mathbf{r})$ is the atomic $1s$ orbital at the origin, and $\mathbf{R}_1$ and $\mathbf{R}_2$ are the positions of the two hydrogen nuclei. The hydrogen $1s$ wavefunction is spherically symmetric and decays exponentially in the radial direction as $\exp(-r/a_0)$, where $a_0 = 0.529$ Å is the Bohr radius. The separation of the hydrogen atoms in the molecule is 0.73 Å. There are then a total of four matrix elements. The matrix elements of the orbitals with itself are

$$\varepsilon = \int d\mathbf{r}\varphi^*(\mathbf{r}-\mathbf{R}_1)\hat{H}\varphi(\mathbf{r}-\mathbf{R}_1) = \int d\mathbf{r}\varphi^*(\mathbf{r}-\mathbf{R}_2)\hat{H}\varphi(\mathbf{r}-\mathbf{R}_2). \tag{3.150}$$

Since the $1s$ orbitals are real, the conjugate is not really necessary. The energy is close to the binding energy of -13.6 eV of the $1s$ orbital. It is not exactly the same due to the presence of the additional nucleus in the molecule. Additionally, if more than one electron is present in the molecule, then this affects the effective potential. However, these corrections are relatively small compared to the binding energy due to the matrix elements between the orbitals on different hydrogen atoms,

$$-t = \int d\mathbf{r}\varphi^*(\mathbf{r}-\mathbf{R}_1)\hat{H}\varphi(\mathbf{r}-\mathbf{R}_2) = \int d\mathbf{r}\varphi^*(\mathbf{r}-\mathbf{R}_2)\hat{H}\varphi(\mathbf{r}-\mathbf{R}_1), \tag{3.151}$$

where t is real, and $t > 0$ for $1s$ orbitals, making the matrix element negative. This is also known as a hopping or transfer matrix element. Note that this matrix element approaches zero if the nuclei are far apart.

The Hamiltonian for the two $1s$ orbitals can be written as

$$\boldsymbol{H} = \sum_{ij=1,2} |\mathbf{R}_i\rangle\langle\mathbf{R}_i|H|\mathbf{R}_j\rangle\langle\mathbf{R}_j| = \begin{pmatrix} \varepsilon & -t \\ -t & \varepsilon \end{pmatrix}. \tag{3.152}$$

There is the complication in that the overlap matrix elements $S = \langle \mathbf{R}_1|\mathbf{R}_2\rangle = \int d\mathbf{r}\varphi^*(\mathbf{r}-\mathbf{R}_1)\varphi(\mathbf{r}-\mathbf{R}_2) \neq 0$ are not equal to zero, since the $1s$ orbitals on different atoms are not orthogonal. It is not too difficult to include the nonzero overlap. However, since it offers little additional insight, we take $S = 0$. The eigenvalues of the Schrödinger equation $\boldsymbol{H}|\psi\rangle = E|\psi\rangle$ can be found by solving the determinant

$$\begin{vmatrix} \varepsilon - E & -t \\ -t & \varepsilon - E \end{vmatrix} = 0 \quad \Rightarrow \quad E_{B/AB} = \varepsilon \mp t, \tag{3.153}$$

where B and AB stand for the bonding and antibonding solutions. The corresponding eigenfunctions are

$$|\psi_{B/AB}\rangle = \frac{1}{\sqrt{2}}(|\mathbf{R}_1\rangle \pm |\mathbf{R}_2\rangle). \tag{3.154}$$

The wavefunctions are schematically shown in Figure 3.1. Note that the signs of the eigenfunctions depend on the sign of the hopping matrix element. The positive sign is

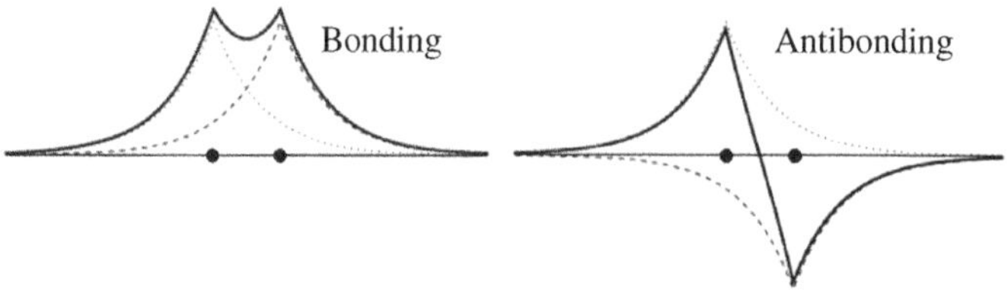

Figure 3.1 Schematic one-dimensional picture of the bonding and antibonding solutions of the H_2 molecule.

lowest for a negative matrix element ($-t < 0$). An electron spends an equal amount of time on atom 1 as it does on atom 2, as it should, because atom 1 is equivalent to atom 2. The electron therefore delocalizes over both hydrogen atoms. This lowers the kinetic energy for the bonding state. It might at first appear counterintuitive that a delocalized orbital has a lower kinetic energy than a localized one. However, a very localized wavefunction, for example the δ function, has a very large derivative and therefore a large kinetic energy. Therefore, the delocalization relaxes the wavefunction. On the flip side, the antibonding has a larger energy, because the sign change between the nuclei increases the derivative and thereby the kinetic energy, see Figure 3.1.

If spin is included, two electrons can reside in the bonding orbital, giving an energy of $2\varepsilon - 2t$. Therefore, an H_2 molecule has a lower energy than two separate hydrogen atoms with an electron each. However, it is not always favorable to form a molecule. For example, a helium atom has two electrons in the $1s$. When bringing two He atoms together to form an He_2 molecule, both bonding and antibonding orbitals need to be filled. This gives a total energy of $2(\varepsilon - t) + 2(\varepsilon + t) = 4\varepsilon$. Therefore, He cannot take any energetic gain from the formation of the bonding and antibonding combinations, and no real bonding is formed. The filled shells is the reason that helium is a noble gas.

We can extend this procedure to chains of arbitrary length N. General wavefunctions for N sites can then be written as

$$\psi(\mathbf{r}) = \sum_{\mathbf{R}} a_{\mathbf{R}} \varphi(\mathbf{r} - \mathbf{R}) \quad \text{or} \quad |\psi\rangle = \sum_{\mathbf{R}} a_{\mathbf{R}} |\mathbf{R}\rangle. \tag{3.155}$$

Again, we want to solve the Schrödinger equation $\boldsymbol{H}|\psi\rangle = E|\psi\rangle$. For simplicity, only matrix elements between nearest-neighbor atoms, given by $-t$ in Eq. (3.151), are included. Next nearest-neighbor and so on become smaller in magnitude compared to the nearest neighbor hopping matrix element. It is straightforward to include these additional terms, but the nearest-neighbor matrix elements offer the most physical insight. The Hamiltonian can be written in matrix form as

$$\boldsymbol{H} = \begin{pmatrix}
\varepsilon & -t & 0 & 0 & \cdots & & & & 0 & 0/-t \\
-t & \varepsilon & -t & 0 & 0 & \cdots & & & 0 & 0 \\
0 & -t & \varepsilon & -t & 0 & 0 & \cdots & & & \\
0 & 0 & -t & \varepsilon & -t & 0 & 0 & \cdots & & \\
\cdots & 0 & 0 & -t & \varepsilon & -t & 0 & 0 & \cdots & \\
 & \cdots & 0 & 0 & -t & \varepsilon & -t & 0 & 0 & \cdots \\
 & & & & \cdot & \cdot & \cdot & \cdot & & \\
0 & 0 & & & \cdot & \cdot & 0 & -t & \varepsilon & -t \\
0/-t & 0 & & & \cdot & \cdot & \cdot & 0 & -t & \varepsilon
\end{pmatrix}. \tag{3.156}$$

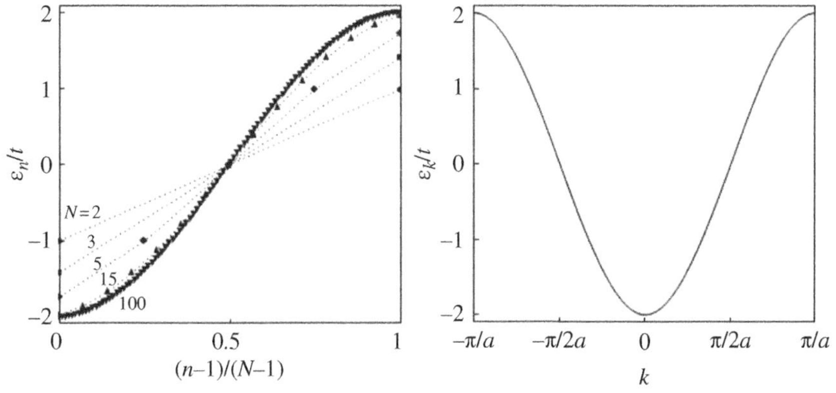

Figure 3.2 The left side shows the eigenenergies ε_n of the Hamiltonian in Eq. (3.156) for different number of sites N in the chain. The eigenenergies ε_n with $n = 1, 2, \ldots, N$ are ordered from small to large. For better comparison, the index n is normalized from 0 to 1 using $(n - 1)/(N - 1)$. The dotted lines are guides to the eye connecting eigenenergies of the same N. The figure on the right shows the eigenenergies for a single tight-binding band $\varepsilon_k = -2t\cos ka$ as a function of k.

The eigenvalues of the above Hamiltonian (with 0 for the bottom-left and top-right matrix elements) can be solved numerically. The values for certain lengths of the chain N are plotted in Figure 3.2. For $N = 2$, the values $\varepsilon_n/t = \pm 1$ for $n = 1, 2$ are found, which are the antibonding/bonding states of the diatomic molecule. When the number of sites N is increased, the eigenvalues are not randomly distributed but quickly approach a particular curve. Compare, for example the eigenvalues for $N = 15$ and $N = 100$. Values of $N > 100$ are virtually indistinguishable from the curve for $N = 100$. The differences occur due to the finite size of the chain. However, these finite size effects are negligibly small for solids, where one is typically dealing with 10^{20}–10^{23} atoms. Note this is only true when we are interested in bulk properties. For certain applications or experiments, surface effects are important. The problem with an open-ended chain is that the atoms in the chain are not equivalent, and therefore, there is no translational symmetry. This can be solved by connecting the beginning and the end of the chain. This amounts to replacing the zeroes at the edges of the matrix by $-t$. For a chain with periodic boundary conditions, the eigenvalues always lie on the large N curve. For $N = 2$, the off-diagonal matrix element in Eq. (3.153) has to be replaced by $-2t$ for hopping from 1 to 2 and hopping from 2 to 1. This gives eigenvalues $\pm 2t$, instead of $\pm t$. The closed loop can, in principle, represent a real system. In two dimensions, a plane with periodic boundary conditions can be represented by a toroid. However, for three dimensions, there is no easy visualization. Therefore, it is better to consider periodic boundary conditions as a mathematical trick to make each site in the chain equivalent, thereby imposing translational symmetry.

Since there is now translational symmetry in real space, one can expect that momentum is a conserved quantity. It is therefore useful to look at the Fourier transform of the system. For the exponential in the Fourier transform, periodic boundary conditions imply that

$$e^{ik(x+L)} = e^{ikx} \;\Rightarrow\; kL = 2n\pi \;\Rightarrow\; k = \frac{2\pi}{L}n, \tag{3.157}$$

where n is an integer, and the length of the chain is $L = Na$, with a the distance between the atoms. This corresponds to an infinite number of k values. However, there are only

N orbitals in the basis, so only a limited numbers are relevant. Restricting ourselves to N values, this gives

$$k = 0, \pm\frac{2\pi}{L}, \pm\frac{4\pi}{L}, \ldots, \pm\frac{\pi}{a}. \tag{3.158}$$

So, we only have k values in the region $[-\frac{\pi}{a}, \frac{\pi}{a}]$. Other k values are not needed since we need only the value at the position of the nuclei $\mathbf{R} = na\mathbf{e}_x$ or $R = na$. In that case, the reciprocal lattice vectors, see Section 1.12,

$$K = 0, \pm\frac{2\pi}{a}, \pm\frac{4\pi}{a}, \cdots = m\frac{2\pi}{a} \quad \text{with} \quad e^{i(k+K)R} = e^{ikR}, \tag{3.159}$$

can be added without changing the eigenfunctions.

The suggested eigenstates for the chain of $1s$ orbitals are

$$\varphi_{\mathbf{k}}(\mathbf{r}) = \langle \mathbf{r}|\mathbf{k}\rangle = \sum_{\mathbf{R}} \langle \mathbf{r}|\mathbf{R}\rangle\langle \mathbf{R}|\mathbf{k}\rangle = \frac{1}{\sqrt{N}}\sum_{\mathbf{R}} e^{i\mathbf{k}\cdot\mathbf{R}}\varphi(\mathbf{r}-\mathbf{R}), \tag{3.160}$$

with $\mathbf{k} = k\mathbf{e}_x$. Note that we are only making a Fourier transform of the lattice points and not of the entire space. The goal is to diagonalize the real space Hamiltonian. We can also write this in bra-ket notation

$$|\mathbf{k}\rangle = \frac{1}{\sqrt{N}}\sum_{\mathbf{R}} e^{i\mathbf{k}\cdot\mathbf{R}}|\mathbf{R}\rangle \quad \text{and} \quad |\mathbf{R}\rangle = \frac{1}{\sqrt{N}}\sum_{\mathbf{k}} e^{-i\mathbf{k}\cdot\mathbf{R}}|\mathbf{k}\rangle. \tag{3.161}$$

The Hamiltonian in real space in bra-ket notation can be written as

$$\boldsymbol{H} = \sum_{\mathbf{R},\mathbf{R}'} |\mathbf{R}'\rangle\langle \mathbf{R}'|H|\mathbf{R}\rangle\langle \mathbf{R}| = -t\sum_{\mathbf{R}\boldsymbol{\delta}} |\mathbf{R}+\boldsymbol{\delta}\rangle\langle \mathbf{R}| = -t\sum_{\mathbf{R}\boldsymbol{\delta}} \mathbf{c}^{\dagger}_{\mathbf{R}+\boldsymbol{\delta}}\mathbf{c}_{\mathbf{R}}, \tag{3.162}$$

where $\boldsymbol{\delta} = \pm a\mathbf{e}_x$ is a vector connecting $\mathbf{R}$ to its nearest neighbors (the atoms on the left and right of the atom in a chain). The last expression in the above equation shows the result in second quantization using Eq. (2.284). Here, the hopping is achieved by first removing the electron from site $\mathbf{R}$ with the annihilation operator $\mathbf{c}_{\mathbf{R}}$. An electron is then created at $\mathbf{R}+\boldsymbol{\delta}$ with the creation operator $\mathbf{c}^{\dagger}_{\mathbf{R}+\boldsymbol{\delta}}$. The action is the same as that of the ket-bra combination $|\mathbf{R}+\boldsymbol{\delta}\rangle\langle \mathbf{R}|$.

The Fourier transform can be made by inserting the transforms into the Hamiltonian:

$$\begin{aligned}\boldsymbol{H} &= -t\sum_{\mathbf{R}\boldsymbol{\delta}} \frac{1}{N}\sum_{\mathbf{k},\mathbf{k}'} e^{-i\mathbf{k}\cdot(\mathbf{R}+\boldsymbol{\delta})}|\mathbf{k}\rangle\langle \mathbf{k}'|e^{i\mathbf{k}'\cdot\mathbf{R}} \\ &= -t\sum_{\mathbf{k},\mathbf{k}'}\left(\frac{1}{N}\sum_{\mathbf{R}} e^{i(\mathbf{k}'-\mathbf{k})\cdot\mathbf{R}}\right)\left(\sum_{\boldsymbol{\delta}} e^{-i\mathbf{k}\cdot\boldsymbol{\delta}}\right)|\mathbf{k}\rangle\langle \mathbf{k}'| = \sum_{\mathbf{k},\mathbf{k}'}\delta_{\mathbf{k},\mathbf{k}'}\varepsilon_{\mathbf{k}}|\mathbf{k}\rangle\langle \mathbf{k}'|.\end{aligned} \tag{3.163}$$

We see that the Hamiltonian is diagonal in $\mathbf{k}$ due to the delta function given by the first term in parentheses. The final result can then be written as

$$\boldsymbol{H} = \sum_{\mathbf{k}} \varepsilon_{\mathbf{k}}|\mathbf{k}\rangle\langle \mathbf{k}|. = \sum_{\mathbf{k}} \varepsilon_{\mathbf{k}}\mathbf{c}^{\dagger}_{\mathbf{k}}\mathbf{c}_{\mathbf{k}}. \tag{3.164}$$

This is a Hamiltonian with only diagonal elements. Therefore, the Hamiltonian, which can in principle be very large, has been diagonalized by making a transform to momentum space. Since the Hamiltonian is diagonal, the electron with momentum $\mathbf{k}$ can only scatter back into the same wavevector.

The diagonal matrix elements (which are directly the eigenvalues) are given by

$$\varepsilon_{\mathbf{k}} = -t\sum_{\boldsymbol{\delta}} e^{-i\mathbf{k}\cdot\boldsymbol{\delta}}. \tag{3.165}$$

For a one-dimensional chain, the nearest-neighbor vectors for a chain in the x-direction are given by $\boldsymbol{\delta} = \pm a\mathbf{e}_x$.

$$\varepsilon_{\mathbf{k}} = -t(e^{ika} + e^{-ika}) = -2t\cos ka, \tag{3.166}$$

where $\mathbf{k} = k\mathbf{e}_x$. Since k goes from $-\frac{\pi}{a}$ to $\frac{\pi}{a}$, the eigenvalues lie on a cosine between $-2t$ and $2t$, see Figure 3.2. The k values are discrete, and there are a total of N eigenstates. However, in solids, N is generally a very large number, so it is plotted as a continuous curve. These eigenvalues reproduce the numerical eigenenergies for a system with periodic boundary conditions. If the band is not entirely filled, the formation of a band lowers the energy of the system and therefore explains, in simple terms, why solids are formed.

Example 3.4 ***The Benzene Molecule***

The benzene molecule (C_6H_6) consists of six carbon atoms that form a hexagon, see Figure 3.3. The hydrogen atoms are attached to the carbons sticking to the outside of the hexagon. Since the atomic number of carbon is 6, we have, including hydrogen, a total of $6 \times (1 + 6) = 42$ electrons. However, 12 of these are strongly bound in the carbon 1s orbitals (with a binding energy of more than 280 eV). For molecular bonding, we can ignore those. An additional 24 electrons are bound in strongly covalent bonds between the carbon atoms and between the carbon and the hydrogen. We want to consider the six remaining electrons that are in the p_z orbitals perpendicular to the plane of the hexagon, see Figure 3.3. The p_z orbitals form a naturally occurring periodic one-dimensional system.

a) Find the eigenvalues numerically using the Hamiltonian in real space. What is the ground-state energy?
b) Find the eigenvalues and eigenvectors analytically using the Hamiltonian in momentum space.

Solution

a) We can define the matrix element between p_z orbitals on carbon atoms on neighboring sites i and j as

$$-t = \langle p_{zi}|H|p_{zj}\rangle = -\int d\mathbf{r}\varphi_{p_z}(\mathbf{r}-\mathbf{R}_i)H\varphi_{p_z}(\mathbf{r}-\mathbf{R}_j). \tag{3.167}$$

The matrix element is negative (making $t > 0$) since the positive lobes are all on the same side of the plane of the hexagon. The 6×6 matrix can be written in the form

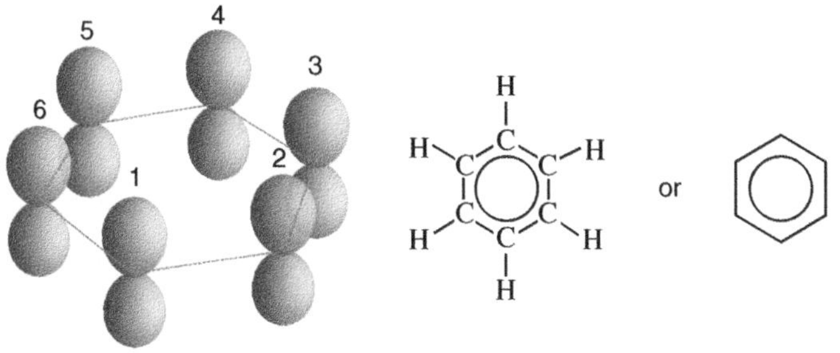

Figure 3.3 The benzene molecule C_6H_6 consists of a hexagon of carbon atoms with hydrogens attached to each carbon. The lowest occupied and unoccupied energy states are formed by the p_z orbitals that stick out of the plane of the hexagon. The chemical notation of benzene is a hexagon with a circle inside.

of Eq. (3.156). Periodic boundary conditions apply for the hexagon, so the top-right and bottom-left matrix elements are $-t$. The eigenvalues can be solved numerically to give

$$E = \varepsilon - 2t, \varepsilon - t, \varepsilon - t, \varepsilon + t, \varepsilon + t, \varepsilon + 2t. \tag{3.168}$$

Including spin, each orbital can be occupied by two electrons. The 6 π electrons go into the lowest molecular orbitals, giving a total energy of $E = 6\varepsilon - 8t$. Note that all the bonding states are occupied. Obviously, the eigenfunctions can be obtained numerically as well.

b) The momenta are given by $k = 2\pi m/L$. Since $L = 6a$ is the total distance around the hexagon, with a the separation between the carbon atoms, this gives $k = \pi m/3a$. However, since there are six carbon atoms, only six momenta are needed,

$$k = 0, \pm\frac{\pi}{3a}, \pm\frac{2\pi}{3a}, \frac{\pi}{a}. \tag{3.169}$$

Note that we could have picked $-\pi/a$ as well, since this value is equivalent to π/a. Inserting the k values into the expression for the energies $\varepsilon_k = \varepsilon - 2t\cos ka$, see Eq. (3.166), reproduces the eigenenergies above. The eigenfunctions are the Fourier transform of the local basis functions,

$$\psi_k = \frac{1}{\sqrt{6}}\sum_{n=1}^{6}\varphi_{p_{nz}} e^{ikna}, \tag{3.170}$$

using the shorthand $\varphi_{p_{nz}} = \varphi_{p_z}(\mathbf{r} - \mathbf{R}_n)$. The wavefunction for the lowest eigenstate with energy $\varepsilon - 2t$ is completely symmetric

$$\psi_0 = \frac{1}{\sqrt{6}}\sum_{n=1}^{6}\varphi_{p_{nz}} = \frac{1}{\sqrt{6}}(\varphi_{p_{1z}} + \varphi_{p_{2z}} + \varphi_{p_{3z}} + \varphi_{p_{4z}} + \varphi_{p_{5z}} + \varphi_{p_{6z}}). \tag{3.171}$$

This eigenfunction contains two electrons (spin up and spin down). This picture differs from that of molecular bonds. An electron in the ψ_0 orbital is delocalized over all the six carbon atoms and not localized in a bond between two carbon atoms. This is indicated as a circle in the chemical notation of a benzene molecule, see Figure 3.3.

The wavefunction with the highest energy $\varepsilon + 2t$ is given by

$$\psi_{\frac{\pi}{a}} = \frac{1}{\sqrt{6}}\sum_{n=1}^{6}(-1)^n\varphi_{p_{nz}} = -\frac{1}{\sqrt{6}}(\varphi_{p_{1z}} - \varphi_{p_{2z}} + \varphi_{p_{3z}} - \varphi_{p_{4z}} + \varphi_{p_{5z}} - \varphi_{p_{6z}}).$$

This is the completely antibonding wavefunction with nodes in the wavefunction between the nuclei.

There are two wavefunctions with energy $\varepsilon - t$. Any linear combination of these two eigenfunctions is also an eigenfunction. Real wavefunctions can be obtained via

$$\begin{aligned}\frac{1}{\sqrt{2}}(\psi_{\frac{\pi}{3a}} + \psi_{-\frac{\pi}{3a}}) &= \frac{1}{\sqrt{3}}\sum_n \varphi_{p_{nz}} \cos n\frac{\pi}{3} \\ &= \frac{1}{\sqrt{3}}\left(\frac{1}{2}\varphi_{p_{1z}} - \frac{1}{2}\varphi_{p_{2z}} - \varphi_{p_{3z}} - \frac{1}{2}\varphi_{p_{4z}} + \frac{1}{2}\varphi_{p_{5z}} + \varphi_{p_{6z}}\right)\end{aligned} \tag{3.172}$$

and

$$\frac{1}{\sqrt{2}i}(\psi_{\frac{\pi}{3a}} - \psi_{-\frac{\pi}{3a}}) = \frac{1}{\sqrt{3}}\sum_n \varphi_{p_{nz}} \sin n\frac{\pi}{3} = \frac{1}{2}(\varphi_{p_{1z}} + \varphi_{p_{2z}} - \varphi_{p_{4z}} - \varphi_{p_{5z}}).$$

Similar wavefunctions can be obtained for $k = \pm 2\pi/3a$, which is left to the reader.

The benzene molecule has two periodicities. One is the periodicity of the hexagon of length $L = 6a$, *i.e.* the size of the system. This leads to the discreteness of the levels with $k = m\pi/3a$. This discreteness is still present for an open chain. The discreteness is a direct consequence of the confinement. Periodicity can be viewed as a type of confinement. For solids, this discretization effectively disappears since the separation $2\pi/L$ is generally very small. The other periodicity is that of the lattice constant a. This limits the region of the momenta from $-\pi/a$ to π/a. Any wavevector $k + m2\pi/a$ produces the same wavefunction as k and therefore offers no new information.

Conceptual summary. – In this section, a one-dimensional solid was created by adding $1s$ orbitals to a chain. The stabilization of the system is a result of the relaxation of the wavefunctions due to the delocalization of the electrons. This allows the electrons to take advantage of the attractive potential of the nuclei but with a lower cost in kinetic energy. For an infinite system, all atoms are equivalent, and there is translational symmetry. For a finite system, the same symmetry can be obtained by applying periodic boundary conditions. The translational symmetry implies that there is conservation in its dual space. This makes the momentum or the wavevector a good quantum number. However, due to the discreteness of the atomic orbitals, only a limited number of momenta are needed. The region of the relevant momenta closest to $\mathbf{k} = 0$ is generally known as the first Brillouin zone.

Although a chain of $1s$ orbitals is relatively simple, the approach can be extended in a straightforward fashion to include different orbitals in a more realistic lattice. This is known as a tight-binding model. Additionally, many materials contain more than one type of atoms, leading to different potentials. Obviously, the problem becomes technically more complicated. For example, the hopping matrix elements from Eq (3.151) are angular dependent for all orbitals other than the s orbital. However, the underlying physics remains similar to the simple one-dimensional problem.

Problems

1 Let us consider an s and a p_x orbital on one site coupled to two s_i orbitals, with $i = 1, 2$ located at $\pm a\mathbf{e}_x$, where a is the distance between the atoms. The matrix elements to the s_1 orbital at $a\mathbf{e}_x$ are $-t_s$ and $-t_p$, with $t_s, t_p > 0$ for s and p_x, respectively. Write down and solve the matrix. The on-site energies can be taken as zero. In chemistry, it is often convenient to make a combination of the s and p_x orbitals, so that the new hybrids predominantly bond in one direction when $t_s \cong t_p$. Find the unitary transformation and write down the Hamiltonian in the new basis.

2 Consider a linear chain in the x-direction of alternating s and p_x orbitals separated by a distance a. The overlap integral is given by

$$t = \int d\mathbf{r}\ \varphi_{p_x}(\mathbf{r} - a\mathbf{e}_x) H \varphi_s(\mathbf{r}) > 0, \tag{3.173}$$

where $\varphi_s(\mathbf{r})$ and $\varphi_{p_x}(\mathbf{r})$ are atomic s and p_x orbitals, respectively. Take the energy of an electron in the s orbital as 0, and the energy of an electron in the p_x orbital as $\Delta > 0$.

a) What are the reciprocal lattice vectors and the first Brillouin zone?
b) Determine the 2×2 matrix for the two tight-binding wavefunctions for a particular k value. Solve for the eigenenergies.
c) Sketch the eigenvalues in the first Brillouin zone.
d) Sketch the orbital arrangements for the lowest eigenstate.

3 Let us consider a one-dimensional chain of equivalent atoms. In the undistorted case, the distance between the atoms is a. Certain materials undergo a distortion at low temperatures that displaces every other atom. This is known as a Peierls distortion. The shorter distance between two neighboring atoms is a'. This also affects the hopping matrix element, which we can take as $t_\pm = t \pm \Delta t$ for the shorter and longer distances, respectively. Calculate the energies and plot the dispersion in the first Brillouin zone.

3.8 Periodic Potential in a Solid

Introduction. – In Section 3.7, (one-dimensional) solids were constructed by aligning $1s$ orbitals in a chain. Electrons, initially bound to a hydrogen atom, are then allowed to hop between different hydrogen atoms. The focus is therefore on obtaining the correct binding to the nucleus and then adding the delocalization. It is also possible to give more emphasis to the kinetic energy by starting with free electrons and then subsequently add the periodic potential.

One might wonder why we approach the same problem from two different angles. Since the end result is the electronic structure of a particular material, the way we describe it should not matter. This is correct, if we did not use any approximations. Unfortunately, physics is full of approximations, and understanding the consequences of making one is important. Additionally, the starting point in this section is free electrons that have a dispersion $\varepsilon_\mathbf{k} \sim k^2$. It is not directly obvious how this can be reconciled with the relation $\varepsilon_\mathbf{k} = -2t \cos ka$ found in the previous section. Furthermore, in Section 3.7, we found that the k values were limited to the region $[-\frac{\pi}{a}, \frac{\pi}{a}]$. There is no limitation on k for free electrons. Another puzzling aspect is how orbital character is formed out of the planes waves.

Since solids are finite, we want to consider particles confined to a volume. However, to retain translational symmetry, periodic boundary conditions are applied, as in Eqs. (1.205) and (1.210). The basis functions are plane waves, see Eq. (1.220),

$$\varphi_\mathbf{k}(\mathbf{r}) = \frac{1}{\sqrt{V}} e^{i\mathbf{k}\cdot\mathbf{r}}. \tag{3.174}$$

The simplifications to the one-dimensional system are made below. The plane waves satisfy the Schrödinger equation for free space

$$-\frac{\hbar^2}{2m}\nabla^2 \varphi_\mathbf{k}(\mathbf{r}) = \varepsilon_\mathbf{k}\varphi_\mathbf{k}(\mathbf{r}), \quad \text{with} \quad \varepsilon_\mathbf{k} = \frac{\hbar^2\mathbf{k}^2}{2m}. \tag{3.175}$$

We now want to add a potential. In a solid, the potential is due to the nuclei in the crystal lattice and the electrons that are strongly bound to the nucleus. This section studies, what is known as, the nearly free electron model. A typical example is aluminum metal.

The $1s$, $2s$, and $2p$ electrons are strongly bound to the nucleus and do not contribute to the low-energy properties, *i.e.* effects of the order of a few electronvolts or less. These electrons are therefore part of the effective nuclear potential. So, the screened nucleus is that of Al^{3+}. The valence electrons, relevant for the low-energy properties, are $3s^2 3p^1$. More advanced calculations also consider an effective negative potential due to those electrons. However, for our purposes, the most important aspect of the effective potential energy is that it is periodic.

We now want to add the periodic potential. Any potential can be described in terms of a Fourier series

$$U(\mathbf{r}) = \sum_{\mathbf{k}} U_{\mathbf{k}} e^{i\mathbf{k}\cdot\mathbf{r}}. \tag{3.176}$$

The $\mathbf{k}$ values are discrete due to the periodic boundary conditions. However, the k values are separated by $2\pi/L$, and this discretization is generally very small. However, we want to impose that the potential satisfies the periodicity of the lattice. Therefore, the potential should be equivalent when moving to a different lattice site with the vectors $\mathbf{R}$ that indicate the periodicity,

$$U(\mathbf{r}+\mathbf{R}) = \sum_{\mathbf{k}} U_{\mathbf{k}} e^{i\mathbf{k}\cdot(\mathbf{r}+\mathbf{R})} = \sum_{\mathbf{k}} e^{i\mathbf{k}\cdot\mathbf{R}} U_{\mathbf{k}} e^{i\mathbf{k}\cdot\mathbf{r}}. \tag{3.177}$$

The wavevectors $\mathbf{K}$ that make the potential periodic satisfy the following condition:

$$e^{i\mathbf{K}\cdot\mathbf{R}} = 1 \quad \Rightarrow \quad U(\mathbf{r}) = \sum_{\mathbf{K}} U_{\mathbf{K}} e^{i\mathbf{K}\cdot\mathbf{r}}. \tag{3.178}$$

We have seen this condition before in Section 1.12. The vectors $\mathbf{K}$ are known as the reciprocal lattice vectors, which were also found in Eq. (1.272) for X-ray diffraction. The vectors can be obtained using Eq. (1.263). The matrix elements between different plane waves are given by

$$\langle \mathbf{k}'|U|\mathbf{k}\rangle = \sum_{\mathbf{K}} \frac{1}{V} \int d\mathbf{r} U_{\mathbf{K}} e^{-i(\mathbf{k}'-\mathbf{K}-\mathbf{k})\cdot\mathbf{r}} = \sum_{\mathbf{K}} U_{\mathbf{K}} \delta_{\mathbf{k}',\mathbf{k}+\mathbf{K}}. \tag{3.179}$$

Therefore, the periodic potential only scatters the electrons between plane waves that differ in wavevector by $\mathbf{K}$. In bra-ket notation or second quantization, we can write the potential as

$$\boldsymbol{U} = \sum_{\mathbf{kK}} U_{\mathbf{K}} |\mathbf{k}+\mathbf{K}\rangle\langle\mathbf{k}| = \sum_{\mathbf{kK}} U_{\mathbf{K}} \mathbf{c}^{\dagger}_{\mathbf{k}+\mathbf{K}} \mathbf{c}_{\mathbf{k}}. \tag{3.180}$$

In a simple one-dimensional chain with equivalent atoms at lattice sites $\mathbf{R} = na\mathbf{e}_x$, where a is the separation between the atoms, this condition is satisfied by

$$e^{iKa} = 1 \;\Rightarrow\; K = \frac{2\pi}{a} m, \tag{3.181}$$

which are the same values as found in Eq. (3.159). Note that since a is much smaller than L, the values of K are much larger than the separation $2\pi/L$ between the k values. The potential has the important consequence that a running wave with wavevector k is coupled to the running waves with wavevectors $k + 2\pi m/a$, with $m = \cdots, -2, -1, 0, 1, 2, \ldots$. However, this implies that k is only a good quantum number in a region in k space of width $2\pi/a$. Conventionally, the region $[-K/2, K/2] = [-\pi/a, \pi/a]$ is chosen. This is known as the first Brillouin zone. Any wavevector outside the first Brillouin zone is coupled to one inside by a reciprocal lattice vector K. However, the wavevectors inside the first Brillouin zone are not coupled to each other by the periodic potential.

Since only the $\mathbf{k}$ values separated by a reciprocal lattice vector $\mathbf{K}$ couple, we can write the eigenfunctions as a linear combination of these running waves

$$\psi_{\mathbf{k}}(\mathbf{r}) = \langle \mathbf{r}|\psi_{\mathbf{k}}\rangle = \sum_{\mathbf{K}} c_{\mathbf{k}-\mathbf{K}} e^{i(\mathbf{k}-\mathbf{K})\cdot\mathbf{r}}, \tag{3.182}$$

where $c_{\mathbf{k}-\mathbf{K}}$ is a coefficient. The Schrödinger equation in matrix form is

$$\boldsymbol{H}|\psi_{\mathbf{k}}\rangle = E|\psi_{\mathbf{k}}\rangle, \tag{3.183}$$

where the Hamiltonian combines the kinetic energy and the periodic potential

$$\boldsymbol{H} = \sum_{\mathbf{k}} \varepsilon_{\mathbf{k}}|\mathbf{k}\rangle\langle\mathbf{k}| + \sum_{\mathbf{kK}} U_{\mathbf{K}}|\mathbf{k}+\mathbf{K}\rangle\langle\mathbf{k}| = \sum_{\mathbf{k}} \varepsilon_{\mathbf{k}} \mathbf{c}_{\mathbf{k}}^{\dagger}\mathbf{c}_{\mathbf{k}} + \sum_{\mathbf{kK}} U_{\mathbf{K}} \mathbf{c}_{\mathbf{k}+\mathbf{K}}^{\dagger}\mathbf{c}_{\mathbf{k}}.$$

For each $\mathbf{k}-\mathbf{K}$, we find a set of coupled equations

$$\varepsilon_{\mathbf{k}-\mathbf{K}} c_{\mathbf{k}-\mathbf{K}} + \sum_{\mathbf{K}'} U_{\mathbf{K}'} c_{\mathbf{k}-\mathbf{K}+\mathbf{K}'} = E c_{\mathbf{k}-\mathbf{K}}. \tag{3.184}$$

Note that this reproduces to $E = \varepsilon_{\mathbf{k}-\mathbf{K}}$ in the absence of a periodic potential.

Since the only good wavevectors k are in the first Brillouin zone $[-\pi/a, \pi/a]$, one generally prefers to plot all eigenstates in that region. This can be done by plotting the dispersions $\varepsilon_{\mathbf{k}-\mathbf{K}}$ or $\varepsilon_{k-\frac{2\pi n}{a}}$ in one dimension, see Figure 3.4. This leads to many parabolas; however, we now only consider k values in $[-\pi/a, \pi/a]$ and ignore everything outside this region. Note that the scattering within one parabola over $K = 2\pi m/a$ (dashed arrow in Figure 3.4) is now replaced by vertical scattering at one k value between different displaced parabolas (solid arrow).

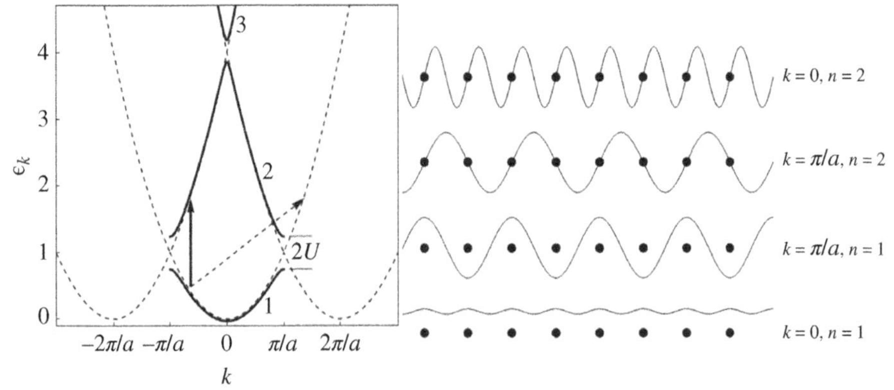

Figure 3.4 The nearly free electron model in one dimension. The parabolas (dashed lines) are free-electron eigenstates with energy $\varepsilon_k = \frac{\hbar^2 k^2}{2m}$. The reduced energy is defined as $\epsilon_k = \varepsilon_k/\varepsilon_{\frac{\pi}{a}}$. The free-electron energies are folded back into the first Brillouin zone $-\frac{\pi}{a} \le k \le \frac{\pi}{a}$ by plotting the parabolas around $K = 2\pi m/a$. The scattering by the periodic potential from k to $k + 2\pi/a$ (dashed arrow) then becomes a vertical transition at k (solid arrow). The solid lines give the eigenenergies for the nearly free electron model. The numbers denote the band index with $n = 1, 2, 3$. The right side shows the eigenfunctions for selected k values. The disks indicate the positions of the nuclei.

This model is known as the nearly free electron model and is generally used only in the case that the periodic potential is weak (such as K, Na, and Al metals). However, even a weak potential can have a large influence if the energy separation of the states that it couples is small (or even zero). For the energy difference to be zero, we require

$$\varepsilon_{\mathbf{k}} = \varepsilon_{\mathbf{k}-\mathbf{K}} \;\Rightarrow\; k^2 = (\mathbf{k}-\mathbf{K})^2 = k^2 - 2\mathbf{k}\cdot\mathbf{K} + K^2 \tag{3.185}$$

or

$$\mathbf{k}\cdot\mathbf{K} = \frac{1}{2}K^2. \tag{3.186}$$

These conditions are the same as the von Laue conditions for X-ray diffraction, see Eq. (1.275). In three dimensions, this condition defines the Bragg planes.

In one dimension, $\mathbf{k}$ and $\mathbf{K}$ are either parallel or antiparallel, and the von Laue condition reduces to $k = \pm\frac{1}{2}K$. In one dimension, there are only Bragg points. Let us consider $K = 2\pi/a$, for which $k = K/2 = \pi/a$ has the same energy as $k - K = -\pi/a$. In this region, we can neglect the coupling to other running waves. Since the potential is real, we have $U_{-\mathbf{K}} = U_{\mathbf{K}}^*$. If in addition, there is inversion symmetry (meaning that the system is unchanged when $\mathbf{r} \to -\mathbf{r}$ around a nucleus), $U_{-\mathbf{K}} = U_{\mathbf{K}}$. This essentially reduces the Fourier series to a cosine series. The energies can be found by solving the determinant

$$\begin{vmatrix} \varepsilon_{\mathbf{k}} - E & U_{\mathbf{K}} \\ U_{\mathbf{K}} & \varepsilon_{\mathbf{k}-\mathbf{K}} - E \end{vmatrix} = 0. \tag{3.187}$$

This leads to the quadratic equation

$$(\varepsilon_{\mathbf{k}} - E)(\varepsilon_{\mathbf{k}-\mathbf{K}} - E) - U_{\mathbf{K}}^2 = 0, \tag{3.188}$$

which has the solutions

$$E_{\mathbf{k}n} = \frac{\varepsilon_{\mathbf{k}} + \varepsilon_{\mathbf{k}-\mathbf{K}}}{2} \mp \sqrt{\left(\frac{\varepsilon_{\mathbf{k}} - \varepsilon_{\mathbf{k}-\mathbf{K}}}{2}\right)^2 + U_{\mathbf{K}}^2}, \tag{3.189}$$

with $n = 1, 2$, respectively. An additional quantum number n is needed to describe the different eigenstates with the same $\mathbf{k}$. This is also known as the band index, see Figure 3.4. In the limit that $\varepsilon_{\mathbf{k}} \cong \varepsilon_{\mathbf{k}-\mathbf{K}}$, this reduces to $E_{\mathbf{k}n} = \varepsilon_{\mathbf{k}} \mp |U_{\mathbf{K}}|$. This is the splitting in the bands observed at the edges of the Brillouin zone at $k = \pm\pi/a$, see Figure 3.4.

We can extend this procedure to include any number of reciprocal lattice vectors K. Let us consider the situation in one dimension for $K = 0$ and $K_\pm = \pm\frac{2\pi}{a}$. Using this we can write down the Hamiltonian,

$$\boldsymbol{H} = \begin{pmatrix} \varepsilon_k & U_K & U_K \\ U_K & \varepsilon_{k-K_-} & U_{2K} \\ U_K & U_{2K} & \varepsilon_{k-K_+} \end{pmatrix}. \tag{3.190}$$

This matrix has to be solved for each k value. The results are given in Figure 3.4. Note that the energies $\varepsilon_{k-K_\pm} = \hbar^2(k - K_\pm)^2/2m$ look in the region $[-K/2, K/2]$ as parabolas centered around $K_\pm$. The eigenvalues of $\boldsymbol{H}$ can be easily solved numerically for a particular choice of U_K and U_{2K}. The values for E_{kn}, where $n = 1, 2, 3$ is the band index, are plotted in the region $[-K/2, K/2]$. Also, drawn in Figure 3.4 are the eigenfunctions at particular kn values. For $k = 0$ and $n = 1$, the eigenfunction is close to a constant, which is the running wave for $k = 0$. There is a small $\cos\frac{2\pi}{a}x$ modulation due to the mixing with $K_\pm$. For $U_K < 0$ this leads to an increased electron density on the nuclei. For $k = \frac{\pi}{a}$, the eigenfunctions are given by $\cos\frac{\pi}{a}a$ and $\sin\frac{\pi}{a}a$. For $U < 0$, the cosine ($n = 1$) is the lowest in energy, leading to maxima on the nuclei. The function for $n = 2$ at $k = \frac{\pi}{a}$ is the sine and has maxima between the nuclei. The eigenfunction for $k = 0$ and $n = 2$ is close to $\sin \frac{2\pi}{a} x$.

Two approaches have been taken to study the electronic structure of a one-dimensional solid, see Figures 3.2 and 3.4. They can be reconciled by looking at the eigenstates of the nearly free electron model, see Figure 3.4. The lowest band ($n = 1$) is s like in nature in the sense that the maxima occur at the nuclei, and the minima between the nuclei. The second band ($n = 2$), on the other hand, shows p character, since there are nodes at the nuclei. Additionally, the bottom of the bands is bonding in nature, *i.e.*, there are no nodes between the nuclei ($k = 0$; $n = 1$ and $k = \frac{\pi}{a}$; $n = 2$). The top of the bands ($k = 0$; $n = 2$ and $k = \frac{\pi}{a}$; $n = 1$) are antibonding in nature and have nodes between the nuclei. The inclusion of the periodic potential therefore leads to the formation of orbital-like character. However, note that the potential $U(x) = -2|U_{\frac{2\pi}{a}}| \cos \frac{2\pi}{a} x$ is simply a cosine if only the lowest Fourier terms are included. This is significantly different from the $1/r$ potential at each of the nuclei expected for atomic potentials.

The tight-binding bands, see Figure 3.2, show the same bonding/antibonding nature since $k = 0$ has all the s orbitals with the same sign, see Eq. (3.160), whereas for $k = \frac{\pi}{a}$ the s orbitals alternate in sign at the nuclei. This behavior is comparable to the $n = 1$ band for the nearly free electron model. The second band is entirely missing from the tight-binding approach. However, this is due to the restriction to only s orbitals at the outset. However, it is straightforward to extend the approach to include p_x orbitals. The coupling of the s orbitals is still given by $\varepsilon_{\mathbf{k}}^{ss} = -2t_{ss} \cos ka$, see Eq. (3.166). The additional subscripts in t_{ss} indicate that this is a hopping matrix element between s orbitals. The coupling between the p orbitals is given by $\varepsilon_{\mathbf{k}}^{pp} = 2t_{pp} \cos ka + \Delta$, where Δ is the difference in onsite energies between the p and s orbitals. The matrix element changes sign. While s orbitals always have the integration over positive wavefunctions, the dominant part of the integral for p orbitals is coming from the lobes of opposite sign that are pointing toward each other. Therefore, the matrix element changes sign. There is an additional off-diagonal matrix element between the s and p orbitals on the neighboring sites. Following Eqs. (3.165) and (3.166), we find

$$\varepsilon_k^{sp} = t_{sp}(e^{ika} - e^{-ika}) = 2it_{sp} \sin ka. \tag{3.191}$$

The sign change is due to the fact that the matrix element of the s orbital is with the negative lobe of the p orbital on the right and the positive lobe of the p orbital on the left. We now have a basis φ_k^s and φ_k^p giving a matrix

$$H = \begin{pmatrix} \varepsilon_k^{ss} & \varepsilon_k^{sp} \\ (\varepsilon_k^{sp})^* & \varepsilon_k^{pp} \end{pmatrix}. \tag{3.192}$$

This matrix needs to be solved for all k values in the region $[-\frac{\pi}{a}, \frac{\pi}{a}]$. Figure 3.5 gives a comparison between the lowest two bands ($n = 1, 2$) in the nearly free electron model and the s and p tight-binding model. The fit has been done by eye. Although a perfect fit cannot be expected due to the different approximations, it is clear that they are giving comparable physics.

Conceptual summary. – We have now derived the electronic structure for a one-dimensional solid from two opposite sides. In Section 3.7, localized s orbitals were used. Bringing the atoms close together led to the formation of bands. In this section,

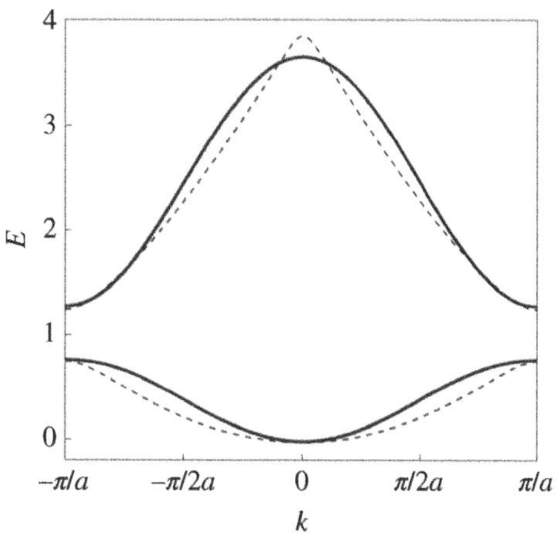

Figure 3.5 Comparison between s and p bands for a one-dimensional system obtained for a tight-binding model (solid) and a nearly free electron model (dashed).

the starting point was free electrons. This directly leads to bands. The scattering due to the periodic potential coupled momenta separated by K led to the formation of orbital character. Even for the simplest potential, *i.e.* a cosine with the minima in the potential on the nuclei, the eigenstates showed a clear s- and p-like characters with maxima and nodes on the nuclei, respectively. Additionally, both approaches found bonding and antibonding-type solutions at the bottom and top of the bands, respectively.

Although we have gained some insight into the electronic structure of solids, the models used are based on parameters. An important area in theoretical condensed-matter physics is dedicated to calculating the electronic structure of solids with the structural properties of the material as the only input parameters. These are called *ab initio* calculations (from the Latin "from the beginning"). The plots of the eigenenergies as a function of momentum are also called the band structure. These plots can only be made if the material can be considered an effective independent-particle system. This does not mean that the Coulomb interactions between the electrons are neglected. However, each electron only feels an effective potential due to the density of the other electrons. The calculation of the effective potential is often done self-consistently, which means that an initial charge density is chosen and the eigenstates are calculated. The electron densities are recalculated, and the procedure is repeated until convergence is obtained. The final eigenstates are effective independent-particle states that can be filled with electrons. This differs from a many-body system, where the eigenstates cannot be described as a product of effective independent-particle states, and the details of the many-particle wavefunction are crucial for understanding the physics. This will be discussed in Chapter 4.

Problems

1 Show that the expression of the wavefunction for a nearly free electron model in Eq. (3.182) with wavevector $\mathbf{k}$ and band index n can be written as $\psi_{n\mathbf{k}}(\mathbf{r}) = e^{i\mathbf{k}\cdot\mathbf{r}}u_{n\mathbf{k}}(\mathbf{r})$, where $u_{n\mathbf{k}}(\mathbf{r})$ is periodic with $\mathbf{R}$. Using this result, derive Bloch's theorem

$$\psi_{n\mathbf{k}}(\mathbf{r}+\mathbf{R}) = e^{i\mathbf{k}\cdot\mathbf{R}}\psi_{n\mathbf{k}}(\mathbf{r}) \tag{3.193}$$

and explain its physical meaning. Demonstrate that Bloch's theorem also holds for tight-binding wavefunctions.

2 Let us consider the nearly free electron model in two dimensions for a square lattice with lattice constant a.

a) Derive an expression for the reciprocal lattice vectors and determine the first Brillouin zone.

b) Draw the free-electron dispersion following the path $(\frac{\pi}{a}, \frac{\pi}{a}) \rightarrow (0,0) \rightarrow (\frac{\pi}{a}, 0) \rightarrow (\frac{\pi}{a}, \frac{\pi}{a})$.

c) Calculate the density of states (the density of states indicates how the number of states changes with the energy).

d) Show qualitatively how the free-electron bands are affected by the presence of a periodic potential.

3.9 Scattering from Local Potential

Introduction. – In Sections 3.7 and 3.8, we considered some fundamental concepts in solids. Although our considerations were restricted to a one-dimensional system, it is clear that many of the fundamental concepts in solids are due to the geometry imposed by the crystal lattice. Conservation of momentum is intimately related to translational symmetry. The wavevector $\mathbf{k}$ is a good quantum number in free space. In a crystal, the periodic potential couples different plane waves to each other with reciprocal lattice vectors $\mathbf{K}$. This leaves only a small region in momentum space, the first Brillouin zone, as proper wavevectors, since all other wavevectors are connected to the first Brillouin zone via reciprocal lattice vectors. Therefore, in a solid, momentum it is still a good quantum number but only for a limited region in momentum space. In this section, we go one step further and consider a potential localized in space. In momentum space, this potential should be completely delocalized, and we therefore expect that momentum is no longer a good quantum number.

A potential localized in space can be written as

$$U(\mathbf{r}) = U\delta(\mathbf{r}). \tag{3.194}$$

Being entirely localized in real space, it is completely delocalized in momentum space. The Hamiltonian is then written as

$$\boldsymbol{H} = \sum_{\mathbf{k}} \varepsilon_{\mathbf{k}} \mathbf{c}_{\mathbf{k}}^{\dagger} \mathbf{c}_{\mathbf{k}} + U \mathbf{c}_{0}^{\dagger} \mathbf{c}_{0} = \sum_{\mathbf{k}} \varepsilon_{\mathbf{k}} \mathbf{c}_{\mathbf{k}}^{\dagger} \mathbf{c}_{\mathbf{k}} + \frac{U}{V} \sum_{\mathbf{k}\mathbf{k}'} \mathbf{c}_{\mathbf{k}'}^{\dagger} \mathbf{c}_{\mathbf{k}}, \tag{3.195}$$

where V is the volume. The energies $\varepsilon_{\mathbf{k}}$ can correspond to free-electron energies or the energies in a band in a solid. The local potential scatters between all possible momenta.

We want to study how the density of states changes in the presence of a local scattering potential using the Green's functions introduced in Section 2.12. Green's functions are particularly powerful for adding an interaction to the Hamiltonian. Depending on the problem, the interaction can be of arbitrary strength. This should be compared with regular perturbation theory, where the perturbation is assumed to be small. In particular, we are interested in the spectral function when placing an electron on the site of the impurity. This can be studied with

$$G_{00}(t) = -\frac{i}{\hbar}\langle 0|\mathbf{c}_0(t)\mathbf{c}_0^\dagger(0)|0\rangle, \quad \text{with} \quad \mathbf{c}_0^\dagger = \frac{1}{\sqrt{V}}\sum_{\mathbf{k}}\mathbf{c}_{\mathbf{k}}^\dagger. \tag{3.196}$$

Note that putting an electron at one particular site requires the summation over all momenta $\mathbf{k}$. The subscript 00 indicates that the propagation starts at $\mathbf{r} = 0$ and ends there as well. Note that there are no electrons in the initial system $|0\rangle$. In the absence of a potential, the local Green's function in momentum space is

$$G^0(E) = \frac{1}{V}\sum_{\mathbf{k}} G_{\mathbf{k}}^0(E) = \frac{1}{V}\sum_{\mathbf{k}} \frac{1}{E - \varepsilon_{\mathbf{k}} + i\eta}, \tag{3.197}$$

using the one-particle Green's function in Eq. (2.286). The superscript 0 indicates that this is the Green's function for free propagation. Therefore, $\mathbf{k}$ is a good quantum number. For the unperturbed Green's function, an electron created with momentum $\mathbf{k}$ stays in that state until it is removed. The imaginary part is directly related to the density of states in the absence of the local potential,

$$-\frac{1}{\pi}\ \mathrm{Im}\ G^0(E) = \frac{1}{V}\sum_{\mathbf{k}} \delta(E - \varepsilon_{\mathbf{k}}) = \rho(E), \tag{3.198}$$

see Eq. (2.287). Note that due to the excitation on one particular site via $\mathbf{c}_0^\dagger$, the density of states is normalized to 1. It is the density of states projected onto $\mathbf{r} = 0$. In this section, we consider a simple square density of states

$$-\frac{1}{\pi}\ \mathrm{Im}\ G^0(E) = \rho(E) = \begin{cases} \dfrac{1}{2W} & \text{for} -W \le E \le W \\ 0 & \text{otherwise,} \end{cases} \tag{3.199}$$

where the integral over the density of states has been normalized to 1. The density of states is shown in Figure 3.6 for $U = 0$.

A perturbation expansion of G_{00} can now be made,

$$G_{00}(E) = G^0(E) + G^0(E)UG_{00}(E). \tag{3.200}$$

The right-hand side can be interpreted as follows. The first term $G^0(E)$ represents the unperturbed motion from $\mathbf{r} = 0$ back to $\mathbf{r} = 0$, which means the electron just stays at the site with $\mathbf{r} = 0$. For the second term, the electron is initially at the local site as described by $G^0(E)$. The electron then scatters off the local potential U into all the other $\mathbf{k}$ states with $(1/\sqrt{V})\sum_{\mathbf{k}}\mathbf{c}_{\mathbf{k}}^\dagger = \mathbf{c}_0^\dagger$, see also Eq. (3.195). However, in the end, we want the electron

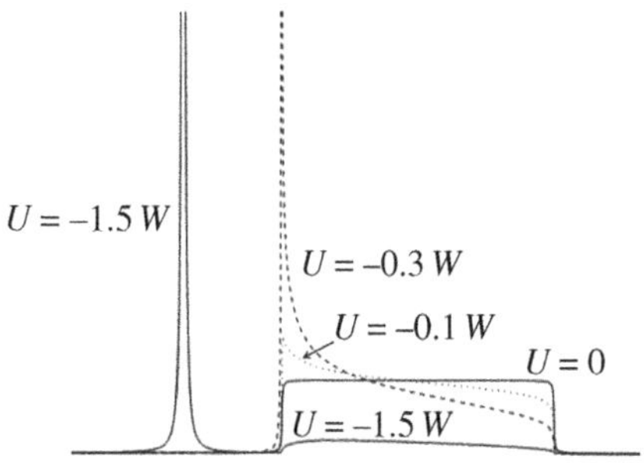

Figure 3.6 The local spectral density $-\frac{1}{\pi}$ Im $G_{00}(E)$ of a square density of states from $-W$ to W in the presence of a local attractive potential $U\delta(\mathbf{r})$. For plotting purposes, a small broadening has been added.

to return to the site at $\mathbf{r} = 0$. This motion is given by the propagator $G_{00}(E)$. This is a closed equation, and we can keep reinserting the Green's function

$$G_{00}(E) = G^0(E) + G^0(E)UG^0(E) + G^0(E)UG^0(E)UG^0(E) + \cdots \tag{3.201}$$

This demonstrates the effects of multiple scattering of the potential. For small U, we can break of the series. This is known as perturbation theory. However, this is not necessary since the problem can be solved exactly. The equation above can be turned into a geometric series

$$G_{00}(E) = G^0(E)\left(1 + UG^0(E) + UG^0(E)UG^0(E) + \cdots\right). \tag{3.202}$$

The series can be summed as

$$G_{00}(E) = \frac{G^0(E)}{1 - UG^0(E)} = \frac{1}{(G^0(E))^{-1} - U}, \tag{3.203}$$

using $\sum_{n=0}^{\infty} x^n = 1/(1-x)$. This result could also have been obtained directly from Eq. (3.200) by bringing $G_{00}(E)$ to the left-hand side.

The unperturbed Green's function in Eq. (3.197) can then be obtained via integration over the density of states in Eq. (3.199),

$$\begin{aligned} G^0(E) &= \int_{-\infty}^{\infty} dE' \frac{\rho(E')}{E - E' + i\eta} = \frac{1}{2W}\int_{-W}^{W} dE' \frac{1}{E - E' + i\eta} \\ &= \frac{1}{2W} \ln \frac{E + W + i\eta}{E - W + i\eta}. \end{aligned} \tag{3.204}$$

Although this can be split again into a real and imaginary part, most modern analytical mathematical software have no problem plotting the imaginary part of this function. Inserting this into Eq. (3.203) allows us to study the effect of a local potential. We take an attractive potential $U < 0$, although the results for $U > 0$ are simply mirrored in $E = 0$. The $U < 0$ case corresponds, for example to a local attractive impurity in a solid.

For $U = 0$, $-\frac{1}{\pi} \operatorname{Im} G_{00}(E) = -\frac{1}{\pi} \operatorname{Im} G^0(E)$ is equal to the unperturbed density of states, see Figure 3.6. Although the local Green's function is studied, the electron scatters equally into each eigenstate. When increasing the strength of the local potential to $U = -0.1W$ and $U = -0.3W$, a resonance is formed at the bottom of the band, see Figure 3.6. For $U < -W$, a bound state appears below the band. Since the potential is so strong, an electron starting at $\mathbf{r} = 0$ has a low probability of leaving the local bound state, although some weak spectral weight is still visible for $-W \leq E \leq W$. It is important to realize that the density of states has not disappeared. If the entire density of states were plotted, the bound state would not be visible, since it is a single state, and there are a macroscopic number of states (order 10^{20}-10^{23}) in the square band. For large U, the energy of the state is close to U (there are some higher order corrections), and the wavefunction is approximately given by $\mathbf{c}_0^\dagger$ in Eq. (3.196). Note that this is a completely symmetrical state: it is a simple summation over all $\mathbf{k}$ values. We will see the appearance of similar states when studying the Coulomb interaction in atoms or the strong interaction for quarks, where a singly degenerate symmetric state is pushed out from the other states (in that case, it is on the high-energy side).

Although the solution with Green's functions produces nice analytical results, it is important to realize that this problem can also be approached numerically. The density of states can be discretized into N states $|i\rangle$ with energies $\varepsilon_i = -W + 2W(i-1)/(N-1)$

for $i = 1, 2, \ldots, N$. The Hamiltonian can then be written in matrix form as

$$\boldsymbol{H} = \begin{pmatrix} \varepsilon_1 + \bar{U} & \bar{U} & \bar{U} & \cdot & \bar{U} & \bar{U} \\ \bar{U} & \varepsilon_2 + \bar{U} & \bar{U} & \cdot & \cdot & \bar{U} \\ \bar{U} & \bar{U} & \varepsilon_3 + \bar{U} & \bar{U} & \cdot & \cdot \\ \cdot & \cdot & \bar{U} & \varepsilon_4 + \bar{U} & \cdot & \cdot \\ \bar{U} & \cdot & \cdot & \cdot & \cdot & \bar{U} \\ \bar{U} & \bar{U} & \cdot & \cdot & \bar{U} & \varepsilon_N + \bar{U} \end{pmatrix}, \tag{3.205}$$

where the local potential $\bar{U} = U/N$ couples all the states $|i\rangle$. This Hamiltonian can be diagonalized numerically, leading to eigenstates $|n\rangle$ with eigenenergies E_n (note that different labels indicate different states). An electron on $\mathbf{r} = 0$ is created via $\mathbf{c}_0^\dagger|0\rangle = \sum_{i=1}^N |i\rangle/\sqrt{N}$, where $|0\rangle$ is the vacuum state. The spectral function $I(E)$ can then be calculated using Fermi's golden rule, see Eq. (2.283),

$$I(E) = \sum_{n=0}^{N} |\langle n|\mathbf{c}_0^\dagger|0\rangle|^2 \delta(E - E_n). \tag{3.206}$$

In order to produce a plot, the δ functions can be written as a Lorentzian, see Eq. (1.212), and a finite broadening η can be used. Alternatively, the δ function can be written as a normalized Gaussian, which reduces the tails in the plot. Although the density of states represents a macroscopic number of states, generally a couple of hundred discretizations is sufficient to produce agreement with the analytical result.

Conceptual summary. – In this section, the important problem of a particle moving in the presence of an additional potential has been described. Note that the interaction was treated to infinite order. The electron interacted with the local potential an arbitrary number of times. If the same problem was treated within the lowest-order perturbation theory, it would only interact once with the potential. However, this approach is not valid when U is large. For a strong potential, a state is separated from all the eigenstates. In the large U limit, this state consists of a linear combination of all the other states. This is also observed in other problems. For example, for many-particle states interacting via the Coulomb or strong interaction, a completely symmetric state is often pushed to high energy away from the other eigenstates.

We will return to the problem of particles moving in the constant presence of a potential in Section 5.3 when we study the generation mass by the Higgs field. However, while here the potential is local and therefore completely delocalized in momentum space, the Higgs field is completely delocalized in real space and therefore localized in momentum space.

Problems

1 Derive the Green's function in Eq. (3.203) for a square density of states in the limit $W \to 0$ and an attractive potential $U < 0$.

2 We want to study the overlap between the lowest states in the absence and presence of an impurity. Use the matrix in Eq. (4.103) to calculate numerically the overlap of the unperturbed states ($U = 0$) and the perturbed states ($U \neq 0$). Look at the center of the band and speculate what happens if these states were filled and a potential was suddenly switched on. This can happen in X-ray spectroscopy where a strongly bound electron can be suddenly removed.

3.10 Single State and a Band

Introduction. – In this section, we study another important example with Green's function techniques. The problem is that of a local state coupled to a continuum. This model, or variations thereof, is applicable to a wide variety of problems. One can think of a local impurity coupled to a band in a solid. It can also apply to an unstable particle or nucleus coupled to a continuum of states that it can decay into. In a somewhat modified form, it can also represent the coupling to a harmonic oscillator potential well. Exact solutions can be found for relatively simple problems.

Additionally, this problem is of interest to see what happens if part of the Hamiltonian is removed from the considerations. The effect of the continuum states then represents itself in terms of the lifetime of a particle or a shift in its energy.

We frame the problem in terms of a local impurity coupled to a band. This system is described by the Hamiltonian $\boldsymbol{H} = \boldsymbol{H}_0 + \boldsymbol{H}_1$ with

$$\boldsymbol{H}_0 = \varepsilon_0 \mathbf{c}_0^\dagger \mathbf{c}_0 + \sum_{\mathbf{k}} \varepsilon_{\mathbf{k}} \mathbf{c}_{\mathbf{k}}^\dagger \mathbf{c}_{\mathbf{k}} \quad \text{and} \quad \boldsymbol{H}_1 = \sum_{\mathbf{k}} V_{\mathbf{k}} (\mathbf{c}_{\mathbf{k}}^\dagger \mathbf{c}_0 + \mathbf{c}_0^\dagger \mathbf{c}_{\mathbf{k}}), \tag{3.207}$$

where $\boldsymbol{H}_1$ is considered the perturbation. The states 0 and $\mathbf{k}$ represent the local impurity and a state in the band, respectively. The combinations $\mathbf{c}_0^\dagger \mathbf{c}_0$ and $\mathbf{c}_{\mathbf{k}}^\dagger \mathbf{c}_{\mathbf{k}}$ are the number operators for the particles, in analogy to Eq. (1.328). They are 1 and 0, depending on whether the state contains a particle or not, respectively. $\boldsymbol{H}_0$ defines the energies of the local impurity and the states in the band. The operator $\mathbf{c}_{\mathbf{k}}^\dagger \mathbf{c}_0$ removes a particle from site 0 and creates a particle in the band with momentum $\mathbf{k}$; $\mathbf{c}_0^\dagger \mathbf{c}_{\mathbf{k}}$ does the opposite. $\boldsymbol{H}_1$ then describes the coupling between the local state and the band.

Again, we would like to study the propagation of a particle from site 0 back to site 0. This corresponds to the spectral function of adding a particle to site 0. A perturbation expansion can be made to the Green's function $G_{00}(E)$

$$G_{00}(E) = G_0^0(E) + G^0(E) \sum_{\mathbf{k}} V_{\mathbf{k}} G_{\mathbf{k}0}(E), \tag{3.208}$$

where $V_{\mathbf{k}} = \langle \mathbf{k} | \boldsymbol{H}_1 | 0 \rangle$ is the matrix element between states $|0\rangle$ and $|\mathbf{k}\rangle$. There are two options. First, the particle can propagate unperturbed from 0 to 0 via the propagator

$$G_0^0(E) = \frac{1}{E - \varepsilon_0 + i\eta}, \tag{3.209}$$

where the subscript indicates that this is the Green's function for the local site, and the 0 in the superscript indicates unperturbed motion. Since the motion is free, the particle remains in the same state. This means that the particle never leaves the impurity. Second, reading from left to right, the particle can start on 0, then hop into the band state with momentum $\mathbf{k}$ via $V_{\mathbf{k}}$, and then find a way to propagate back to 0 via $G_{\mathbf{k}0}$. Note that $G_{\mathbf{k}0}$

can still contain many different hoppings between 0 and different band states. It is just as complicated as $G_{00}(E)$. However, let us expand it

$$G_{\mathbf{k}0}(E) = G^0_{\mathbf{k}}(E)V_{\mathbf{k}}G_{00}(E). \tag{3.210}$$

First, note that there is no unperturbed way to go from $\mathbf{k}$ to 0. This propagation always involves $\boldsymbol{H}_1$. This is given in the second term, where the particle initially propagates freely in the band state with

$$G^0_{\mathbf{k}}(E) = \frac{1}{E - \varepsilon_{\mathbf{k}} + i\eta}, \tag{3.211}$$

where the 0 in the superscript indicates unperturbed motion. There is no scattering in the Hamiltonian between different $\mathbf{k}$ values, as was, for example discussed in Section 3.8. In fact, this would make the problem much harder to solve. Note that there is no summation over $\mathbf{k}$ since the particle starts in state $\mathbf{k}$. The particle then scatters with $V_{\mathbf{k}}$. The particle is now back on site 0 and can start the process all over again, as indicated by the full propagator $G_{00}(E)$. Although it appears that no progress has been made, let us insert $G_{\mathbf{k}0}(E)$ back into Eq. (3.208). This gives

$$G_{00}(E) = G^0_0(E) + G^0_0(E)\sum_{\mathbf{k}} V_{\mathbf{k}}G^0_{\mathbf{k}}(E)V_{\mathbf{k}}G_{00}(E). \tag{3.212}$$

The only full propagator that is left is now $G_{00}(E)$, and we can solve for this Green's function, giving

$$G_{00}(E) = \frac{G^0_0(E)}{1 - G^0_0(E)\sum_{\mathbf{k}} V_{\mathbf{k}}G^0_{\mathbf{k}}(E)V_{\mathbf{k}}} = \frac{1}{(G^0_0(E))^{-1} - \sum_{\mathbf{k}} V_{\mathbf{k}}G^0_{\mathbf{k}}(E)V_{\mathbf{k}}}.$$

Inserting the free propagators gives

$$G_{00}(E) = \frac{1}{E - \varepsilon_0 + i\eta - \sum_{\mathbf{k}} \dfrac{V^2_{\mathbf{k}}}{E - \varepsilon_{\mathbf{k}} + i\eta}}. \tag{3.213}$$

To understand what this Green's function means, let us simplify the problem somewhat. We take $\varepsilon_0 = 0$ and use the square density of states from Section 3.9. By taking the hopping matrix element constant, $V_{\mathbf{k}} = V$, the integral of the unperturbed Green's function for the band $G^0(E)$ in Eq. (3.204) can be used again. The Green's function is then

$$G_{00}(E) = \frac{1}{E + i\eta - V^2G^0(E)}. \tag{3.214}$$

The imaginary part of $G_{00}(E)$, *i.e.* the spectral function, is shown in Figure 3.7 for several values of V. The spectral function looks like a simple Lorentzian, apart from the fact that the edges of the band are becoming apparent for large V. Note that the width of the Lorentzian is much larger than the finite value taken for η for plotting purposes. The broadening can be understood as follows. The Green's function $G^0(E)$ contains a real and an imaginary part. The real part causes an energy shift. However, since the local state was placed at the center of the band, there are just as many band states pushing it up or down in energy. Therefore, there is no shift in energy. This only leaves the imaginary part of $G^0(E)$. However, this is directly related to the density of states of the band, see

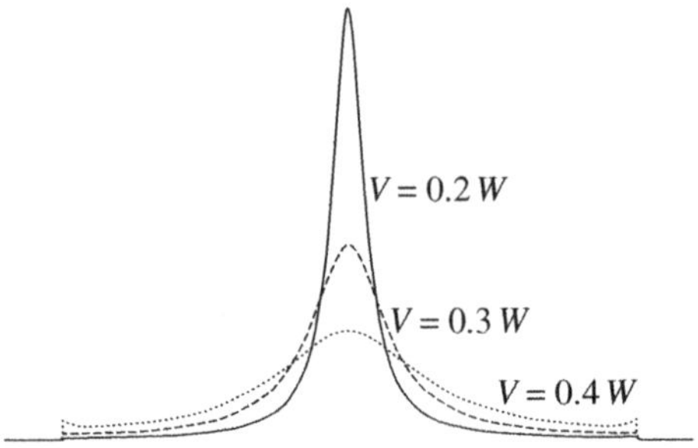

Figure 3.7 The local spectral density $-\frac{1}{\pi}\,\text{Im}\,G_{00}(E)$ of a state at $E = 0$ coupled to a square density of states from $-W$ to W for different values of the coupling constant V (indicated in the figure). The edges of the band become visible for large V.

Eq. (3.198). The whole effect of $G^0(E)$ can then be effectively replaced by a broadening parameter

$$\Gamma \cong -V^2\,\text{Im}\,G^0(\varepsilon_0) = \pi V^2 \rho(\varepsilon_0) = \frac{\pi V^2}{2W}, \tag{3.215}$$

where $\rho = 1/2W$ for the square density of states. In general, the density of states is not as smooth as the one used in this example. However, if V is not too large, one can assume that the density of states $\rho(\varepsilon_0)$ around $E = \varepsilon_0$ is approximately constant. Equation (3.214) then becomes

$$G_{00}(E) = \frac{1}{E + i\Gamma} \quad \Rightarrow \quad -\frac{1}{\pi}\,\text{Im}\,G_{00}(E) = \frac{\Gamma}{\pi}\frac{1}{E^2 + \Gamma^2}. \tag{3.216}$$

The spectral function of a state coupled to a continuum is therefore a Lorentzian with half-width-half-maximum of Γ. The integrated intensity equals 1.

The Green's function can also be transformed back into the time domain. Following Eq. (2.285), we find

$$G_{00}(t) = -\frac{i}{\hbar} e^{-\frac{i}{\hbar}\varepsilon_0 t - \frac{\Gamma}{\hbar} t}\theta(t). \tag{3.217}$$

This is a decaying exponential, and the state 0 has now a finite lifetime. Therefore, the complex quantity η introduced for convergence of the integral now takes on a physical meaning. For the initial Hamiltonian in Eq. (3.207), the particle had an infinite lifetime. However, although the particle starts out at site 0, it dephases exponentially into the continuum. The decay time increases for a large density of states and a larger coupling, see Eq. (3.215). Therefore, broadening of features in experiments (*i.e.* in addition to experimental broadenings) provides information on the stability of the state. A large broadening implies that the state is unstable. As we will see, in particle physics, the lifetime is often related to the mass of the particle.

The propagation becomes decaying because the continuum states were projected out of the problem. This does not mean that the particle is destroyed. It simply scattered in a continuum of different states and propagates in those states. The exponential decay implies that it never returns to site 0. However, this is in principle an approximation. It is correct if the coupling occurs with a density of states with a very small level splitting. However, if the level splitting is large, the particle returns to the original state. Let us look at this in more detail. Instead of having a dense continuum, the coupling of a single state at energy $\varepsilon_0 = 0$ to a set of states is considered. The coupling strength is taken as $v = V/\sqrt{N}$. The N states are equally spaced from $-W$ to W, giving an energy $\varepsilon_n = -W + 2W(n-1)/(N-1)$ for $n = 1, 2, \ldots, N$. For the limiting case, $n = 1$, we take

$\varepsilon_1 = 0$. In general, the Hamiltonian in matrix form can be written as

$$\boldsymbol{H} = \begin{pmatrix} 0 & v & v & v & \cdots & v \\ v & \varepsilon_1 & 0 & 0 & & \\ v & 0 & \varepsilon_2 & 0 & & \cdot \\ v & 0 & 0 & \varepsilon_3 & & \cdot \\ \cdot & & & & \cdot & \cdot \\ v & 0 & 0 & 0 & \cdots & \varepsilon_N \end{pmatrix}. \tag{3.218}$$

The matrix can be diagonalized numerically, giving $N+1$ eigenstates $|E_n\rangle$ with energies $E_n = \hbar\omega_n$. Let us take as the initial state $|0\rangle$ with energy $\varepsilon_0 = 0$. The wavefunction at a particular time t is then given by

$$|\psi(t)\rangle = \sum_{n=1}^{N+1} e^{-i\omega_n t}|E_n\rangle\langle E_n|0\rangle. \tag{3.219}$$

We now want to track the occupation of the state zero. This can be done using the operator $\boldsymbol{n}_0 = |0\rangle\langle 0|$. This gives

$$P_0(t) = \langle\psi(t)|\boldsymbol{n}_0|\psi(t)\rangle = \left|\sum_{n=1}^{N+1} e^{-i\omega_n t}|\langle E_n|0\rangle|^2\right|^2. \tag{3.220}$$

Let us first look at $N = 1$. This is a two-level system with two states at energy zero (the expression for ε_n does not work for $N = 1$, so we take $\varepsilon_1 = 0$). This can be easily solved to give eigenvalues $E_{1,2} \pm V$ and eigenstates $|E_{1,2}\rangle = (|0\rangle \mp |1\rangle)/\sqrt{2}$ if $V > 0$. The time dependence then simplifies to $\cos 2\pi t/T$, where the oscillation period is given by $T_0 = 2\pi\hbar/(E_2 - E_1) = \pi\hbar/V$. The two-level system therefore gives rise to a sinusoidal behavior, see Figure 3.8, known as Rabi oscillations. For $N = 3$, the level couples to states at $-W, 0, W$ (skipping the even N that have no state at zero energy). There is an increase in the oscillation period. However, the $N = 3$ is not really different from $N = 1$.

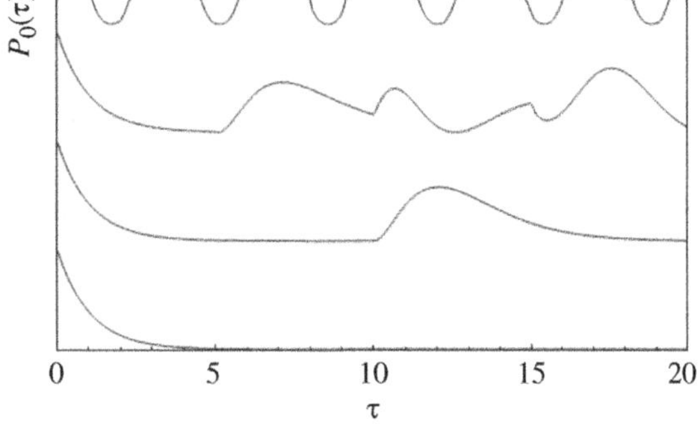

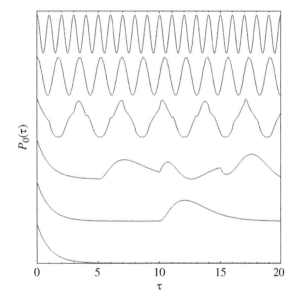

Figure 3.8 Time dependence of the occupation of an initial state coupled to a set of N equally spaced states in a band from $-W$ to W. The coupling constant is $V/\sqrt{N}$, with $V = 0.1\ W$. The number of states is $N = 1, 3, 11, 51, 101, 201$ from top to bottom. All the probability curves are between 0 and 1. The normalized time is given by $\tau = t/T_0$, with $T_0 = \pi\hbar/V$ the oscillation period for $N = 1$.

The states at $-W$ and W couple only weakly to $|0\rangle$. Therefore, effectively, this is still close to a two-level system except that the coupling has decreased to $V/\sqrt{3}$. For $N = 11$, the behavior still looks periodic with a further decrease in oscillation period, but higher order terms start to play a role.

For $N > 50$, the initial time dependence takes on the exponential decay from Eq. (3.217) expected for the coupling of a state to a continuum. However, a recurrence is still observed. In this limit, the problem starts to look more like that of a displaced harmonic oscillator, see Section 3.2. The discrete states effectively act like the levels of a quantum harmonic oscillator with energies $n\hbar\omega_0$ with the oscillation energy $\hbar\omega_0 = 2W/N$ given by the separation between the levels. Starting at the state $|0\rangle$, the particle effectively falls into the harmonic oscillator well, swings through the well, and returns to its original point after an oscillation period of $T = 2\pi/\omega_0 = \pi\hbar N/W$. Note that the period increases linearly with N, so the recurrence time increases as the continuum becomes denser. In the reduced units of the plot in Figure 3.8, the expected occurrence is at $\tau = T/T_0 = VN/W = 0.1N = 5.1, 10.1, 20.1$ for $N = 51, 101, 201$ for the used value of $V/W = 0.1$. Note that the oscillation period was initially proportional to $T \sim \sqrt{N}/V$ for few states. The recurrence period becomes proportional to $T \sim N/W$ when states are dense, and the coupling now determines how fast the system exponentially delocalizes into the continuum, see Eq. (3.215).

Let us now consider the case when the state lies outside the density of states, see Fig 3.9. Since the imaginary part of $G_{00}(E)$ is plotted, almost all the spectral weight is in the single state, and the density of states is barely visible. The imaginary part in the denominator of $G_{00}(E)$ is very small, and there is almost no lifetime broadening (the small width of the single state is due to a finite η for plotting purposes). However, there is now a substantial change in energy. Focusing on the single state, the problem can also be simplified by replacing the density of states by an effective level in Eq. (3.213),

$$G_{00}(E) = \frac{1}{E - \varepsilon_0 + i\eta - \dfrac{V^2}{E - \Delta_{\text{eff}} + i\eta}}. \tag{3.221}$$

The poles of this Green's function occur at the eigenenergies E_n, with $n = 1, 2$ of the 2×2 matrix

$$\boldsymbol{H}_{\text{eff}} = \begin{pmatrix} \varepsilon_0 & V \\ V & \Delta_{\text{eff}} \end{pmatrix}. \tag{3.222}$$

The weights are given by $|\langle E_n|0\rangle|^2$. In the limit, $\Delta_{\text{eff}} \gg V$, the change in energy becomes V^2/Δ_{eff}. Therefore, even though the particle couples to an entire continuum of states, the net effect is a shift in the energy of the particle.

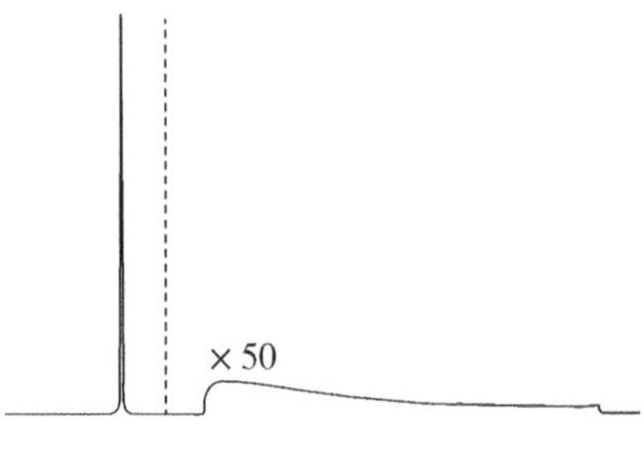

Figure 3.9 The coupling of a single state to a rectangular density of states $\rho = 1/2\ W$ from $-W$ to W. The state is located at $\varepsilon_0 = -1.2W$ (indicated by the dashed line), and the coupling is $V = 0.5W$. The intensity of the band states is multiplied by 50.

Conceptual summary. – The lineshape of a single state coupled to a continuum corresponds to a Lorentzian with a width proportional to the coupling constant to the continuum and the number of states in the continuum at energies close to the level of the single state. When the continuum states are effectively removed from the problem, this becomes comparable to a state with a finite lifetime. This is what is often observed when looking at highly excited states. These states often show a very large broadening since they couple strongly to a continuum of other eigenstates, and they decay therefore quickly into other states. The situation becomes more complex when the continuum is not dense, and recurrences can be observed. This is comparable to a particle oscillating in a potential well with a period $T = 2\pi/\omega_0$, where $\hbar\omega_0$ is the separation between the states. When the level is not in the continuum, the presence of the continuum can be represented by a shift in energy.

Although this problem is relatively easy, it is important when discussing particle physics. We prefer to visualize a particle as moving independently in space. However, a particle is continuously subjected to all kinds of interactions, represented by the continuum here. When focusing solely on the behavior of the particle, these interactions manifest themselves as a finite lifetime of the particle and shifts in energy. Now since $E = mc^2$, the latter can also be represented in terms of the mass of a particle. Therefore, lifetime and mass are often manifestations of other interactions that we are treating in a very simplified way. Compare this with the change in the speed of light inside a solid. This is just the most basic effective model to account for the very complicated interactions of the electromagnetic radiation with the charges in the solid.

Problems

1 The Green's function techniques in Sections 3.9 and 3.10 can be applied to many problems and are generally preferable to typical perturbation theory approaches. Let us consider a three-level system given by the Hamiltonian

$$\boldsymbol{H} = \sum_{i=1}^{3} \varepsilon_i \mathbf{c}_i^\dagger \mathbf{c}_i + V_{12}(\mathbf{c}_2^\dagger \mathbf{c}_1 + \mathbf{c}_1^\dagger \mathbf{c}_2) + V_{23}(\mathbf{c}_3^\dagger \mathbf{c}_2 + \mathbf{c}_2^\dagger \mathbf{c}_3). \tag{3.223}$$

Determine the Green's function $G_{11}(E)$. Plot the imaginary part of G_{11} for $\varepsilon_1 = 0$, $\varepsilon_2 = 1$, and $\varepsilon_3 = 2$ and $V_{12} = 1$, $V_{23} = 2$. Obtain the same results via exact diagonalization.

4

Many-Particle Systems

Introduction

Our attention so far has focused on obtaining the function spaces for a variety of different systems. The goal was to understand the behavior of a single particle in a potential. Generally, one has to deal with systems consisting of many particles that interact with each other. In this case, one has to distinguish two broad classes: independent-particle and many-body systems.

When a system can be treated within the independent-particle limit, one can first calculate all the energy levels and then fill them with particles. This does not mean that the interactions between the particles are ignored, but they can be treated as an effective potential. For an atomic problem, this would imply that an electron in a certain orbital feels a different effective nucleus. For example, the electron in the 1s orbital is close to the nucleus, and the effective charge of the nucleus is close to the real charge. For orbitals, with increasing nl, the nucleus becomes increasingly screened by the other electrons, and the effective nuclear charge decreases. In the end, we are left with effective atomic levels that are filled with electrons.

For many-body systems, the above procedure no longer works. The interactions between the particles additionally split the many-particle energy levels, and the details of the wavefunction formed by the particles become important. Although this should in principle be the way to treat many-particle systems, the number of possible states increases exponentially with the number of particles. Therefore, these treatments are only feasible for systems with a limited number of particles.

To better understand how to deal with systems containing many particles, the chapter is set up as follows.

4.1 In order to construct many-particle wavefunctions, we have to distinguish different types of particles. For one class, the particles behave similar to (higher order) vectors: they anticommute when they are interchanged. The other class behaves similarly to the commuting quantities used to build functions, such as indistinguishable spinors and step operators. The particles are called fermions and bosons, respectively. The many-body wavefunctions associated with these types of particles differ as well.

4.2 Fermions and bosons also have different statistics. Fermions have to obey Pauli's principle, which states that they cannot occupy the same quantum state. Bosons do not have this restriction. This also strongly affects their behavior at different temperatures.

Geometric Quantum Mechanics, First Edition. Michel van Veenendaal.

4.3 This section studies how the Coulomb interaction between electrons affects the lowest independent-particle state in a system of free electrons. The end result is somewhat unphysical, indicating that the interaction cannot be treated as an effective potential, and that the electron occupations need to be adjusted to minimize the mutual interactions.

4.4 In order to see the splitting of the energies of many-particle states, it is convenient to express the interactions in the appropriate symmetry in terms of tensors. This section extends the ideas used for angular momentum, a tensor of rank 1 (a vector), to higher ranks. These tensors are needed to describe the Coulomb and strong interactions.

4.5 The ideas of tensors are used to describe the Coulomb interactions between electrons on an atom. It is shown that the interaction splits the many-particle states into distinct groups of eigenstates, known as Coulomb multiplets.

4.6 Hadrons are composite particles consisting of several quarks that are held together by the strong interaction. The strong interaction between the quarks splits the many-particle states into distinct groups. This section considers mesons that consist of a quark and an antiquark. This problem has strong similarities to the Coulomb interaction between electrons on an atom, but the symmetry of the problem is different.

4.7 The ideas from the preceding section are extended to the more complicated baryons that consist of three quarks. The best known baryons are the proton and the neutron.

4.8 The proton and the neutron bind together under the residual strong force to form the nucleus of atoms. This problem is treated within the limit of the nuclear shell model that treats the protons and neutrons as independent particles in an effective three-dimensional quantum harmonic oscillator well. Adding the spin–orbit interaction allows us to explain the increased stability of the nucleus for certain atomic numbers.

4.1 Wavefunctions for Many-Body Systems

Introduction. – Up to this point, we have found two distinct ways to change the order of unit vectors and their higher order composites. To incorporate the properties of space, unit vectors in three dimensions had to anticommute with each other, see Eq. (1.102). The anticommutation also holds in spacetime, although the metric changed, see Eq. (2.15). Although unit vectors are sufficient to locate a point in space, they are insufficient to describe each point in space independently. This required the construction of a function space. For free particles or a relatively simple potential, the function spaces could be constructed with commuting quantities, see Eq. (1.340), such as indistinguishable spinors or step operators. Increasing the number of commuting operators can be viewed as changing the amplitude of the vector. The best known example is the quantum harmonic oscillator, see Section 3.1, where $|n\rangle$ represents a different oscillation amplitude. For the spherical harmonics, see Section 1.15, increasing the number of indistinguishable spinors increases the angular momentum. Amplitude should be thought of in the most general terms, *i.e.* it can be an increase in the relevant quantum numbers or, for systems with a very low symmetry, the energy (which is always a good quantum number in a nonrelativistic time-invariant system).

The same distinction occurs when combining particles into, what is known as, a many-body wavefunction. One type of particles, such as protons, neutrons, and electrons, behave like anticommuting vectors. These are known as fermions. The other type of particles commute with each other, and adding a particle to a many-body state appears like an effective increase in the amplitude of the function. These particles are known as bosons. All known force carriers, such as photons and gluons, behave like bosons. For example, adding more photons to a many-photon state effectively increases the amplitude of the electromagnetic wave. Although a change in sign when changing the order of the particles might appear minor, it actually has dramatic consequences on the properties of the many-body system.

Let us start with a simple example of two particles in two distinct levels indicated by the generic quantum numbers k and k' (k can be the wavevector $\mathbf{k}$ for free particles, n for the quantum harmonic oscillator, nlm for the atomic orbitals, etc.). This gives a product wavefunction $\varphi_k(\mathbf{r}_1)\varphi_{k'}(\mathbf{r}_2)$. However, the particles are now not indistinguishable, since particles 1 and 2 are in eigenstates k and k', respectively. This can be solved by exchanging the particles given $\varphi_{k'}(\mathbf{r}_1)\varphi_k(\mathbf{r}_2)$. There are now two ways of combining the two permutations

$$\Phi^{\pm}_{kk'}(\mathbf{r}_1, \mathbf{r}_2) = \frac{1}{\sqrt{2}}(\varphi_k(\mathbf{r}_1)\varphi_{k'}(\mathbf{r}_2) \pm \varphi_{k'}(\mathbf{r}_1)\varphi_k(\mathbf{r}_2)). \tag{4.1}$$

The interchange of the particles produces a sign change $\Phi^{\pm}_{kk'}(\mathbf{r}_2, \mathbf{r}_1) = \pm\Phi^{\pm}_{kk'}(\mathbf{r}_1, \mathbf{r}_2)$, so we have symmetric (+) and antisymmetric (−) two-particle wavefunctions. This corresponds to a wavefunction for bosons and fermions, respectively. This sign change might appear as a minor difference. However, note what happens when we try to put two fermions in the same quantum state: $\Phi^{-}_{kk}(\mathbf{r}_1, \mathbf{r}_2) = 0$. The wavefunction is zero, and therefore, two fermions cannot occupy the same quantum state. This is known as Pauli's principle and follows directly from the anticommutative properties of the fermions.

Let us now jump to the statistics of particles. The behavior of the particles can therefore be distinguished by their commutation properties. However, as mentioned above, these same properties were found when studying the geometry of space and spacetime. In a many-body wavefunction, the fermions appear to behave as if they were different vectors. Bosons behave as if they are changing the effective amplitude of the many-body wavefunction. This behavior can be captured in the commutation relations. Let us start with bosons that satisfy the commutation relations

$$[\boldsymbol{a}_k, \boldsymbol{a}^\dagger_{k'}] = \delta_{kk'}, \quad [\boldsymbol{a}^\dagger_k, \boldsymbol{a}^\dagger_{k'}] = 0, \quad [\boldsymbol{a}_k, \boldsymbol{a}_{k'}] = 0. \tag{4.2}$$

In the context of particles, $\boldsymbol{a}^\dagger_k$ and $\boldsymbol{a}_k$ are called creation and annihilation operators that add or remove a boson from the quantum state labeled k, respectively. However, note that the distinction between a particle and the change of the amplitude of a field can become rather blurred.

For bosons, the quantum states k_i can be occupied by more than one particle at a time. For each quantum state k_i, the wavefunction is effectively that of a quantum harmonic oscillator filled with n_{k_i} bosons

$$|n_{k_1} \cdots n_{k_N}\rangle = \prod_{i=1}^{N} |n_{k_i}\rangle = \prod_{i=1}^{N} \frac{1}{\sqrt{n_{k_i}!}} (\boldsymbol{a}^\dagger_{k_i})^{n_{k_i}} |0\rangle, \tag{4.3}$$

where N is the total number of quantum states. The product wavefunction can also be expressed in real space by taking the inner product with the real space product wavefunction

$$\langle \mathbf{r}^N| = \frac{1}{\sqrt{N!}}\langle \mathbf{r}_1 \ \cdots \ \mathbf{r}_N|. \tag{4.4}$$

Since particles are indistinguishable, a $1/\sqrt{N!}$ normalization factor appears. However, different indices are attached to $\langle \mathbf{r}_i|$ to allow us to distinguish the different wavefunctions. Permutations over all the possible position vectors need to be made in order to make the particles indistinguishable. Taking the inner product with Eq. (4.4) gives

$$\begin{aligned}\Phi_{n_{k_1}\cdots n_{k_N}}(\mathbf{r}_1 \ \cdots \ \mathbf{r}_N) &= \langle \mathbf{r}^N | n_{k_1} \ \cdots \ n_{k_N}\rangle \\ &= \frac{\prod_{i=1}^{N}\sqrt{n_{k_i}!}}{\sqrt{N!}}\sum_{P_{k_i}}\varphi_{k_{i_1}}(\mathbf{r}_1)\cdots \ \varphi_{k_{i_N}}(\mathbf{r}_N).\end{aligned} \tag{4.5}$$

The summation P_{k_i} now goes over all the permutations of different k values. For example, for $N = 2$, there are two permutations when the k values are different,

$$\Phi_{1_k 1_{k'}}(\mathbf{r}_1, \mathbf{r}_2) = \frac{1}{\sqrt{2}}(\varphi_k(\mathbf{r}_1)\varphi_{k'}(\mathbf{r}_2) + \varphi_{k'}(\mathbf{r}_1)\varphi_k(\mathbf{r}_2)), \tag{4.6}$$

but only one permutation when the quantum numbers are equal,

$$\Phi_{2_k}(\mathbf{r}_1, \mathbf{r}_2) = \frac{\sqrt{2!}}{\sqrt{2!}}\varphi_k(\mathbf{r}_1)\varphi_k(\mathbf{r}_2) = \varphi_k(\mathbf{r}_1)\varphi_k(\mathbf{r}_2). \tag{4.7}$$

Example 4.1 ***The Photon Field***

So far, we have seen that adding and removing photons from the field is comparable to a harmonic oscillator, with the understanding that each mode $\mathbf{q}\alpha$ corresponds to a different oscillator. Although the photon is an elementary particle, the mathematics is very similar to that of the harmonic oscillator. The vector potential is the equivalent of the displacement $\boldsymbol{x}$ for the quantum harmonic oscillator, see Eq. (3.41),

$$\boldsymbol{A}_{\mathbf{q}\alpha} = \boldsymbol{\epsilon}_{\mathbf{q}\alpha}\sqrt{\frac{\hbar}{2\varepsilon_0\omega_{\mathbf{q}}}}(\langle \mathbf{q}\alpha| + |\mathbf{q}\alpha\rangle) = \boldsymbol{\epsilon}_{\mathbf{q}\alpha}\sqrt{\frac{\hbar}{2\varepsilon_0\omega_{\mathbf{q}}}}(\boldsymbol{a}_{\mathbf{q}\alpha} + \boldsymbol{a}_{\mathbf{q}\alpha}^{\dagger}). \tag{4.8}$$

The unit vectors for a particular wavevector $\mathbf{q}$ and polarization α are given by

$$|n_{\mathbf{q}\alpha}\rangle = \frac{1}{\sqrt{n_{\mathbf{q}\alpha}!}}|\mathbf{q}\alpha\rangle|\mathbf{q}\alpha\rangle\cdots|\mathbf{q}\alpha\rangle = \frac{1}{\sqrt{n_{\mathbf{q}\alpha}!}}(\boldsymbol{a}_{\mathbf{q}\alpha}^{\dagger})^{n_{\mathbf{q}\alpha}}|0\rangle, \tag{4.9}$$

where there are $n_{\mathbf{q}\alpha}$ kets of the type $|\mathbf{q}\alpha\rangle$. We have already seen in Eq. (2.161) that the vacuum permittivity ε_0 plays the role of the mass. There are some additional differences. Since the oscillation occurs in three dimensions, we need a polarization vector $\boldsymbol{\epsilon}_{\mathbf{q}\alpha}$ to indicate the direction of the oscillation. Note that this directional component also needs to be present to describe a one-dimensional quantum harmonic oscillator in three dimensions. Since the oscillation lies in a plane perpendicular to the wavevector $\mathbf{q}$, there are two polarization directions indicated by α. Each oscillation mode is associated with a different oscillator.

In Section 2.12, we looked at the free propagator for fermionic particles. We want to do the same for the propagation of the vector potential. In analogy to Eq. (2.278), the Green's function is given by

$$\begin{aligned} G^0_{\mathbf{q}\alpha}(t) &= -\frac{i}{\hbar}\langle 0|\boldsymbol{A}_{\mathbf{q}\alpha}(t)\boldsymbol{A}_{\mathbf{q}\alpha}(0)|0\rangle \\ &= -\frac{i}{\hbar}\frac{\hbar}{2\varepsilon_0\omega_{\mathbf{q}}}\langle 0|[\boldsymbol{a}^\dagger_{\mathbf{q}\alpha}(t)+\boldsymbol{a}_{\mathbf{q}\alpha}(t)][\boldsymbol{a}^\dagger_{\mathbf{q}\alpha}(0)+\boldsymbol{a}_{\mathbf{q}\alpha}(0)]|0\rangle. \end{aligned} \tag{4.10}$$

However, this expression is somewhat problematic. This expression contains four terms. However, since the expectation value is in the vacuum state, the terms $\boldsymbol{a}^\dagger_{\mathbf{q}\alpha}\boldsymbol{a}^\dagger_{\mathbf{q}\alpha}$ and $\boldsymbol{a}_{\mathbf{q}\alpha}\boldsymbol{a}_{\mathbf{q}\alpha}$ are zero. The product $\boldsymbol{a}_{\mathbf{q}\alpha}(t)\boldsymbol{a}^\dagger_{\mathbf{q}\alpha}(0)$ creates a photon at $t=0$ and removes it at a time t. This is a proper process as long as $t>0$. The term $\boldsymbol{a}^\dagger_{\mathbf{q}\alpha}(t)\boldsymbol{a}_{\mathbf{q}\alpha}(0)$ is unphysical since it removes a photon before creating it. This can be fixed by changing the order giving $\boldsymbol{a}_{\mathbf{q}\alpha}(0)\boldsymbol{a}^\dagger_{\mathbf{q}\alpha}(t)$. However, this requires that for this term $t<0$ since the photon needs to be created before it is removed. Inserting the appropriate step functions for the time gives

$$G^0_{\mathbf{q}\alpha}(t) = -\frac{i}{2\varepsilon_0\omega_{\mathbf{q}}}\langle 0|[\boldsymbol{a}_{\mathbf{q}\alpha}(t)\boldsymbol{a}^\dagger_{\mathbf{q}\alpha}(0)\theta(t)+\boldsymbol{a}_{\mathbf{q}\alpha}(0)\boldsymbol{a}^\dagger_{\mathbf{q}\alpha}(t)\theta(-t)]|0\rangle.$$

The photons are now effectively traveling in both directions between the particles, making the interaction equivalent for both charges. Following the discussion above Eq. (2.280), it is straightforward to calculate the Fourier transform

$$G^0_{\mathbf{q}}(\omega) = \frac{1}{2\varepsilon_0\omega_{\mathbf{q}}}\frac{1}{\omega-\omega_{\mathbf{q}}+i\eta} - \frac{1}{2\varepsilon_0\omega_{\mathbf{q}}}\frac{1}{\omega+\omega_{\mathbf{q}}-i\eta}, \tag{4.11}$$

and we drop the subscript α, since the final result is independent of the polarization. The total Green's function then becomes

$$G^0_{\mathbf{q}}(\omega) = \frac{1}{\varepsilon_0(\omega^2-\omega^2_{\mathbf{q}}+i\eta)}. \tag{4.12}$$

This is the same propagator as was found earlier in Eq. (2.167). This Green's function has poles at $\omega=\pm\omega_{\mathbf{q}}$. The Green's function is the same as that needed to describe a classically driven harmonic oscillator.

In the preceding discussion, we have demonstrated that the vector potential can be effectively viewed as a quantum harmonic oscillator. The creation and annihilation operators for the photons, the force carriers, take the role as step operators. Equation (4.8) expresses the vector potential in function space. However, the electric and magnetic fields are related to the vector potential by derivatives in space and time, see Eq. (2.139). So, we also want to express the vector potential in real space. A field at a point in space can be expanded into plane waves as

$$\langle \mathbf{r},t| = \sum_\alpha \int d\mathbf{q}\,\langle \mathbf{r},t|\underline{\mathbf{q}\alpha}\rangle\langle\underline{\mathbf{q}\alpha}| = \sum_\alpha \int d\mathbf{q}\,\varphi_{\mathbf{q}}(\mathbf{r},t)\boldsymbol{d}_{\mathbf{q}\alpha}, \tag{4.13}$$

where the underline in $|\underline{\mathbf{q}\alpha}\rangle$ indicates that we are dealing with a displacement operator

$$|\underline{\mathbf{q}\alpha}\rangle = \boldsymbol{d}^\dagger_{\mathbf{q}\alpha} = \boldsymbol{\epsilon}_{\mathbf{q}\alpha}\sqrt{\frac{\hbar}{\varepsilon_0\omega_{\mathbf{q}}}}\boldsymbol{a}^\dagger_{\mathbf{q}\alpha} \quad \Rightarrow \quad \boldsymbol{A}_{\mathbf{q}\alpha} = \frac{1}{\sqrt{2}}(\boldsymbol{d}_{\mathbf{q}\alpha}+\boldsymbol{d}^\dagger_{\mathbf{q}\alpha}), \tag{4.14}$$

using the vector potential from Eq. (4.8). This is similar to the introduction of dimensionless coordinates for the quantum harmonic oscillator, see Section 3.1. So far, we have mainly dealt with scalar fields, which means that each point in space is associated with a number. However, for vector fields, each point in space is associated with a vector, indicating the direction in which the field is oscillating. Therefore, the displacement operator is a vector. In free space, the field is expressed in plane waves

$$\varphi_{\mathbf{q}}(\mathbf{r},t) = \langle \mathbf{r}, t|\underline{\mathbf{q}\alpha}\rangle = \frac{1}{(2\pi)^{\frac{3}{2}}} e^{i\mathbf{q}\cdot\mathbf{r}-i\omega_{\mathbf{q}}t}, \tag{4.15}$$

with $\omega_{\mathbf{q}} = cq$. The vector potential in real space can then be expanded in terms of the function space of plane waves as

$$\boldsymbol{A}(\mathbf{r},t) = \frac{1}{\sqrt{2}}(\langle \mathbf{r},t| + |\mathbf{r},t\rangle) = \frac{1}{\sqrt{2}}\sum_{\alpha}\int d\mathbf{q}\left(\langle \mathbf{r},t|\underline{\mathbf{q}\alpha}\rangle\langle\underline{\mathbf{q}\alpha}| + |\underline{\mathbf{q}\alpha}\rangle\langle\underline{\mathbf{q}\alpha}|\mathbf{r},t\rangle\right).$$

Or, expressed in the original creation and annihilation operators,

$$\boldsymbol{A}(\mathbf{r},t) = \sum_{\alpha}\int d\mathbf{q}\sqrt{\frac{\hbar}{2\varepsilon_0\omega_{\mathbf{q}}}}\left(\boldsymbol{\epsilon}_{\mathbf{q}\alpha}\varphi_{\mathbf{q}}(\mathbf{r},t)\boldsymbol{a}_{\mathbf{q}\alpha} + \boldsymbol{\epsilon}^*_{\mathbf{q}\alpha}\varphi^*_{\mathbf{q}}(\mathbf{r},t)\boldsymbol{a}^\dagger_{\mathbf{q}\alpha}\right). \tag{4.16}$$

This is the quantum field operator for the vector potential.

Now that we have an expression for the vector potential field operator in space and time, we can find expressions for the electric and magnetic fields using the expressions in Eq. (2.139),

$$\mathbf{E} = -\frac{\partial \mathbf{A}}{\partial t} = \sum_{\alpha}\int d\mathbf{q}\sqrt{\frac{\hbar}{2\varepsilon_0\omega_{\mathbf{q}}}}i\omega_{\mathbf{q}}\left(\boldsymbol{\epsilon}_{\mathbf{q}\alpha}\varphi_{\mathbf{q}}(\mathbf{r},t)\boldsymbol{a}_{\mathbf{q}\alpha} - \boldsymbol{\epsilon}^*_{\mathbf{q}\alpha}\varphi^*_{\mathbf{q}}(\mathbf{r},t)\boldsymbol{a}^\dagger_{\mathbf{q}\alpha}\right)$$

and

$$\mathbf{B} = \nabla\times\mathbf{A} = \sum_{\alpha}\int d\mathbf{q}\sqrt{\frac{\hbar}{2\varepsilon_0\omega_{\mathbf{q}}}}i\left(\mathbf{q}\times\boldsymbol{\epsilon}_{\mathbf{q}\alpha}\varphi_{\mathbf{q}}(\mathbf{r},t)\boldsymbol{a}_{\mathbf{q}\alpha} - \mathbf{q}\times\boldsymbol{\epsilon}^*_{\mathbf{q}\alpha}\varphi^*_{\mathbf{q}}(\mathbf{r},t)\boldsymbol{a}^\dagger_{\mathbf{q}\alpha}\right).$$

The Hamiltonian then follows from the conventional energy density of the electromagnetic field, but now expressed in the quantized field operators

$$\mathrm{H} = \frac{1}{2}\varepsilon_0\int d\mathbf{r}\ \mathbf{E}^\dagger\mathbf{E} + \frac{1}{2\mu_0}\int d\mathbf{r}\ \mathbf{B}^\dagger\mathbf{B}. \tag{4.17}$$

Looking at the term with the electric field:

$$\begin{aligned}\frac{1}{2}\varepsilon_0\int d\mathbf{r}\ |\mathbf{E}|^2 = \frac{1}{4}\sum_{\mathbf{q}\alpha}\hbar\omega_{\mathbf{q}}&\Big[(a^\dagger_{\mathbf{q}\alpha}a_{\mathbf{q}\alpha} + a_{\mathbf{q}\alpha}a^\dagger_{\mathbf{q}\alpha})\\ &- (-1)^\alpha(a^\dagger_{\mathbf{q}\alpha}a^\dagger_{-\mathbf{q}\alpha} + a_{\mathbf{q}\alpha}a_{-\mathbf{q}\alpha})\Big],\end{aligned} \tag{4.18}$$

where the different $\mathbf{q}$ values arise from the integration over the plane waves, giving, for example $\int d\mathbf{r}\ \varphi^*_{\mathbf{q}'}(\mathbf{r})\varphi_{\mathbf{q}}(\mathbf{r}) = \delta_{\mathbf{q}'\mathbf{q}}$ and $\int d\mathbf{r}\ \varphi^*_{\mathbf{q}'}(\mathbf{r})\varphi^*_{\mathbf{q}}(\mathbf{r}) = \delta_{\mathbf{q}',-\mathbf{q}}$. The factor $(-1)^\alpha$ comes from the different orientation of the second polarization vector for a right-handed system when the momentum $\mathbf{q}$ is inversed, *i.e.* $\boldsymbol{\epsilon}_{\pm\mathbf{q}\alpha}\cdot\boldsymbol{\epsilon}_{\mathbf{q}\alpha'} = (\pm 1)^\alpha\delta_{\alpha\alpha'}$. For example, for a wave traveling in the positive z-direction, $\mathbf{e}_x$ and $\mathbf{e}_y$ form a right-handed pair of polarization vectors. However, when traveling in the negative z-direction, $\mathbf{e}_x$ and $-\mathbf{e}_y$ form a right-handed system. Similarly, we find for the magnetic field

$$\begin{aligned}\frac{1}{2\mu_0}\int d\mathbf{r}\ |\mathbf{B}|^2 = \frac{1}{4}\sum_{\mathbf{q}\alpha}\hbar\omega_{\mathbf{q}}&\Big[(a^\dagger_{\mathbf{q}\alpha}a_{\mathbf{q}\alpha} + a_{\mathbf{q}\alpha}a^\dagger_{\mathbf{q}\alpha})\\ &+(-1)^\alpha(a^\dagger_{\mathbf{q}\alpha}a^\dagger_{-\mathbf{q}\alpha} + a_{\mathbf{q}\alpha}a_{-\mathbf{q}\alpha})\Big].\end{aligned} \tag{4.19}$$

Note the sign change in the second term, which is due to $((-\mathbf{q}) \times \boldsymbol{\epsilon}_{-\mathbf{q}\alpha'}) \cdot (\mathbf{q} \times \boldsymbol{\epsilon}_{\mathbf{q}\alpha}) = (\mp 1)^{\alpha} \delta_{\alpha\alpha'}$. Therefore, the second terms cancel, and we are left with

$$\mathrm{H} = \frac{1}{2} \sum_{\mathbf{q}\alpha} \hbar\omega_{\mathbf{q}} (a^{\dagger}_{\mathbf{q}\alpha} a_{\mathbf{q}\alpha} + a_{\mathbf{q}\alpha} a^{\dagger}_{\mathbf{q}\alpha}) = \sum_{\mathbf{q}\alpha} \hbar\omega_{\mathbf{q}} \left(a^{\dagger}_{\mathbf{q}\alpha} a_{\mathbf{q}\alpha} + \frac{1}{2} \right). \tag{4.20}$$

Therefore, the modes in the photon field behave like independent harmonic oscillators.

Let us now return to the commutation relationship for fermions, where the creation and annihilation operators anticommute with each other,

$$\{\mathbf{c}_k, \mathbf{c}^{\dagger}_{k'}\} = \delta_{kk'}, \quad \{\mathbf{c}^{\dagger}_k, \mathbf{c}^{\dagger}_{k'}\} = 0, \quad \{\mathbf{c}_k, \mathbf{c}_{k'}\} = 0, \tag{4.21}$$

where k represents the quantum number(s) describing the state. Their behavior is therefore similar to vectors. All elementary particles that are not force carriers are fermions. This includes quarks and leptons. We follow the conventional notation of indicating fermions with the letter c, but using boldface to indicate its anticommuting character; $\mathbf{c}^{\dagger}_k$ creates a particle in the quantum state k, and $\mathbf{c}_k$ removes the particle from this state. When looking at two-particle wavefunctions, see Eq. (4.1), we saw that it was not possible to put two fermions in the same quantum state. This is known as Pauli's exclusion principle. In terms of creation operators, this directly follows from the anticommutation relation,

$$\mathbf{c}^{\dagger}_k \mathbf{c}^{\dagger}_k = \frac{1}{2}(\mathbf{c}^{\dagger}_k \mathbf{c}^{\dagger}_k + \mathbf{c}^{\dagger}_k \mathbf{c}^{\dagger}_k) = \frac{1}{2}\{\mathbf{c}^{\dagger}_k, \mathbf{c}^{\dagger}_k\} = 0. \tag{4.22}$$

However, many particles are composites of other fermions. The commutation properties of composite particles can differ from that of a single fermion. Let us consider the commutation of a pair of fermions

$$\mathbf{c}^{\dagger}_{k_1} \mathbf{c}^{\dagger}_{k'_1} \mathbf{c}^{\dagger}_{k_2} \mathbf{c}^{\dagger}_{k'_2} = \mathbf{c}^{\dagger}_{k_2} \mathbf{c}^{\dagger}_{k_1} \mathbf{c}^{\dagger}_{k'_1} \mathbf{c}^{\dagger}_{k'_2} = \mathbf{c}^{\dagger}_{k_2} \mathbf{c}^{\dagger}_{k'_2} \mathbf{c}^{\dagger}_{k_1} \mathbf{c}^{\dagger}_{k'_1}, \tag{4.23}$$

where k_i and k'_i, with $i = 1, 2$, are the quantum numbers of the composite particles labeled 1 and 2, respectively. Due to Pauli's principle all these quantum numbers are different. From Eq. (4.23), we see that pairs of fermions effectively commute with each other. This is found for mesons that are particles consisting of a quark and an antiquark bound together by the strong interaction, see Section 4.6. However, sometimes the bosonic character shows up unexpectedly. In some materials, an effective attractive interaction between electrons mediated by lattice vibrations causes pair formation. This dramatically changes the properties of the material giving rise to superconductivity, see Section 5.2.

For three particles, we find

$$\begin{aligned} \mathbf{c}^{\dagger}_{k_1} \mathbf{c}^{\dagger}_{k'_1} \mathbf{c}^{\dagger}_{k''_1} \mathbf{c}^{\dagger}_{k_2} \mathbf{c}^{\dagger}_{k'_2} \mathbf{c}^{\dagger}_{k''_2} &= -\mathbf{c}^{\dagger}_{k_2} \mathbf{c}^{\dagger}_{k_1} \mathbf{c}^{\dagger}_{k'_1} \mathbf{c}^{\dagger}_{k''_1} \mathbf{c}^{\dagger}_{k'_2} \mathbf{c}^{\dagger}_{k''_2} \\ &= \mathbf{c}^{\dagger}_{k_2} \mathbf{c}^{\dagger}_{k'_2} \mathbf{c}^{\dagger}_{k_1} \mathbf{c}^{\dagger}_{k'_1} \mathbf{c}^{\dagger}_{k''_1} \mathbf{c}^{\dagger}_{k''_2} = -\mathbf{c}^{\dagger}_{k_2} \mathbf{c}^{\dagger}_{k'_2} \mathbf{c}^{\dagger}_{k''_2} \mathbf{c}^{\dagger}_{k_1} \mathbf{c}^{\dagger}_{k'_1} \mathbf{c}^{\dagger}_{k''_1}. \end{aligned} \tag{4.24}$$

Therefore, composites of three fermions anticommute again. This is the case for baryons, which are subatomic particles that contain three quarks. The best known examples are the proton and the neutron, see Section 4.7. Note that, generally, protons and neutrons are not described as a composite particle but as a single fermion.

A fermionic many-body product wavefunction can be written as

$$|k_1 \cdots k_N\rangle = \mathbf{c}^{\dagger}_{k_N} \cdots \mathbf{c}^{\dagger}_{k_1} |0\rangle, \quad \langle k_1 \cdots k_N| = \langle 0| \mathbf{c}_{k_1} \cdots \mathbf{c}_{k_N}, \tag{4.25}$$

where the conventional vacuum state $|0\rangle$ has been added. Note that there are many different combinations of quantum numbers k_i possible. The state above is not necessarily an eigenfunction of the Hamiltonian of the system. In general, the eigenfunctions are linear combinations of these basis functions above. Any approach that describes a many-body system only in terms of single product wavefunctions is an effective independent-particle model. This is still the case for an incoherent occupation of different product wavefunctions, *i.e.* not a linear combination. Examples are the thermal occupation of different product wavefunctions or mean-field-type approaches.

The real space equivalent of the product wavefunction can be obtained by taking the inner product with the position product function from Eq. (4.4)

$$\Phi_{k_1 \cdots k_N}(\mathbf{r}_N \cdots \mathbf{r}_1) = \langle \mathbf{r}^N | k_1 \cdots k_N \rangle = \frac{1}{\sqrt{N!}} \sum_{P_{i_1 \cdots i_N}} \epsilon_{i_1 \cdots i_N} \varphi_{k_1}(\mathbf{r}_{i_1}) \cdots \varphi_{k_N}(\mathbf{r}_{i_N}),$$

where all the possible inner products are taken. The single-particle wavefunctions are given by $\varphi_{k_j}(\mathbf{r}_i) = \langle \mathbf{r}_i | k_j \rangle$. The summation goes over all the possible permutations. For example, for $N = 3$, $i_1 i_2 i_3 = 123, 132, 231, 213, 312, 321$, giving a total of $3! = 6$ permutations. Due to the anticommutativity of the creation operators, the interchange of particles causes a sign change. This is taken care of by the Levi-Civita symbol $\epsilon_{i_1 \cdots i_N}$ that is ± 1 for an even/odd number of permutations needed to bring $i_1 \cdots i_N$ back to $1, 2, \ldots, N$. For the three-fermion example, this gives $\epsilon_{i_1 i_2 i_3} = 1, -1, 1, -1, 1, -1$. The above wavefunction can also be written as a determinant,

$$\Phi_{k_1 \cdots k_N}(\mathbf{r}_1 \cdots \mathbf{r}_N) = \frac{1}{\sqrt{N!}} \begin{vmatrix} \varphi_{k_1}(\mathbf{r}_1) & \cdots & \varphi_{k_1}(\mathbf{r}_N) \\ \cdot & & \cdot \\ \varphi_{k_N}(\mathbf{r}_1) & \cdots & \varphi_{k_N}(\mathbf{r}_N) \end{vmatrix}. \tag{4.26}$$

This is known as a Slater determinant. However, although the real-space coordinates are needed, one rarely writes out the full wavefunction in real space since the permutations over all the coordinates become cumbersome. The properties of fermions also follow from the properties of determinants: there is a sign change when two neighboring rows or columns are interchanged, and the determinant is zero when two columns or rows are equivalent.

For many-body systems, it is generally preferable to work with the anticommuting and commuting vectors to represent the different particles. This is also known as second quantization. Expressions for wavefunctions expressed in real space generally become very long. However, the coordinates are often still needed when evaluating matrix elements. Let us discuss the matrix elements for one- and two-particle interaction terms. A general one-particle interaction can be written as

$$\boldsymbol{H}_0 = \sum_{kk'} |k'\rangle \langle k' | \boldsymbol{H}_0 | k \rangle \langle k| = \sum_{kk'} H_{k'k} \mathbf{c}^\dagger_{k'} \mathbf{c}_k. \tag{4.27}$$

The scalar $H_{k'k}$ tells how the functions represented by $\mathbf{c}^\dagger_{k'}$ and $\mathbf{c}_k$ transform into each other under a particular operation. For some of the function spaces that we have derived, the matrix elements can be derived analytically. However, for a space with a complicated potential landscape, the coefficients have to be determined numerically. The matrix element is given by

$$H_{k'k} = \langle k' | \boldsymbol{H}_0 | k \rangle = \int d\mathbf{r}' \int d\mathbf{r} \; \langle k' | \mathbf{r}' \rangle \langle \mathbf{r}' | \boldsymbol{H}_0 | \mathbf{r} \rangle \langle \mathbf{r} | k \rangle, \tag{4.28}$$

where two spatial completeness relationships have been inserted on the right-hand side. All known Hamiltonians are local, *i.e.*

$$\langle \mathbf{r}' | \boldsymbol{H}_0 | \mathbf{r} \rangle = \hat{H}_0(\mathbf{r}) \delta(\mathbf{r} - \mathbf{r}'). \tag{4.29}$$

In general, $\hat{H}_0(\mathbf{r})$ is an operator, such as the momentum or angular momentum operator, or a function in space, such as a potential (which is, in principle, still an operator). The matrix element then becomes

$$H_{k'k} = \int d\mathbf{r}\, \varphi^*_{k'}(\mathbf{r})\hat{H}_0(\mathbf{r})\varphi_k(\mathbf{r}). \tag{4.30}$$

Once the matrix elements $H_{k'k}$ are evaluated, there is no need for the real-space coordinates, and the transformations for a single particle can be used. Since $\boldsymbol{H}_0$ can only change a single quantum number, there are only two options for finite many-particle matrix elements. Let us discuss the situation for fermions. The many-particle wavefunctions can be the same on both sides

$$\langle k_1 \cdots k_N|\boldsymbol{H}_0|k_1 \cdots k_N\rangle = \sum_{i=1}^{N} H_{k_ik_i}. \tag{4.31}$$

The fermion from a state k_i scatters back into k_i. For a diagonal matrix element, this can be done for all k_i, so there is a summation over i. The second possibility is that the bra and ket differ by one quantum number

$$\langle k_1 \cdots k' \cdots k_{N-1}|\boldsymbol{H}_0|k_1 \cdots k \cdots k_{N-1}\rangle = P_{k'k}H_{k'k}. \tag{4.32}$$

This is an off-diagonal matrix element. The remaining $N-1$ quantum numbers $k_1, \ldots,$ k_{N-1} are equal in the bra and the ket. Pauli's principle ensures that they are all different from k and k'. The phase factor $P_{k'k}$ is ± 1 for an even/odd number of permutations needed to bring k/k' to the left/right of the ket/bra, respectively. All matrix elements between basis functions that differ by more than one quantum number are zero. Obtaining the correct sign is annoying but important. Mistakes in the sign of the matrix elements can lead to incorrect energies and wavefunctions.

Let us look at an example with three fermions. A matrix element for the situation where the bra and ket differ by one quantum number is

$$\begin{aligned}\langle k_1k_3k_4|\boldsymbol{H}_0|k_1k_2k_3\rangle &= \langle 0|\mathbf{c}_{k_1}\mathbf{c}_{k_3}\mathbf{c}_{k_4}\boldsymbol{H}_0\mathbf{c}^\dagger_{k_3}\mathbf{c}^\dagger_{k_2}\mathbf{c}^\dagger_{k_1}|0\rangle \\ &= -\langle 0|\mathbf{c}_{k_1}\mathbf{c}_{k_3}\mathbf{c}_{k_4}\boldsymbol{H}_0\mathbf{c}^\dagger_{k_2}\mathbf{c}^\dagger_{k_3}\mathbf{c}^\dagger_{k_1}|0\rangle \\ &= -\langle k_1k_3|k_1k_3\rangle\langle k_4|\boldsymbol{H}_0|k_2\rangle = -H_{k_4k_2}.\end{aligned} \tag{4.33}$$

In order to obtain the correct sign, the operators for k_2 and k_4 need to be permuted until they are right next of the interaction.

Let us now consider a two-particle interaction. In general, this can be written as

$$\begin{aligned}\boldsymbol{H}_1 &= \sum_{k_1k_2k_3k_4} |k_3k_4\rangle\langle k_4k_3|\boldsymbol{H}_1|k_1k_2\rangle\langle k_2k_1| \\ &= \sum_{k_1k_2k_3k_4} H_{k_4k_3k_2k_1}\mathbf{c}^\dagger_{k_4}\mathbf{c}^\dagger_{k_3}\mathbf{c}_{k_2}\mathbf{c}_{k_1}.\end{aligned} \tag{4.34}$$

The indices have been chosen such that the subscripts in the operators are ascending from right to left. However, note that k_1 and k_4 are in a similar position, and so are k_2 and k_3. The matrix element is given by

$$\begin{aligned}H_{k_4k_3k_2k_1} &= H_{k_3k_4k_1k_2} = \langle k_3k_4|\boldsymbol{H}_1|k_2k_1\rangle = \langle k_4k_3|\boldsymbol{H}_1|k_1k_2\rangle \\ &= \frac{1}{2}\int d\mathbf{r}'\int d\mathbf{r}\, \langle k_3k_4|\mathbf{r}'\mathbf{r}\rangle\langle \mathbf{r}\mathbf{r}'|\boldsymbol{H}_1|\mathbf{r}\mathbf{r}'\rangle\langle \mathbf{r}\mathbf{r}'|k_2k_1\rangle,\end{aligned} \tag{4.35}$$

where the locality of the Hamiltonian has already been included. Generally, the interaction between particles only depends on the distance between them $\langle \mathbf{r}\mathbf{r}'|\boldsymbol{H}_1|\mathbf{r}\mathbf{r}'\rangle = \hat{H}_1(\mathbf{r}-\mathbf{r}')$. There are four combinations of the bra-kets inside the matrix element of which two are the same. The particle with position vector $\mathbf{r}$ can be put into k_1 and k_4, and the particle with position vector $\mathbf{r}'$ into k_2 and k_3. Due to the indistinguishability, the same occurs for $\mathbf{r} \leftrightarrow \mathbf{r}'$. The other option is $\mathbf{r} \rightarrow k_1, k_3$ and $\mathbf{r}' \rightarrow k_2, k_4$. This also has the equivalent permutation $\mathbf{r} \leftrightarrow \mathbf{r}'$. The $\mathbf{r} \leftrightarrow \mathbf{r}'$ permutation simply gives a factor 2 when changing the integration variables, which cancels with the factor $\frac{1}{2}$ in Eq. (4.35). This leaves two distinct terms

$$H_{k_4k_3k_2k_1} = U_{k_4k_3k_2k_1} - U_{k_3k_4k_2k_1}. \tag{4.36}$$

Note the minus sign in front of the second term. The matrix elements can be written explicitly as

$$U_{k_4k_3k_2k_1} = \int d\mathbf{r}' \int d\mathbf{r}\, \varphi^*_{k_4}(\mathbf{r})\varphi^*_{k_3}(\mathbf{r}')\hat{H}_1(\mathbf{r}-\mathbf{r}')\varphi_{k_2}(\mathbf{r}')\varphi_{k_1}(\mathbf{r}). \tag{4.37}$$

The two matrix elements in Eq. (4.36) are often known as the direct and exchange interactions, respectively. Again, once the matrix element $H_{k_4k_3k_2k_1}$ has been calculated, the transformation properties are known, and there is no need to deal with the positions in real space.

There are three possibilities for finite matrix elements between product wavefunctions. If the bra and ket are equal, we have

$$\langle k_1 \cdots k_N|\boldsymbol{H}_1|k_1 \cdots k_N\rangle = \sum_{j>i} H_{k_ik_jk_jk_i}. \tag{4.38}$$

When one quantum number is different, and $N-1$ quantum number are equal, the matrix element is given by

$$\langle k_1 \cdots k' \cdots k_{N-1}|\boldsymbol{H}_0|k_1 \cdots k \cdots k_{N-1}\rangle = P_{k'k}\sum_{i=1}^{N-1} H_{k'k_ik_ik}, \tag{4.39}$$

where the sign is comparable to that in Eq. (4.32). Since $\boldsymbol{H}_1$ is a two-particle interaction, there is still a summation over the remaining quantum numbers, since another particle needs to participate in the interaction without changing its quantum number. When two quantum numbers are different, and $N-2$ are the same, the matrix element is

$$\langle k_1 \cdots k'' \cdots k''' \cdots k_{N-2}|\boldsymbol{H}_1|k_1 \cdots k \cdots k' \cdots k_{N-2}\rangle = P_{k'''k''k'k}H_{k'''k''k'k},$$

where $P_{k'''k''k'k}$ is ± 1 for an even/odd number of permutations to bring the quantum numbers that are involved in the interaction next of $\boldsymbol{H}_1$. The inner product for the other quantum numbers is one. Any matrix element where the bra and ket differ by more than two quantum numbers is zero.

Conceptual summary. – The different commutative properties of particles have drastic consequences for the behavior of a many-body system. One of the most important differences is the occupation of eigenfunctions. For fermions, only one particle can occupy an eigenstate. This is known as Pauli's principle. This is why all electrons on an atom do not reside in the 1s orbital, which would be the lowest energy state, but each orbital (including spin) is only filled with a single electron. There is no such restriction for bosons, and one can have, for example many photons in the same state.

As we will see later, the commutative properties of composite particles can differ from their components. For example, an attractive interaction between two electrons, which are fermions, can effectively create a boson. These pairs can now occupy the same quantum state leading to dramatically different properties. We also considered in detail the photon field. Apart from the increase in complexity, its behavior is remarkably similar to the quantum harmonic oscillator.

Problems

1 Determine the matrix elements $\langle k_1 k_3 k_4 | \boldsymbol{H}_1 | k_1 k_2 k_3 \rangle$ for the two-fermion interaction $\boldsymbol{H}_1$.

2 An electron also interacts with itself by emitting a photon and then absorbing it again via the $\mathbf{p} \cdot \mathbf{A}$ interaction.

a) Write down the Green's function for the self-interaction of an electron with a plane wave with momentum $\mathbf{k}$ using the techniques from Section 3.10. Only consider states with zero and one photon.

b) Evaluate the matrix element in the dipole limit, *i.e.* $e^{i\mathbf{q}\cdot\mathbf{r}} \cong 1$. As we will see, this approximation leads to unphysical results.

c) Derive the energy shift and show that it diverges.

d) Argue that the energy shift should be finite and give an estimate of its magnitude.

e) Provide a reason that the gyromagnetic ratio $g > 2$ and provide an estimate of g.

4.2 Quantum Statistics

Introduction. – In the nineteenth century, Ludwig Boltzmann demonstrated that thermodynamics is intimately related to statistics. The fact that bosons can occupy the same quantum state but fermions cannot leads to different thermodynamics properties. In fact, the thermodynamic properties of blackbodies led Max Planck to his hypothesis that electromagnetic waves are quantized in energy units $\hbar\omega$. In this section, we discuss some basic elements of statistical thermodynamics of many-particle systems relevant for quantum physics.

In most of our problems, we are simply considering very small systems. However, this system is generally not isolated from its surroundings. We therefore want to connect it to a much larger system, generally called the reservoir. The reservoir is not necessarily in its lowest energy state, it can contain excitations with respect to this state. We are interested in finding the probability that a state in our local system is occupied when the local system is coupled to this reservoir. The probability $p(E)$ of finding a particular state with energy E is directly proportional to the number of states with energy E, also known as the multiplicity $g(E)$. Note that we are not considering the nonequilibrium dynamics that tell us how these states become occupied but simply assume that nature finds a way. Since we will not be able to calculate the entire multiplicity of the reservoir, we end up with only relative probabilities of finding a state with energy E occupied. For a particular energy E of the system under consideration, the multiplicity g of the system and the reservoir is the multiplicity g_s of the system times the multiplicity of the reservoir g_R,

$$g(E_{\text{tot}}, E) = g_s(E) g_R(E_{\text{tot}} - E). \tag{4.40}$$

The total energy E_{tot} of the system and the reservoir is considered a conserved quantity. We want to find the energy E of the system such that the maximum multiplicity, *i.e.* the maximum probability, is obtained,

$$\frac{\partial g}{\partial E} = \frac{\partial g_s}{\partial E} g_R - g_s(E)\frac{\partial g_R}{\partial E} = 0. \tag{4.41}$$

This gives

$$\frac{1}{g_s}\frac{\partial g_s}{\partial E} = \frac{1}{g_R}\frac{\partial g_R}{\partial E} \quad \Rightarrow \quad \frac{\partial \ln g_s}{\partial E} = \frac{\partial \ln g_R}{\partial E}. \tag{4.42}$$

The logarithm of the multiplicity is related to the thermodynamic quantity of entropy S whose derivative defines the temperature

$$S \equiv k_B \ln g \quad \text{with} \quad \frac{1}{T} \equiv \frac{\partial S}{\partial E} = k_B \frac{\partial \ln g}{\partial E}, \tag{4.43}$$

where $k_B = 1.38 \times 10^{-23}$ J/K is Boltzmann's constant. Therefore, an increase in entropy simply means that the system is going toward a state with a higher multiplicity, *i.e.* a state that has a higher probability of occurring. Inserting the definitions into the above equation gives

$$\frac{\partial S_s}{\partial E} = \frac{\partial S_R}{\partial E} \quad \Rightarrow \quad \frac{1}{T_s} = \frac{1}{T} \quad \text{or } T_s = T. \tag{4.44}$$

Therefore, the temperature of the system T_s is the same as the reservoir $T_R = T$. Since the much larger reservoir determines the actual temperature, the system is in thermal equilibrium with the reservoir.

We are not interested in the reservoir but more in relative probabilities of finding states in the local system. The total entropy of the system and the reservoir can be divided into two terms

$$S(E_{\text{tot}}, E) = k_B \ln g_s(E) g_R(E_{\text{tot}} - E) = k_B \ln g_s(E) + k_B \ln g_R(E_{\text{tot}} - E).$$

However, since the multiplicity of the reservoir is much larger than that of the system, the entropy can be approximated by

$$S(E_{\text{tot}}, E) \cong S_R(E_{\text{tot}} - E) = S_R(E_{\text{tot}}) - \left.\frac{\partial S_R}{\partial E}\right|_{E_{\text{tot}}} E = S_R(E_{\text{tot}}) - \frac{E}{T}.$$

Since the system is small, the relative probabilities are entirely determined by the relative multitudes of the reservoir

$$p(E) = \frac{g_R(E_{\text{tot}} - E)}{g_R(E_{\text{tot}})} = \frac{e^{\frac{1}{k_B} S_R(E_{\text{tot}} - E)}}{e^{\frac{1}{k_B} S_R(E_{\text{tot}})}} = e^{-\frac{E}{k_B T}}. \tag{4.45}$$

This relative probability is known as the Boltzmann factor.

Since the probabilities are relative and not normalized, it is useful to define the partition function

$$Z = \sum_n p(E_n) = \sum_n e^{-\frac{E_n}{k_B T}}, \tag{4.46}$$

where the sum goes over all the possible states n. Z is essentially a total probability for all the states. This allows us to normalize the probability of a particular state to the total probability.

Let us first consider the case of bosons where the particle number is not fixed. The energy of a particular state is given by $E_n = n\hbar\omega_k$, where k is a generic index identifying the different modes. For a quantum harmonic oscillator, the energy can be replaced by $E_n = n\hbar\omega_0$. For photons, the energy is given by $E_n = n\hbar\omega_{\mathbf{q}}$, see Eq. (4.20). The zero-point energy has been left out in both cases. The partition function can be summed analytically

$$Z = \sum_{n=0}^{\infty} e^{-\frac{n\hbar\omega_k}{k_BT}} = \frac{1}{1-e^{-\frac{\hbar\omega_k}{k_BT}}} = \frac{1}{1-e^{-x}}, \tag{4.47}$$

with $x = \hbar\omega_k/k_BT$. The average value of n_k for a particular ω_k can be obtained:

$$\begin{aligned}\langle n_k\rangle &= \frac{1}{Z}\sum_{n=0}^{\infty} ne^{-nx} = -\frac{1}{Z}\frac{\partial}{\partial x}\sum_{n=0}^{\infty} e^{-nx} = -\frac{1}{Z}\frac{\partial Z}{\partial x} = -\frac{\partial \ln Z}{\partial x} \\ &= \frac{\partial \ln(1-e^{-x})}{\partial x} = \frac{e^{-x}}{1-e^{-x}} = \frac{1}{e^x-1} = \frac{1}{e^{\frac{\hbar\omega_k}{k_BT}}-1}.\end{aligned} \tag{4.48}$$

The average energy for a particular mode k is then

$$\langle E_{\omega_k}\rangle = \langle n_k\rangle\hbar\omega_k = \frac{\hbar\omega_k}{e^{\frac{\hbar\omega_k}{k_BT}}-1}. \tag{4.49}$$

This is known as Planck's distribution law.

Fermi–Dirac distribution. – In the above calculation of $\langle n_k\rangle$, the number of particles was free, *i.e.* the magnitude of the oscillation or the photon field can adjust itself to the temperature. However, often the number of particles is a fixed quantity. Let us start with fermions, where a particular state can only contain zero or one fermion due to Pauli's principle in Eq. (4.22). Let us first consider the Boltzmann factor $\exp(-E/k_BT)$ at zero temperature. If the energy is positive (which is the case for free particles), then this is 1 for $E = 0$ and 0 otherwise. Therefore, the number of fermions that can be put in this system is equal to the number of states with energy 0. If there are more fermions in the system, then they start to occupy states with zero probability. This means that if the system is connected to a reservoir, these fermions will leave the system. In order for the system to be stable, there needs to be an energy level up to which the states can be filled without fermions leaving the system. This level is known as the chemical potential μ. Therefore, in the reservoir, the states are also filled up to this energy level. The chemical potential effectively shifts the "zero" energy level in the Boltzmann factor, $\exp(-(E-\mu)/k_BT)$. To obtain a normalized probability for a particular state with energy E, we have to divide by the total probability or the distribution function Z which now only contains the probabilities of the state being empty or occupied. This gives

$$f(E) = \frac{e^{-\frac{E-\mu}{k_BT}}}{Z} = \frac{e^{-\frac{E-\mu}{k_BT}}}{1+e^{-\frac{E-\mu}{k_BT}}}. \tag{4.50}$$

The final result is known as the Fermi–Dirac distribution,

$$f(E) = \frac{1}{1+e^{\frac{E-\mu}{k_BT}}}. \tag{4.51}$$

The chemical potential is the energy where the occupation switches from $f(E) > \frac{1}{2}$ to $f(E) < \frac{1}{2}$. Note that the chemical potential needs to adjust as a function of temperature to maintain a constant number of particles. This effect is generally relatively small.

Bose–Einstein distribution and condensation. – A distribution can also be derived for bosons. A boson state with energy E can contain an arbitrary number of bosons n giving an energy nE. A chemical potential is now added to fix the number of bosons. The derivation is the same as for Planck's distribution law in Eq. (4.48), giving

$$f(E) = \frac{1}{e^{\frac{E-\mu}{k_B T}} - 1}. \tag{4.52}$$

This is known as the Bose–Einstein distribution. The position of the chemical potential is now fixed by the total number of particles

$$N = \sum_k f(E_k) = \sum_k \frac{1}{e^{\frac{E_k-\mu}{k_B T}} - 1}, \tag{4.53}$$

where the summation goes over all the possible boson states. Note that the summation in Eq. (4.48) only converges if the exponential is less than 1. This implies that the chemical potential for bosons is less than 0.

When the temperature goes to zero, all the bosons condense into the lowest state. Taking the lowest energy $E_0 = 0$, its occupation in the limit $T \to 0$ is given by

$$f(0) = \frac{1}{e^{\frac{-\mu}{k_B T}} - 1} \cong \frac{1}{1 - \frac{\mu}{k_B T} - 1} = -\frac{k_B T}{\mu}, \tag{4.54}$$

with $\mu < 0$ and $|\mu| \ll k_B T$ in order to obtain a finite occupation. If all particles are in the lowest state, the chemical potential can be found via

$$N = f(0) = -\frac{k_B T}{\mu} \quad \Rightarrow \quad \mu = -\frac{k_B T}{N}. \tag{4.55}$$

Since the number of particles is generally very large, the chemical potential is very close to the lowest level $E_0 = 0$. When increasing the temperature, higher lying states will be excited. When considering the case of free bosons in three dimensions, with an energy $E = \hbar^2 \mathbf{k}^2/2M$, where M is the mass of the boson, surfaces of equal energy in momentum space are determined by the norm k of the wavevector. The total number of states with a momentum less than k is given by

$$N = \left(\frac{L}{2\pi}\right)^3 \frac{4}{3}\pi k^3 = \frac{V}{6\pi^2} k^3 = \frac{V}{6\pi^2}\left(\frac{2M}{\hbar^2}E\right)^{\frac{3}{2}}, \tag{4.56}$$

using the fact that the k values are separated in momentum space by $2\pi/L$ in each direction due to the periodic boundary conditions. The density of states, *i.e.* the change in the number of states at a particular energy, is given by

$$D(E) = \frac{dN}{dE} = \frac{V}{4\pi^2}\left(\frac{2M}{\hbar^2}\right)^{\frac{3}{2}} \sqrt{E} = A\sqrt{E} \text{ with } A = \frac{V}{4\pi^2}\left(\frac{2M}{\hbar^2}\right)^{\frac{3}{2}}, \tag{4.57}$$

where V is the volume. Even though the energy splitting between the states can be small, there is only a microscopic number of low-lying energy levels, reflected in the fact that the density of states goes to zero. Therefore, even though there is only one lowest energy state, it is still relatively stable with respect to thermal excitations. The number of bosons in excited states can be calculated using the density of states

$$N_{\text{exc}} = A \int_0^\infty dE \frac{\sqrt{E}}{e^{\frac{E}{k_B T}} - 1} = A(k_B T)^{\frac{3}{2}} \int_0^\infty dx \frac{\sqrt{x}}{e^x - 1} = A(k_B T)^{\frac{3}{2}} 1.306\sqrt{\pi},$$

with $x = E/k_B T$. The chemical potential has been taken as zero in this evaluation. The Bose–Einstein condensation temperature is defined as the temperature where all the bosons are in the excited states, *i.e.* $N_{\text{exc}} = N$. This can be written as

$$T_{\mathrm{BE}} = \frac{2\pi\hbar^2}{k_B M}\left(\frac{N}{2.612V}\right)^{\frac{2}{3}}, \tag{4.58}$$

where N/V is the density of the bosons. With some rewriting, the relative number of bosons in excited states is then

$$\frac{N_{\mathrm{exc}}}{N} = \left(\frac{T}{T_{\mathrm{BE}}}\right)^{\frac{3}{2}}. \tag{4.59}$$

For the well-known free boson ^{4}He, the condensation temperature is estimated to be $T_{\mathrm{BE}} = 3$ K, which is close to the actual temperature of 2.17 K.

Example 4.2 ***Blackbody Radiation***

The quantum theory started with the interpretation of blackbody radiation. All matter emits electromagnetic radiation when its temperature is above absolute zero. The radiation results from the conversion of thermal energy into electromagnetic energy. A blackbody is a system that absorbs all incident radiation. The blackbody is an idealization, but the general ideas hold for any system. The intensity depends on the frequency and has a maximum that is dependent on the temperature T. The maximum is in the visible region for the Sun ($T = 5500$ K) and in the infrared region for a human body. Using purely thermodynamic arguments, Wilhelm Wien showed in 1893 that the spectral distribution must be given by

$$I(\omega) = \omega^3 f\left(\frac{\omega}{T}\right). \tag{4.60}$$

For a blackbody, we can assume standing waves inside a cavity.

a) Using the equipartition theorem, determine the spectral distribution in the classical limit. Show that this satisfies Wien's law but explain why this calculation fails.
b) Show that Planck's distribution law gives the correct physical result in both the low- and high-frequency limits.

Solution

a) The eigenfunctions for a cavity are standing waves, for example $\sin k_x x$ in the x-direction. This leads to the quantization of momenta as $k_i = n\pi/L$ (assuming a cubic cavity). The number of states in a shell of thickness dk for a given wavenumber k is

$$dN = 2\frac{1}{8}\left(\frac{L}{\pi}\right)^3 4\pi k^2 dk = \frac{V}{\pi^2 c^3}\omega^2 d\omega = V D(\omega) d\omega, \tag{4.61}$$

where $(L/\pi)^3$ is the volume in momentum space occupied by one $\mathbf{k}$ point. The total volume is $V = L^3$. The factor 2 accounts for the two polarization directions. The factor $\frac{1}{8}$ is needed since there are only positive k values. The equipartition theorem states that the energy for a classical oscillator is $k_B T$. The spectral distribution is then the average energy times the number of oscillator states,

$$I(\omega) = D(\omega) k_B T = \frac{k_B T}{\pi^2 c^3}\omega^2. \tag{4.62}$$

This result is known as the Rayleigh–Jeans law. It satisfies Wien's law with $f(\omega/T) \sim T/\omega$. However, it fails physically since $I(\omega)$ diverges for $\omega \to \infty$, since the energy for each oscillator remains constant, but the number of oscillator states increases. This is known as the ultraviolet catastrophe.

b) The classical energy is now replaced by Planck's distribution law from Eq. (4.49),

$$I(\omega) = D(\omega)\langle E_\omega \rangle = \frac{\hbar\omega^3}{c^3\pi^2}\frac{1}{e^{\frac{\hbar\omega}{k_BT}} - 1}. \tag{4.63}$$

It is easily checked that this still satisfies Wien's law. In the limit $\hbar\omega \ll k_BT$, $e^{\frac{\hbar\omega}{k_BT}} - 1 \cong \frac{\hbar\omega}{k_BT}$, and the Rayleigh–Jeans law is reproduced. Therefore, for relatively large temperatures, quantum effects are washed out. However, in the opposite limit $\hbar\omega \gg k_BT$, we obtain $I(\omega) \cong \frac{\hbar\omega^3}{c^3\pi^2}\exp(-\frac{\hbar\omega}{k_BT})$. The spectral intensity of the high-frequency states becomes suppressed, since the temperature is insufficient to excite these oscillators.

Conceptual summary. – The different statistics of fermions and bosons lead to different thermodynamic properties. Statistical thermodynamics is very well suited to quantum mechanics, since we are looking at the occupation of different states. For a particular system, a state with the lowest energy can be identified. Excitations with respect to this state can be made leading to an excess energy. As we saw in Section 3.6, creating excitations is a complicated process. Additionally, there are, in principle, an infinite number of different excited states. However, thermodynamics tells us that, in thermal equilibrium with the surroundings, the excess energy and the related occupations of different levels can be identified by a single parameter, namely k_BT, where T is the temperature. This differs from Section 3.6, where a system is excited, but it is not in equilibrium.

The assumption in statistical thermodynamics is that the only thing that matters is the number of states and their energies. How these states become occupied is not considered. Under these assumptions, the possible occupation of a level can be described solely in terms of the temperature and the energy of the state. These functions are known as the Fermi–Dirac and Bose–Einstein distributions for fermions and bosons, respectively. The Bose–Einstein distribution has the interesting characteristic that, below a certain energy, all particles condense into the lowest state. This gives rise to several interesting phenomena. We return to this in later sections.

Problems

1 Consider a system of spins with two states (↑ and ↓). Show that the probability of different configurations can be written as a Gaussian in terms of the magnetization $\overline{M} = (N_\uparrow - N_\downarrow)/N$.

4.3 The Fermi Sea in Solids

Introduction. – A particle in a box is one of the first examples treated in quantum mechanics. If the box is small, *i.e.* on the order of nanometers or less, the quantization effects can become apparent. This means that the quantization of the energy levels becomes larger than the temperature. However, for a large box, the quantization of the energy levels becomes so small that it does not affect the physical properties. This is the limit generally assumed for solids. In Section 3.8, we studied the effect of a periodic potential in such a system. In this section, we revisit the problem of a solid as electrons in a three-dimensional potential well. The periodic potential due to the atoms in the solid will be neglected. However, this time we consider the effect of the Coulomb interactions between electrons in a solid.

The Hamiltonian for free particles is given by

$$\boldsymbol{H}_0 = \sum_{\mathbf{k}\sigma} \varepsilon_{\mathbf{k}} \mathbf{c}^{\dagger}_{\mathbf{k}\sigma} \mathbf{c}_{\mathbf{k}\sigma}, \quad \text{with} \quad \varepsilon_{\mathbf{k}} = \frac{\hbar^2 \mathbf{k}^2}{2m}. \tag{4.64}$$

Since electrons are fermions, they have to obey Pauli's principle and cannot all occupy the $\mathbf{k} = 0$ state as the bosons as discussed in the previous section. Therefore, for free electrons, the levels are filled up to a certain momentum, known as the Fermi wavevector k_F. The calculation of the number of states is analogous to that of bosons, see Eq. (4.56), except that a factor 2 has to be included for the spin degrees of freedom

$$N = 2\left(\frac{L}{2\pi}\right)^3 \frac{4}{3}\pi k^3 = \frac{V}{3\pi^2} k^3 \quad \Rightarrow \quad k_F = \left(\frac{3\pi^2 N_e}{V}\right)^{\frac{1}{3}} = \left(3\pi^2 n_e\right)^{\frac{1}{3}}, \tag{4.65}$$

where N_e is the total number of electrons in the system, and n_e is the electron density. The ground state at $T = 0$ is the Fermi sphere, *i.e.* the levels are filled starting from the lowest energy and up to k_F

$$|F\rangle = \prod_{|\mathbf{k}|<k_F} \mathbf{c}^{\dagger}_{\mathbf{k}\downarrow} \mathbf{c}^{\dagger}_{\mathbf{k}\uparrow} |0\rangle, \tag{4.66}$$

where $\mathbf{c}^{\dagger}_{\mathbf{k}\sigma}$ creates an electron with momentum $\mathbf{k}$ and spin $\sigma = \uparrow, \downarrow$.

The total kinetic energy is then

$$\langle \boldsymbol{H}_0 \rangle = \langle F | \boldsymbol{H}_0 | F \rangle = \frac{\hbar^2}{2m} \sum_{|\mathbf{k}|<k_F, \sigma} \mathbf{k}^2. \tag{4.67}$$

The energy is more easily evaluated by replacing the summation over momentum by an integral over the energy. To perform the integration, the density of states at a particular energy is needed. This was already obtained for free bosons in Eq. (4.57), and we only need to multiply by 2 to account for the spin

$$D(E) = \frac{dN}{dE} = \frac{V}{2\pi^2}\left(\frac{2m}{\hbar^2}\right)^{\frac{3}{2}} \sqrt{E}. \tag{4.68}$$

The integration is then done up to the Fermi energy,

$$E_F = \frac{\hbar^2 k_F^2}{2m} = \frac{\hbar^2}{2m}\left(\frac{3\pi^2 N_e}{V}\right)^{2/3}. \tag{4.69}$$

The total kinetic energy can now be evaluated

$$\begin{aligned} \langle \boldsymbol{H}_0 \rangle &= \int_0^{E_F} dE\, D(E) E = \frac{V}{2\pi^2}\left(\frac{2m}{\hbar^2}\right)^{3/2} \int_0^{E_F} dE\, E^{\frac{3}{2}} \\ &= \frac{V}{2\pi^2}\left(\frac{2m}{\hbar^2}\right)^{3/2} E_F^{\frac{5}{2}} = \frac{3}{5} N_e E_F, \end{aligned} \tag{4.70}$$

where $\frac{3}{5}E_F$ is the average kinetic energy per electron in the Fermi sphere.

Obviously, electrons are not free but interact with each other via the Coulomb interaction. The Coulomb potential between the electrons is given by

$$H_1(\mathbf{r}) = \frac{1}{4\pi\varepsilon_0} \sum_{j>i} \frac{e^2}{|\mathbf{r}_i - \mathbf{r}_j|}, \tag{4.71}$$

where the summation i, j goes over the electrons in the system. However, since the eigenstates of the kinetic energy are planes waves, we prefer to write the Coulomb interaction in momentum space. The matrix element can be evaluated as follows:

$$U_{\mathbf{k}'''\sigma'''\mathbf{k}''\sigma''\mathbf{k}'\sigma'\mathbf{k}\sigma} = \frac{1}{4\pi\varepsilon_0 V^2}\delta_{\sigma\sigma'''}\delta_{\sigma'\sigma''}$$
$$\times \int d\mathbf{r}_1 \int d\mathbf{r}_2 e^{-i\mathbf{k}'''\cdot\mathbf{r}_1} e^{-i\mathbf{k}''\cdot\mathbf{r}_2} \frac{e^2}{|\mathbf{r}_1 - \mathbf{r}_2|} e^{i\mathbf{k}'\cdot\mathbf{r}_2} e^{i\mathbf{k}\cdot\mathbf{r}_1}.$$

The Coulomb interaction does not change the translational symmetry of the problem since it only depends on the relative positions $\mathbf{r}_i - \mathbf{r}_j$ of the electrons, and additionally, it is just a scalar. Therefore, we should be able to write this matrix element in terms of a part reflecting the symmetry of the problem (*i.e.* leading to momentum conservation) and a coupling strength. To bring this out, let us look at the spatial part and take $\mathbf{r}_2 = \mathbf{r}_1 - \mathbf{r}$:

$$U_{\mathbf{k}'''\sigma'''\mathbf{k}''\sigma''\mathbf{k}'\sigma'\mathbf{k}\sigma} = \frac{1}{4\pi\varepsilon_0 V}\delta_{\sigma\sigma'''}\delta_{\sigma'\sigma''}\left[\frac{1}{V}\int d\mathbf{r}_1 e^{i(-\mathbf{k}'''-\mathbf{k}''+\mathbf{k}'+\mathbf{k})\cdot\mathbf{r}_1}\right]$$
$$\times \int d\mathbf{r} \frac{e^2}{r} e^{i(\mathbf{k}''-\mathbf{k}')\cdot\mathbf{r}}.$$

The integral in the brackets equals the Kronecker δ function $\delta_{\mathbf{k}'''+\mathbf{k}'',\mathbf{k}'+\mathbf{k}}$ imposing momentum conservation. We can always satisfy this by taking $\mathbf{k}'' = \mathbf{k}' - \mathbf{q}$ and $\mathbf{k}''' = \mathbf{k} + \mathbf{q}$. This gives

$$U_{\mathbf{k}+\mathbf{q},\sigma''',\mathbf{k}'-\mathbf{q},\sigma''\mathbf{k}'\sigma'\mathbf{k}\sigma} = \frac{U_\mathbf{q}}{V}\delta_{\sigma\sigma'''}\delta_{\sigma'\sigma''}. \tag{4.72}$$

The Coulomb interaction between the electrons can be written as

$$\boldsymbol{H}_1 = \sum_{\substack{\mathbf{k}\mathbf{k}'\mathbf{q}\\ \sigma\sigma'}} U_\mathbf{q} c^\dagger_{\mathbf{k}+\mathbf{q},\sigma} c^\dagger_{\mathbf{k}'-\mathbf{q},\sigma'} c_{\mathbf{k}'\sigma'} c_{\mathbf{k}\sigma}. \tag{4.73}$$

The Fourier component of the Coulomb interaction between two electrons can be directly obtained from Gauss's law for a point charge

$$\nabla\cdot\mathbf{E} = -\nabla^2 V = \nabla^2\frac{U}{e} = -\frac{e\delta(\mathbf{r})}{\varepsilon_0} \Rightarrow -q^2 U_\mathbf{q} = -\frac{e^2}{\varepsilon_0} \Rightarrow U_\mathbf{q} = \frac{e^2}{\varepsilon_0 q^2},$$

where $V = -U/e$ is the potential, and U the potential energy. Note that the δ function in real space is completely delocalized in momentum space. However, for other purposes, it is actually instructive to make a Fourier transform of the real-space potential. The nonunitary Fourier transform, see also Eq. (1.248), is given by

$$U(r) = \frac{1}{4\pi\varepsilon_0}\frac{e^2}{r}e^{-\kappa r} \Rightarrow U_\mathbf{q} = \langle\mathbf{q}|U\rangle = \int d\mathbf{r}\, \langle\mathbf{q}|\mathbf{r}\rangle\langle\mathbf{r}|U\rangle = \int d\mathbf{r}\, e^{-i\mathbf{q}\cdot\mathbf{r}}U(\mathbf{r}).$$

A decaying component $e^{-\kappa r}$ has been added to the potential for convergence. Without lack of generality, we can take $\mathbf{q}$ along the z-axis. This gives

$$\begin{aligned} U_\mathbf{q} &= \frac{e^2}{4\pi\varepsilon_0}\int_0^{2\pi} d\varphi \int_0^\infty dr \int_0^\pi d\theta\, r^2 \sin\theta\, e^{-iqr\cos\theta-\kappa r}\frac{1}{r} \\ &= \frac{e^2}{2\varepsilon_0}\int_0^\infty dr \int_{-1}^1 d\cos\theta\; re^{-iqr\cos\theta-\kappa r} = \frac{e^2}{2iq\varepsilon_0}\int_0^\infty dr(e^{iqr} - e^{-iqr})e^{-\kappa r} \\ &= \frac{e^2}{2iq\varepsilon_0}\left(-\frac{1}{iq-\kappa} + \frac{1}{-iq-\kappa}\right) = \frac{e^2}{\varepsilon_0}\frac{1}{q^2+\kappa^2}. \end{aligned} \tag{4.74}$$

In the limit that $\kappa \to 0$, this reduces to the result obtained with Gauss's law. However, in Section 5.5, we will see that κ becomes finite when considering the dynamical response of the electrons to a potential.

We would now like to know the ground state energy. We do this in lowest order perturbation theory, which assumes that the Coulomb interactions leave the Fermi sphere unchanged. One might question the validity of this procedure since the Coulomb interaction is by no means a small perturbation. However, we start with this and see where it gets us. The expectation value of the Coulomb interaction in the Fermi sphere is

$$\langle \boldsymbol{H}_1 \rangle = \langle F|\boldsymbol{H}_1|F\rangle = \sum_{\sigma\sigma'}\sum_{\mathbf{kk'q}} \frac{U_{\mathbf{q}}}{V}\langle F|\mathbf{c}^{\dagger}_{\mathbf{k+q},\sigma}\mathbf{c}^{\dagger}_{\mathbf{k'-q},\sigma'}\mathbf{c}_{\mathbf{k'}\sigma'}\mathbf{c}_{\mathbf{k}\sigma}|F\rangle. \tag{4.75}$$

Since there is only one configuration (the Fermi sea), there are only diagonal matrix elements. One way to obtain a finite expectation value is to take $\mathbf{q} = 0$. Note that the Coulomb matrix element in Eq. (4.74) diverges for $q = 0$.

Let us consider the contribution of the $q = 0$ term. The matrix element is

$$\begin{aligned}\langle \boldsymbol{H}_1 \rangle_{\text{direct}} &= \frac{1}{2}\frac{U_0}{V}\sum_{\sigma\sigma'}\sum_{\mathbf{kk'}}\langle F|\mathbf{c}^{\dagger}_{\mathbf{k}\sigma}\mathbf{c}^{\dagger}_{\mathbf{k'}\sigma'}\mathbf{c}_{\mathbf{k'}\sigma'}\mathbf{c}_{\mathbf{k}\sigma}|F\rangle \\ &= -\frac{1}{2}\frac{U_0}{V}\sum_{\sigma\sigma'}\sum_{\mathbf{kk'}}\langle F|\mathbf{c}^{\dagger}_{\mathbf{k'}\sigma'}(\delta_{\mathbf{kk'}}\delta_{\sigma\sigma'} - \mathbf{c}_{\mathbf{k'}\sigma'}\mathbf{c}^{\dagger}_{\mathbf{k}\sigma})\mathbf{c}_{\mathbf{k}\sigma}|F\rangle \\ &= \frac{1}{2}\frac{U_0}{V}\sum_{\sigma\sigma'}\sum_{\mathbf{kk'}}\langle F|(\mathrm{N}_{\mathbf{k'}\sigma'}\mathrm{N}_{\mathbf{k}\sigma} - \mathrm{N}_{\mathbf{k}\sigma}\delta_{\mathbf{kk'}}\delta_{\sigma\sigma'})|F\rangle,\end{aligned} \tag{4.76}$$

using the anticommutation from Eq. (4.21), and where the number operator is defined as $\mathrm{N}_{\mathbf{k}\sigma} = \mathbf{c}^{\dagger}_{\mathbf{k}\sigma}\mathbf{c}_{\mathbf{k}\sigma}$. This result simplifies to

$$\langle \boldsymbol{H}_1 \rangle_{\text{direct}} = \frac{1}{2}\frac{U_0}{V}N(N-1) \cong \frac{1}{2}\frac{U_0}{V}N^2 = \frac{1}{2}\frac{e^2N^2}{V\varepsilon_0\kappa^2}, \tag{4.77}$$

which is a very large repulsive interaction. Note that $\frac{1}{2}N(N-1)$ is the number of pairs of electrons. This term is the static Coulomb repulsion between the electrons. However, generally, we are dealing with charge-neutral systems. Let us consider the situation for a solid. We have left out, so far, the positive background of the nuclei that make the total system neutral. Suppose we can describe the nuclei by a uniform background density $\rho(\mathbf{r}) = N/V$. We then have the following two contributions:

$$\langle \boldsymbol{H}_1 \rangle_{\text{background}} = \frac{1}{2}\frac{e^2}{4\pi\varepsilon_0}\int \rho(\mathbf{r}_1)\rho(\mathbf{r}_2)\frac{e^{-\kappa|\mathbf{r}_1-\mathbf{r}_2|}}{|\mathbf{r}_1-\mathbf{r}_2|}d\mathbf{r}_1 d\mathbf{r}_2 = \frac{1}{2}\frac{e^2N^2}{V\varepsilon_0\kappa^2}, \tag{4.78}$$

describing the interaction between the nuclei, and

$$\langle \boldsymbol{H}_1 \rangle_{\text{e-background}} = -\frac{e^2}{4\pi\varepsilon_0}\sum_i\int \rho(\mathbf{r})\frac{e^{-\kappa|\mathbf{r}-\mathbf{r}_i|}}{|\mathbf{r}-\mathbf{r}_i|}d\mathbf{r} = -\frac{e^2N^2}{V\varepsilon_0\kappa^2}, \tag{4.79}$$

describing the interaction of the electrons with the nuclei. The integrals are essentially equivalent to that in Eq. (4.74). Note the absence of the factor 1/2 in the latter, since we do not have to correct for double counting the interactions here. The contributions from Eqs. (4.77)–(4.79) cancel each other. Therefore, the static contribution of the Coulomb interactions is zero for a neutral system. From now on, we implicitly assume that the summation in the Coulomb interaction excludes $\mathbf{q} = 0$.

There is, however, an additional way to get a finite expectation value. Note that there were two terms in Eq. (4.36). The matrix element above combined the terms with the same spin. This is the direct term. However, we can also exchange the particles in that case $\mathbf{c}^\dagger_{\mathbf{k}+\mathbf{q},\sigma}$ annihilates $\mathbf{c}_{\mathbf{k}'\sigma'}$, and $\mathbf{c}^\dagger_{\mathbf{k}'-\mathbf{q},\sigma'}$ annihilates $\mathbf{c}_{\mathbf{k}\sigma}$. This can be satisfied by taking $\mathbf{k}' = \mathbf{k}+\mathbf{q}$. Additionally, the exchange term is only nonzero if the spins are parallel, *i.e.* $\sigma = \sigma'$. The sum of the expectation values of the direct and the exchange terms within the Fermi sea is called the Hartree–Fock approximation. The exchange term can be written as

$$\begin{aligned}\langle \boldsymbol{H}_1\rangle_{\text{exch}} &= \sum_\sigma\sum_{\mathbf{kq}} \frac{\mathbf{q}}{V}\langle F|\mathbf{c}^\dagger_{\mathbf{k}+\mathbf{q},\sigma}\mathbf{c}^\dagger_{\mathbf{k}\sigma}\mathbf{c}_{\mathbf{k}+\mathbf{q},\sigma}\mathbf{c}_{\mathbf{k}\sigma}|F\rangle \\ &= -\sum_\sigma\sum_{\mathbf{kq}} \frac{U_\mathbf{q}}{V}\langle F|\mathrm{N}_{\mathbf{k}+\mathbf{q},\sigma}\mathrm{N}_{\mathbf{k}\sigma}|F\rangle. \end{aligned} \tag{4.80}$$

We want to write this in a somewhat different form. We can write the kinetic energy per particle as

$$\langle \boldsymbol{H}_0\rangle = \frac{1}{N}\sum_{\mathbf{k}\sigma}\varepsilon_\mathbf{k}\langle F|\mathrm{N}_{\mathbf{k},\sigma}|F\rangle. \tag{4.81}$$

Rewriting the exchange term in the same fashion gives

$$\langle \boldsymbol{H}_1\rangle_{\text{exch}} = \frac{1}{N}\sum_{\mathbf{k}\sigma}\Sigma_\mathbf{k}\langle F|\mathrm{N}_{\mathbf{k},\sigma}|F\rangle \quad \text{with} \quad \Sigma_\mathbf{k} = -\frac{1}{V}\sum_\mathbf{q} U_\mathbf{q}\langle F|\mathrm{N}_{\mathbf{k}+\mathbf{q}}|F\rangle.$$

$\Sigma_\mathbf{k}$ is known as the self-energy of the particle. We see that it is negative. This is a result of Pauli's principle. Since electrons with the same spin cannot be in the same quantum state, they tend to avoid each other. This effectively leads to a somewhat lower Coulomb interaction between electrons with parallel spins. We can evaluate the integral in the self-energy as

$$\begin{aligned}\frac{1}{V}\sum_{|\mathbf{k}'|<k_F}\frac{1}{|\mathbf{k}-\mathbf{k}'|^2} &= \frac{1}{(2\pi)^2}\int_0^{k_F} dk' k'^2\int_{-1}^{1} d\cos\theta \frac{1}{k^2+k'^2-2kk'\cos\theta} \\ &= \frac{1}{(2\pi)^2}\int_0^{k_F} dk' k' \ln\left|\frac{k'+k}{k'-k}\right| \\ &= \frac{k_F}{2\pi^2}\left(\frac{1}{2}+\frac{k_F^2-k^2}{4kk_F}\ln\left|\frac{k_F+k}{k_F-k}\right|\right) = \frac{k_F}{2\pi^2}F(k). \end{aligned} \tag{4.82}$$

The total energy for an electron is then

$$E_\mathbf{k} = \varepsilon_\mathbf{k} + \Sigma_\mathbf{k} = \varepsilon_\mathbf{k} - \frac{e^2 k_F}{2\pi^2\varepsilon_0}F(k) \tag{4.83}$$

Figure 4.1 shows the energies as a function of k. Note that the total electron–electron interaction is positive. Only the reduction of the Coulomb energy due to the exchange term is shown here.

When looking at the total energy in Figure 4.1, we see that it is no longer a parabola. For free electrons, the curvature of the energy dispersion directly corresponds to the mass,

$$\frac{1}{m} = \frac{1}{\hbar^2}\frac{\partial^2\varepsilon_\mathbf{k}}{\partial k^2}. \tag{4.84}$$

We now want to reinterpret the curvature of the total energy as an effective mass $m^*_\mathbf{k}$,

$$E_\mathbf{k} = E_{\text{shift}} + \frac{\hbar^2k^2}{2m^*_\mathbf{k}} \qquad \Rightarrow \qquad dE_\mathbf{k} = \frac{\hbar^2 k}{m^*_\mathbf{k}}dk. \tag{4.85}$$

Figure 4.1 The energies $\varepsilon_{\mathbf{k}}$ and $\Sigma_{\mathbf{k}}^{\text{exch}}$ plus the total energy for $a_0 k_F/\pi = 1$, where $E_F = \hbar^2 k_F^2/2m$. The solid line are the results for $\kappa = 0$, and the dashed line for $\kappa = 0.1$.

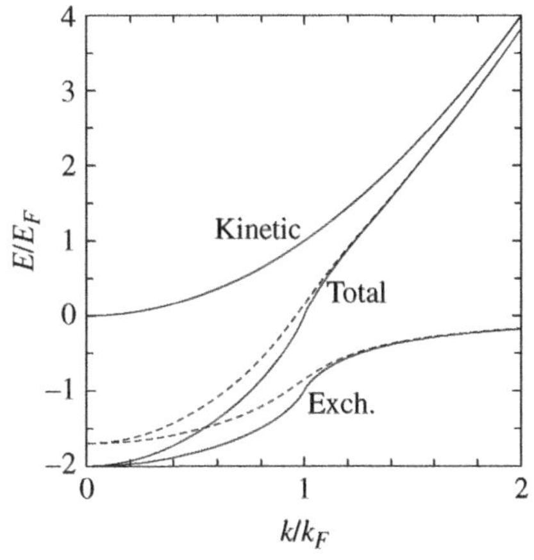

Applying this to Eq. (4.83) and dividing by $1/m$ gives

$$\frac{m}{m_{\mathbf{k}}^*} = 1 + \frac{\partial \Sigma_{\mathbf{k}}^{\text{exch}}}{\partial \varepsilon_{\mathbf{k}}} = \frac{e^2 m}{2\pi} \frac{k_F}{k^2} \left(\frac{k^2 + k_F^2}{k k_F} \ln \left| \frac{k + k_F}{k - k_F} \right| - 2 \right). \tag{4.86}$$

Therefore, $1/m_{\mathbf{k}}^*$ has a logarithmic divergence at the Fermi energy when $k/k_F \to 1$. This would have serious consequences at low temperature, where the Fermi gas would become unstable. For example, the specific heat would diverge as $T/\ln T$, which is not observed experimentally. The divergence is a result of the long range of the Coulomb interaction. It can be removed by a nonzero κ. In Figure 4.1, one sees that, although the divergence is removed, the energies change relatively little. The effective Coulomb interaction is then $U(\mathbf{r}) = e^2/(4\pi\varepsilon_0)(e^{-\lambda r}/r)$. The limited range of the Coulomb interaction is known as screening.

Conceptual summary. – In this section, we considered a first example of a many-body problem. Obviously, we looked at problems with several charges before, when looking at atoms, molecules, and solids. However, here the protons were treated as fixed in space and therefore were treated as an effective potential. Here, we are effectively doing the same thing for the Coulomb interaction between electrons. However, the electrons are not considered fixed in real space but in momentum space. The end goal is the same. We want to treat the electron–electron interaction as an effective potential. The eigenfunctions are not adjusted to account for the interactions between electrons. This method fails and leads to unphysical results such as a diverging effective mass at the Fermi energy.

This can be fixed by taking a screened Coulomb potential. However, we will see in Section 5.5 that screening is an effective model to describe rearrangements of the electron density.

The state used in this section, see Eq. (4.66), is a single product of eigenfunctions or a single Slater determinant. Even when including the effects of screening, we are only still adjusting occupations of states. This means that we are still framing the many-body system as an effective independent-particle model, where an electron feels a potential due to its interaction with other electrons. In the forthcoming sections, we will be looking at problems where such an approach is not possible, and we have to consider combinations of Slater determinants or product states.

Since we are working in terms of an effective independent-electron approach, it still makes sense to plot the energy of an electron as a function of momentum, see Figure 4.1 (this no longer makes sense for real many-body wavefunctions). However, reinterpreting the dispersion curve for the effective "independent" electron leads to an energy-dependent mass. Note that we could have done the same exercise in Sections 3.7 and 3.8. This leads to the following question: if the change of dispersion of a particle can be interpreted as an effective mass, what about the rest mass itself? Is the rest mass an intrinsic quantity of the particle or also just an effective model for other interactions? We return to this in Section 5.3.

Problems

1 Although the Coulomb interactions treated within the Hartree–Fock limit change the effective mass at the Fermi level, the system remains a metal. A model where Coulomb interactions can turn a metallic system into an insulator is the Hubbard model,

$$\boldsymbol{H} = -t \sum_{\mathbf{R}\delta\sigma} \mathbf{c}^{\dagger}_{\mathbf{R}+\delta,\sigma} \mathbf{c}_{\mathbf{R}\sigma} + U \sum_{\mathbf{R}} n_{\mathbf{R}\uparrow} n_{\mathbf{R}\downarrow}. \tag{4.87}$$

This is essentially a system with a single orbital per site with nearest-neighbor hopping. This part is essentially the tight-binding system from Section 3.7 extended with spin $\sigma = \uparrow, \downarrow$. Additionally, when two electrons are on the same site, they experience a Coulomb repulsion U.

a) Generally, there is no simple exact solution for this model. To turn this into an effective one-particle model, let us consider the motion of an upspin through the system. For a particular site, the orbital with opposite spin can be either occupied or empty. For half-filling, determine the local Green's function, *i.e.* for $t = 0$ (to obtain a symmetric Green's function $G^0(E)$, take the on-site energy $\varepsilon_0 = -U/2$).

b) Determine the full Green's function $G_{\mathbf{RR}'}(E)$ in the presence of hopping. Make a Fourier transform to obtain $G_{\mathbf{k}}(E)$ (include only nearest-neighbor hopping).

c) Show that the result can be written as a 2×2 matrix.

d) Sketch the eigenenergies and weights for a one-dimensional system.

4.4 Tensors

Introduction. – In this chapter, we are looking at many-body problems. The vast majority of interactions can be viewed as occurring between two particles. It is convenient to view a two-particle interaction as two processes. Let us consider the Coulomb interaction in Eq. (4.73). One of the electrons scatters from $\mathbf{k}$ to $\mathbf{k} + \mathbf{q}$, whereas the other electron scatters from $\mathbf{k}'$ to $\mathbf{k}' - \mathbf{q}$. In the process, momentum $\mathbf{q}$ is exchanged between the particles. Note that the Coulomb interaction is a scalar, meaning that momentum is only exchanged in the process; no momentum is created by the Coulomb interaction. Viewing the two-particle interaction as two separate one-particle interactions coupled via an exchange is useful when considering other problems. In the forthcoming sections, we will look at the many-body interactions between electrons on an atom and between quarks inside mesons and baryons. Although these look like rather different problems, the underlying ideas are actually surprisingly similar.

The advantage of splitting the two-particle interaction into two one-particle interactions is that we can use results obtained in Chapter 1 regarding one-particle interactions. For electrons on an atom, we are still looking at the Coulomb interaction, but the different symmetry of the problem dictates what is exchanged. For free space, the photons carry momentum, which is a good quantum number. However, for an atom, a spherically symmetric problem, the photons carry angular momentum. The difference is a result of the eigenfunctions necessary to describe the problem, not because the photon decides to exchange different quantities.

In Chapter 1, it was shown that higher order vectors could be expressed in terms of indistinguishable spinors. For atoms, we restrict the discussion to s and p orbitals; d and f orbitals are more complex, but the underlying physics is the same. The threefold p orbitals also have nice parallels with quarks.

In order to understand the interaction between two particles, let us first look at the scattering of a single particle. For a p orbital with $l = 1$, the basis $|1m\rangle$ in terms of spinors is given by $|11\rangle = \boldsymbol{e}_1 = \boldsymbol{a}_\uparrow^\dagger \boldsymbol{a}_\uparrow^\dagger/\sqrt{2}$, $|10\rangle = \boldsymbol{e}_0 = \boldsymbol{a}_\uparrow^\dagger \boldsymbol{a}_\downarrow^\dagger$, and $|1,-1\rangle = \boldsymbol{e}_{-1} = \boldsymbol{a}_\downarrow^\dagger \boldsymbol{a}_\downarrow^\dagger/\sqrt{2}$. This is the same as the unit vectors $\boldsymbol{e}_q = |1q\rangle$ in a three-dimensional spherical basis written in terms of pairs of spinors, see Eq. (1.351). One way that a particle can scatter is given by the angular momentum operators $\mathbf{L}_q$, with $q = 1, 0, -1$. The $q = 0$ component, $\mathbf{L}_0 = \mathbf{L}_z = \frac{1}{2}(\boldsymbol{a}_\uparrow^\dagger \boldsymbol{a}_\uparrow - \boldsymbol{a}_\downarrow^\dagger \boldsymbol{a}_\downarrow)$, see Eq. (1.330), gives the scattering where no projected angular momentum is transferred. For the operators $\mathbf{L}_{\pm 1} = \mp\frac{1}{\sqrt{2}} \boldsymbol{a}_{\pm\frac{1}{2}}^\dagger \boldsymbol{a}_{\mp\frac{1}{2}}$, see Eq. (1.385), the projected angular momentum is changed by ± 1. For p orbitals, we express these operators in matrix form in the basis $|1m\rangle$. For example,

$$\boldsymbol{W}_1^1 = \sum_{m,m'=-1}^{1} |1m'\rangle\langle m'|\mathbf{L}_1|1m\rangle\langle 1m| = \begin{pmatrix} 0 & -1 & 0 \\ 0 & 0 & -1 \\ 0 & 0 & 0 \end{pmatrix}, \tag{4.88}$$

where the notation $\boldsymbol{W}_q^k$ has been introduced to allow for further generalization. The indices in the matrix are ordered $m = 1, 0, -1$. The index k gives the rank of the tensor, where $k = 1$ indicates vectors; $q = k, k-1, \ldots, -k$ are the $2k+1$ components for a particular values of k. To avoid too many indices, the values of $l = l' = 1$ are implicitly assumed. For different l and l', the matrices have different dimensions.

Although spherical tensors are perfect for spherical symmetry, we also introduce the tensors in real form. This produces Hermitian matrices that connect better to those used in the discussion of quarks. Equation (1.348) gives $\boldsymbol{W}_x^1 = \mathbf{L}_x = (-\boldsymbol{W}_1^1 + \boldsymbol{W}_{-1}^1)/\sqrt{2}$, $\boldsymbol{W}_y^1 = \mathbf{L}_y = i(\boldsymbol{W}_1^1 + \boldsymbol{W}_{-1}^1)/\sqrt{2}$, $\boldsymbol{W}_z^1 = \mathbf{L}_z = \boldsymbol{W}_0^1$. Written in matrix form for the p orbitals, the operators are

$$\boldsymbol{W}_x^1 = \mathbf{L}_x = \begin{pmatrix} 0 & \frac{1}{\sqrt{2}} & 0 \\ \frac{1}{\sqrt{2}} & 0 & \frac{1}{\sqrt{2}} \\ 0 & \frac{1}{\sqrt{2}} & 0 \end{pmatrix}, \quad \boldsymbol{W}_y^1 = \mathbf{L}_y = \begin{pmatrix} 0 & -\frac{i}{\sqrt{2}} & 0 \\ \frac{i}{\sqrt{2}} & 0 & -\frac{i}{\sqrt{2}} \\ 0 & \frac{i}{\sqrt{2}} & 0 \end{pmatrix},$$

$$\boldsymbol{W}_z^1 = \mathbf{L}_z = \begin{pmatrix} 1 & 0 & 0 \\ 0 & 0 & 0 \\ 0 & 0 & -1 \end{pmatrix}. \tag{4.89}$$

Note that the basis for the orbitals is spherical, whereas the basis for the operators is real or tesseral. However, these are not all the possible operations on this basis. For example, we have no operator that brings $|11\rangle = \boldsymbol{e}_1 = \boldsymbol{a}_\uparrow^\dagger \boldsymbol{a}_\uparrow^\dagger/\sqrt{2}$ into $|1,-1\rangle = \boldsymbol{e}_{-1} = \boldsymbol{a}_\downarrow^\dagger \boldsymbol{a}_\downarrow^\dagger/\sqrt{2}$. This transition requires the change of two spinors. Furthermore, there are nine elements in the matrix but only three matrices. For spherical operators, all matrix elements are real. For real operators, the upper and lower triangles are related by conjugation. However, the matrix elements can now be complex, leading to the same number of elements. There are therefore six other independent matrices. One of those is the identity matrix $\mathbb{1}_3$. This leaves five matrices. These can be described by matrices that change two spinors. In principle, one can make operators that change three spinors. However, these do not lead to any new operations for $l = 1$, although they do for higher l values.

The evaluation of the matrices is somewhat laborious, so let us restrict ourselves to an example and consider the following products:

$$\boldsymbol{L}_{x/y}\boldsymbol{L}_{x/y} = \frac{1}{2}\left(\pm \boldsymbol{a}_\downarrow^\dagger \boldsymbol{a}_\uparrow \boldsymbol{a}_\downarrow^\dagger \boldsymbol{a}_\uparrow + \boldsymbol{a}_\downarrow^\dagger \boldsymbol{a}_\uparrow \boldsymbol{a}_\uparrow^\dagger \boldsymbol{a}_\downarrow + \boldsymbol{a}_\uparrow^\dagger \boldsymbol{a}_\downarrow \boldsymbol{a}_\downarrow^\dagger \boldsymbol{a}_\uparrow \pm \boldsymbol{a}_\uparrow^\dagger \boldsymbol{a}_\downarrow \boldsymbol{a}_\uparrow^\dagger \boldsymbol{a}_\downarrow\right). \tag{4.90}$$

The result for $l = 1$ can be written in matrix form as

$$\boldsymbol{L}_{x/y}\boldsymbol{L}_{x/y} = \begin{pmatrix} \frac{1}{2} & 0 & \pm\frac{1}{2} \\ 0 & 1 & 0 \\ \pm\frac{1}{2} & 0 & \frac{1}{2} \end{pmatrix}. \tag{4.91}$$

Although the different combinations are proper operators, conventionally these are written in the real combinations of spherical harmonics

$$\frac{r^2}{\sqrt{2}}(C_{22}(\hat{\boldsymbol{r}}) + C_{2,-2}(\hat{\boldsymbol{r}})) = \frac{\sqrt{3}}{2} r^2 \sin^2\theta \cos 2\varphi = \frac{\sqrt{3}}{2}(x^2 - y^2). \tag{4.92}$$

The tesseral combinations correspond to the real d orbitals of an atom: $x^2 - y^2$ and xy (combinations of $m = \pm 2$), yz and zx ($m = \pm 1$), and $3z^2 - r^2$ ($m = 0$). Taking the same combinations for the operator above gives

$$\boldsymbol{W}^2_{x^2-y^2} = \sqrt{3}(\boldsymbol{L}_x\boldsymbol{L}_x - \boldsymbol{L}_y\boldsymbol{L}_y) = \sqrt{3}\begin{pmatrix} 0 & 0 & 1 \\ 0 & 0 & 0 \\ 1 & 0 & 0 \end{pmatrix}. \tag{4.93}$$

This gives the operator that can change m by ± 2. There is some arbitrariness in the relative scaling of the W_i^k for different k. Here, the normalization is such that the diagonal element with $m = l$ is one for the diagonal tensors, *i.e.* $i = z$, $3z^2 - r^2$ for $k = 1, 2$, respectively. Note that this corresponds to the 0 component in spherical symmetry. This is a convenient normalization for spherical tensors.

The remaining four matrices, we give without derivation:

$$\boldsymbol{W}^2_{xy} = \sqrt{3}(\boldsymbol{L}_x\boldsymbol{L}_y + \boldsymbol{L}_y\boldsymbol{L}_x) = \sqrt{3}\begin{pmatrix} 0 & 0 & -i \\ 0 & 0 & 0 \\ i & 0 & 0 \end{pmatrix},$$

$$\boldsymbol{W}^2_{yz} = \sqrt{3}(\boldsymbol{L}_y\boldsymbol{L}_z + \boldsymbol{L}_z\boldsymbol{L}_y) = \sqrt{\frac{3}{2}}\begin{pmatrix} 0 & -i & 0 \\ i & 0 & i \\ 0 & -i & 0 \end{pmatrix},$$

$$\boldsymbol{W}^2_{zx} = \sqrt{3}(\boldsymbol{L}_z\boldsymbol{L}_x + \boldsymbol{L}_x\boldsymbol{L}_z) = \sqrt{\frac{3}{2}}\begin{pmatrix} 0 & 1 & 0 \\ 1 & 0 & -1 \\ 0 & -1 & 0 \end{pmatrix},$$

$$\boldsymbol{W}^2_{3z^2-r^2} = 2\boldsymbol{L}_z\boldsymbol{L}_z - \boldsymbol{L}_x\boldsymbol{L}_x - \boldsymbol{L}_y\boldsymbol{L}_y = \begin{pmatrix} 1 & 0 & 0 \\ 0 & -2 & 0 \\ 0 & 0 & 1 \end{pmatrix}. \tag{4.94}$$

The $\boldsymbol{W}^2_i$ are tensors of rank 2. Note that the components correspond to the real d atomic orbitals $x^2 - y^2, xy, yz, zx, 3z^2 - r^2$. The sum of the squared matrix elements of $\boldsymbol{W}^2_i$ is 6. Together with the tensor of rank zero, $\boldsymbol{W}^0_s = \mathbb{1}_3$, these nine matrices can describe any operator acting on a p orbital. The matrices form a basis for all possible Hermitian 3×3 matrices. There are many different ways to form an orthogonal basis for Hermitian 3×3 matrices. This basis is convenient for spherical symmetry, since each matrix belongs to a particular rank k in spherical symmetry.

Just as orbitals have a particular angular momentum, the operators that scatter between different orbitals can also be assigned an angular momentum or rank. If the initial and final orbitals have an angular momentum l and l', respectively, the rank ranges from $k = l + l', l + l' - 1, \ldots, |l - l'|$. For example, for the scattering between two p orbitals ($l = l' = 1$), as described above, this gives possible ranks $k = 2, 1, 0$. An operator of rank k has $2k + 1$ components. This gives a total of $5 + 3 + 1 = 9$ matrices. The number of matrices corresponds directly to the number of matrix elements $(2l + 1)(2l + 1) = 3 \times 3 = 9$.

The above matrices describe the scattering of a single particle, in this case, the scattering of an electron between p orbitals. A similar matrix is needed to describe the scattering of the other electron in the Coulomb interaction. Since the Coulomb interaction is a scalar, the product of the tensors also needs to be a scalar. It is therefore given by the inner product

$$H_1 = \sum_k A_k \boldsymbol{W}'^k \cdot \boldsymbol{W}^k = \sum_{ki} A_k \boldsymbol{W}'^k_i \boldsymbol{W}^k_i, \tag{4.95}$$

where $i = s$ for $l = 0$, $i = x, y, z$ for $l = 1$, and $i = x^2 - y^2, xy, yz, zx, 3z^2 - r^2$ for $l = 2$. The prime indicates that $\boldsymbol{W}'^k$ is operating on a different particle than $\boldsymbol{W}^k$. To maintain spherical symmetry, the coefficient A_k can only depend on the rank k of the tensor and not on the coefficients i. Although the matrices look very different, they will only transform into matrices of the same rank k when the system is rotated. This is similar to orbitals: for example, a rotation will never change a p orbital into a d orbital and vice versa. Of course, it is possible to create an operator where the coefficient depends on i. For example, one can have equal coefficients for the x and y components but a different coefficient for the z-direction. However, this will create a planar system and therefore breaks spherical symmetry. There is an extensive literature on what matrices to use for systems with a particular symmetry. This is known as representation theory.

The group of matrices is an example of, what is known as, the special orthogonal group in three dimensions or SO(3). The matrices describe all the properties for rotation for $l = 1$. Since the rotation can be over any angle, it is a continuous group. This should be contrasted with, for example solids, where only rotations over a particular angle (say, 90° or 120°) are symmetry operations of the system. This is known as a point group. Additionally, the rotations can also be obtained via differentiation as we saw in Chapter 1. This makes SO(3) a Lie group.

Although spherical symmetry is already a very high symmetry, it is possible to make an even higher symmetry for scattering between the different $l = 1$ states. In spherical symmetry, the matrices split into three groups indicated by their rank $k = 0, 1, 2$. In the end, we are interested to see how a two-particle interaction splits the many-body states into different clusters of eigenstates. For this splitting, the $k = 0$ matrix is irrelevant. Since it is effectively a scalar term, it only depends on the number of particles N in the many-body states. The matrix element of the $k = 0$ term is diagonal and proportional to the number of pairs $\frac{1}{2}N(N-1)$. It therefore shifts all the many-body states equally. This leaves the $k = 1, 2$ terms. The symmetry operations of the systems tell us what has to be the same. Therefore, all the orbitals for a certain nl have the same energy since rotations transform the different components indicated by m into each other. This does not directly imply that they have a different energy. For example, the potential of the nucleus in a hydrogen atom does not split the 3p and 3d electrons in energy since the energy is proportional to $1/n^2$. It is also important to note that symmetry tells you how states might split, it does not tell you the ordering of these states in energy. It also tells you nothing about how much they split. Tensors are "symmetry" operators, they do not provide information about the underlying "physical" nature of the interaction (there is some arbitrariness in the designation, and one can, obviously, consider momentum and angular momentum physical operators. This is fine, as long as you understand their intimate connection to the symmetry of the system).

Symmetry therefore tells you what tensors are effectively equivalent. When the symmetry is lowered, more tensors become inequivalent, and more coefficients (or interactions strengths) are needed to describe the problem. For example, in spherical symmetry (SO(3)), all tensor $\boldsymbol{W}^1_\mu$ are equivalent. However, in cylindrical symmetry (SO(2)), one of the matrices will have a different coefficient. For example, for a magnetic field in the z-direction, $\boldsymbol{W}^1_z$ has a nonzero coefficient. This will split the components of a p orbital into $1, 0, -1$ (the Zeeman splitting).

On the other hand, for a higher symmetry, more matrices become effectively equivalent. For spherical symmetry, which is already a very high symmetry, the matrices still split into $\boldsymbol{W}^1$ and $\boldsymbol{W}^2$. In order to make an even higher symmetry, we want to make all matrices effectively equivalent. This is somewhat comparable to the local potential treated in Section 3.9. For the local potential, there are no restrictions imposed by the symmetry (such as momentum conservation), and all the states with different momenta scatter into each other in an equivalent fashion. We want to create a similar situation for the 3×3 matrices. This is known as special unitary or SU(3).

In order to make all basis functions equivalent, the tensors have to be scaled in the same fashion. The easiest way to achieve this is by looking at the square of the elements of the tensors. For $k = 1, 2$, the sum of the matrix elements squared are 2 and 6, respectively. In order to make them comparable, $\boldsymbol{W}^2$ can be scaled by $1/\sqrt{3}$. Although the combined matrices of $\boldsymbol{W}^1$ and $\boldsymbol{W}^2/\sqrt{3}$ form a good set of operators for SU(3), the matrices are usually given in a somewhat different form

$$\boldsymbol{\lambda}_1 = \frac{1}{\sqrt{3}}\boldsymbol{W}^2_{x^2-y^2}, \quad \boldsymbol{\lambda}_2 = \frac{1}{\sqrt{3}}\boldsymbol{W}^2_{xy}, \quad \boldsymbol{\lambda}_3 = \boldsymbol{W}^1_z,$$

$$\boldsymbol{\lambda}_4 = \frac{1}{\sqrt{2}}\left(\boldsymbol{W}^1_x + \frac{1}{\sqrt{3}}\boldsymbol{W}^2_{zx}\right) = \begin{pmatrix} 0 & 1 & 0 \\ 1 & 0 & 0 \\ 0 & 0 & 0 \end{pmatrix},$$

$$\boldsymbol{\lambda}_5 = \frac{1}{\sqrt{2}}\left(\boldsymbol{W}^1_y + \frac{1}{\sqrt{3}}\boldsymbol{W}^2_{yz}\right) = \begin{pmatrix} 0 & -i & 0 \\ i & 0 & 0 \\ 0 & 0 & 0 \end{pmatrix}, \tag{4.96}$$

$$\boldsymbol{\lambda}_6 = \frac{1}{\sqrt{2}}\left(\boldsymbol{W}^1_x - \frac{1}{\sqrt{3}}\boldsymbol{W}^2_{zx}\right) = \begin{pmatrix} 0 & 0 & 0 \\ 0 & 0 & 1 \\ 0 & 1 & 0 \end{pmatrix},$$

$$\boldsymbol{\lambda}_7 = \frac{1}{\sqrt{2}}\left(\boldsymbol{W}^1_y - \frac{1}{\sqrt{3}}\boldsymbol{W}^2_{yz}\right) = \begin{pmatrix} 0 & 0 & 0 \\ 0 & 0 & -i \\ 0 & i & 0 \end{pmatrix}, \quad \boldsymbol{\lambda}_8 = \frac{1}{\sqrt{3}}\boldsymbol{W}^2_{3z^2-r^2}$$

These matrices are known as the Gell-Mann matrices. They are usually given in the basis $m = 1, -1, 0$ instead of $1, 0, -1$. Note that transitions with $\Delta m = \pm 2$ are now treated in a similar way to $\Delta m = \pm 1$. These matrices are important for describing the strong interaction between quarks. Note that there are only eight independent Hermitian matrices. The one that is missing is the identity matrix $\boldsymbol{W}^0_s = \mathbb{1}_3$, which we could denote as $\boldsymbol{\lambda}_0$. Since all components are now equivalent, the use of expressing everything in terms of one and two boson operators no longer makes sense. New operators can be introduced that have three components $\boldsymbol{a}^\dagger_q$ as opposed to the two components $(\uparrow, \downarrow)$ that were used for the spinors in three-dimensional space. The index $q = r, g, b$ is often given as red, green, and blue. However, these are simply labels and chosen such that the neutral object is white. The Gell-Mann matrices can then be expressed in terms of the new operators. For example,

$$\boldsymbol{\lambda}_4 = \boldsymbol{a}^\dagger_r \boldsymbol{a}_g + \boldsymbol{a}^\dagger_g \boldsymbol{a}_r. \tag{4.97}$$

However, we will be using the matrix form to express the strong interaction.

Conceptual summary. – In this section, we looked at the properties of matrices of dimension n, restricting ourselves to $n \times n = 3 \times 3$ matrices. Although often more difficult to visualize, matrices obey similar symmetry properties as the wavefunctions. In spherical symmetry, the orbitals for $n = 3$ split into $l = 0, 1, 2$. In the same fashion, the 3×3 matrices split into three distinct groups of rank $k = 0, 1, 2$. The symmetry operations tell us which matrices are equivalent to each other. The 3×3 matrices describe the symmetry properties of scattering between threefold-degenerate states (for example, a p orbital). It does not say anything of the physical aspects underlying the interaction. This is all contained in the coefficients in front of the matrices. The coefficients provide the information on the strength and the sign of the interaction. For spherical symmetry, this is generally encapsulated in the Wigner–Eckart theorem, see Eq. (3.135). However, the results are more general, leading to representation theory, which connects the geometry of space to algebraic quantities, such as matrices.

In the forthcoming sections, we combine the tensors, which describe the scattering of a single particle, to obtain two-particle interactions. We look at the Coulomb interaction in spherical symmetry (SO(3)) and the strong interaction in special unitary symmetry (SU(3)). Our focus lies on how these two-particle interactions split the many-body states into different groups. Again, this splitting is entirely determined by the symmetry of the problem and not by the detailed nature of the interaction. For example, for an atom this splitting occurs because the system is rotationally symmetric and happens regardless of the radial dependence of the interaction.

Problems

1 Tensors also occur when describing the local surroundings of an atom in a molecule or a solid. Suppose an atom is surrounded by six point charges in the x, y, and z directions. The additional potential on the p orbitals is then $U(\mathbf{r}) = \sum_{\delta=\pm 1}\sum_{i=x,y,z} U_0 (\mathbf{r} - a_i\delta\mathbf{e}_i)$, where $U_0(\mathbf{r} - \mathbf{R})$ is the potential of a point charge at position $\mathbf{R}$. The effect of this potential on the p can always be expressed in terms of renormalized spherical harmonics $C_q^k(\hat{\mathbf{r}})$ and radial wavefunctions $R_k(r)$, *i.e.* $U(\mathbf{r}) = \sum_{kq} C_q^k(\hat{\mathbf{r}})R_k(r)$. Using symmetry arguments, show what k and q remain for $a_i = a$ and for $a_x = a_y = a$ and $a_z = a'$. How do the p orbitals split?

2 Extend the discussion of the preceding problem for d orbitals for the case $a_i = a$ (octahedral symmetry). The matrix elements due to the six surrounding point charges between the d orbitals can be written as

$$\boldsymbol{U} = \Delta \begin{pmatrix} 1 & 0 & 0 & 0 & 5 \\ 0 & -4 & 0 & 0 & 0 \\ 0 & 0 & 6 & 0 & 0 \\ 0 & 0 & 0 & -4 & 0 \\ 5 & 0 & 0 & 0 & 1 \end{pmatrix}, \tag{4.98}$$

for the basis $m = 2, 1, 0, -1, -2$; Δ is the strength of the interaction. Explain from symmetry what matrix elements can be finite. How do the orbitals split for this interaction?

4.5 Electron Interactions on an Atom

Introduction. – We will now apply the results from the previous section on spherical tensors to the Coulomb interaction between electrons on an atom. Let us consider an orbital given by the quantum numbers nl containing N electrons. Including spin, there are $M = 2(2l + 1)$ different states. This gives $M!/(M - N)!N!$ different ways of placing N electrons into M states. When only considering the interaction of the electrons with the nucleus, all $(nl)^N$ configurations have the same energy. However, in the presence of the Coulomb interaction between electrons, the configurations interact with each other and split into, what is known as, Coulomb multiplets with different energy. Some eigenfunctions correspond to a single configuration. However, many eigenfunctions are linear combinations of different electron configurations.

We first consider more closely the Coulomb interaction between electrons. This interaction is inversely proportional to the distance between the electrons $1/|\mathbf{r} - \mathbf{r}'|$, where $\mathbf{r}$ and $\mathbf{r}'$ are the coordinates of the electrons. Note that the potential of the nucleus is proportional to $1/r$ and is properly written in spherical polar coordinates. However, the Coulomb interaction between electrons contains two vectors and therefore depends on the angle between the position vectors of the electrons. We therefore first have to expand the interaction in terms of spherical harmonics.

The potential between the electrons is given by

$$H_1(\mathbf{r}-\mathbf{r}') = \frac{1}{4\pi\varepsilon_0}\frac{e^2}{|\mathbf{r}-\mathbf{r}'|} = \frac{1}{4\pi\varepsilon_0}\frac{e^2}{r_>}\left(1+\frac{r_<^2}{r_>^2}-2\frac{r_<}{r_>}\cos\theta\right)^{-1/2}, \tag{4.99}$$

where θ is the angle between $\mathbf{r}$ and $\mathbf{r}'$, and $r_>$ and $r_<$ are the greater and the smaller of r and r', respectively. The expansion of the square root is the generating function of the Legendre polynomial, giving

$$H_1 = \frac{1}{4\pi\varepsilon_0}\frac{e^2}{r_>}\sum_{k=0}^{\infty}\left(\frac{r_<}{r_>}\right)^k P_k(\cos\theta). \tag{4.100}$$

In the above expression, the coordinates of both electrons are still intimately coupled, and we want to separate the scattering of the particles. This can be done as follows. The Legendre polynomial $P_k(\cos\theta) = C_0^k(\hat{\mathbf{r}})$ is the zero component of the renormalized spherical harmonics, see Eq. (1.363). The lowest terms are $P_k(z) = 1, z, \frac{3}{2}z^2 - r^2$ for $k = 0, 1, 2$. Without loss of generality, we can pick $\mathbf{r}' = \mathbf{e}_z$ and take the other vector with respect to the z-axis at an angle θ, and let us call that $\mathbf{r} \rightarrow \mathbf{r}_0$. This gives

$$P_k(\cos\theta) = C_0^k(\hat{\mathbf{r}}_0) = C_0^k(\hat{\mathbf{r}}_0)C_0^{k*}(\mathbf{e}_z). \tag{4.101}$$

The other spherical harmonic can be added using $C_q^k(\mathbf{e}_z) = \delta_{q,0}$ or $C_0^k(\mathbf{e}_z) = 1$, since for $\mathbf{r}' = \mathbf{e}_z$, the angle θ' is zero. Since all the other components contain $\sin\theta'$, the spherical harmonics are zero for $q \neq 0$. However, if $C_q^k(\mathbf{e}_z) = 0$ for $q \neq 0$, the other components can be added to the summation. The result can then be written as an inner product

$$P_k(\cos\theta) = \sum_{q=-k}^{k} C_q^k(\hat{\mathbf{r}}_0)C_q^{k*}(\mathbf{e}_z) = \mathbf{C}^k(\hat{\mathbf{r}}_0)\cdot\mathbf{C}^k(\mathbf{e}_z) = \mathbf{C}^k(\hat{\mathbf{r}})\cdot\mathbf{C}^k(\hat{\mathbf{r}}'). \tag{4.102}$$

However, since an inner product is invariant under rotation, we can now pick $\mathbf{r}'$ in any direction. The above equation is known as the spherical addition theorem. For $k = 1$, this reduces to $\cos\theta = \hat{\mathbf{r}}\cdot\hat{\mathbf{r}}'$, so we can view this as a generalization of the inner product for arbitrary rank k. This allows us to split $P_k(\cos\theta)$, which depends on the position of both particles into the inner product of spherical harmonics that only depend on the position of one particle.

We now want to obtain the matrix element $H_{\nu'''\nu''\nu'\nu}$ for the Coulomb interaction between electrons in hydrogen-like atomic orbitals $\psi_\nu(\mathbf{r})$ from Eq. (3.113), where $\nu = n_\nu l_\nu m_\nu \sigma_\nu$ is a shorthand for all the quantum numbers (including spin). The entire matrix element, given by $H_{\nu'''\nu''\nu'\nu} = U_{\nu'''\nu''\nu'\nu} - U_{\nu''\nu'''\nu'\nu}$, contains a direct and an exchange term as in Eq. (4.36). The matrix element can be split into different parts as follows:

$$\begin{aligned} U_{4321} &= \int d\mathbf{r}\int d\mathbf{r}'\psi_4^*(\mathbf{r})\psi_3^*(\mathbf{r}')U\psi_2(\mathbf{r}')\psi_1(\mathbf{r}) \\ &= \sum_k R_{4321}^k a_{l_4 l_1 k} a_{l_3 l_2 k}\, \boldsymbol{W}^k(4,1)\cdot \boldsymbol{W}^k(3,2)\delta_{\sigma_4,\sigma_1}\delta_{\sigma_3,\sigma_2}. \end{aligned} \tag{4.103}$$

The first term is the radial matrix element,

$$R_{4321}^k = \int dr\, r^2 \int dr'\, r'^2 R_{n_4 l_4}(r)R_{n_3 l_3}(r')\frac{e^2 r_<^k}{4\pi\varepsilon_0 r_>^{k+1}}R_{n_2 l_2}(r')R_{n_1 l_1}(r),$$

where $r_>$ and $r_<$ are the greater and the lesser, respectively, of r and r'. This matrix element contains most of the physical aspects of the interaction, *i.e.* the actual dependence of the potential on the distance. It determines the strength of the interaction. Although it is possible to evaluate these matrix elements, we will take them as a parameter since they do not affect the level structure.

The angular part of the matrix element is predominantly determined by the spherical symmetry of the problem. Since the Coulomb interaction only depends on the distance between the electrons, the integrals over the unit vectors $\hat{\mathbf{r}}$ and $\hat{\mathbf{r}}'$ can be separated. For example, for the electron with coordinate $\hat{\mathbf{r}}$

$$\int d\hat{\mathbf{r}}\; Y^*_{l_4 m_4}(\hat{\mathbf{r}}) C^k_q(\hat{\mathbf{r}}) Y_{l_1 m_1}(\hat{\mathbf{r}}) = a_{l_4 l_1 k} W^k_{q,m_4 m_1}, \tag{4.104}$$

where $W^k_{q,m'm}$ is the matrix element of the tensor $\boldsymbol{W}^k_q$ from Section 4.4. The renormalized spherical harmonic C^k_q comes from the Coulomb interaction, see Eq. (4.102), and $Y_{l_1 m_1}$ and $Y_{l_4 m_4}$ from the atomic orbitals. There is a similar matrix element for the other angular coordinate. Since C^k_q is a spherical tensor, the Wigner–Eckart theorem in Eq. (3.135) applies, and the matrix elements can be described in terms of indistinguishable spinors, see Eq. (3.138) and the preceding section. The factor $a_{l_4 l_1 k}$ contains what is known as the reduced matrix element, *i.e.* everything in the matrix element that cannot be described by the tensors from Section 4.4. This is essentially the equivalent of $(l_4||C^k||l_1)$ in Eq. (3.135) (some factors have been moved from the $3j$ symbol into the reduced matrix element to avoid dealing with square roots from the normalization of the $3j$ symbols that are not needed when dealing with operators). The derivation is somewhat laborious and will be omitted. The factor is given by

$$a_{l_4 l_1 k} = \sqrt{(2l_4+1)(2l_1+1)} \begin{pmatrix} l_4 & k & l_1 \\ 0 & 0 & 0 \end{pmatrix} \begin{pmatrix} l_4 & k & l_1 \\ -l_4 & 0 & l_1 \end{pmatrix}. \tag{4.105}$$

Note that it does not depend on m_1 and m_4, and $a_{l_4 l_1 k}$ is generally simply a constant. The reduced matrix element contains the physical aspects of the Coulomb interaction. Note that many of the physical aspects are in the radial matrix element; however, a couple remain in the angular part. The value of $a_{l_4 l_1 k}$ is zero when $l_1 + l_4 + k$ is odd. This reflects the fact that the integral is zero when the function that is integrated over is odd. Note that the tensors $\boldsymbol{W}^k_q$ themselves are not necessarily zero when $l_1 + l_4 + k$ is odd.

The only part that depends on m_1 and m_4 is given by a matrix element $(\boldsymbol{W}^k_q)_{m_4,m_1}$ of the spherical tensors. The spherical tensors describe how orbitals are transformed into each other when angular momentum is added to them, *i.e.* when the orbitals are rotated. They are entirely determined by the symmetry of the problem. The final part of the matrix element in Eq. (4.103) is spin conservation since the Coulomb interaction, being a scalar, cannot change the spin.

Both the radial and reduced matrix elements only depend on the n and l quantum numbers. Therefore, when looking at a particular configuration, they are constant scaling factors. The splitting of the energy levels is then entirely due to $\boldsymbol{W}^k_q$. Therefore, the "physical" aspect, the radial and reduced matrix elements, determines the size and order of the splitting. However, how the states split is determined by the "symmetry" part of the interaction, *i.e.* the spherical tensors. The Coulomb interaction between two electrons can then be written as

$$U_{4321} = \sum_k F^k\, \boldsymbol{W}'^k \cdot \boldsymbol{W}^k \delta_{\sigma_4,\sigma_1} \delta_{\sigma_3,\sigma_2}, \qquad F^k = R^k_{4321} a_{l_4 l_1 k} a_{l_3 l_2 k}.$$

Since we are generally looking at specific orbitals, the nl quantum numbers are assumed implicitly. The tensors are assumed to work on different electrons. Let us treat some examples to see how this works in practice.

Example 4.3 ***Two Electrons in an s Orbital (s^2)***
Let us look at the simplest example: two electrons in an s orbital. The matrix element in Eq. (4.103) reduces to $F^0\delta_{\sigma_1,\sigma_4}\delta_{\sigma_2,\sigma_3}$, with $F^0 = R^0_{1111}$ being the direct radial matrix element. Note that the tensor operator is simply $\boldsymbol{W}^0 = 1$. If the two electrons are in the same s orbital, the only configuration is $|n00\uparrow n00\downarrow\rangle$. The matrix element is

$$\langle n00\uparrow, n00\downarrow|\boldsymbol{H}_1|n00\uparrow, n00\downarrow\rangle = F^0. \tag{4.106}$$

This is the repulsion between two electrons in an s orbital.

Example 4.4 ***Two Electrons in Two s Orbitals ($ns^1n's^1$)***
Things become more interesting when we put the electrons in different s shells (for example, 1s and 2s on an atom). Since there are two different shells involved, we have, in addition to the direct term $F^0 = R^0_{1221}$, the exchange term $G^0 = R^0_{2121}$, where $1 \equiv nlm\sigma = n00\sigma$ and $2 \equiv n'00\sigma'$. The $2 \times 2 = 4$ two-electron configurations are $|n00\uparrow; n'00\uparrow\rangle$, $|n00\uparrow; n'00\downarrow\rangle$, $|n00\downarrow; n'00\uparrow\rangle$, and $|n00\downarrow; n'00\downarrow\rangle$. Since the Coulomb interaction does not change the spin, the states can be classified by the total spin $M_S = \sigma + \sigma'$, giving $M_S = 1, 0, 0, -1$ for the four states. When the spins are parallel ($M_S = 1$), the exchange term needs to be taken into account

$$\langle n00\uparrow; n'00\uparrow|\boldsymbol{H}_1|n00\uparrow; n'00\uparrow\rangle = F^0 - G^0. \tag{4.107}$$

The same energy is found for $M_S = -1$. Since we are dealing with a problem with spherical symmetry, these are two components of a total spin $S = 1$. Since $S = 1$ forms a triplet with components $M_S = 1, 0, -1$, we also anticipate the same energy for $M_S = 0$.

The situation for $M_S = 0$ is more complicated, since there are two states in that case. The diagonal matrix element for each state is

$$\langle n00, \pm\frac{1}{2}; n'00, \mp\frac{1}{2}|\boldsymbol{H}_1|n00, \pm\frac{1}{2}; n'00, \mp\frac{1}{2}\rangle = F^0. \tag{4.108}$$

Additionally, there is a coupling between the two states

$$\langle n00, \pm\frac{1}{2}n'00, \mp\frac{1}{2}|\boldsymbol{H}_1|n00, \mp\frac{1}{2}n'00, \pm\frac{1}{2}\rangle = -G^0. \tag{4.109}$$

Since the spin changes for each particle, the direct term is zero. However, for the exchange term, the spin is conserved. The matrix elements can be written as a 2×2 matrix:

$$\begin{pmatrix} F^0 & -G^0 \\ -G^0 & F^0 \end{pmatrix} \begin{matrix} |n00\uparrow; n'00\downarrow\rangle \\ |n00\downarrow; n'00\uparrow\rangle \end{matrix}, \tag{4.110}$$

leading to eigenvalues $F^0 \pm G^0$. The value $F^0 - G^0$ with eigenvector $(|n00\uparrow; n'00\downarrow\rangle + |n00\downarrow; n'00\uparrow\rangle)/\sqrt{2}$ comes from the $M_S = 0$ term of the $S = 1$ triplet. The $F^0 + G^0$ energy has only an $M_S = 0$ component. It is, therefore, the $M_S = 0$ term of the singlet with $S = 0$ with eigenvector $(|n00\uparrow; n'00\downarrow\rangle - |n00\downarrow; n'00\uparrow\rangle)/\sqrt{2}$.

Since there is no orbital angular momentum, the problem can also be written as the coupling of two $s = \frac{1}{2}$ spins. The eigenvalues of the $\mathbf{s}_1 \cdot \mathbf{s}_2$ interaction can be obtained analogously to Eq. (2.245). Taking $\hbar \equiv 1$ gives

$$\mathbf{s}_1 \cdot \mathbf{s}_2 = \frac{1}{2}(\mathbf{S}^2 - \mathbf{s}_1^2 - \mathbf{s}_1^2) = \frac{1}{2}\left(S(S+1) - \frac{3}{2}\right) = -\frac{3}{4}, \frac{1}{4} \quad \text{for } S = 0, 1, \tag{4.111}$$

where the different total S values are one- and threefold degenerate for $S = 0$ and 1, respectively. The Coulomb interaction between two electrons in different s shells can therefore also be written as

$$\boldsymbol{H}_1 = F^0 - 2G^0\left(\mathbf{s}_1 \cdot \mathbf{s}_2 + \frac{1}{4}\right) = F^0 \pm G^0, \quad \text{for } S = 0, 1. \tag{4.112}$$

Example 4.5 ***Two Electrons in a p Orbital (p^2)***
Let us consider next the example of two electrons in a p shell. The different electron configurations are shown in Table 4.1. Including spin, there are a total of $6 \times 5/2 = 15$ different ways of arranging two electrons in six states. Since both electrons are in the same p orbital, the notation can be simplified to $|m_1\sigma_1; m_2\sigma_2\rangle$. The splitting of the configurations can already be determined before calculating the energies. For the Coulomb interaction, L, M_L, S, and M_S are good quantum numbers. It is straightforward to arrange the configurations in terms of M_L and M_S, see Table 4.1. The different LS terms can be found by looking at the extremal M_L and M_S. The configuration $|1\uparrow; 1\downarrow\rangle$ belongs to the term with $L = 2$ and $S = 0$, since there is no configuration with $M_L = 2$ and $M_S = 1$. The $L = 2$ and $S = 0$ is denoted by ^1D (say singlet D). The letter indicates the value of L, which corresponds to the notation for atomic orbitals. The different values for L are denoted by

$$\begin{array}{llllll} L: & 0 & 1 & 2 & 3 & 4 \\ & S & P & D & F & G \end{array} \tag{4.113}$$

The superscript is $2S + 1$. Note that the ^1D term appears in the configurations with $M_L = 2, 1, 0, -1, -2$ and $M_S = 0$. This energy level is $(2L + 1)(2S + 1) = 5$-fold degenerate. The configuration $|1\uparrow; 0\uparrow\rangle$ belongs to a ^3P term (triplet P), which is ninefold degenerate, with $M_L = 1, 0, -1$ and $M_S = 1, 0, -1$. Then, there is only one configuration unaccounted for with $M_L = M_S = 0$, which must be an ^1S term.

Since $a_{11k} = 1, 0, \frac{1}{5}$ for $k = 0, 1, 2$, the Coulomb interaction for $l = 1$ only contains the $k = 0, 2$ terms,

$$\begin{aligned} \boldsymbol{H}_1 = {} & \left(F^0 \mathrm{W}^0(4,1)\mathrm{W}^0(3,2) + \frac{F^2}{25}\mathrm{W}^2(4,1) \cdot \mathrm{W}^2(3,2)\right)\delta_{\sigma_1,\sigma_4}\delta_{\sigma_2,\sigma_3} \\ & -\frac{F^2}{25}\delta_{\sigma_1,\sigma_3}\delta_{\sigma_2,\sigma_4}\mathrm{W}^2(3,1) \cdot \mathrm{W}^2(4,2), \end{aligned} \tag{4.114}$$

Table 4.1 Product basis $|m_1\sigma_1\ m_2\sigma_2\rangle$ for a p^2 configuration arranged for different M and M_S values

p^2	$M_S = 1$	$M_S = 0$	$M_S = -1$
$M = 2$		$\|1\uparrow; 1\downarrow\rangle$	
$M = 1$	$\|1\uparrow; 0\uparrow\rangle$	$\|1\uparrow; 0\downarrow\rangle, \|1\uparrow; 0\downarrow\rangle$	$\|1\downarrow; 0\downarrow\rangle$
$M = 0$	$\|1\uparrow; -1\uparrow\rangle$	$\|1\uparrow; -1\downarrow\rangle, \|-1\uparrow; 1\downarrow\rangle, \|0\uparrow; 0\downarrow\rangle$	$\|1\downarrow; -1\downarrow\rangle$
$M = -1$	$\|0\uparrow; -1\uparrow\rangle$	$\|-1\uparrow; 0\downarrow\rangle, \|-1\uparrow; 0\downarrow\rangle$	$\|0\downarrow; -1\downarrow\rangle$
$M = -2$		$\|-1\uparrow; -1\downarrow\rangle$	

where the last term is the exchange contribution. The radial wavefunction $R_{n1}(r)$ is identical for all electrons. This also implies that the radial matrix element for the exchange is equal to that of the direct interaction. The tensor for $k = 0$ is equal to the identity matrix $\mathrm{W}^0 = \mathbb{1}_3$. This implies that for $k = 0$, m cannot change. However, the only different states with equal m have opposite spin; therefore, the exchange term for $k = 0$ is zero.

For ^1D, the only nonzero terms in the interaction are

$$E(^1D) = E(1\uparrow;1\downarrow) = F^0 + \frac{1}{25}\left(W^2_{3z^2-r^2,11}\right)^2 F^2 = F^0 + \frac{1}{25}F^2, \tag{4.115}$$

where $W^k_{q,m'm}$ is the matrix element of the tensor $\boldsymbol{W}^k_q$ from Section 4.4. Note that $\boldsymbol{W}^2_{3z^2-r^2}$ is the only tensor of rank 2 with diagonal matrix elements. From Eq. (4.94), see that $W^2_{3z^2-r^2,11} = 1$. In the case of ^3P with equal spin, there is also an exchange contribution

$$\begin{aligned} E(^3P) = E(1\uparrow;0\uparrow) &= F^0 + W^2_{3z^2-r^2,11}W^2_{3z^2-r^2,00}\frac{F^2}{25} - \sum_{q=yz,zx} W^2_{q,01}W^2_{q,10}\frac{F^2}{25} \\ &= F^0 + \left(1\times(-2) - 2\times\left(\sqrt{\frac{3}{2}}\right)^2\right)\frac{F^2}{25} = F^0 - \frac{1}{5}F^2. \end{aligned} \tag{4.116}$$

The matrices $\boldsymbol{W}^2_q$, with $q = yz, xz$, allow for the scattering $m = 1$ to $m' = 0$ and vice versa, see Eq. (4.94).

The eigenenergies so far could be easily obtained since we were able to find configurations that were directly eigenstates of the Coulomb interaction. Symmetry tells us that all the LS terms with different M_L and M_S have the same energy. The ^3P and ^1D eigenstates account for the eigenstates of all the M_L and M_S values except for $M_L = M_S = 0$ which contains three basis functions and should therefore lead to three different eigenstates. This gives the remaining eigenstate, which is the nondegenerate LS term ^1S. To calculate the eigenenergy, one needs to consider all the matrix elements between the product functions $|1\uparrow;-1\downarrow\rangle$, $|0\uparrow;0\downarrow\rangle$, and $|-1\uparrow;1\downarrow\rangle$. This also leads to off-diagonal matrix elements, such as

$$\begin{aligned} \langle\pm1\uparrow;\mp1\downarrow|\boldsymbol{H}_1|0\uparrow;0\downarrow\rangle &= \sum_{q=yz,zx} W^2_{q,10}W^2_{q,-10}\frac{F^2}{25} \\ &= \left(\sqrt{\frac{3}{2}}\right)^2((-i)\times(-i) + 1\times(-1))\frac{F^2}{25} = -\frac{3}{25}F^2. \end{aligned} \tag{4.117}$$

After working out all the matrix elements, the total matrix for $M_L = M_S = 0$ can be written as

$$\boldsymbol{H}_1 = F^0\mathbb{1}_3 - \frac{F^0}{25}\begin{pmatrix} 1 & -3 & 6 \\ -3 & 4 & -3 \\ 6 & -3 & 1 \end{pmatrix}, \tag{4.118}$$

which has the following eigenvalues:

$$F^0 - \frac{1}{5}F^2, \quad F^0 + \frac{1}{25}F^2, \quad F^0 + \frac{2}{5}F^2. \tag{4.119}$$

The first two are the eigenenergies of ^3P and ^1D, respectively. This leaves the latter as the energy for ^1S. The singlet state has the highest energy. The corresponding eigenstate $|LM, SM_S\rangle = |00,00\rangle$ is given by

$$|00,00\rangle = \frac{1}{\sqrt{3}}(|1\uparrow;-1\downarrow\rangle - |0\uparrow;0\downarrow\rangle + |-1\uparrow;1\downarrow\rangle). \tag{4.120}$$

In this wavefunction, all configurations contribute equally. This is reminiscent of the local potential where the δ-function potential caused a symmetric state to split off from the continuum states, see Section 3.9. This effect is observed regularly.

The eigenenergies follow Hund's rules, that is the state with the highest spin S has the lowest energy. The reason underlying this rule is that the maximization of the spin forces electrons into different orbitals since they have to satisfy Pauli's exclusion principle. This increases the separation between the electrons and minimizes the Coulomb energy. When there are several states with the same S, then the state among them with the highest L is the lowest. Here, again, maximizing L forces electrons into different orbitals. The minimization of the Coulomb interaction therefore leads to a magnetic moment. This explains why magnetism is often strongest in materials where the orbitals have a small radial extent, such as first-row transition metals (3d) and rare earths (4f), which causes a strong electron repulsion.

Conceptual summary. – Although this section was restricted to some relatively simple configurations, the underlying physics is the same for more complex problems. When considering only the Coulomb interaction by the nucleus, p^2 configurations would have an energy of $2\varepsilon_{\mathrm{p}}$, where ε_{p} is the binding energy of the p electron. However, the Coulomb interaction between the electrons splits the 15-fold-degenerate states into three groups, which are nine-, five-, and onefold-degenerate multiplets. Note that the quantum numbers lm and $\frac{1}{2}\sigma$ are no longer good quantum numbers. Only the total angular momentum L and spin S and their projections M and M_S are conserved quantities. This is comparable to free space, where for a many-particle system, momentum $\mathbf{k}$ is no longer a good quantum number but only the total momentum $\sum_{i=1}^{N} \mathbf{k}_i$ is. In the presence of electron–electron interactions, momentum is continuously exchanged between the electrons. The Coulomb interaction can effectively be viewed as electrons "colliding" with each other. The same happens in atoms. However, now angular momentum is transferred between the electrons.

Another important aspect of the eigenstates is that they are linear combinations of different configurations of single-particle solutions of the Hamiltonian, *i.e.* in the absence of electron–electron interactions. Note that different linear combinations of the same configurations can lead to very different eigenstates. This is a feature of a many-body system. It is no longer possible to treat the electron–electron interactions as an effective potential, calculate the energy levels, and then fill the levels with electrons. It becomes essential to treat all the different configurations for a particular number of electrons explicitly.

Additional interactions can further split the multiplets. The values of the total angular momentum are given by $J = L + S, L + S - 1, \dots, |L - S|$. For the singlet states with $S = 0$, ^1D and ^1S, where $J = L = 2$ and $J = L = 0$, respectively, there is no additional splitting. However, for the ^3P multiplet, the total angular momentum values are $J = 2, 1, 0$ with a degeneracy of $2J + 1$. These levels will split in the presence of spin–orbit interaction. Note, however, that this additional splitting does not represent a lowering of the symmetry.

Problems

1 Find the different eigenenergies for a p^3 configuration. Label the eigenstates by their spectral notation.

2 The spin–orbit interaction can additionally split the LS terms of the Coulomb interaction.

a) Calculate the splitting due to the spin–orbit interaction for a p^2 configuration in the limit that its strength ζ is much weaker than the Coulomb interaction, *i.e.* $\zeta \ll F^k$.

b) Calculate the energies in the absence of the Coulomb interaction.

c) Show why the energy of the $M = 2$ state of the ^1D multiplet is unchanged by the spin–orbit interaction when the Coulomb interaction is large.

4.6 Strong Interaction: Mesons

Introduction. – Obviously, it is beyond the scope of this book to describe the standard model in detail. However, we can gain insight into some of the underlying fundamental concepts. One of the impressive feats of high-energy physics was the prediction of particles. To do this, it is important to understand the symmetry of the interactions between elementary particles. For a certain number of particles, the symmetry then determines how the states split. For example, for the electron–electron interactions for a p^2 configuration in the previous section, one expects the Coulomb multiplets ^3P, ^1D, and ^1S. One cannot have, for example the ^1S missing or only 8 states in the ^3P instead of 9. The same is true for particle physics. If one assumes that particles are composites of a fixed number of interacting elementary particles (say, neutrons and protons consisting of three quarks), then the symmetry of the interaction entirely determines into what groups the composite particles are split. Therefore, if a certain theory predicts that there is a group of nine composite particles, then finding a number unequal to 9 means that the theory is incorrect.

In the previous section, we considered the Coulomb interactions between electrons on an atom. This is a spherically symmetric problem in three dimensions (denoted as SO(3) or special orthogonal in three dimensions). In SO(3), transitions between different projected angular momentum values m are not necessarily equivalent. For example, for a p orbital, there is a clear difference between scattering from $m = 1$ to $m = 0$ ($\Delta m = -1$) and from $m = 1$ to $m = -1$ ($\Delta m = -2$). For baryons, such as neutrons and protons, it was suggested that they are composite particles consisting of three quarks. The quarks are very strongly bound together by the strong interaction. The suggested symmetry of the interaction was SU(3), where SU stands for special unitary. SU(3) is a high symmetry, and all scattering processes become effectively equivalent. In Section 4.4, the spherical tensors of rank 1 and 2 could be scaled in such a way that all three components $m = 1, 0, -1$ are treated on an equal footing. This led to the eight Gell-Mann matrices, see Eq. (4.96). Since the strong interaction is a different interaction, the three components are generally not labeled by m. To bring out the equivalent nature of the basis, they are usually labeled as red, green, and blue. Just as electrons on an atom exchange angular momentum, quarks exchange color.

Since the strong force is several orders of magnitude larger than the other fundamental interactions, the quarks combine to form neutral objects. Quarks are almost never observed independently, and particle accelerators are needed to split the neutral objects. The neutral composite objects are known as hadrons (from the Greek word for large, massive). There are two different groups of hadrons. Quarks can be combined in an odd number to form a baryon. Usually these consists of three quarks, and the best known examples are protons and neutrons. Note that the combination of red, green, and blue creates white, which is a neutral color. (This has nothing to do with actual color. It is just a theoretical way of describing that the combination of three different quantities gives something neutral.) Quarks can also be combined in an even number to form mesons. This is usually a quark and an antiquark. Here, the combination of a color and its anticolor forms a neutral object.

Although there are a total of six quarks, early theories were often restricted to the three lightest quarks: the up, down, and strange quarks. Virtually, all matter one deals with in everyday life consist only of the up and down quarks, which have comparable masses. The strange quark is an order of magnitude heavier. The three remaining quarks, charm, bottom, and top, are several orders of magnitude heavier than the up and down quarks and are not considered here. Since not all three quarks are equivalent, SU(3) is only an approximate symmetry. However, initially, we will treat the three quarks as equivalent. The goal of this and the following section is to gain an understanding of how the symmetry of an interaction splits many-body states into different groups in a fashion similar to the Coulomb interaction on an atom.

Let us start with mesons, which are the composite particles of a quark and an antiquark. The interaction between the quarks and antiquarks can be written as

$$\boldsymbol{H}_1 = \boldsymbol{\lambda} \cdot \overline{\boldsymbol{\lambda}}, \tag{4.121}$$

using Eq. (4.96). The overline indicates that these Gell-Mann matrices describe the scattering of the antiquark. In principle, this interaction should have a strength. However, since we are only interested in the splitting, this has been set to 1. Therefore, we are only looking at the symmetry of the problem. The basis for the Gell-Mann matrices can now be changed from $m = 1, 0, -1$ to $|u\rangle$, $|s\rangle$, and $|d\rangle$. The basis is written in terms of the flavor, up (u), down (d), and strange (s), of the particle. Each quark also has a color at a specific point in time. However, colors are continuously exchanged between the quarks via the strong interaction. The quark–antiquark basis then has nine components $|u\overline{u}\rangle$, $|u\overline{s}\rangle$, $|u\overline{d}\rangle$, $|s\overline{u}\rangle$, $|s\overline{s}\rangle$, $|s\overline{d}\rangle$, $|d\overline{u}\rangle$, $|d\overline{s}\rangle$, and $|d\overline{d}\rangle$. This leads to a 9×9 matrix.

The diagonal components are given by

$$\langle q\overline{q}|\boldsymbol{H}_1|q\overline{q}\rangle = \lambda_{3,qq}\lambda^*_{3,\overline{q}\,\overline{q}} + \lambda_{8,qq}\lambda^*_{8,\overline{q}\,\overline{q}} = \begin{cases} \dfrac{4}{3} & \text{for } q = \overline{q}, \\ -\dfrac{2}{3} & \text{for } q \neq \overline{q}, \end{cases} \tag{4.122}$$

where $\lambda_{i,qq'}$ is the matrix element of $\boldsymbol{\lambda}_i$, with $i = 1, 2, \ldots, 8$ at position (q, q'), with $q = u, s, d$ or $\overline{q} = \overline{u}, \overline{s}, \overline{d}$. The conjugate is necessary for antiquarks. Note that $\boldsymbol{\lambda}_3$ and $\boldsymbol{\lambda}_8$ are the only matrices with diagonal components.

As it turns out, the only nonzero off-diagonal matrix elements occur between the states with $q = \overline{q}$. All the states with $q \neq \overline{q}$ are eigenstates with an eigenvalue $-\frac{2}{3}$, see Eq. (4.122). The off-diagonal matrix can be obtained in the same fashion. For example,

$$\langle s\overline{s}|\boldsymbol{H}_1|u\overline{u}\rangle = \lambda_{4,su}\lambda^*_{4,\overline{s}\,\overline{u}} + \lambda_{5,su}\lambda^*_{5,\overline{s}\,\overline{u}} = 1 \times 1^* + (-i) \times (-i)^* = 2, \tag{4.123}$$

see Eq. (4.96). The matrix for the $q = \overline{q}$ states is then

$$\boldsymbol{H}_1 = \begin{pmatrix} \frac{4}{3} & 2 & 2 \\ 2 & \frac{4}{3} & 2 \\ 2 & 2 & \frac{4}{3} \end{pmatrix} \begin{matrix} u\overline{u} \\ s\overline{s} \\ d\overline{d} \end{matrix}, \tag{4.124}$$

where the basis of the matrix is indicated on the right. This matrix can be diagonalized by hand or numerically and gives eigenvalues $-\frac{2}{3}$, $-\frac{2}{3}$, and $\frac{16}{3}$. In total, there are eight eigenstates with an eigenvalue $-\frac{2}{3}$ and one with an eigenvalue $\frac{16}{3}$. This splitting is usually summarized as

$$\overline{3} \otimes 3 = 8 \oplus 1. \tag{4.125}$$

We again see that a symmetric singlet state is separated from the other states, as we have seen in Section 3.9 for a local potential and in the preceding sections for the Coulomb interactions between electrons on an atom. The "color" singlet wavefunction with eigenvalue $\frac{16}{3}$ can be written out as

$$|\eta'\rangle = |\varphi_0\rangle = \frac{1}{\sqrt{3}}(|u\overline{u}\rangle + |d\overline{d}\rangle + |s\overline{s}\rangle). \tag{4.126}$$

This is the wavefunction of what is commonly known as the η' meson. The $|\varphi_0\rangle$ notation indicates that the coefficients follow the diagonal elements of the identity matrix ($\boldsymbol{\lambda}_0$). The other wavefunctions that are part of the meson octet with eigenvalue $-\frac{2}{3}$ are

$$|\eta\rangle = |\varphi_8\rangle = \frac{1}{\sqrt{6}}(|u\overline{u}\rangle + |d\overline{d}\rangle - 2|s\overline{s}\rangle), \tag{4.127}$$

for the η meson, and

$$|\pi^0\rangle = |\varphi_3\rangle = \frac{1}{\sqrt{2}}(|u\overline{u}\rangle - |d\overline{d}\rangle), \tag{4.128}$$

for the π^0 meson. The coefficients follow the diagonal elements of the Gell-Mann matrices $\boldsymbol{\lambda}_8$ and $\boldsymbol{\lambda}_3$, respectively.

For particles, an important property is the rest mass, which is directly related to the energy since $E = m_0c^2$. Although a calculation of the mass is beyond the scope of this book, we see from the eigenvalues of an interaction with SU(3) symmetry, $-\frac{2}{3}$ and $\frac{16}{3}$, that the singlet state must be separated from the other mesons. The mass for the η' meson is 958 MeV/c^2. This is significantly higher than the other mesons that have masses between 135 and 549 MeV/ c^2. However, from our SU(3) interaction, we had expected their masses to be all the same. This is because SU(3), while very important in recognizing the constituents of hadrons, is only an approximate symmetry. Our interaction was constructed such that all components were equivalent. This is largely true, at least for their masses, for the up and down quarks, but the rest mass of the strange quark is significantly larger. Now since the quarks are relativistic particles, mass is not a conserved quantity, but the important aspect is that the high SU(3) symmetry is broken. Therefore, one often likes to classify heavy particles by their components. For this, two useful numbers are introduced

$$I_3 = -\overline{I}_3 = \frac{1}{2}\boldsymbol{\lambda}_3 = \frac{1}{2}(n_u - n_d),$$
$$Y = -\overline{Y} = \frac{1}{\sqrt{3}}\boldsymbol{\lambda}_8 = \frac{1}{3}(n_u + n_d - 2n_s), \qquad (4.129)$$

which are known as the isospin and hypercharge, respectively. For the antiquarks (given by $\overline{I}_3$ and $\overline{Y}$), the numbers are reversed. The quantum numbers can be directly related to the diagonal Gell-Mann matrices. As an example, the meson $|u\overline{s}\rangle$ has a total isospin $I_3 + \overline{I}_3 = \frac{1}{2} - 0 = \frac{1}{2}$ and a total hypercharge of $Y + \overline{Y} = \frac{1}{3} - (-\frac{2}{3}) = 1$. This classification allows the placement of each of the mesons in a two-dimensional plane, see Figure 4.2.

Due to the different mass of the strange quark, the SU(3) symmetry is lifted. The masses of the mesons can be described phenomenologically. However, for this we additionally need to consider the fact that all quarks have a spin of $\frac{1}{2}$. The interaction between the spin moments $\boldsymbol{\mu} \sim \mathbf{s}/m$ significantly affects the energy and therefore the mass. The effective mass can then be written as

$$M_{12} = m_1 + m_2 + a\frac{\mathbf{s}_1 \cdot \mathbf{s}_2}{m_1 m_2}. \qquad (4.130)$$

The mass of the meson M_{12} is therefore the sum of its components plus an additional term due to the spin–spin interaction, where a is a phenomenological coupling strength. The $\mathbf{s}_1 \cdot \mathbf{s}_2$ term can be evaluated in the same fashion as Eq. (4.112) (taking $\hbar \equiv 1$). Let us first discuss the pseudoscalar ($S = 0$) mesons, before discussing the vector mesons ($S = 1$) in more detail. The masses and coupling strengths are taken as free parameters. Their values are approximately $m_u = m_d \cong 300$–$308\,\mathrm{MeV/c^2}$, $m_s \cong 478$–$483\,\mathrm{MeV/c^2}$, and $a = 628$–$636 m_u^2$. For the pseudoscalar mesons with $\mathbf{s}_1 \cdot \mathbf{s}_2 = -\frac{3}{4}$, this gives masses of 139–141 and 561–564 MeV/c^2 for the π and η mesons, respectively. The values should be compared with experimental values of 135–140 and 549 MeV/c^2, respectively. For K mesons, see Figure 4.2, where one the quarks is the strange quark, the 487–490 MeV/c^2 value also agrees well with the 494–498 MeV/c^2 experimental values. However, note that the predicted value of the mass for the singlet meson η' of 350 MeV/c^2 is significantly

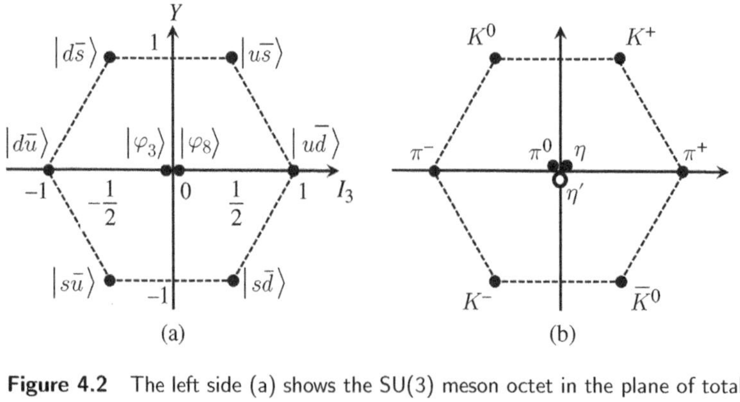

Figure 4.2 The left side (a) shows the SU(3) meson octet in the plane of total isospin I_3 and total hypercharge Y with the corresponding quark–antiquark wavefunctions, with $|\varphi_3\rangle = (|u\overline{u}\rangle - |d\overline{d}\rangle)/\sqrt{2}$ and $|\varphi_8\rangle = (|u\overline{u}\rangle + |d\overline{d}\rangle - 2|s\overline{s}\rangle)/\sqrt{6}$. The right (b) gives the names of the associated pseudoscalar mesons. The singlet η' meson is also indicated.

smaller than the experimentally observed value of 958 MeV/c^2, due to the significantly larger strong interaction between the quarks.

Let us now consider the vector mesons. For $S = 1$, $\mathbf{s}_1 \cdot \mathbf{s}_2 = \frac{1}{4}$, and the masses are expected to be higher. For the K^* mesons, the estimated mass is 874–892 MeV/c^2, which is close to twice the mass of the comparable K pseudoscalar mesons. The values agree well with the 892–896 MeV/c^2 experimental values. An even larger increase in mass is seen for the ρ mesons that consist of up–down pairs of quark and antiquark and should therefore be compared to the π pseudoscalar mesons. Here, the mass increases from 135–140 to 770 MeV/c^2. The latter is estimated by the model to be 753–775 MeV/c^2.

However, the simple model seems to fail when calculating the weight of a vector meson with the wavefunction in Eq. (4.127). The estimated weight is 928 MeV/c^2, which is significantly larger than the observed 782 MeV/c^2 of the ω vector meson. However, what is going wrong is not so much the phenomenological model as the breaking of the SU(3) symmetry. Mixing of the different configurations only occurs for configurations with zero isospin I_3 and hypercharge Y, see Figure 4.2. For pure SU(3) mixing, the wavefunctions in Eqs. (4.126)–(4.128) are found. Although this works well for the pseudoscalar mesons, the different behavior of the $s\bar{s}$ configuration changes affects the wavefunctions. The interaction in Eq. (4.124) can be modified as follows:

$$U = \begin{pmatrix} \frac{4}{3} & 2\alpha & 2 \\ 2\alpha & \frac{4}{3} + \Delta & 2\alpha \\ 2 & 2\alpha & \frac{4}{3} \end{pmatrix} \begin{matrix} u\bar{u} \\ s\bar{s} \\ d\bar{d}. \end{matrix} \tag{4.131}$$

The term $\Delta > 0$ reflects the higher mass of the strange quark. Additionally, when including higher order terms, the coupling between the strange quark and the up/down quarks becomes suppressed. This is reflected in the factor $0 \leq \alpha \leq 1$. Solving the matrix and using Eqs. (4.126)–(4.128) gives the eigenfunctions $|\varphi_3\rangle$ and

$$\begin{aligned} |\omega\rangle &= \cos\theta|\varphi_0\rangle + \sin\theta|\varphi_8\rangle & &\rightarrow \frac{1}{\sqrt{2}}(|u\bar{u}\rangle + |d\bar{d}\rangle), \\ |\phi\rangle &= \cos\theta|\varphi_8\rangle - \sin\theta|\varphi_0\rangle & &\rightarrow |s\bar{s}\rangle, \end{aligned} \tag{4.132}$$

for the ω and ϕ vector mesons, respectively. For $\alpha = 1$, $\arctan 2\theta = 2\sqrt{2}\Delta/(\Delta - 18)$. Therefore, for $\Delta = 0$ and $\alpha = 1$, we have $\theta = 0$ and $|\omega\rangle = |\varphi_0\rangle$ and $|\phi\rangle = |\varphi_8\rangle$, as expected, since the matrix is unchanged. The limiting values are given for the case of "ideal" mixing, *i.e.* the limit $\Delta \rightarrow \infty$ or $\alpha \rightarrow 0$. This corresponds to $\theta = \arctan(1/\sqrt{2}) \cong 35.3°$. As it happens, the wavefunctions appear to be close to ideal mixing. Therefore, the wavefunctions are substantially different from what is expected from SU(3) symmetry. Recalculating the masses in this limit gives 753–775 and 1016–1031 MeV/c^2 for the ω and ϕ mesons, respectively. This is again remarkably close to the experimental values of 782 and 1020 MeV/c^2, respectively.

Conceptual summary. – The theory of particle physics can be daunting, and this section only touches upon one particular, but important, aspect, namely how the symmetry of the strong interaction splits the quark–antiquark states into two distinct groups. We will see this again in the following section on baryons, which consist of three quarks. This classification of particles is an important step in the validation of the underlying theory. For the mesons described in this chapter, one expects nine mesons consisting of a group of eight mesons and a singlet meson separated from the other eight by having a substantially higher mass. If these mesons are not found, then the underlying quark theory is incorrect. This allowed the prediction of other particles. Although this was a good starting point, nature turns out to be more complicated than this, and SU(3) symmetry is only approximate due to the different masses of the quarks and additional interactions.

Problems

1 In order to simplify the math, but still get a good idea of how hadrons are constructed from quarks, let us forget about the strange quark. The problem for the up and down quarks becomes equivalent to that of $ns^1n's^1$ example in Section 4.5, *i.e.* u and d play the same role as $\uparrow$ and $\downarrow$.

a) Give the eigenfunctions for two quarks with effective $S = 0$ and $S = 1$.

b) Eigenfunctions for three quarks with effective $S = \frac{1}{2}$ can be found by adding up and down quarks to the two-quark $S = 0$ state. Give the eigenvectors.

c) Find the three-quark state with an effective $S = \frac{3}{2}$. Find the other components using step operators.

d) Determine the total number of states and find the missing eigenvectors.

2 Express the Gell-Mann matrices relevant for the u and d quarks in terms of $\boldsymbol{a}_{\pm\frac{1}{2}}$ and $\boldsymbol{a}^\dagger_{\pm\frac{1}{2}}$, where u/d replace $\uparrow$ / $\downarrow$. Use the dual spinors $\boldsymbol{a}_{\pm\frac{1}{2}} = \pm\underline{\boldsymbol{a}}^\dagger_{\pm\frac{1}{2}}$ to write the operation in π mesons.

4.7 Strong Interaction: Baryons

Introduction. – In the preceding section, we looked at the formation of a neutral composite particle via the combination of a quark and an antiquark. This is known as a meson. In this section, a neutral object is obtained by combining three quarks. These composite particles are known as baryons and are of particular interest since the proton and the neutron belong to this group of particles. We restrict our considerations to the up, down, and strange quarks. This leads to a total of $3^3 = 27$ three-quark configurations. This leads to a total of $27^2 = 729$ matrix elements describing the interaction between the different states. Although the evaluation becomes somewhat tedious, the goal is the same as in the previous section: how does an interaction with SU(3) symmetry (the strong force) combine these combinations into groups of eigenstates.

A neutral object with respect to the strong interaction can be obtained by combining all the colors. Combining red, green, and blue creates white, *i.e.* a neutral color. Before combining three quarks, let us first look at the interaction between two quarks

$$\boldsymbol{H}_1 = -\boldsymbol{\lambda} \cdot \boldsymbol{\lambda}', \tag{4.133}$$

where the Gell-Mann matrices operate on different particles (as indicated with the prime). The minus sign is chosen such that the singlet will have the highest eigenvalue. Since we are predominantly interested in the splitting of the levels under an SU(3) interaction, the interaction strength has been taken unity.

For two quarks, this leads again to a 9×9 matrix with a basis $|qq'\rangle$. Since the interaction is now no longer between a quark and an antiquark, there is no longer a conjugate in the matrix elements. This does not affect the diagonal matrix elements from Eq. (4.122) apart from the minus sign. However, without conjugate the off-diagonal matrix elements in Eq. (4.123) become zero, and the states $|uu\rangle$, $|ss\rangle$, and $|dd\rangle$ are now eigenstates of $\boldsymbol{H}_1$ with eigenvalues $-\frac{4}{3}$. However, other off-diagonal matrix elements are now nonzero. For example,

$$\langle su|\boldsymbol{H}_1|us\rangle = -\lambda_{4,su}\lambda_{4,us} - \lambda_{5,su}\lambda_{5,us} = -1 \times 1 - i \times (-i) = -2. \tag{4.134}$$

Since these states do not couple to others, it gives an effective 2×2 matrix

$$U = \begin{pmatrix} \frac{2}{3} & -2 \\ -2 & \frac{2}{3} \end{pmatrix} \begin{matrix} |su\rangle \\ |us\rangle \end{matrix} . \tag{4.135}$$

The eigenvalues are $-\frac{4}{3}$ and $\frac{8}{3}$ with the eigenvectors

$$\frac{1}{\sqrt{2}}(|su\rangle \pm |us\rangle), \tag{4.136}$$

respectively. Equivalent matrices are obtained for $|sd\rangle$ and $|ds\rangle$ and for $|ud\rangle$ and $|du\rangle$. This gives a total of six eigenvalues of $-\frac{4}{3}$ and three eigenvalues of $\frac{8}{3}$. Therefore, the coupling of two quarks leads to a different level splitting than a quark and an antiquark,

$$3 \otimes 3 = 6 \oplus \overline{3}, \tag{4.137}$$

where $\overline{3}$ indicates that the states at $\frac{8}{3}$ are antisymmetric states.

Let us now extend this to three quarks and restrict ourselves to the light quarks, up, down, and strange. There are then $3^3 = 27$ different combinations of three quarks. However, there are only matrix elements between equivalent quark configurations, *i.e.* qqq (3 times), qqq' (6), and $qq'q''$ (1), where each q, q', q'' can be u, d, s. The 27×27 matrix therefore splits into 10 smaller matrices. Let us start with the states where all the quarks are equal $|uuu\rangle$, $|ddd\rangle$, and $|sss\rangle$. These states do not couple to any other states. Using Eq. (4.122), the eigenvalues are then $3 \times (-4/3) = -4$. Next, we consider the states with two equal quarks. There are three configurations for each combination of q and q': $|qqq'\rangle$, $|qq'q\rangle$, and $|q'qq\rangle$. The diagonal matrix elements are the same, $-\frac{4}{3} + \frac{2}{3} + \frac{2}{3} = 0$. There are only off-diagonal matrix elements between unequal quarks, see Eq. (4.134). Since this is an interaction between two quarks, the other quark cannot change. This gives, for example $\langle qqq'|\boldsymbol{H}_1|qq'q\rangle = \delta_{qq}\langle qq'|\boldsymbol{H}_1|q'q\rangle = -2$. The total matrix is then

$$\boldsymbol{H}_1 = \begin{pmatrix} 0 & -2 & -2 \\ -2 & 0 & -2 \\ -2 & -2 & 0 \end{pmatrix} \begin{matrix} |qqq'\rangle \\ |qq'q\rangle \\ |q'qq\rangle \end{matrix} . \tag{4.138}$$

The eigenvalues are easily found numerically giving -4, 2, and 2. This occurs six times for the combinations *uud*, *uus*, *ddu*, *dds*, *ssu*, and *ssd*.

The last matrix is for the states with all unequal quarks. There are six possible states. The diagonal matrix elements are $3 \times \frac{2}{3} = 2$. The off-diagonal matrix elements are -2

or 0, depending on the number of permutations. For example, $\langle qq'q''|\boldsymbol{H}_1|qq''q'\rangle = -2$ and $\langle qq'q''|\boldsymbol{H}_1|q''qq'\rangle = 0$. The 6×6 matrix is then

$$U = \begin{pmatrix} 2 & -2 & -2 & -2 & 0 & 0 \\ -2 & 2 & 0 & 0 & -2 & -2 \\ -2 & 0 & 2 & 0 & -2 & -2 \\ -2 & 0 & 0 & 2 & -2 & -2 \\ 0 & -2 & -2 & -2 & 2 & 0 \\ 0 & -2 & -2 & -2 & 0 & 2 \end{pmatrix} \begin{matrix} |dsu\rangle \\ |sdu\rangle \\ |dus\rangle \\ |usd\rangle \\ |uds\rangle \\ |sud\rangle \end{matrix} . \tag{4.139}$$

The eigenvalues can be obtained numerically and are -4, 2 (4 times), and 8.

In summary, we have found eigenvalues -4 (10 times), 2 (16 times), and 8 (once). However, it can be shown that the eigenvalue 2 consists of two different groups of eigenvalues. This can be found by slightly changing the interaction strength between different pairs of quarks. Since each interaction between a pair of quarks has SU(3) symmetry, this does not break the symmetry. So, a better way of grouping the eigenvalues is -4 (10 times), 2 (8 times), 2 (8 times), and 8 (once). Or, in group theoretical language,

$$3 \otimes 3 \otimes 3 = (6 \oplus \overline{3}) \otimes 3 = 10 \oplus 8 \oplus 8 \oplus 1, \tag{4.140}$$

which are called decuplet, octet, and singlet.

For the eigenstates, one also needs to consider the $\frac{1}{2}$ spin of the quarks. Since the derivation of the wavefunctions becomes quite involved, only a few examples are given. The decuplet and octets, see Figure 4.3, are states with $J = \frac{3}{2}, \frac{1}{2}$, respectively. It turns out that only particles with a completely symmetric wavefunction are stable. This implies that the flavor and spin functions are either both symmetric or both antisymmetric. Based on the discussion in the previous section, one could expect that the states with a higher J value have a higher mass. This indeed is the case; the masses in the decuplet range from 1232 to 1672 MeV/c^2, whereas the octet states have masses from 939 to 1318 MeV/c^2. Therefore, the more interesting low-energy baryons are in the octet.

The wavefunctions for all equal quarks are $|qqq\rangle$, with $q = u, d, s$. The spin function for $M_J = \frac{3}{2}$ is $|\uparrow\uparrow\uparrow\rangle$. The total wavefunction is then

$$|qqq_{10}; \frac{3}{2}\frac{3}{2}\rangle = |qqq\rangle|\uparrow\uparrow\uparrow\rangle = |q\uparrow q\uparrow q\uparrow\rangle. \tag{4.141}$$

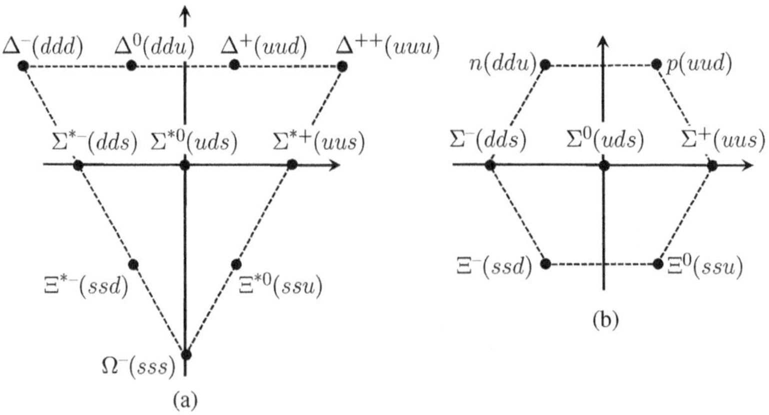

Figure 4.3 The eigenstates of the up, down, and strange quarks under an interaction with SU(3) symmetry split into four different manifolds of which the baryon decuplet (a) and octet (b) contain the most important particles. The particles are shown in the plane of isospin I_3 and hypercharge Y.

Note that the content qqq only occurs in the decuplet indicated by the subscript 10. The eigenfunctions for the 3 × 3 matrices in Eq. (4.138) are

$$|qqq'_{10}\rangle = \frac{1}{\sqrt{3}}(|q'qq\rangle + |qq'q\rangle + |qqq'\rangle), \tag{4.142}$$

$$|qqq'_{8_1}\rangle = \frac{1}{\sqrt{6}}(|q'qq\rangle + |qq'q\rangle - 2|qqq'\rangle), \tag{4.143}$$

$$|qqq'_{8_2}\rangle = \frac{1}{\sqrt{2}}(|q'qq\rangle - |qq'q\rangle). \tag{4.144}$$

The latter two are antisymmetric wavefunctions. In order to find a completely symmetric wavefunction, they need to be multiplied by $J = \frac{1}{2}$ wavefunctions $|\uparrow\uparrow\downarrow_{8_i}\rangle$, with $i = 1, 2$ of the same symmetry. This gives

$$|qqq'_8; \frac{1}{2}\frac{1}{2}\rangle = \frac{1}{\sqrt{2}}(|qqq'_{8_1}\rangle|\uparrow\uparrow\downarrow_{8_1}\rangle + |qqq'_{8_2}\rangle|\uparrow\uparrow\downarrow_{8_2}\rangle). \tag{4.145}$$

Expanding gives

$$\begin{aligned} |qqq'_8; \frac{1}{2}\frac{1}{2}\rangle = \frac{1}{\sqrt{18}}(2|q\uparrow q\uparrow q'\downarrow\rangle - |q\uparrow q\downarrow q'\uparrow\rangle - |q\downarrow q\uparrow q'\uparrow\rangle \\ + 2|q\uparrow q'\downarrow q\uparrow\rangle - |q\uparrow q'\uparrow q\downarrow\rangle - |q\downarrow q'\uparrow q\uparrow\rangle \quad (4.146) \\ + 2|q'\downarrow q\uparrow q\uparrow\rangle - |q'\uparrow q\uparrow q\downarrow\rangle - |q'\uparrow q\downarrow q\uparrow\rangle). \quad (4.147) \end{aligned}$$

This is the wavefunction for the proton and the neutron for $qqq' = uud, ddu$, respectively.

Another important property of the quarks is their charge. The up, down, and strange quarks have a charge of $\frac{2}{3}e$, $-\frac{1}{3}e$, and $-\frac{1}{3}e$, respectively. This gives the proton a charge of $+e$ and makes the neutron a particle with zero charge.

Conceptual summary. – In this section, it was shown how the SU(3) symmetry of the strong interaction splits the 27 different configurations of the up, down, and strange quarks into four different groups of eigenstates containing 10, 8(2×), and 1 composite particles. The proton and the neutron are part of an octet of baryons. Obviously, many details have been skimmed over or have been entirely omitted, since the complexity of the theory rapidly increases. However, the take-home message is that symmetry strongly affects the level structure of many-body states. By writing the strong interaction in terms of the Gell-Mann matrices, the level structure of the baryon states is directly imposed. The splitting into different groups of particles is a result of the SU(3) symmetry and independent of the detailed "physical" nature of the interaction. Again, we note that SU(3) symmetry is only an approximation since the quarks are not entirely equivalent degrees of freedom, and additional interactions come into play.

Problems

1 In Eq. (4.138), we see that the quark configuration uud occurs three times. Let us compare the masses of the well-known proton, part of the octet, and Δ^+, which is part of the decuplet, see Figure 4.3. For $m = m_u = m_d \cong 0.36\,\text{GeV}$, the mass of the baryon can be estimated using

$$M = 3m + A\sum_{i<j}\frac{\mathbf{S}_i \cdot \mathbf{S}_j}{m^2}, \tag{4.148}$$

where the interaction between the spins of the quarks additionally affects the mass. Calculate the masses for the proton and Δ^+, which have a total spin of $S = \frac{1}{2}, \frac{3}{2}$, respectively. The experimental value of $A = 0.026\,\text{GeV}^3$.

2 The wavefunctions for the proton and the neutron are given in Eq. (4.147). The magnetic moment of a spin $\frac{1}{2}$ particle with charge q and mass m is given by $\mu = q\hbar/2m$, giving $\mu_u = \frac{2}{3}\mu_N$ and $\mu_d = -\frac{1}{3}\mu_N$, for the up and down quarks, respectively; $\mu_N = e\hbar/2m$ is the nuclear magneton. Calculate the total magnetic moments for the proton and neutron.

4.8 Nuclear Structure

Introduction. – In the preceding sections, we have seen how the strong interaction binds quarks together to form heavy particles, known as hadrons, that are essentially neutral with respect to the strong interaction. However, this then leaves the problem how the nucleus that consists of protons and neutrons is stable when the protons experience a strong Coulomb repulsion. The answer lies in the size of the nucleus. A single proton has a diameter $d_{\text{proton}} = 1.76\,\text{fm}{=}1.76{\times}10^{-15}\,\text{m}$. However, for uranium, the diameter of the nucleus is still only about 11.71 fm. A naive calculation shows that each nucleon (proton or neutron) occupies a sphere with diameter 1.9 fm, which is not too different from the size of a single proton. Therefore, the nucleons are so close together that quarks in the neighboring nucleons can still interact with each other. Additionally, quarks can be interchanged between different nucleons. For example, a proton (*uud* in quark composition) can exchange an up quark with a down quark from a neutron (*ddu*). Note that after the interaction, the proton has turned into a neutron and vice versa. The interaction can also occur without the nucleons changing character and can therefore also occur between equivalent nucleons.

A somewhat different way of looking at the same interaction above is that the proton emits an up quark and a down antiquark or that the neutron emits a down quark and an up antiquark. From Figure 4.2, we see that such quark–antiquark combinations correspond to charged $\pi^{\pm}$ mesons, respectively. If the nucleons do not change character, they exchange neutral π^0 mesons. We have now created an effective interaction with massive π mesons as force carriers. However, note that this is not a fundamental force but only an effective description of the strong interaction. This is also called the residual strong force.

In this section, we first have a closer look at the residual strong force. However, to understand the nuclear structure, the interaction will be strongly simplified.

Although we have studied extensively the Coulomb force, where the force carrier, the photon, is massless, we have no understanding how the mass affects the interaction. From Eq. (2.27), we know that the norm of the four-wavevector for a free massive particle is given by

$$\mathbf{k}'^2 = \frac{m^2c^2}{\hbar^2}. \tag{4.149}$$

Using Eq. (2.55) and making the change to nonrelativistic notation, this can be written as the following wave equation:

$$-\nabla'^2\phi(\mathbf{r}') = \nabla^2\phi(\mathbf{r},t) - \frac{1}{c^2}\frac{\partial\phi(\mathbf{r},t)}{\partial t^2} = \frac{m^2c^2}{\hbar^2}\phi(\mathbf{r},t). \tag{4.150}$$

This is known as the Klein–Gordan equation. Let us now look at the limit of a steady state with a δ function source

$$\nabla^2\phi(\mathbf{r}) - \frac{m^2c^2}{\hbar^2}\phi(\mathbf{r}) = 4\pi g\delta(\mathbf{r}), \tag{4.151}$$

where the time derivatives have been removed. Essentially, we have a point charge of strength g emitting massive pions. In momentum space, the point charge becomes completely delocalized, and this can be written as

$$-\mathbf{q}^2\phi_{\mathbf{q}} - \frac{m^2c^2}{\hbar^2}\phi_{\mathbf{q}} = 4\pi g \quad \Rightarrow \quad \phi_{\mathbf{q}} = -\frac{4\pi g}{\mathbf{q}^2 + \dfrac{m^2c^2}{\hbar^2}}. \tag{4.152}$$

However, we have seen this Fourier transform before in Eq. (4.74), where the additional term in the denominator was introduced for convergence of the integral. Here, we see that this term takes on a physical meaning. In real space, the potential energy associated with this potential is

$$U(\mathbf{r}) = g\phi(\mathbf{r}) = -g^2\frac{e^{-\frac{r}{R}}}{r}, \quad \text{with} \quad R = \frac{\hbar}{mc}. \tag{4.153}$$

The potential is effectively screened by the lifetime of the massive particle. This is known as a Yukawa potential. Taking the diameter of the proton as the effective range of the residual strong force gives a mass $mc^2 = \hbar c/R = \hbar c/d_{\text{proton}} \cong 112\,\text{MeV}$. This simple estimate is close to the masses of 140 and 135 MeV of the $\pi^\pm$ and π^0 mesons, respectively.

Although we have found a way to keep the nucleons together, this still leaves a complex many-body problem of many nucleons. It does not directly help in explaining why certain nuclei are stabler than others. Experimentally, it was found that nuclei are particularly stable when the number of protons or neutrons is equal to one of the so-called magic numbers 2, 8, 20, 28, 50, 82, and 126. Particularly, the most common isotope of lead $^{208}_{82}\text{Pb}$ has a particularly stable nucleus and contains two magic numbers: 82 protons and 126 neutrons. In order to understand this, we switch to an effective one-particle model, known as the nuclear shell model.

The first step in the nuclear shell model is to view the protons and neutrons as particles bound in a spherical harmonic potential well, $U(r) = \frac{1}{2}m\omega_0^2 r^2 = \frac{1}{2}m\omega_0^2(x^2 + y^2 + z^2)$. The nucleus is obviously not a potential well, but it is a reasonable starting point. Note that we are approximating a complex many-body problem into an effective one-particle problem. A nucleon feels an effective potential due to the residual strong force of all the other nucleons, thereby keeping it inside the nucleus. Just as in Section 3.3, for the harmonic oscillator in two dimensions, the problem can be solved by separation of variables, giving the Hamiltonian

$$\boldsymbol{H} = \hbar\omega_0\left(\boldsymbol{a}_x^\dagger\boldsymbol{a}_x + \boldsymbol{a}_y^\dagger\boldsymbol{a}_y + \boldsymbol{a}_z^\dagger\boldsymbol{a}_z + \frac{3}{2}\right). \tag{4.154}$$

However, in order to obtain a better insight into the spherical nature of the problem, the following operators are introduced:

$$\boldsymbol{a}_{\pm1}^\dagger = \frac{1}{\sqrt{2}}(\boldsymbol{a}_x^\dagger \pm \boldsymbol{i}\boldsymbol{a}_y^\dagger) \quad \text{and} \quad \boldsymbol{a}_0^\dagger = \boldsymbol{a}_z^\dagger, \tag{4.155}$$

following Eq. (3.103). Note that in Eq. (3.103), for two dimensions, there were only two different oscillators, $\boldsymbol{a}_{\pm1}^\dagger$, whereas there are three components in three dimensions.

This changes the Hamiltonian into

$$\boldsymbol{H} = \hbar\omega_0 \left(\boldsymbol{a}_1^\dagger \boldsymbol{a}_1 + \boldsymbol{a}_0^\dagger \boldsymbol{a}_0 + \boldsymbol{a}_{-1}^\dagger \boldsymbol{a}_{-1} + \frac{3}{2} \right). \tag{4.156}$$

In order to study different configurations of the same energy, we look at the products $(\boldsymbol{a}_1^\dagger + \boldsymbol{a}_0^\dagger + \boldsymbol{a}_{-1}^\dagger)^N$. The energy of each product state for a particular N is $E = (N + \frac{3}{2})\hbar\omega_0$. For $N = 0$, this gives only one state with zero step operators. This corresponds to $l = 0$. In analogy to atomic orbitals, this can also be indicated as 1s. For $N = 1$, there are three states $\boldsymbol{a}_1^\dagger + \boldsymbol{a}_0^\dagger + \boldsymbol{a}_{-1}^\dagger$, giving $l = 1$. This is usually denoted as 1p. Things becomes a little more interesting for $N = 2$,

$$(\boldsymbol{a}_1^\dagger + \boldsymbol{a}_0^\dagger + \boldsymbol{a}_{-1}^\dagger)^2 = \boldsymbol{a}_1^\dagger \boldsymbol{a}_1^\dagger + 2\boldsymbol{a}_1^\dagger \boldsymbol{a}_0^\dagger + 2\boldsymbol{a}_1^\dagger \boldsymbol{a}_{-1}^\dagger + \boldsymbol{a}_0^\dagger \boldsymbol{a}_0^\dagger + 2\boldsymbol{a}_0^\dagger \boldsymbol{a}_{-1}^\dagger + \boldsymbol{a}_{-1}^\dagger \boldsymbol{a}_{-1}^\dagger.$$

This gives a total of six configurations, but among those, $\boldsymbol{a}_0^\dagger \boldsymbol{a}_0^\dagger$ and $\boldsymbol{a}_1^\dagger \boldsymbol{a}_{-1}^\dagger$ have a projected angular momentum $m = 0$. One combination of those forms a $l = 0$ state. Since this is the second $l = 0$, the notation is given by 2s. The other five states correspond to $(l = 2)$. This is called the 1d.

Although the angular momentum quantum can already be determined from the product states, the eigenstates require a little more work. As with spherical harmonics, we can also define angular momentum operators

$$\boldsymbol{L}_0 = \boldsymbol{a}_1^\dagger \boldsymbol{a}_1 - \boldsymbol{a}_{-1}^\dagger \boldsymbol{a}_{-1}, \quad \boldsymbol{L}_{-1} = \boldsymbol{a}_0^\dagger \boldsymbol{a}_1 + \boldsymbol{a}_{-1}^\dagger \boldsymbol{a}_0, \quad \boldsymbol{L}_1 = \boldsymbol{a}_1^\dagger \boldsymbol{a}_0 + \boldsymbol{a}_0^\dagger \boldsymbol{a}_{-1}, \tag{4.157}$$

omitting the Condon–Shortley phase factor. The normalized $m = 2$ component of the 1d states is

$$|22\rangle = \frac{1}{\sqrt{2}} \boldsymbol{a}_1^\dagger \boldsymbol{a}_1^\dagger |0\rangle = |2;0;0\rangle, \tag{4.158}$$

using the notation $|lm\rangle$ for angular momentum quantum numbers and the notation $|n_1;n_0;n_{-1}\rangle$ for quantum numbers of the oscillators. Since this is the only state with $m = 2$, it is directly an eigenstate. Eigenstates of the same l with lower m can be obtained by applying the step-down operator. Operating with $\boldsymbol{L}_{-1}$ on the right-hand side of Eq. (4.158) gives

$$\boldsymbol{L}_{-1}|2;0;0\rangle = \sqrt{2 \times 1}|1;1;0\rangle = \sqrt{2}\boldsymbol{a}_1^\dagger \boldsymbol{a}_0|0\rangle. \tag{4.159}$$

We can perform a similar operation on the angular momentum states on the left-hand side in Eq. (4.158). Following Eq. (1.361), the states $|lm\rangle$ can be expressed in terms of spinors $\boldsymbol{a}_{\pm\frac{1}{2}}^\dagger = \boldsymbol{a}_\uparrow^\dagger, \boldsymbol{a}_\downarrow^\dagger$ and therefore $|lm\rangle = |n_\uparrow; n_\downarrow\rangle$. Therefore, the state with $m = l = 2$ is given by $|22\rangle = |4;0\rangle = \frac{1}{\sqrt{4!}}(\boldsymbol{a}^\dagger)^4$. The properly normalized angular momentum operators for the spinors are

$$\boldsymbol{L}_0 = \frac{1}{2}(\boldsymbol{a}_\uparrow^\dagger \boldsymbol{a}_\uparrow - \boldsymbol{a}_\downarrow^\dagger \boldsymbol{a}_\downarrow), \quad \boldsymbol{L}_{-1} = \frac{1}{\sqrt{2}} \boldsymbol{a}_\downarrow^\dagger \boldsymbol{a}_\uparrow, \quad \boldsymbol{L}_1 = \frac{1}{\sqrt{2}} \boldsymbol{a}_\uparrow^\dagger \boldsymbol{a}_\downarrow, \tag{4.160}$$

again omitting the Condon–Shortley phase factor. Stepping down the left-hand side of Eq. (4.158) gives

$$\boldsymbol{L}_{-1}|22\rangle = \boldsymbol{L}_{-1}|4;0\rangle = \frac{1}{\sqrt{2}}\sqrt{4 \times 1}|3;1\rangle = \sqrt{2}|21\rangle. \tag{4.161}$$

Comparing Eqs. (4.159) and (4.161) shows that $|21\rangle = |1;1;0\rangle$. This is not entirely surprising since it is the only state with $m = 1$. Stepping down again gives

$$\begin{aligned} \boldsymbol{L}_{-1}|21\rangle = \boldsymbol{L}_{-1}|3;1\rangle &= \frac{1}{\sqrt{2}}\sqrt{3\times 2}|2;2\rangle = \sqrt{3}|20\rangle \\ &= \boldsymbol{L}_{-1}|1;1;0\rangle = \sqrt{1\times 2}|0;2;0\rangle + \sqrt{1\times 1}|1;0;1\rangle, \end{aligned} \tag{4.162}$$

making sure to use the correct step-down operators for the left- and right-hand side. This gives the eigenfunction

$$|20\rangle = \sqrt{\frac{2}{3}}|0;2;0\rangle + \frac{1}{\sqrt{3}}|1;0;1\rangle = \frac{1}{\sqrt{3}}\left(\boldsymbol{a}_0^\dagger\boldsymbol{a}_0^\dagger|0\rangle + \boldsymbol{a}_1^\dagger\boldsymbol{a}_{-1}^\dagger|0\rangle\right). \tag{4.163}$$

This eigenstate contains both $m = 0$ configurations. However, this implies that there should be another eigenstate perpendicular to this one. Since $m = 0$ is the only projected angular momentum with two configurations, the eigenstate must belong to an $l = 0$ term

$$|00\rangle = \frac{1}{\sqrt{3}}|0;2;0\rangle - \sqrt{\frac{2}{3}}|1;0;1\rangle. \tag{4.164}$$

Note that $\boldsymbol{L}_{\pm 1}|00\rangle = 0$, showing that this is indeed an $l = 0$ term. The other eigenstates of $l = 2$ are $|2,-1\rangle = |0,1,1\rangle$ and $|2,-2\rangle = |0,0,2\rangle$.

All the possible states with three bosons are given by

$$\begin{aligned} (\boldsymbol{a}_1^\dagger + \boldsymbol{a}_0^\dagger + \boldsymbol{a}_{-1}^\dagger)^3 &= (\boldsymbol{a}_1^\dagger)^3 + 3(\boldsymbol{a}_1^\dagger)^2\boldsymbol{a}_0^\dagger + 3(\boldsymbol{a}_1^\dagger)^2\boldsymbol{a}_{-1}^\dagger + 3\boldsymbol{a}_1^\dagger(\boldsymbol{a}_0^\dagger)^2 + 6\boldsymbol{a}_1^\dagger\boldsymbol{a}_0^\dagger\boldsymbol{a}_{-1}^\dagger \\ &+(\boldsymbol{a}_0^\dagger)^3 + 3(\boldsymbol{a}_0^\dagger)^2\boldsymbol{a}_{-1}^\dagger + 3\boldsymbol{a}_1^\dagger(\boldsymbol{a}_{-1}^\dagger)^2 + 3\boldsymbol{a}_0^\dagger(\boldsymbol{a}_{-1}^\dagger)^2 + (\boldsymbol{a}_{-1}^\dagger)^3. \end{aligned}$$

There are a total of 10 states, with projected angular momentum ranging from 3 to -3. For $m = 1, 0, -1$, there are two configurations. Therefore, the angular momentum states are $l = 1, 3$ or 2p and 1f in spectral notation. The m values for which there is only one configuration are directly eigenstates for $l = 3$: $|33\rangle = |3;0;0\rangle$, $|32\rangle = |2;1;0\rangle$, $|3,-2\rangle = |0;1;2\rangle$, and $|3,-3\rangle = |0;0;3\rangle$. The state for $l = 3$ and $m = 1$ can be found by stepping down the $m = 2$ state

$$\begin{aligned} \boldsymbol{L}_{-1}|32\rangle = \boldsymbol{L}_{-1}|5;1\rangle &= \frac{1}{\sqrt{2}}\sqrt{5\times 2}|4;2\rangle = \sqrt{5}|31\rangle \\ &= \boldsymbol{L}_{-1}|2;1;0\rangle = \sqrt{2\times 2}|1;2;0\rangle + \sqrt{1\times 1}|2;0;1\rangle. \end{aligned} \tag{4.165}$$

This gives an eigenfunction

$$|31\rangle = \frac{2}{\sqrt{5}}|1;2;0\rangle + \frac{1}{\sqrt{5}}|2;0;1\rangle. \tag{4.166}$$

The $l = 1$ eigenstate has to be orthogonal to that

$$|11\rangle = \frac{1}{\sqrt{5}}|1;2;0\rangle - \frac{2}{\sqrt{5}}|2;0;1\rangle. \tag{4.167}$$

The procedure can be continued to find the other eigenstates for $N = 3$.

For $N = 4$, the different combinations of spinors are

$$\begin{aligned} (\boldsymbol{a}_1^\dagger + \boldsymbol{a}_0^\dagger + \boldsymbol{a}_{-1}^\dagger)^4 &= (\boldsymbol{a}_1^\dagger)^4 + 4(\boldsymbol{a}_1^\dagger)^3\boldsymbol{a}_0^\dagger + 4(\boldsymbol{a}_1^\dagger)^3\boldsymbol{a}_{-1}^\dagger + 6(\boldsymbol{a}_1^\dagger)^2(\boldsymbol{a}_0^\dagger)^2 \\ &+4\boldsymbol{a}_1^\dagger(\boldsymbol{a}_0^\dagger)^3 + 12(\boldsymbol{a}_1^\dagger)^2\boldsymbol{a}_0^\dagger\boldsymbol{a}_{-1}^\dagger \\ &+6(\boldsymbol{a}_1^\dagger)^2(\boldsymbol{a}_{-1}^\dagger)^2 + 12\boldsymbol{a}_{-1}^\dagger(\boldsymbol{a}_0^\dagger)^2\boldsymbol{a}_1^\dagger + (\boldsymbol{a}_0^\dagger)^4 + \cdots, \end{aligned} \tag{4.168}$$

where only the configurations with $m \geq 0$ are indicated. Looking again at the projected angular momentum m, we see that m ranges from 4 to -4. More than one configuration exist for $m \leq 2$. The last line shows the three different states with $m = 0$. This total angular momentum quantum numbers are therefore $l = 4, 2, 0$ or in spectral notation 1g, 2d, and 3s.

In summary, we have found so far

N	l	Spectral notation	Degeneracy	Total number
0	0	$1s$	2	2
1	1	$1p$	6	8
2	$0, 2$	$2s, 1d$	12	20
3	$1, 3$	$2p, 1f$	20	40
4	$0, 2, 4$	$3s, 2d, 1g$	30	70

This gives the magic numbers 2, 8, and 20 but does not reproduce the higher magic numbers. Therefore, there are still some aspects missing in the model. The first step to recognize is that the potential cannot be harmonic. This would mean that it approaches infinity when a nucleon moves away from the nucleus. However, if a nucleon detaches itself from the nucleus, it would end up in the vacuum where we generally take the potential zero. A potential that works better is the Woods-Saxon potential

$$U(r) = -\frac{U_0}{1 + \exp\dfrac{r - R}{a}}, \tag{4.169}$$

where r is the distance to the nucleus. This potential is essentially a smoothed potential well. An estimate of the width of the potential well is $R = r_0 A^{\frac{1}{3}}$, where $r_0 \cong 1.25$–$1.30\,\text{fm}$, and A is the mass number, *i.e.* the total number of protons and neutrons in the nucleus. The surface thickness is of the order of $a = 0.5$–$0.7\,\text{fm}$. The depth of the potential well is $U_0 = 50\,\text{MeV}$. Approximating this by an harmonic-oscillator potential gives an energy of approximately $\hbar\omega_0 \cong 41/A^{\frac{1}{3}}$. Therefore, the potential becomes flatter as the mass of the nucleus increases. Unfortunately, since the Woods–Saxon potential becomes complex to evaluate, we approximate the higher order effects. If the nucleus is still spherical, the corrections can only depend on the angular momentum l. We take here a correction proportional to $-l^2$, *i.e.* higher l values are lower in energy, which is in agreement with experiment. However, as is clear from Figure 4.4, this does not solve the magic number problem, since there is no clear gap in the energy spectrum around an occupancy of 28, 50, and 82.

The next ingredient is to take into account that the nucleons have a spin of $\frac{1}{2}$, and that the levels can be split due to the spin–orbit interaction as described that arises from the potential, see Section 2.11. This leads to a total angular momentum $j = l \pm \frac{1}{2}$, which gives energy shifts of $-\zeta l/2, \zeta(l+1)/2$, respectively, see Eq. (2.245). The total results are shown by the dots in Figure 4.4. Note that the figure is qualitative, and no extensive attempts to fit experimental results have been made. However, as is clear from the figure, the spin–orbit interaction opens up clear gaps in the energies around the magic numbers 28, 50, and 82 by splitting the large-l levels 1f, 1g, and 1h, respectively.

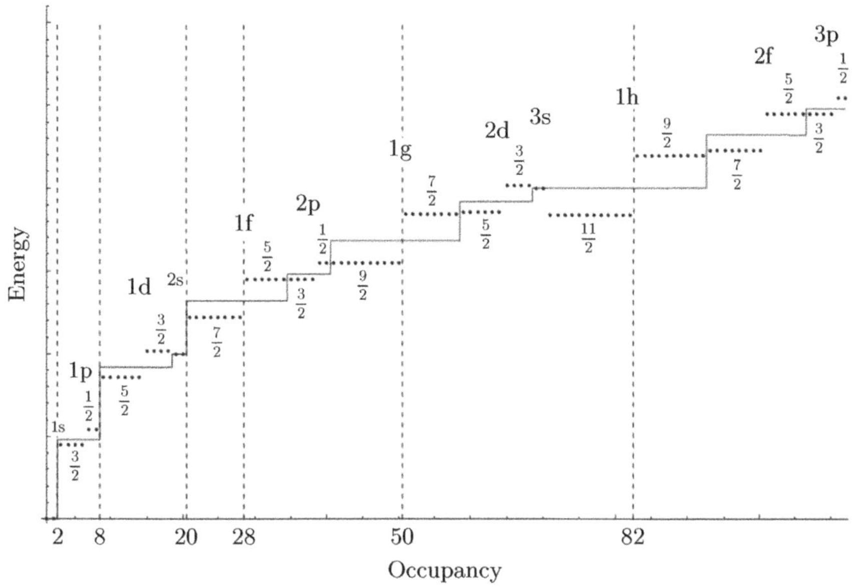

Figure 4.4 Energies in the nuclear shell model as a function of occupancy. The solid line gives the energies of the nuclear shells for a spherical harmonic oscillator potential with a correction proportional to $-l^2$ to account for the fact that the potential looks more like a smoothed potential well instead of a harmonic oscillator. Each horizontal line corresponds to a particular nl level. The spectral notation is given above the horizontal line. The spin–orbit interaction splits each level with a particular l into $j = l \pm \frac{1}{2}$ total angular momentum states indicated by dots. The $l + \frac{1}{2}$ and $l - \frac{1}{2}$ are indicated below and above their states, respectively. The axis indicates the magic numbers, where relatively large jumps in energy occur.

Conceptual summary. – In this section, we looked at the interaction between protons and neutrons inside the nucleus. Although the proton and neutron are both color neutral particles, their proximity inside the nucleus still allows them to interact via the strong force. The interaction can be viewed as an exchange of quarks, which, alternatively, can be viewed as an exchange of π mesons. Since the effective force carrier is now massive, the interaction becomes short ranged.

To understand the energy structure of the nucleus, the system is treated as an effective three-dimensional harmonic oscillator. Note that this simplifies the interaction between the nucleon to an effective single-particle potential. The protons and neutrons just want to stay inside the nucleus. The effective potential due to the other nucleons is taken to be similar to a harmonic potential. The level structure is more easily understood by rewriting the step operators in spherical coordinates. These are further split by higher order corrections when taking a more accurate shape of the potential. A final ingredient is the addition of a spin–orbit interaction term. The rather phenomenological nuclear shell model gives a reasonable understanding of the nuclear structure and the existence of magic numbers, *i.e.* it predicts certain stable isotopes.

5

Collective and Emergent Phenomena

Introduction

In this chapter, we look at phenomena called collective and emergent. Essentially, we study problems where the behavior of many quantum objects is different from that of a single quantum object. The definition of what to call an emergent phenomenon is obviously somewhat subjective. For example, there are several examples from Chapters 3 and 4 where new properties emerge when quantum objects are combined. For example, molecules and solids behave differently from single atoms; baryons, such as protons and neutrons, have different characteristics than the quarks that they are made off; and the statistics of fermions and bosons only make sense when there is more than a single quantum particle. However, in this chapter, we focus more on phenomena that occur when the behavior of many quantum objects is dramatically different from what we expect.

5.1 In Section 4.5, we saw that the minimization of electron–electron interactions on an atom can lead to the formation of magnetic moments. Similar effects also occur in solids. However, additionally, the magnetic moments on different sites can interact with each other. This can lead to long-range order below a critical temperature.

5.2 The nuclei have mainly been treated as fixed in space and giving rise to a potential. It is shown that the interplay between the electrons and the nuclear motion in solids can give rise to an effective attractive electron–electron interaction. The resulting pair formation of electrons leads to effective bosons that condense below a critical temperature. This effect can lead to superconductivity where electrical currents flow without resistance.

5.3 In classical mechanics, there is a clear distinction between mass and energy. However, we have seen that effective mass can result from interactions, and relativity tells us that mass and energy are equivalent. This section shows that the mass of leptons, for example electrons and quarks, can also be viewed as arising from the interaction with a scalar field, known as the Higgs field.

5.4 Although all magnetization directions are in principle equivalent, the system picks one particular direction, and excitations now occur with respect to this broken symmetry state. The theory underlying this symmetry breaking shows some interesting overlaps between magnetism, superconductivity, and the Higgs field.

5.5 In Section 4.3, we saw that calculating the Coulomb interaction between electrons for the Fermi sea led to unphysical results. Here, we see that this can be resolved by

Geometric Quantum Mechanics, First Edition. Michel van Veenendaal.

allowing electron–hole pair excitations with respect to the Fermi sea. This leads to an effective screened potential, comparable to the Yukawa potential.

5.6 The response of the Fermi sea to electromagnetic fields leads to collective excitations known as plasmons. These excitations play an important role in understanding the optical properties of materials. For example, metals are reflective for light with a frequency below the plasmon frequency.

5.2 Liquid helium-4 atoms are bosons, and it is therefore expected that the particles undergo a Bose–Einstein condensation. However, this lowest state has zero momentum. In order to have the condensate move without resistance at a finite momentum, a collective phenomenon known as superfluidity, interactions between the bosons need to be included.

5.1 Magnetism

Introduction. – A nice example of a collective phenomenon is magnetism. For many materials, the atoms have finite magnetic moments. This can be the case with one electron per site, but also for many-electron atoms finite moments are often found, see Section 4.5. Electrons with parallel spin are not allowed to occupy the same orbitals. This results in a lowering of the total electron repulsion. A local magnetic moment is therefore often a result of the minimization of the Coulomb interaction and common in materials where the orbitals have a small radial extent, such as (first-row) transition metals and rare earths and their compounds. If a material has noninteracting spins, then it is a paramagnet and only becomes magnetic in the presence of an external magnetic field. However, often the spins interact with each other and show long-range order. This can lead to a spontaneous magnetization where the magnetic moments align parallel or antiparallel, below a critical temperature. This is called ferromagnetism or antiferromagnetism, respectively.

Let us start with paramagnetism where the moments align in a magnetic field. The interaction with the field is given by the Zeeman interaction, see Eq. (2.224),

$$H = g\mu_B \mathbf{B}\cdot\mathbf{S} = -\mathbf{B}\cdot\boldsymbol{\mu}, \quad \text{with} \quad \boldsymbol{\mu} = -g\mu_B\mathbf{S}, \tag{5.1}$$

where μ_B is the Bohr magneton. The magnetic moments $\boldsymbol{\mu}$ align parallel to the magnetic field. We restrict the discussion here to $S = \frac{1}{2}$ spins. This simplifies the mathematics, but the essence of the physics remains the same for arbitrary S. The direction of the magnetic field is taken as the z-axis, *i.e.* $\mathbf{B} = B\mathbf{e}_z$. The expectation value of the spin is then (in units of $\hbar$)

$$\langle S\rangle = \frac{\sum_{\sigma=\pm\frac{1}{2}} \sigma e^{-\frac{g\mu_B B\sigma}{k_BT}}}{\sum_{\sigma=\pm\frac{1}{2}} e^{-\frac{g\mu_B B\sigma}{k_BT}}} = \frac{1}{2}\frac{e^{\frac{-\mu_B B}{k_BT}} - e^{\frac{\mu_B B}{k_BT}}}{e^{\frac{-\mu_B B}{k_BT}} + e^{\frac{\mu_B B}{k_BT}}} = -\frac{1}{2}\tanh\frac{\mu_B B}{k_BT}, \tag{5.2}$$

taking the gyromagnetic factor $g = 2$ and the partition function from Eq. (4.46). The spins therefore align antiparallel to the applied field. For $T \to 0$, only the $\sigma = -\frac{1}{2}$ is occupied. Note that the temperature scale associated with the Zeeman splitting is very small since

$\mu_B B$ corresponds to 0.058 meV for a field of $B = 1$ T. This should be compared to room temperature, which is $k_B T_{\text{room}} = 25$ meV. The total magnetization of the material is then

$$M = n\mu = -g\mu_B n\langle S\rangle = n\mu_B \tanh\frac{\mu_B B}{k_B T} \cong \frac{n\mu_B^2 B}{k_B T}, \tag{5.3}$$

where n is the density of spins, and where the last step gives the approximation in the high-temperature limit. The susceptibility $\chi = \mu_0 \mathbf{M}/\mathbf{B}$ is then the response of the material to the external field $\mathbf{B}$

$$\chi = \frac{\mu_0 M}{B} \cong \frac{n\mu_0\mu_B^2}{k_B T} = \frac{C}{T}, \quad \text{with} \quad C = \frac{n\mu_0\mu_B^2}{k_B}, \tag{5.4}$$

where C is the Curie constant for $S = \frac{1}{2}$. For low temperatures, the hyperbolic tangent approaches the asymptotic level of 1, and the magnetization levels out at $n\mu_B$. The magnetic susceptibility then approaches zero.

Let us now consider a material where the spins interact ferromagnetically. The spins by themselves generate magnetic fields, and this can cause a dipole interaction between the spins. However, this is not what generally causes the ferromagnetic interaction between spins in solids. Usually, the exchange is there to minimize the Coulomb interaction between the electrons. For two electrons in two different s orbitals on the same atom, the Coulomb interaction can be written as a spin–spin interaction, see Eq. (4.112). The lowest Coulomb interaction occurred for parallel spins. The interaction can be written in the same form if the spins are on orbitals on the neighboring sites. The Hamiltonian for the spins related to the magnetic sites in a solid is given by

$$H = -\sum_{ij} J_{ij} \mathbf{S}_i \cdot \mathbf{S}_j - g\mu_0\mu_B \sum_i \mathbf{H} \cdot \mathbf{S}_i. \tag{5.5}$$

This Hamiltonian is known as the Heisenberg model. The parameter $J_{ij} = J$ for the nearest neighbors i and j, and zero otherwise. The minus sign in front of this term ensures that the exchange is ferromagnetic. The second term is the applied magnetic field $\mathbf{H}$. In the absence of a response of the solid $\mathbf{B} = \mu_0\mathbf{H}$. However, for magnetic materials the presence of a field $\mathbf{H}$ causes a magnetization $\mathbf{M}$, which is generally assumed proportional to $\mathbf{H}$, *i.e.* $\mathbf{M} = \chi\mathbf{H}$, where χ is the magnetic susceptibility. The total magnetic field is then $\mathbf{B} = \mu_0(\mathbf{H} + \mathbf{M}) = \mu_0(1+\chi)\mathbf{H} = \mu\mathbf{H}$, where $\mu = \mu_0(1+\chi)$. We want to write the first term in the same form as the Zeeman interaction. Let us assume that the exchange interactions with the neighboring spins can be written as an effective field

$$-g\mu_0\mu_B \mathbf{H}_{\text{int}} = -2\sum_{j\in \text{n.n.}} J\langle \mathbf{S}_j\rangle = 2zJ\langle \mathbf{S}\rangle, \tag{5.6}$$

where the summation only runs over the nearest neighbors, and z is the number of nearest neighbors. The factor 2 arises from the double counting of the interaction when summing over i and j. This approximation is generally known as mean-field theory. It is not a great approximation for $S = \frac{1}{2}$ spins since fluctuations are always large due to the small size of the spin. However, it can still provide important insights into the physics of the problem. Equation (5.2) can now be used to determine the magnetization. Neglecting the external field for the moment gives

$$\langle \mathbf{S} \rangle = \frac{1}{2} \tanh \frac{zJ\langle \mathbf{S} \rangle}{k_B T} \quad \Rightarrow \quad M = \mu_B \tanh \frac{zJM}{2\mu_B k_B T}, \tag{5.7}$$

where $\mathbf{M} = 2\mu_B \mathbf{S}$ is the magnetization on each site. The magnetization is now no longer simply a function of the external field but has to be determined self-consistently.

When only considering the magnitude, Equation (5.7) is essentially the intersection between a straight line M and a hyperbolic tangent. For large x (low temperature), $\tanh x \to 1$; for small x (high temperature), $\tanh x \cong x$. Therefore, in the high-temperature limit, both sides of the equation can be approximated by a straight line. If the right-hand side has a slope less than 1, the lines never cross, and no spontaneous magnetization occurs. The critical temperature occurs when the slopes are both 1. Making a series of the magnetization gives

$$M = \frac{zJM}{2k_B T_c} \quad \Rightarrow \quad k_B T_c = \frac{1}{2} Jz. \tag{5.8}$$

Using these results, the susceptibility above the critical temperature (where the material is a paramagnet) can be calculated. The magnetic field due to the interaction between the spins in Eq. (5.6) can be written as

$$H_{\text{int}} = \frac{4k_B T_c \langle S_z \rangle}{g\mu_0 \mu_B} = \frac{k_B T_c M}{\mu_0 \mu_B^2 n} = \frac{T_c}{C} M, \tag{5.9}$$

using Eqs. (5.3) and (5.4). The susceptibility for a ferromagnet is then

$$M = \frac{C}{T}(H_{\text{int}} + H) = \frac{C}{T}\left(\frac{T_c}{C} M + H\right) \quad \Rightarrow \chi = \frac{M}{H} = \frac{C}{T - T_c}, \tag{5.10}$$

which is known as the Curie–Weiss law.

Let us derive this result in a different fashion, which will be important in Section 5.4. We want to determine the total free energy, which consists of the internal energy U and the entropy S. The energy density $u = U/V$ of the system is given by

$$u = -zJ\langle S \rangle^2 n = -\frac{1}{4} zJ\overline{M}^2 n = -\frac{1}{2} k_B T_C \overline{M}^2 n, \tag{5.11}$$

using Eq. (5.8) and where the normalized magnetization is $\overline{M} = M/M_0 = 2\langle S \rangle$, where $M_0 = n\mu_B$ is the saturation magnetization for an $S = \frac{1}{2}$ system. The entropy of a certain state is related to its multiplicity, see Eq. (4.43). For a particular configuration, the multiplicity is given by

$$g = \frac{N!}{N_\downarrow ! N_\uparrow !}, \quad \text{with} \quad N_{\pm\frac{1}{2}} = \frac{N}{2}(1 \pm \overline{M}). \tag{5.12}$$

The maximum multiplicity occurs for $N_\downarrow = N_\uparrow$, *i.e.* for a nonmagnetic system. On the other hand, the multiplicity g is 1 for the magnetic systems, *i.e.* the entropy, is zero for the fully magnetized state. For large N, Stirling's approximation $\ln n! \cong n \ln n - n$ can be used to express the entropy $S = k_B \ln g$ from Eq. (4.43) as

$$S = k_B \left\{ N \ln N - \frac{N}{2}(1 + \overline{M}) \ln \frac{N}{2}(1 + \overline{M}) - \frac{N}{2}(1 - \overline{M}) \ln \frac{N}{2}(1 - \overline{M}) \right\}.$$

Close to the critical temperature, the entropy can be expanded in terms of $\overline{M}$, giving for the entropy density

$$s = \frac{S}{V} \cong nk_B \left(\ln 2 - \frac{1}{2}\overline{M}^2 - \frac{1}{12}\overline{M}^4 + \cdots \right). \tag{5.13}$$

Omitting the constant term in the entropy, the free-energy density can now be written as

$$f = u - Ts = \frac{1}{2}nk_BT\left\{\left(1 - \frac{T_c}{T}\right)\overline{M}^2 + \frac{1}{6}\overline{M}^4\right\} - \mu_0 MH, \tag{5.14}$$

using Eq. (5.11) and where the energy due to the external magnetic field H has been added. In the paramagnetic state close to T_c, the free energy can be minimized by taking the derivative with respect to the magnetization,

$$\frac{\partial f}{\partial M} \cong \frac{nk_BT}{M_0^2}\left(1 - \frac{T_c}{T}\right)M - \mu_0 H = 0, \tag{5.15}$$

where the higher order term in the magnetization has been neglected. The susceptibility can then be obtained as

$$\chi = \frac{M}{H} = \frac{n\mu_0\mu_B^2}{k_B}\frac{1}{T - T_c} = \frac{C}{T - T_c}, \tag{5.16}$$

with $M_0 = n\mu_B$ and using Eq. (5.4). This result reproduces Eq. (5.10).

The approach above is a version of the Ginzburg–Landau theory of second-order phase transitions, where there is no discontinuity in the free energy when crossing the transition temperature, unlike a first-order phase transition. If we are interested in the spontaneous magnetization, we can write

$$f = \alpha\mathbf{M}^2 + \frac{1}{2}\beta\mathbf{M}^4, \tag{5.17}$$

where $\alpha = a(T - T_c)$, with $a = \mu_0/2C$ and $\beta \cong \frac{1}{6}k_BT_c/n^3\mu_B^4$, where in the region of the critical temperature, we take $T \cong T_c$ for the constant a. The magnetization can be found by minimizing the free energy,

$$\frac{\partial f}{\partial \mathbf{M}} = 2\alpha\mathbf{M} + 2\beta\mathbf{M}^3 = 0. \tag{5.18}$$

This equation has two solutions $\mathbf{M} = 0$ and, more interesting,

$$\mathbf{M}^2 = -\frac{\alpha}{\beta} = \frac{a(T_c - T)}{\beta} \quad \Rightarrow \quad M = \sqrt{\frac{a(T_c - T)}{\beta}}. \tag{5.19}$$

This solution only exists for $T < T_c$, *i.e.* below the critical temperature.

In Eq. (5.5), the interaction between the spin is ferromagnetic. To minimize the Coulomb repulsion, one prefers electrons to occupy different states or orbitals, which simply increases their average distance. This can be achieved by aligning their spins in which cases Pauli's principle forbids the occupation of the same state. This leads to a ferromagnetic coupling. However, it is also possible to have an antiferromagnetic coupling. This is generally related to the relaxation of the wavefunctions. For simplicity, let us consider two neighboring s orbitals (or any other singly degenerate state with spin). If the orbitals are separated from each other, then the lowest configuration is to put an electron on each orbital. Putting two electrons in the same orbital causes a Coulomb interaction between the two electrons. Let us denote that interaction by U. However, let us now introduce a hopping term between the two states. The Hamiltonian is then given by

$$\boldsymbol{H} = -t\sum_{\langle ij\rangle\sigma} \mathbf{c}_{i\sigma}^{\dagger}\mathbf{c}_{j\sigma} + U\sum_i \boldsymbol{n}_{i\uparrow}\boldsymbol{n}_{i\downarrow}, \tag{5.20}$$

where $\langle ij\rangle$ indicates hopping between the nearest neighbors. Here, we limit ourselves to only two sites, *i.e.* $i, j = 1, 2$, but for an extended system, this is known as the Hubbard model. For two sites, this problem is that of an H_2 molecule, Section 3.7. However, in that case the bonding and antibonding eigenstates were calculated first, and then, the molecule was filled with electrons. Now, we look at the molecule and calculate the electronic structure including explicitly the number of electrons. There are a total of six configurations. The states with parallel spins $|\uparrow;\uparrow\rangle$ and $|\downarrow;\downarrow\rangle$, where the semicolon separates the sites, are at energy zero since Pauli's principle does not allow hopping between the sites. For $S_z = 0$, the problem can be written as a 4×4 matrix,

$$\boldsymbol{H} = \begin{pmatrix} 0 & -t & -t & 0 \\ -t & U & 0 & -t \\ -t & 0 & U & -t \\ 0 & -t & -t & 0 \end{pmatrix} \begin{matrix} |\uparrow;\downarrow\rangle \\ |\uparrow\downarrow;0\rangle \\ |0;\uparrow\downarrow\rangle \\ |\downarrow;\uparrow\rangle \end{matrix} \quad \rightarrow \quad \boldsymbol{H} = \begin{pmatrix} 0 & -2t \\ -2t & U \end{pmatrix}. \tag{5.21}$$

The 4×4 matrix can be diagonalized exactly or transformed into a 2×2 matrix and two nonbonding states at energies 0 and U. The 2×2 matrix can be diagonalized to give the eigenenergies

$$E_\pm = \frac{U}{2} \pm \frac{1}{2}\sqrt{U^2 + 16t^2} \cong \begin{cases} -2t + \dfrac{U}{2}, 2t + \dfrac{U}{2}, & t \gg U, \\ -\dfrac{4t^2}{U}, U + \dfrac{4t^2}{U}, & t \ll U, \end{cases} \tag{5.22}$$

with eigenfunctions

$$|E_\pm\rangle = \frac{\cos\theta}{\sqrt{2}}(|\uparrow;\downarrow\rangle + |\downarrow;\uparrow\rangle) + \frac{\sin\theta}{\sqrt{2}}(|\uparrow\downarrow;0\rangle + |0;\uparrow\downarrow\rangle). \tag{5.23}$$

In the limit $t \gg U$, the solutions are similar to what is expected for an H_2 molecule. The two electrons settle in the bonding states, giving an energy of $-2t$. Additionally, $\theta \cong 45°$, so the electrons spend approximately equal amount of time on a site with and without the other electron giving an average Coulomb repulsion of $U/2$. In the limit $t \ll U$, the gain due to the delocalization of the orbital is significantly less. However, the energy of the singlet state $-4t^2/U$ is still less than putting the spins parallel, giving an effective antiferromagnetic coupling

$$H = \sum_{\langle ij\rangle} J\left(\mathbf{S}_i \cdot \mathbf{S}_j - \frac{1}{4}\right) \quad \text{with} \quad J = \frac{2t^2}{U}, \tag{5.24}$$

using the spin–spin interaction from Eq. (4.112). This gives a singlet state with $S = 0$ at energy $-4t^2/U$ and a triplet state with $S = 1$ and $S_z = 1, 0, -1$ at zero energy. (Note that the summation over $\langle ij\rangle$ counts each interaction twice).

Conceptual summary. – In a magnetic system, there are two competing phenomena. On one hand, there are interactions of the local moments with an external field or with each other. These terms often want to increase the total magnetization. On the other hand, the entropy decreases the total magnetization, since it is much more likely to have a state that is not fully magnetized. The magnetization therefore only appears below a particular temperature, known as the critical temperature.

Another interesting aspect occurs in magnetism, which we will encounter again in Section 5.4. This is the concept of broken symmetry. In principle, all magnetization directions are equivalent. However, the total magnetization does not continuously rotate between those directions. The system picks one particular direction for the magnetization. Now the material has a particular magnetization axis, and the symmetry of the system is broken.

Problems

1 Let us consider two equal neighboring spins with relatively large S that can rotate freely. An additional electron can hop between the two sites with the spins in an s-like orbital and prefers to be parallel with the large spin. Derive a Hamiltonian for this system, solve for the eigenenergies, and consider the limit that the coupling between the large spin and the spin of the free electron is much larger than the hopping matrix element (Hint: take one of the sites as the reference axis).

5.2 Superconductivity

Introduction. – In solids, the electronic and nuclear degrees of freedom are often treated separately. This makes sense since the electrons move significantly faster than the nuclei and are generally no longer around to feel the response of the lattice to their presence. However, other electrons can feel the retarded response of the lattice. Since the electron that distorted the lattice has moved on, its Coulomb interaction has reduced. In that case, the interaction with the distorted lattice dominates, and this can lead to an effective attractive interaction between electrons mediated by the lattice distortions. The effective attractive interaction between the electrons leads to the formation of a pair of electrons that is an effective boson. Since bosons follow different statistics, the pairs can condensate, which dramatically changes the properties of the material. The most surprising effect is that a current can now move without resistance, a phenomenon known as superconductivity.

A detailed derivation of this interaction using canonical transformations is long, so a somewhat simplified version is presented, which conveys the spirit of the derivation. This is also justified in light of additional approximations.

In Section 3.2, the problem of the displaced harmonic oscillator was studied. The Coulomb interaction between the electrons and the oscillator leads to a displacement of the oscillator that lowers the energy. It does not matter whether the interaction is attractive or repulsive (depending on the charge of the nuclei), this only affects the direction of the displacement. The problem for a solid is very similar in idea but has several added complexities. First, there is no longer a single harmonic oscillator, but the lattice vibrations, the phonons, have a dispersion giving energies $\hbar\omega_{\mathbf{q}}$, where $\mathbf{q}$ is the wavevector of the phonon. Second, the lattice is not responding to a fixed charge but to an electron density with dispersion $\varepsilon_{\mathbf{k}}$. Third, as in the displaced harmonic oscillator, the interaction is still taken linear in the displacement, but the coupling constants now depend on the momentum and the electron density. The total Hamiltonian is then

$$\boldsymbol{H} = \sum_{\mathbf{q}} \hbar\omega_{\mathbf{q}} \boldsymbol{a}^{\dagger}_{\mathbf{q}} \boldsymbol{a}_{\mathbf{q}} + \sum_{\mathbf{k}} \varepsilon_{\mathbf{k}} \mathbf{c}^{\dagger}_{\mathbf{k}\sigma} \mathbf{c}_{\mathbf{k}\sigma} - \sum_{\mathbf{kq}} \lambda_{\mathbf{q}} \mathbf{c}^{\dagger}_{\mathbf{k}+\mathbf{q},\sigma} \mathbf{c}_{\mathbf{k}\sigma} (\boldsymbol{a}_{\mathbf{q}} + \boldsymbol{a}^{\dagger}_{-\mathbf{q}}), \tag{5.25}$$

where $\sigma = \uparrow, \downarrow$ is the spin.

Despite the additional detail, the Hamiltonian is still similar in structure to the displaced harmonic oscillator in Eq. (3.31). This Hamiltonian was diagonalized by displacing the step operators, effectively shifting to the minimum of the displaced parabolic potential. We again introduce displaced oscillators $\boldsymbol{a}_{\mathbf{q}}^{\dagger} = \boldsymbol{a}_{\mathbf{q}}^{\prime\dagger} - \Delta_{\mathbf{q}}^{\dagger}$ and $\boldsymbol{a}_{\mathbf{q}} = \boldsymbol{a}_{\mathbf{q}}^{\prime} - \Delta_{\mathbf{q}}$. The Hamiltonian can be rewritten as

$$\begin{aligned} \boldsymbol{H}_{\mathrm{p}} &= \sum_{\mathbf{q}} \hbar\omega_{\mathbf{q}} \boldsymbol{a}_{\mathbf{q}}^{\prime\dagger} \boldsymbol{a}_{\mathbf{q}}^{\prime} + \sum_{\mathbf{q}} \hbar\omega_{\mathbf{q}} \Delta_{\mathbf{q}}^{\dagger} \Delta_{\mathbf{q}} + \sum_{\mathbf{kq}} \lambda_{\mathbf{q}} \mathbf{c}_{\mathbf{k}+\mathbf{q},\sigma}^{\dagger} \mathbf{c}_{\mathbf{k}\sigma} (\Delta_{-\mathbf{q}}^{\dagger} + \Delta_{\mathbf{q}}) \\ &\quad - \sum_{\mathbf{kq}} \left(\lambda_{\mathbf{q}} \mathbf{c}_{\mathbf{k}+\mathbf{q},\sigma}^{\dagger} \mathbf{c}_{\mathbf{k}\sigma} (\boldsymbol{a}_{\mathbf{q}}^{\prime} + \boldsymbol{a}_{-\mathbf{q}}^{\prime\dagger}) + \hbar\omega_{\mathbf{q}} (\Delta_{\mathbf{q}}^{\dagger} \boldsymbol{a}_{\mathbf{q}}^{\prime} + \Delta_{-\mathbf{q}} \boldsymbol{a}_{-\mathbf{q}}^{\prime\dagger}) \right), \end{aligned} \tag{5.26}$$

where the Hamiltonian for the phonons $\boldsymbol{H}_{\mathrm{p}}$ leaves out the energies of the electrons. Inversion symmetry has been assumed, so that $\omega_{-\mathbf{q}} = \omega_{\mathbf{q}}$ and $\lambda_{-\mathbf{q}} = \lambda_{\mathbf{q}}$. We want to cancel the linear displacement term in the second line. This can in principle be done exactly by taking $\Delta_{\mathbf{q}}^{\dagger} = \Delta_{-\mathbf{q}} = -\lambda_{\mathbf{q}} \mathbf{c}_{\mathbf{k}+\mathbf{q},\sigma}^{\dagger} \mathbf{c}_{\mathbf{k}\sigma} / \hbar\omega_{\mathbf{q}}$. However, it has to be expressed in terms of real physical excitations. In $\Delta_{\mathbf{q}}^{\dagger} \boldsymbol{a}_{\mathbf{q}}^{\prime}$, $\Delta_{\mathbf{q}}^{\dagger}$ is coupled to the annihilation of a phonon with momentum $\mathbf{q}$. This can be done by exciting an electron from $\mathbf{k} \to \mathbf{k} + \mathbf{q}$, giving an excitation with total energy $\varepsilon_{\mathbf{k}+\mathbf{q}} - \varepsilon_{\mathbf{k}} - \hbar\omega_{\mathbf{q}}$. Therefore, unlike the displacement harmonic oscillator where the coupling was to a fixed local object that could absorb any momentum, here excitations with respect to the Fermi sphere are created. The suggested displacement is

$$\Delta_{\mathbf{q}}^{\dagger} = \sum_{\mathbf{k}} \frac{\lambda_{\mathbf{q}} \mathbf{c}_{\mathbf{k}+\mathbf{q},\sigma}^{\dagger} \mathbf{c}_{\mathbf{k}\sigma}}{\varepsilon_{\mathbf{k}+\mathbf{q}} - \varepsilon_{\mathbf{k}} - \hbar\omega_{\mathbf{q}}}, \quad \Delta_{\mathbf{q}} = -\sum_{\mathbf{k}} \frac{\lambda_{\mathbf{q}} \mathbf{c}_{\mathbf{k}-\mathbf{q},\sigma}^{\dagger} \mathbf{c}_{\mathbf{k}\sigma}}{\hbar\omega_{\mathbf{q}} + \varepsilon_{\mathbf{k}-\mathbf{q}} - \varepsilon_{\mathbf{k}}}. \tag{5.27}$$

The denominator in $\Delta_{\mathbf{q}}$ corresponds to the creation of a phonon via an electron excitation from $\mathbf{k} \to \mathbf{k} - \mathbf{q}$. The displacements cancel the interaction term for $|\varepsilon_{\mathbf{k}+\mathbf{q}} - \varepsilon_{\mathbf{k}}| \ll \hbar\omega_{\mathbf{q}}$. When this is not satisfied, higher order corrections occur that we ignore.

We are more interested in the additional terms that occur in the first line of Eq. (5.26). For the displaced harmonic oscillator in Eq. (3.33), these terms simply caused a shift in energy. Due to the presence of the creation and annihilation operators, they are now an effective interaction between electrons. Let us consider the terms containing $\lambda_{\mathbf{q}}$ and $\Delta_{\mathbf{q}}$ in Eq. (5.26). The electron–phonon interaction is then

$$\boldsymbol{H}_{\text{e-p}} = \sum_{\mathbf{kk'q}} \lambda_{\mathbf{q}} \mathbf{c}_{\mathbf{k}+\mathbf{q},\sigma}^{\dagger} \mathbf{c}_{\mathbf{k}\sigma} \left(\frac{\lambda_{\mathbf{q}} \mathbf{c}_{\mathbf{k'}-\mathbf{q},\sigma'}^{\dagger} \mathbf{c}_{\mathbf{k'}\sigma'}}{\varepsilon_{\mathbf{k'}-\mathbf{q}} - \varepsilon_{\mathbf{k'}} - \hbar\omega_{\mathbf{q}}} - \frac{\lambda_{\mathbf{q}} \mathbf{c}_{\mathbf{k'}-\mathbf{q},\sigma'}^{\dagger} \mathbf{c}_{\mathbf{k'}\sigma'}}{\hbar\omega_{\mathbf{q}} + \varepsilon_{\mathbf{k'}-\mathbf{q}} - \varepsilon_{\mathbf{k'}}} \right).$$

Note that primes need to be added in the summations in $\Delta_{\mathbf{q}}$. The terms can be brought under the same denominator,

$$\boldsymbol{H}_{\text{e-p}} = \sum_{\mathbf{kk'q}} \frac{2\lambda_{\mathbf{q}}^2 \hbar\omega_{\mathbf{q}}}{(\varepsilon_{\mathbf{k'}} - \varepsilon_{\mathbf{k'}-\mathbf{q}})^2 - (\hbar\omega_{\mathbf{q}})^2} \mathbf{c}_{\mathbf{k}+\mathbf{q},\sigma}^{\dagger} \mathbf{c}_{\mathbf{k'}-\mathbf{q},\sigma'}^{\dagger} \mathbf{c}_{\mathbf{k'}\sigma'} \mathbf{c}_{\mathbf{k}\sigma}. \tag{5.28}$$

The term $\Delta_{\mathbf{q}}^{\dagger} \Delta_{\mathbf{q}}$ in Eq. (5.26) contains similar terms, and the factor 2 disappears in the total interaction. Since the interaction will be simplified further, let us focus on the physics. Since the phonon operators have effectively been removed, Equation (5.28) looks like an effective interaction between electrons. For small excitation energies, $\varepsilon_{\mathbf{k'}} - \varepsilon_{\mathbf{k'}-\mathbf{q}} \ll \hbar\omega_{\mathbf{q}}$, the interaction is attractive. Therefore, as in Section 3.2, the interaction of the charges with the oscillators lowers the energy. For large excitation energies, there is no net benefit, and the interaction becomes repulsive.

We have now obtained an effective interaction that can bind electrons together. In Eq. (4.23), we saw that pairs of particles behave effectively like bosons. Note that the individual electrons are still fermions and occupy different states. These electron pairs, known as Cooper pairs, can undergo a Bose–Einstein condensation at low temperatures. However, there is a big difference between free bosons and pairs of electrons in a solid. For free bosons, a macroscopic number of particles condenses into a single lowest state. The lowest excited states have a very small energy, but there is only a microscopic number of them. Therefore, the condensate is stable. For electrons, the lowest state is the Fermi sea. Therefore, if the lowest excited states were simply electron–hole pairs, the condensate would be unstable, since there is a macroscopic number of low-energy excited states. However, this is resolved by having a finite gap in the excitation spectrum. Therefore, there is still a macroscopic number of excited states, but they are no longer at zero energy. The condensate then becomes stable if the temperature is below the gap in the excitation spectrum.

In the theory by Bardeen, Cooper, and Schrieffer (BCS), the electron–phonon interaction is further simplified to understand the excitation spectra of the electron pairs. The attractive interaction is assumed to be a constant $-U$. Acoustic phonons only exist for a finite frequency range. Therefore, a cut-off ω_D, known as the Debye frequency, is introduced. The potential then becomes

$$U_{\mathbf{q}} = \begin{cases} -U, & \text{for } \hbar\omega \leq \hbar\omega_D, \\ 0, & \text{otherwise.} \end{cases} \tag{5.29}$$

Since, additionally, we are looking for a boson condensate, the expectation is that this state has zero momentum and zero spin. This implies that $\mathbf{k}' = -\mathbf{k}$ and $\sigma' = -\sigma$. The entire electronic Hamiltonian then becomes

$$\boldsymbol{H} = \sum_{\mathbf{k}} \xi_{\mathbf{k}} \mathbf{c}^{\dagger}_{\mathbf{k}\sigma} \mathbf{c}_{\mathbf{k}\sigma} + \sum_{\mathbf{k}\mathbf{k}'} U_{\mathbf{k}'-\mathbf{k}} \mathbf{c}^{\dagger}_{\mathbf{k}'\uparrow} \mathbf{c}^{\dagger}_{-\mathbf{k}'\downarrow} \mathbf{c}_{-\mathbf{k}\downarrow} \mathbf{c}_{\mathbf{k}\uparrow}, \tag{5.30}$$

where $\xi_{\mathbf{k}} = \varepsilon_{\mathbf{k}} - \mu$ is the electron energy with respect to the chemical potential. The electron–electron interaction term now effectively describes the scattering of a Cooper pair with quantum numbers $\mathbf{k} \uparrow$ and $-\mathbf{k} \downarrow$ to a pair with $\mathbf{k}' \uparrow$ and $-\mathbf{k}' \downarrow$.

It is important to note that the expectation value of the electron–phonon interaction in the Fermi sea is zero, since the electrons are unable to scatter into different states due to Pauli's principle. Therefore, the electron density close to the chemical potential will be redistributed to allow the scattering. We want to see what happens at a particular $\mathbf{k}$ value. It is useful to take the Fermi sea as the starting point. If $\xi_{\mathbf{k}} \leq 0$, the states at both $\mathbf{k}$ and $-\mathbf{k}$ are occupied. The electron–phonon coupling removes the electron pair with $\mathbf{k} \uparrow; -\mathbf{k} \downarrow$ and scatters them somewhere else. If we only look at occupation of the wavevector $\mathbf{k}$, the particle number appears not conserved since electrons scatter in and out of this momentum due to the interaction. However, for the entire system, the particle number is still conserved since electron pairs are only scattered to a different wavevector. Likewise, if $\xi_{\mathbf{k}} > 0$, the state is initially empty but can be filled by a scattered electron pair. The scattering rate is determined by an effective field

$$\Delta_{\mathbf{k}} = -\sum_{\mathbf{k}'} U_{\mathbf{k}'-\mathbf{k}} \langle \mathbf{c}^{\dagger}_{\mathbf{k}'\uparrow} \mathbf{c}^{\dagger}_{-\mathbf{k}'\downarrow} \rangle, \tag{5.31}$$

combining the strength of the scattering and the probability of creating an electron pair at $\mathbf{k}'$. A minus sign has been included since we want $\Delta_{\mathbf{k}}$ to be a positive number. The effective Hamiltonian now becomes

$$\boldsymbol{H} = \sum_{\mathbf{k}} \xi_{\mathbf{k}} \mathbf{c}^{\dagger}_{\mathbf{k}\sigma} \mathbf{c}_{\mathbf{k}\sigma} - \sum_{\mathbf{k}} \Delta_{\mathbf{k}} (\mathbf{c}^{\dagger}_{-\mathbf{k}\downarrow} \mathbf{c}^{\dagger}_{\mathbf{k}\uparrow} + \mathbf{c}_{-\mathbf{k}\downarrow} \mathbf{c}_{\mathbf{k}\uparrow}). \tag{5.32}$$

This leads to two 2×2 matrices for states below and above the chemical potential:

$$\begin{pmatrix} 2\xi_{\mathbf{k}} & -\Delta_{\mathbf{k}} \\ -\Delta_{\mathbf{k}} & 0 \end{pmatrix} \begin{matrix} \mathbf{k}\uparrow; -\mathbf{k}\downarrow \\ 0_{\mathbf{k}\uparrow}; 0_{-\mathbf{k}\downarrow} \end{matrix}, \quad \begin{pmatrix} 0 & -\Delta_{\mathbf{k}} \\ -\Delta_{\mathbf{k}} & 2\xi_{\mathbf{k}} \end{pmatrix} \begin{matrix} 0_{\mathbf{k}\uparrow}; 0_{-\mathbf{k}\downarrow} \\ \mathbf{k}\uparrow; -\mathbf{k}\downarrow \end{matrix},$$

for $\xi_{\mathbf{k}} \leq 0$ and $\xi_{\mathbf{k}} > 0$, respectively, where for $\xi_{\mathbf{k}} \leq 0$, the electron pair $|\mathbf{k}\uparrow; -\mathbf{k}\downarrow\rangle$ scatters to somewhere else, leaving an empty state $|0_{\mathbf{k}\uparrow}; 0_{-\mathbf{k}\downarrow}\rangle$. Above the Fermi level, an initially empty state $|0_{\mathbf{k}\uparrow}; 0_{-\mathbf{k}\downarrow}\rangle$ can be filled by a scattered electron pair giving $|\mathbf{k}\uparrow; -\mathbf{k}\downarrow\rangle$. These are, apart from shift in energy, equivalent matrices. For $\xi_{\mathbf{k}} \leq 0$, the solutions for the lowest states can be written as

$$|\varphi_{\mathbf{k}}^{-}\rangle = \cos\theta_{\mathbf{k}}|\mathbf{k}\uparrow; -\mathbf{k}\downarrow\rangle + \sin\theta_{\mathbf{k}}|0_{\mathbf{k}\uparrow}; 0_{-\mathbf{k}\downarrow}\rangle. \tag{5.33}$$

The angle $\theta_{\mathbf{k}}$ is given by

$$\tan 2\theta_{\mathbf{k}} = \frac{\Delta_{\mathbf{k}}}{\xi_{\mathbf{k}}} \quad \text{or} \quad \cos 2\theta_{\mathbf{k}} = \frac{\xi_{\mathbf{k}}}{E_{\mathbf{k}}} \quad \text{with} \quad E_{\mathbf{k}} = \sqrt{\xi_{\mathbf{k}}^2 + \Delta_{\mathbf{k}}^2}. \tag{5.34}$$

The probability of the occupied part of the wavefunction in Eq. (5.33) is then

$$v_{\mathbf{k}}^2 = \cos^2\theta_{\mathbf{k}} = \frac{1}{2} - \frac{1}{2}\cos 2\theta_{\mathbf{k}} = \frac{1}{2} - \frac{\xi_{\mathbf{k}}}{2E_{\mathbf{k}}} = \frac{1}{2}\left(1 - \frac{\xi_{\mathbf{k}}}{\sqrt{\xi_{\mathbf{k}}^2 + \Delta_{\mathbf{k}}^2}}\right), \tag{5.35}$$

where $v_{\mathbf{k}}$ is the conventional notation. With $u_{\mathbf{k}}^2 = 1 - v_{\mathbf{k}}^2$, the BCS ground state then can be written as

$$|\psi_0\rangle = \prod_{\mathbf{k}} (u_{\mathbf{k}} + v_{\mathbf{k}} \mathbf{c}_{\mathbf{k}\uparrow}^{\dagger} \mathbf{c}_{-\mathbf{k}\downarrow}^{\dagger})|0\rangle. \tag{5.36}$$

Therefore, even for $T = 0$, there is a broadened electron distribution around the Fermi level. Note that this does not conflict with the Fermi–Dirac distribution, since we are looking at the occupation of the free-electron levels $\varepsilon_{\mathbf{k}}$. The eigenstates $E_{\mathbf{k}}$ do follow the Fermi–Dirac distribution.

The value of $\Delta_{\mathbf{k}}$ needs to be determined self-consistently,

$$\begin{aligned} \Delta_{\mathbf{k}} &= -\sum_{\mathbf{k}'} U_{\mathbf{k}'-\mathbf{k}} \langle \varphi_{\mathbf{k}}^{-} | \mathbf{c}_{\mathbf{k}'\uparrow}^{\dagger} \mathbf{c}_{-\mathbf{k}'\downarrow}^{\dagger} | \varphi_{\mathbf{k}}^{-} \rangle = -\sum_{\mathbf{k}'} U_{\mathbf{k}'-\mathbf{k}} \sin\theta_{\mathbf{k}} \cos\theta_{\mathbf{k}} \\ &= -\frac{1}{2}\sum_{\mathbf{k}'} U_{\mathbf{k}'-\mathbf{k}} \sin 2\theta_{\mathbf{k}} = -\frac{1}{2}\sum_{\mathbf{k}'} \frac{U_{\mathbf{k}'-\mathbf{k}}\Delta_{\mathbf{k}}}{E_{\mathbf{k}}}, \end{aligned} \tag{5.37}$$

with $\sin 2\theta_{\mathbf{k}} = \Delta_{\mathbf{k}}/E_{\mathbf{k}}$ using Eq. (5.34). Using Eq. (5.29), a constant value of $\Delta = \Delta_{\mathbf{k}}$ is found for $\xi_{\mathbf{k}} \leq \hbar\omega_D$. The above equation then becomes

$$1 = \frac{U}{2}\sum_{\mathbf{k}} \frac{1}{E_{\mathbf{k}}}. \tag{5.38}$$

Since $\hbar\omega_D$ is relatively small, we can consider the density of states $\rho(E)$ close to the chemical potential $\mu = 0$ as constant. Replacing the summation by an integration gives

$$\frac{1}{\rho(0)U} = \int_0^{\hbar\omega_D} d\xi \frac{1}{\sqrt{\xi^2 + \Delta^2}} = \sinh^{-1}\frac{\hbar\omega_D}{\Delta}. \tag{5.39}$$

This gives for Δ,

$$\Delta = \frac{\hbar\omega_D}{\sinh(1/\rho(0)U)} \cong 2\hbar\omega_D e^{-\frac{1}{\rho(0)U}}, \tag{5.40}$$

where the weak-coupling limit was taken in the last step.

The goal was to increase the energy of the lowest excitation to stabilize the bosons and the Bose–Einstein condensate. The energies $E_{\mathbf{k}}^{\pm}$ at a particular $\mathbf{k}$ value can be determined from Eq. (5.33). The excitation energy per electron is half the difference of the two energies,

$$E_{\mathbf{k}}^{\pm} = \xi_{\mathbf{k}} \pm \sqrt{\xi_{\mathbf{k}}^2 + \Delta_{\mathbf{k}}^2} \quad \Rightarrow \quad E_{\mathbf{k}} = \frac{1}{2}(E_{\mathbf{k}}^{+} - E_{\mathbf{k}}^{-}) = \sqrt{\xi_{\mathbf{k}}^2 + \Delta_{\mathbf{k}}^2}. \tag{5.41}$$

Therefore, the electron–phonon interaction has significantly increased the stability of the Cooper pairs by creating an energy gap in the excitation spectrum. The size of this gap depends on the temperature, since the expectation value $\langle \varphi_{\mathbf{k}}^{-} | \mathbf{c}_{\mathbf{k}'\uparrow}^{\dagger} \mathbf{c}_{-\mathbf{k}'\downarrow}^{\dagger} | \varphi_{\mathbf{k}}^{-} \rangle$ is reduced by the excitations across the gap. This modifies Eq. (5.38) to

$$1 = \frac{U}{2} \sum_{\mathbf{k}} \frac{1}{E_{\mathbf{k}}} (1 - 2f(E_{\mathbf{k}})) = \frac{U}{2} \sum_{\mathbf{k}} \frac{1}{E_{\mathbf{k}}} \tanh \frac{E_{\mathbf{k}}}{2k_B T}, \tag{5.42}$$

where $f(E)$ is the Fermi–Dirac distribution function from Eq. (4.51). The factor 2 accounts for the fact that a Cooper pair consists of two electrons. To find the critical temperature, we take the limit $\Delta \to 0$ when $T \to T_c$. This means $E_{\mathbf{k}} \to \xi_{\mathbf{k}}$. Replacing the summation by an integral and again assuming that the density of states is more or less constant close to the chemical potential gives

$$\frac{1}{\rho(0)U} = \int_0^{\hbar\omega_D/2k_B T_c} \frac{\tanh x}{x} dx, \tag{5.43}$$

taking $x = \xi/2k_B T_c$. The evaluation of the integral is not straightforward but yields

$$k_B T_c = 1.13\hbar\omega_D e^{-\frac{1}{\rho(0)U}} \quad \Rightarrow \quad \frac{\Delta(0)}{k_B T_c} = 1.764, \tag{5.44}$$

using Eq. (5.40). This ratio has been tested experimentally and found in satisfactory agreement for conventional superconductors.

Conceptual summary. – The physics underlying the effective attractive electron–electron interaction that creates superconductivity is similar to that of a displaced harmonic oscillator. The displaced harmonic oscillator gives the response of the system to a fixed charge. For the phonons in a superconductor, the interaction occurs with the electron density, which causes a displacement. However, the displaced phonons then interact with the electron density. Removing the phonons from the problem leads to an effective attractive interaction between electrons.

The attractive interaction disturbs the electron distribution close to the Fermi level most strongly. A gap occurs in the electron spectrum due to the formation of pairs. This stabilizes the effective bosons created by the pairing of the electrons. The pairs condense below a certain temperature, creating a superconductor.

Problems

1 Calculate the density of states in the superconducting state. Assume that the density of states for the normal state is approximately constant.

5.3 Mass Generation

Introduction. – In Sections 5.1 and 5.2 of this chapter, we looked at the phenomena that emerged from interactions between magnetic moments and between electrons and the crystal lattice. Now we consider a property, the mass of a particle, that in the classical world is viewed as an intrinsic characteristic of matter and that was demonstrated to actually result from interactions between particles and a field. Up to this point, we already had to revise the classical notions of mass several times. For solids, it was useful to relate the effective mass m^* directly to the curvature in the energy dispersion $E_{\mathbf{k}}$, *i.e.* $1/m^* = (1/\hbar^2)(\partial^2 E_{\mathbf{k}}/\partial k^2)$, see Eq. (4.84). This gives indeed the rest mass m for a free particle but can affect the mass when the dispersion changes in a solid, see Sections 3.7 and 3.8. Additionally, electron–electron interactions can further modify the effective mass, see Section 4.3. We also saw that in relativity, mass is directly related to the energy, see Eq. (2.39). The majority of the mass for hadrons is due to the binding energy of the quarks via the strong interaction, see Sections 4.6 and 4.7. However, this still leaves the rest masses of several particles, such as the electron and the W and Z bosons, unexplained. In this section, it is shown that the rest mass is not an inherent property of the particle but also effectively describes an interaction, namely the coupling of particles to, what is known as, the Higgs field.

Here, a simple understanding is given of the mass generation. The problem is similar to that of the local potential studied in Section 3.9. A potential localized in real space is completely delocalized in momentum space. Therefore, for the local potential, momentum is not a good quantum number since the potential scatters a plane wave to any other plane wave. However, the idea of the Higgs field is that particles couple to a scalar field that is completely delocalized in space. This implies that it is localized in momentum space; therefore, momentum is a conserved quantity. However, the continuous interaction with the Higgs field changes the norm of the four-vector, thereby producing mass. The value of the mass depends on the strength of the coupling to the Higgs field.

Let us revisit the Dirac equation for a free particle, see Eq. (2.27),

$$\mathbf{p}' - mc = 0, \tag{5.45}$$

where $\mathbf{p}'$ is the four-momentum with the Minkowski metric, m is the rest mass, and c is the speed of light. There are two ways to interpret this. We can think of this as a four-momentum with a finite mass $\mathbf{p}' = mc$. However, we can also draw an analogy with $\mathbf{p}' - q\mathbf{A}'$, which is a particle with four-momentum $\mathbf{p}'$ moving in a four-potential $\mathbf{A}'$. Therefore, the expression in (5.45) is also the Dirac equation for a four-momentum $\mathbf{p}'$ with norm zero in the presence of a scalar field mc. This field is known as the Higgs field. The mass is then the coupling strength to the Higgs field.

Let us start with a massless particle. The unperturbed Green's function needs to satisfy

$$\mathbf{p}'c = 0 \;\Rightarrow\; \mathbf{p}'cG^0(\mathbf{p}') = 1 \;\Rightarrow\; G^0(\mathbf{p}') = \frac{1}{\mathbf{p}'c}. \tag{5.46}$$

The mass is now considered as a pertubation. The equation that we want to solve is then

$$\mathbf{p}'c = mc^2. \tag{5.47}$$

The full Green's function in the presence of a field causing mass is then

$$\begin{aligned} G(\mathbf{p}') &= G^0(\mathbf{p}') + G^0(\mathbf{p}')mc^2G(\mathbf{p}') \\ &= G^0(\mathbf{p}') + G^0(\mathbf{p}')mc^2G^0(\mathbf{p}') + G^0(\mathbf{p}')mc^2G^0(\mathbf{p}')mc^2G^0(\mathbf{p}') + \cdots \end{aligned} \tag{5.48}$$

The second line is not really needed, but it shows nicely the repeated scattering of the particle by the mass-generating field. However, the Green's function already closes on itself in the first line. This gives for the full Green's function

$$G(\mathbf{p}') = \frac{G^0(\mathbf{p}')}{1 - G^0(\mathbf{p}')mc^2} = \frac{1}{(G^0(\mathbf{p}'))^{-1} - mc^2} = \frac{1}{\mathbf{p}'c - mc^2}. \tag{5.49}$$

The denominator is now a four-vector. This can also be made real

$$G(\mathbf{p}') = \frac{\mathbf{p}'c + mc^2}{(\mathbf{p}'c)^2 - (mc^2)^2} = \frac{2mc^2}{(\mathbf{p}'c)^2 - (mc^2)^2} = \frac{2mc^2}{E^2 - \mathrm{p}^2c^2 - m^2c^4}, \tag{5.50}$$

using Eq. (5.47). The poles of the Green's function satisfy Einstein's energy-momentum relation in Eq. (2.49).

We could have saved ourselves all the trouble by considering Eq. (5.47) as the free propagation of a particle with mass, *i.e.* $\mathbf{p}c - mc^2 = 0$. The Green's function for this is

$$(\mathbf{p}'c - mc^2)G(\mathbf{p}') = 1 \;\; \Rightarrow \;\; G(\mathbf{p}') = \frac{1}{\mathbf{p}'c - mc^2}, \tag{5.51}$$

reproducing the same result as Eq. (5.49).

This exercise can be repeated in the Dirac form. Multiplying Eq. (5.47) from the right side by $\mathbf{e}_0'$ gives

$$\mathbf{p}c = mc^2\mathbf{e}_0' \;\; \Rightarrow \;\; \mathbf{p}cG_D^0(\mathbf{p}) = 1 \;\; \Rightarrow \;\; G_D^0(\mathbf{p}) = \frac{1}{\mathbf{p}c}, \tag{5.52}$$

where $\mathbf{p}c = E + \mathrm{p}c$. Note that the interaction is now a vector. This can again be expressed in a series with multiple scattering of the field

$$G_D(\mathbf{p}) = \frac{1}{\mathbf{p}c} + \frac{1}{\mathbf{p}c}mc^2\mathbf{e}_0'\frac{1}{\mathbf{p}c} + \frac{1}{\mathbf{p}c}mc^2\mathbf{e}_0'\frac{1}{\mathbf{p}c}mc\mathbf{e}_0'\frac{1}{\mathbf{p}c} + \cdots \tag{5.53}$$

Although not necessary for the summation, let us shift the unit vector $\mathbf{e}_0'$ to the right

$$G_D(\mathbf{p}) = \frac{1}{\mathbf{p}c} + \frac{1}{\mathbf{p}c}mc^2\frac{1}{\underline{\mathbf{p}}c}\mathbf{e}_0' + \frac{1}{\mathbf{p}c}mc^2\frac{1}{\underline{\mathbf{p}}c}mc^2\frac{1}{\mathbf{p}c} + \cdots ,$$

where $\underline{\mathbf{p}}c = E - \mathrm{p}c$. The change $\mathbf{p} \leftrightarrow \underline{\mathbf{p}}$ is a change in chirality. This occurs when the interaction is written in terms of vectors but does not happen when expanding in terms of scalars. The series in Eq. (5.53) can be summed to give

$$G_D(\mathbf{p}) = \frac{1}{\mathbf{p}c - mc^2\mathbf{e}_0'} = \frac{1}{E + \mathrm{p}c - mc^2\mathbf{e}_0'}. \tag{5.54}$$

This can be made real via

$$G_D(\mathbf{p}') = \frac{1}{E + \mathrm{p}c - mc^2\mathbf{e}_0'}\frac{E - \mathrm{p}c + mc^2\mathbf{e}_0'}{E - \mathrm{p}c + mc^2\mathbf{e}_0'} = \frac{2E}{E^2 - \mathrm{p}^2c^2 - m^2c^4}. \tag{5.55}$$

From Eq. (5.52), $E + \mathrm{p}c = mc^2$ or $-\mathrm{p}c + mc^2 = E$. This is the same Green's function as in Eq. (5.50), except that the norm has changed from mc^2 to E.

The propagator for the relativistic particle in Eq. (5.50) looks rather different from the nonrelativistic free-particle propagator in Eq. (2.286). In fact, it looks a lot closer to the Green's function for a photon in Eq. (2.167) or a harmonic oscillator, see Eq. (2.159). This implies that we should be able to find a correlation function that looks similar to Eq. (4.10). A relativistic particle can therefore be interpreted as an effective displacement at time t,

$$|\boldsymbol{x}(t)\rangle = \sum_{\mathrm{p},\nu=\pm}\sqrt{\frac{\hbar}{2m\omega_{\mathrm{p}}}}|\mathrm{p}\nu\rangle\langle\mathrm{p}\nu|\boldsymbol{x}(t)\rangle = \sum_{\mathrm{p}}\sqrt{\frac{\hbar}{2m\omega_{\mathrm{p}}}}(\boldsymbol{a}_{\mathrm{p}-}^{\dagger}e^{i\omega_{\mathrm{p}}t} + \boldsymbol{a}_{\mathrm{p}+}^{\dagger}e^{-i\omega_{\mathrm{p}}t}),$$

with $|\mathbf{p}\pm\rangle = \boldsymbol{a}^\dagger_{\mathbf{p}\pm}$ being the effective spinor for counterclockwise and clockwise rotations in spacetime. Instead of $\boldsymbol{a}^\dagger$ and $\boldsymbol{a}$ used for the oscillator, the operators are now $\boldsymbol{a}^\dagger_{\mathbf{p}+}$ and $\boldsymbol{a}^\dagger_{\mathbf{p}-}$ that create particles and antiparticles both with a positive energy $\hbar\omega_\mathbf{p} = \sqrt{m^2c^4 + \mathbf{p}^2c^2}$. This avoids the issues of removing particles from a vacuum and negative energy states. Adding the spatial part of the function space of complex exponential gives

$$\phi(\mathbf{r}, t) = \sum_\mathbf{p} \sqrt{\frac{\hbar}{2m\omega_\mathbf{p}}} (\boldsymbol{a}^\dagger_{\mathbf{p}-} e^{-\frac{i}{\hbar}\mathbf{p}\cdot\mathbf{r} - i\omega_\mathbf{p} t} + \boldsymbol{a}^\dagger_{\mathbf{p}+} e^{\frac{i}{\hbar}\mathbf{p}\cdot\mathbf{r} - i\omega_\mathbf{p} t}). \tag{5.56}$$

This is generally known as a quantum field operator.

For independent free particles, the different $\mathbf{p}$ components behave independently in Eq. (5.56). We take the displacement for a particular momentum as $|\boldsymbol{x}_\mathbf{p}(t)\rangle$, with $|\boldsymbol{x}(t)\rangle = \sum_\mathbf{p} |\boldsymbol{x}_\mathbf{p}(t)\rangle$. The correlation or Green's function is given by

$$G_\mathbf{p}(t) = -\frac{i}{\hbar} \langle \boldsymbol{x}_\mathbf{p}(t) | \boldsymbol{x}^0_\mathbf{p} \rangle \theta(t). \tag{5.57}$$

Care should be taken with $\boldsymbol{x}^0_\mathbf{p}$. For $t = 0$, we have $|\boldsymbol{x}(0)\rangle \sim \boldsymbol{a}^\dagger_{\mathbf{p}-} + \boldsymbol{a}^\dagger_{\mathbf{p}+}$, which gives a finite value at $t = 0$. Although giving an oscillator a finite displacement at $t = 0$ is a perfectly fine way to initiate an oscillation, it does not work for our purposes since we want the $\boldsymbol{x}(0)$ to be zero for $t < 0$. Using $|\boldsymbol{x}(0)\rangle$ would lead to a discontinuity at $t = 0$. We therefore prefer to start the oscillation by having a finite derivative and zero amplitude. This is essentially the response to a δ function impulse. Therefore, the oscillation starts at the finite phase $\omega_\mathbf{p} t_0 = \frac{\pi}{2}$, giving

$$|\boldsymbol{x}^0_\mathbf{p}\rangle = \sqrt{\frac{\hbar}{2m\omega_\mathbf{p}}} (\boldsymbol{a}^\dagger_{\mathbf{p}-} - \boldsymbol{a}^\dagger_{\mathbf{p}+}). \tag{5.58}$$

Or, in even simpler terms, to ensure continuity, we want to start a sine at $t = 0$ and not a cosine, see Eq. (2.173). The Green's function is then

$$\begin{aligned} G_\mathbf{p}(t) &= -\frac{i}{\hbar} \frac{\hbar}{2m\omega_\mathbf{p}} \left(e^{-i\omega_\mathbf{q} t} \langle \boldsymbol{a}_{\mathbf{p}-} \boldsymbol{a}^\dagger_{\mathbf{p}-} \rangle - e^{i\omega_\mathbf{q} t} \langle \boldsymbol{a}_{\mathbf{p}+} \boldsymbol{a}^\dagger_{\mathbf{p}+} \rangle \right) \theta(t) \\ &= -\frac{i}{2m\omega_\mathbf{p}} \left(e^{-i\omega_\mathbf{p} t} - e^{i\omega_\mathbf{p} t} \right) \theta(t) = -\frac{1}{m\omega_\mathbf{p}} \sin \omega_\mathbf{p} t \, \theta(t), \end{aligned} \tag{5.59}$$

omitting the cross terms. This propagator is continuous at $t = 0$.

We are also interested in the Fourier transform of the Green's function

$$\begin{aligned} G_\mathbf{p}(\omega) = \langle \omega | G_\mathbf{p} \rangle &= \int_{-\infty}^{\infty} dt \, \langle \omega | t \rangle \langle t | G_\mathbf{p} \rangle = \int_{-\infty}^{\infty} dt \, e^{i\omega t} G_\mathbf{p}(t) \\ &= -\frac{i}{2m\omega_\mathbf{p}} \int_0^\infty dt \, \left(e^{i(\omega - \omega_\mathbf{p})t} - e^{i(\omega_\mathbf{p} + \omega)t} \right) e^{-\eta t} \\ &= \frac{1}{2m\omega_\mathbf{p}} \left(\frac{1}{\omega - \omega_\mathbf{p} + i\eta} - \frac{1}{\omega + \omega_\mathbf{p} + i\eta} \right), \end{aligned} \tag{5.60}$$

using the nonunitary Fourier transform, see Eq. (1.248). A small decaying exponential with $\eta \to 0$ has been introduced to remove the oscillatory behavior in the limit $t \to \infty$. The two terms can be put under the same denominator

$$G_\mathbf{p}(\omega) = \frac{1}{m} \frac{1}{\omega^2 - \omega_\mathbf{p}^2 + i\eta}. \tag{5.61}$$

The Green's function has the same form as that of a classical oscillator, see Eq. (2.159), except that many different frequencies $\omega_{\mathbf{p}}$ can be excited. It also has the same form as the propagator of the photon, see Eq. (2.167), except that the permittivity of space ε_0 replaces the mass, and that, $m = 0$ has to be taken for the energy $\hbar\omega_{\mathbf{p}} = \sqrt{\mathbf{p}^2c^2 + m^2c^4} = pc$.

The propagator in Eq. (5.61) should be compared with those in Eqs. (5.50) and (5.55). The latter were obtained directly from the Dirac equation. In the derivation of Eq. (5.61), the underlying assumption was that we are talking about some real displacement in space. However, the complex exponentials $e^{\mp i\omega_{\mathbf{p}}t}$ are related to the time-dependent part of the function space. Obviously, for a quantum oscillator, this is directly related to the displacement $\boldsymbol{x}$. However, removal of the factor $\sqrt{\hbar/2m\omega_{\mathbf{p}}}$ squared gives

$$\frac{2m\omega_{\mathbf{p}}}{\hbar} G_{\mathbf{p}}(E) = -\frac{i}{\hbar}\langle \boldsymbol{\xi}_{\mathbf{p}}(t)|\boldsymbol{\xi}^0_{\mathbf{p}}\rangle\theta(t) = G_D(\mathbf{p}) = \frac{mc^2}{E_{\mathbf{p}}} G(\mathbf{p}'), \tag{5.62}$$

where $\boldsymbol{\xi} = \sqrt{2m\omega_{\mathbf{p}}/\hbar}\, \boldsymbol{x}$ is the dimensionless displacement. Therefore, using $\boldsymbol{\xi}$ gives the Dirac propagator $G_D(\mathbf{p})$ obtained using four-vectors. The Green's function using the Minkowski metric $G(\mathbf{p}')$ differs from the Dirac propagator by a factor related to the norms of the four-momenta in the different bases.

Up to this point the mass has been considered to be a real number. However, we have seen that mass is the result of the coupling to the Higgs field. Therefore, mass is a way to effectively treat a more complicated interaction. As we saw in Section 3.10, coupling to a continuum of states can lead to two different effects: a shift in energy, which can be identified as the mass, and a broadening that can be interpreted as the inverse lifetime of the particle. This is often indicated as the self-energy $\Sigma_{\mathbf{q}}(\omega)$, which depends on the energy and momentum of the particle. Let us ignore momentum dependence and assume that the interaction has a complex component $\Sigma = mc^2 - i\Gamma$. Inserting this into Eq. (5.49) gives

$$G(\mathbf{p}') = \frac{1}{\mathbf{p}'c - mc^2 + i\Gamma}. \tag{5.63}$$

The Green's function can be written in scalar form as

$$G(\mathbf{p}') = \frac{\mathbf{p}'c + mc^2}{\mathbf{p}'c + mc^2}\frac{1}{\mathbf{p}'c - mc^2 + i\Gamma} \cong \frac{2mc^2}{\mathbf{p}'^2c^2 - m^2c^4 + i2mc^2\Gamma}, \tag{5.64}$$

using that $\mathbf{p}'c + mc^2 = 2mc^2$. The imaginary part is given by

$$-\frac{1}{\pi}\,\mathrm{Im}G(\mathbf{p}') = \frac{1}{\pi}\frac{(2mc^2)^2\Gamma}{(E^2 - E_{\mathbf{p}}^2)^2 + (2mc^2\Gamma)^2}, \tag{5.65}$$

with $E_{\mathbf{p}}^2 = \mathbf{q}^2c^2 + m^2c^4$. This expression is related to the Breit–Wigner distribution. Therefore, the spectral width of the distribution is also related to the mass of the particle. It also shows why heavier particles often have a shorter lifetime.

Let us now turn our attention to the electroweak force, as discussed in Section 2.13. In that section, the four-potential $\mathbf{A}'$ that only works on a single particle was extended with additional four-vector terms that coupled different particles. In principle, this could be very many terms, but the interaction turns out to be the strongest between certain fermion pairs, such as the up/down quark and the electron/neutrino. This leads to a 2×2 matrix of vector interactions. Using the Weinberg angle this was then split into a diagonal 2×2 matrix with unequal matrix elements, which was identified with the electromagnetic interaction, and another 2×2 matrix that also contained the off-diagonal interactions, which was identified as the weak interaction. However, there was no good reason to split it in this fashion. The reason for the split is that the bosons mediating the interaction propagate differently in space since the bosons related to the weak interaction will acquire mass through the Higgs field.

The assumption is that the vector bosons interact with the Higgs field in the same fashion as neutrino/electron and down/up quark pairs. Therefore, there is a pair of Higgs bosons that can scatter into each other under the electroweak interaction. Now this approach fails if both Higgs bosons are equally important. So, we assume for the moment that one of the bosons dominates the interaction. This is a broken symmetry, and it will be discussed in Section 5.4 how this can happen. Let us now adjust the Higgs field, such that it agrees with the experimental observations. The dominant Higgs field is a scalar field and should not interact via the electromagnetic four-potential. These conditions can be obtained by taking a hypercharge $Y = 1$. Inserting the hypercharge into Eq. (2.292) gives for the electroweak interaction for the Higgs pair

$$\boldsymbol{H}_{\text{EW}} = \begin{pmatrix} e\mathbf{A}' + G\cos 2\theta\,\mathbf{Z}' & \sqrt{2}g'\mathbf{W}'_- \\ \sqrt{2}g'\mathbf{W}'_+ & G\mathbf{Z}' \end{pmatrix} \begin{matrix} \phi_+ \\ \phi_0 \end{matrix} . \tag{5.66}$$

This interaction is very similar to Eq. (2.295). The scalar Higgs field, indicated by ϕ_0, couples to the electroweak interaction in a similar fashion as the neutrino, *i.e.* it does not experience the four-potential $\mathbf{A}'$ and only interacts with the $\mathbf{Z}'$ and $\mathbf{W}'_\pm$ bosons. Therefore, only the charge carriers associated with the weak interaction gain mass. It is the mass generation that turns the electroweak interaction into two distinct interactions. Note that photons would also obtain mass if there was a large positively charged ϕ_+ field. However, the existence of ϕ_+ bosons has not been demonstrated, indicating that the ϕ_0 field dominates.

Finally, it is important to note that the Higgs mechanism is not the only source of mass. Due to the mass-energy equivalence, additional energy terms can also be viewed effectively as mass. Therefore, the mass of the proton and neutron is predominantly caused by the strong interaction between the quarks. The Higgs field plays therefore an important role for noncomposite particles, such as leptons (electrons) and force carriers (W and Z bosons).

Conceptual summary. – The Dirac equation sets the norm of the four-momentum equal to mc. In this section, we considered the mc term as an effective interaction, which can give a massless particle a finite mass. Since the interaction only changes the norm of the four-momentum but not the momentum, we conclude that the interaction is a scalar field delocalized in space. This is known as the Higgs field.

Due to its hypercharge, the scalar Higgs field ϕ_0 interacts with the electroweak force in a similar fashion as the neutrino. This means that the scalar Higgs field does not interact with the electromagnetic four-potential, leaving the photons massless. However, the Higgs field does affect the W and Z bosons and gives them a finite mass. When initially describing the electroweak interaction in Eq. (2.13), we rewrote it to bring out better the electromagnetic and weak interactions. If both force carriers were massless, we could have rewritten the interaction in many different ways. However, it is the finite mass of the force carriers that makes the weak interaction different from the electromagnetic interaction.

Problems

1 Find the ratio between the masses of the Z and $W_\pm$ bosons taking a Weinberg angle of 28°.

5.4 Symmetry Breaking

Introduction. – In Section 5.3, it was demonstrated how the interaction with the Higgs field created mass for the leptons and the W and Z bosons. However, a crucial element is that the Higgs field is dominated by the scalar Higgs bosons ϕ_0, and that the charged Higgs fields $\phi_\pm$ are very weak. This essentially breaks the symmetry between the scalar and charged Higgs field. In this section, we consider how a finite expectation value of a field can be obtained.

This is not the first case that we have seen symmetry breaking. One of the most intuitive examples of symmetry breaking happens in magnetism, see Section 5.1. Here, it was found that below a critical temperature the system orders. The magnetization of the system occurs in a particular direction. This causes a spontaneous symmetry breaking, since excitations now occur with respect to the magnetization direction of the material. Also, in superconductors, we observe the appearance of an order parameter below a critical temperature.

Let us first consider the cost of making an excitation in a ferromagnet. For superconductors, there was a gap in the excitation spectrum, since it required breaking up a Cooper pair, see Eq. (5.41). At first, one might think that an excitation in a ferromagnet is gapped as well since it requires breaking ferromagnetic pairs that are coupled via $J\mathbf{S}_i \cdot \mathbf{S}_j$, where $\mathbf{S}_i = \mathbf{S}_{ix}\boldsymbol{e}_x + \mathbf{S}_{iy}\boldsymbol{e}_y + \mathbf{S}_{iz}\boldsymbol{e}_z$ is the Pauli vector of spin operators for site i, see Eq. (5.5). For $S = \frac{1}{2}$, the energy of the triplet and singlet states are $-\frac{1}{4}J$ and $\frac{3}{4}J$, respectively. This would then require a cost zJ, where z is the number of nearest neighbors, to create an excitation, *i.e.* flip a spin. This turns out not to be the case. For the ground state $|0\rangle$, all the spins are pointing in a particular direction. However, all directions are equivalent. Therefore, the lowest excitations correspond to a very small rotation of the direction of magnetization. Since the new direction has the same energy as the previous direction, this excitation spectrum should be gapless.

To understand the excitation, let us consider, for simplicity, a one-dimensional system with spins of magnitude S. The Heisenberg Hamiltonian, see Eq. (5.5), can be written as

$$\boldsymbol{H} = -J\sum_{\langle ij\rangle}\mathbf{S}_j \cdot \mathbf{S}_i = -J\sum_{\langle ij\rangle}(\mathbf{S}_{j0}\mathbf{S}_{i0} + \mathbf{S}_{j,-1}\mathbf{S}_{i1} + \mathbf{S}_{j1}\mathbf{S}_{i,-1}), \tag{5.67}$$

where $\mathbf{S}_{i,\pm 1} = (\mathbf{S}_{ix} \pm i\mathbf{S}_{iy})/\sqrt{2}$ are the step operators for the spin (omitting the Condon–Shortley phase), see Eq. (1.372), and the summation $\langle ij\rangle$ goes over the nearest-neighbor pairs. The ground state energy for an ordered system is given by

$$E_0 = \langle 0|\boldsymbol{H}|0\rangle = -J\sum_{\langle ij\rangle}\langle 0|\mathbf{S}_j \cdot \mathbf{S}_i|0\rangle = -2JNS^2, \tag{5.68}$$

where N is the number of sites. This energy only involves the $\mathbf{S}_{j0}\mathbf{S}_{i0}$ term. The factor 2 reflects the number of nearest neighbors in a one-dimensional chain. Using Eq. (1.372), a spin at a particular site i can be lowered with respect to the ferromagnetic ground state via

$$\mathbf{S}_{i,-1}|0\rangle = \frac{1}{\sqrt{2}}\sqrt{(S+S)(S-S+1)}|i\rangle = \sqrt{S}|i\rangle,$$

omitting any Condon–Shortley phase factors. The ket $|i\rangle$ indicates that all sites have a projected spin $M_S = S$, except site i, which has a projected spin $M_S = S - 1$. Similarly, we can show that $S_{i1}|i\rangle = \sqrt{S}|0\rangle$. Working with the Hamiltonian on the state $|i\rangle$ gives

$$\begin{aligned} \boldsymbol{H}|i\rangle &= -J\sum_{j,\delta=\pm 1}\left(\mathbf{S}_{j+\delta,0}\mathbf{S}_{j0} + \mathbf{S}_{j+\delta,-1}\mathbf{S}_{j1}\right)|i\rangle \\ &= -J\sum_{\delta=\pm 1}(S(S-1)|i\rangle + S|i+\delta\rangle) - 2J(N-1)S^2|i\rangle, \end{aligned} \tag{5.69}$$

where $\delta = \pm 1$ goes over the nearest-neighbor sites. The contribution $-J\sum_{\delta=\pm 1}S^2|i\rangle + 2JS^2|i\rangle = 0$. The spin-flip term $\mathbf{S}_{j+\delta,-1}\mathbf{S}_{j1}$ is only nonzero for $j = i$. The other spin-flip term is always zero. The Hamiltonian effectively describes the hopping of the $M_S = S - 1$ state through the chain. This excitation is called a magnon. This can also be written in a different fashion. The diagonal term $JS\sum_i \boldsymbol{a}_i^\dagger \boldsymbol{a}_i|i'\rangle = JS|i'\rangle$ only gives a contribution for the excited spin on site i. The spin-flip term $-JS\sum_{i\delta}\boldsymbol{a}_{i+\delta}^\dagger \boldsymbol{a}_i|i'\rangle = -JS\sum_\delta |i'+\delta\rangle$ moves the excited spin to a neighboring site. This is comparable to the tight-binding Hamiltonian in Eq. (3.162), which describes the hopping of an electron on a lattice. We can describe this as the effective hopping of a particle

$$\boldsymbol{H}_{\text{eff}} = JS\sum_{i,\delta}(\boldsymbol{a}_i^\dagger \boldsymbol{a}_i - \boldsymbol{a}_{i+\delta}^\dagger \boldsymbol{a}_i) - 2JNS^2, \tag{5.70}$$

where $\boldsymbol{a}_i^\dagger|0\rangle = |i\rangle$ creates an $S_z = S - 1$ spin at site i in a background of $S_z = S$ spins. This is known as a Holstein–Primakoff transformation. The first term is the energy change JS with respect to the ground-state energy for $|i\rangle$ and its neighbor $|i+\delta\rangle$. The second term describes the hopping of the excitation to a neighboring site. Just like the tight-binding model, the Hamiltonian can be diagonalized by going to momentum space, giving

$$\boldsymbol{H}_{\text{eff}} = 2JS\sum_k(1-\gamma_k)\boldsymbol{a}_k^\dagger \boldsymbol{a}_k - 2JS^2, \tag{5.71}$$

with

$$\gamma_k = \frac{1}{2}\sum_{\delta=\pm 1} e^{ik\delta} \quad \Rightarrow \quad \gamma_k = \cos k, \tag{5.72}$$

taking the distance between the spins equal to 1. The relevant k values are in the region $[-\pi, \pi]$. The dispersion relation in one dimension is given by

$$\hbar\omega_k = 2JS(1-\cos k) = 4JS\sin^2\frac{k}{2} \cong JSk^2, \tag{5.73}$$

where the last step takes the limit $k \to 0$. It is important to note that the excitation spectrum is gapless. In the long wavelength limit, $k \to 0$, the excitation corresponds to a rotation of all the spins in the same direction. Since this does not affect the angle between the spins, the change in energy should be zero.

Another common element between magnetism, superconductivity, and the Higgs field is the presence of a finite order parameter, which is related to the magnetization, the superconducting gap, and the size of the scalar Higgs field ϕ_0, respectively. When only focusing on the order parameter, there are some remarkable similarities between the theories, since they are all essentially second-order phase transitions.

An approach similar to magnetism was used to describe superconductivity by Ginzburg and Landau. Following Eq. (5.17), a difference in free-energy density of the superconducting state relative to the normal state is written as

$$f = \alpha|\psi|^2 + \frac{\beta}{2}|\psi|^4, \tag{5.74}$$

where ψ is a complex order parameter that is nonzero in the superconducting state, and $\alpha = a(T - T_c)$. As in magnetism, there are two contributions to the free energy term: $-aT_c|\psi|^2$ is an attractive energy dependent on the superconducting density, and $a|\psi|^2 + b|\psi|^4$, with $\beta = bT \cong bT_c$ near the critical temperature, is due to the entropy that resists the order. Since ψ is a complex order parameter, the field equation for ψ is obtained by minimizing with respect to ψ^*

$$\frac{\partial f}{\partial \psi^*} = \alpha\psi + \beta|\psi|^2\psi = U\psi = 0, \quad \text{with} \quad U = \alpha + \beta|\psi|^2. \tag{5.75}$$

This gives for the order parameter

$$|\psi|^2 = -\frac{\alpha}{\beta} = \frac{a}{b}\left(1 - \frac{T}{T_c}\right) \quad \Rightarrow \quad |\psi| = \sqrt{\frac{a}{b}\left(1 - \frac{T}{T_c}\right)}, \tag{5.76}$$

giving the same temperature dependence as Eq. (5.19). This square-root temperature dependence is characteristic of mean-field approaches to second-order phase transitions.

As we saw above in the discussion of spin waves in ferromagnets, excitations can occur with respect to the order. The energy of these excitation is proportional to k^2, see Eq. (5.73), and can be viewed as comparable to the kinetic energy. For superconductors, a comparable term related to slow variations $\nabla\psi$ in the superconducting order parameter is added. This leads to the Ginzburg-Landau equations

$$\frac{(-i\hbar\nabla - 2e\mathbf{A})^2}{2m^*}\psi + \alpha\psi + \beta|\psi|^2\psi = 0, \tag{5.77}$$

where the vector potential $\mathbf{A}$ has been added. Since we know from BCS theory, see Section 5.2, that the superconducting order is due to the condensation of pairs of electrons, the effective charge has been taken as $2e$; m^* is the effective mass of the Cooper pair.

Let us look at the behavior in the absence of an electromagnetic field. Since the coefficients are real, let us assume that ψ is real. The value of the order parameter is then $\psi_0^2 = |\alpha|/\beta$, where $\alpha < 0$ for $T < T_c$. Multiplying the Ginzburg–Landau equation by $\beta^{\frac{1}{2}}/|\alpha|^{\frac{3}{2}}$ gives

$$-\frac{\hbar^2\beta^{\frac{1}{2}}}{2m^*|\alpha|^{\frac{3}{2}}}\nabla^2\psi + \alpha\frac{\beta^{\frac{1}{2}}}{|\alpha|^{\frac{3}{2}}}\psi + \left(\frac{\beta}{|\alpha|}\right)^{\frac{3}{2}}\psi^3 = 0 \tag{5.78}$$

or $\overline{\psi} = \psi/\psi_0$

$$\frac{\hbar^2}{2m^*|\alpha|}\nabla^2\overline{\psi} + \overline{\psi} - \overline{\psi}^3 = 0 \quad \Rightarrow \quad \xi^2\frac{d^2\overline{\psi}}{dx^2} + \overline{\psi} - \overline{\psi}^3 = 0, \tag{5.79}$$

where a characteristic length $\xi = \hbar^2/2m^*|\alpha|$ has been introduced. In the last step, we limit ourselves to a one-dimensional problem. Let us assume that there is a relatively small fluctuation of the order parameter $\overline{\psi}(x) = 1 + \chi(x)$. In that limit, we can write to the leading order

$$\xi^2 \frac{d^2\chi(x)}{dx^2} + 1 + \chi(x) - (1 + \chi(x))^3 \cong \xi^2 \frac{d^2\chi(x)}{dx^2} - 2\chi(x) = 0, \tag{5.80}$$

keeping only the leading terms. The differential equation has as solutions

$$\chi(x) = e^{\pm \frac{\sqrt{2}x}{\xi}} = \exp\left(\pm \frac{2\sqrt{2}m^* a|T_c - T|x}{\hbar^2}\right). \tag{5.81}$$

Note that the characteristic length ξ, also known as the coherence length, diverges for $T \to T_c$. This implies that the system becomes very sensitive to fluctuations. A small perturbation will propagate throughout the system. Since, close to T_c, the superconducting gap is small, these excitations are comparable to the delocalized magnon excitations in ferromagnets.

The approach used in magnetism and superconductivity is a very general approach to phase transitions, leading to a broken symmetry. In high-energy physics, similar ideas have been used to explain the presence of a large scalar field, known as the Higgs field. In Section 5.3, the successful split into massless photons and massive W and Z bosons depended on the presence of a large massive scalar field. In order to understand this somewhat better, let us first have a look at what a massive scalar field looks like. A massive relativistic particle must satisfy $\mathbf{p}' = mc$. This can be written in a differential equation for the field using Eqs. (2.55) and (2.58),

$$\hat{\mathbf{p}}'^2 c^2 \phi(\mathbf{r}') = m^2 c^4 \phi(\mathbf{r}') \;\Rightarrow\; -\hbar^2 c^2 \Box^2 \phi = -\hbar^2 \frac{\partial^2 \phi}{\partial t^2} + \hbar^2 c^2 \nabla^2 \phi = m^2 c^4 \phi.$$

This equation is known as the Klein–Gordan equation. Due to the mass-energy equivalence, the mass term can also be thought of as a potential energy. In analogy to the energy term in the Ginzburg–Landau equation, the potential energy density is given by

$$u = m^2 c^4 |\phi|^2 = m^2 c^4 \phi^* \phi \;\Rightarrow\; \frac{\partial u}{\partial \phi^*} = m^2 c^4 \phi. \tag{5.82}$$

Since ϕ is a complex quantity, the derivative with respect to ϕ^* gives the mass term in the Klein–Gordan equation.

Unfortunately, the above equation for the massive field does not satisfy our criteria since it would produce a relatively small complex field and not the large real field. Following the results in magnetism and superconductivity, the potential energy density is rewritten as

$$u(\mathbf{r}') = -a|\phi(\mathbf{r}')|^2 + \frac{b}{2}|\phi(\mathbf{r}')|^4. \tag{5.83}$$

The quadratic term wants to impose a long-range order, whereas the quartic term prefers $\phi = 0$. Note that the temperature dependence has been removed. We are therefore looking at the ordered state. This is purely a phenomenological theory. There is, as of yet, no fundamental explanation for this potential. This result looks strange. Taking $-a = m^2c^4$ for the coefficient of the quadratic term seems to imply an imaginary mass.

The minimum in the potential-energy density can be obtained by taking the derivative with respect to ϕ^*,

$$\frac{du}{d\phi^*} = -a\phi + b|\phi|^2\phi = 0 \;\Rightarrow\; \phi_0^2 = \frac{a}{b}. \tag{5.84}$$

This is the large scalar field that produces the mass of the W and Z bosons. Although we will choose the field ϕ_0 real, any direction in the complex plane works. The important aspect is that the field is pointing in a particular direction.

Although the original potential was symmetric around $\phi = 0$ in all directions in the complex plane, this is no longer the case around the minimum $\phi_0 = a/b$. For convenience, we now shift the value of the energy density at the minimum to zero,

$$u = \frac{b}{2}(|\phi|^2 - \phi_0^2)^2 = \frac{b}{2}(|\phi|^4 - 2|\phi|^2\phi_0^2 + \phi_0^4) = \frac{b}{2}|\phi|^4 - a|\phi|^2 + \frac{a^2}{2b}.$$

The field is now expanded around ϕ_0,

$$\phi(\mathbf{r}') \cong \phi_0 + \chi_1(\mathbf{r}') + i\chi_2(\mathbf{r}'), \tag{5.85}$$

where χ_1 and χ_2 are assumed small with respect to ϕ_0. Inserting the expansion into the potential and retaining only the lowest term gives

$$u(\mathbf{r}') \cong \frac{b}{2}(2\phi_0\chi_1 + \chi_1^2 + \chi_2^2)^2 \cong 2b\phi_0^2\chi_1^2 = 2a\chi_1^2(\mathbf{r}') = 2m^2c^4\chi_1^2(\mathbf{r}'), \tag{5.86}$$

again using $a = mc^2$. So, what initially appeared to be a negative mass term $-m^2c^4$ for ϕ has now become a positive mass term $2m^2c^4$ for the real field χ_1. The field equations are now given by

$$\nabla^2\chi_1(\mathbf{r}') - \frac{2mc^2}{\hbar^2}\chi_1(\mathbf{r}') + i\nabla^2\chi_2(\mathbf{r}') = 0. \tag{5.87}$$

Therefore, χ_1 behaves like a massive particle following a Klein–Gordan equation, whereas χ_2 has a massless wave equation. In high-energy physics, this is known as the Nambu–Goldstein boson.

Conceptual summary. – The physics underlying spontaneous symmetry breaking can be described with a relatively simple phenomenological model describing second-order phase transitions. Interactions between particles lead to the formation of a different ground state with a finite expectation value of some order parameter (the local magnetization, the superconducting gap, or the Higgs field). If the order has a direction, one particular direction is usually chosen for the order. This breaks the symmetry of the system. Additionally, excitations in the system now occur with respect to the new ground state, which changes the properties. If the field has a directional component, excitations can occur with zero energy, since each direction is still equivalent. However, since the order involves many particles, a limited number of excitations does not affect the broken symmetry, unless the system is close to its critical temperature.

Although some of these models are phenomenological, for magnetism and superconductivity, these findings are supported by microscopic theories. However, for the Higgs field, the detailed physics underlying the formation of a finite scalar field is not well understood.

Problems

1 The magnon dispersion can also be derived for an antiferromagnet, *i.e.* a system where the sign of the nearest-neighbor couplings is $J > 0$, leading to an ordering of alternating up and down spins. This can be solved by creating two sublattices. Since we are interested in excitations with respect to the ordered lattice, we take for the sublattice with the down spins $\boldsymbol{S}_{i0} \to -\boldsymbol{S}_{i0}$ and $\boldsymbol{S}_{i,\pm 1} \to \boldsymbol{S}_{i,\mp 1}$.

a) Show the change in the Hamiltonian in Eq. (5.67) when making the substitution. Consider a one-dimensional system.
b) Write the Hamiltonian in $\boldsymbol{a}_i$ and $\boldsymbol{a}_i^\dagger$.
c) Make a Fourier transform of the Hamiltonian to momentum space.
d) Calculate the magnon dispersion.

5.5 Screening in Solids

Introduction. – Up till now, we have encountered several physical phenomena that turned out to be effective models for more complex underlying interactions. In Section 5.3 on the generation of mass, we saw that the presence of a rest mass can also be viewed as a massless particle moving in a delocalized scalar field, known as the Higgs field. This leads to a constant shift which we can also include by giving the four-momentum a finite norm, *i.e.* a mass. For force carriers, the presence of a mass changes the nature of the interaction. The mass of the exchange particles introduces an exponential decay, making the interaction short ranged as we saw for the Yukawa potential in Eq. (4.153). The finite mass of the W and Z bosons due to their interaction with the Higgs field makes the weak interaction distinct from the electromagnetic interaction.

The mass of the photon, the force carrier of the electromagnetic field, is 0. However, when looking at the electron–electron interactions in a solid, unphysical results were obtained, leading to an infinite effective mass at the Fermi level, see Section 4.3. These problems could be resolved using a screened Coulomb interaction, which is comparable to the Yukawa potential. In this section, we see that this interaction effectively takes into account the complex response of the solid to the presence of a potential. The underlying theory has important practical applications. For example, the metallic appearance of materials, such as silver, platinum, and aluminum, is directly related to a collective response of the electrons to the incoming electromagnetic radiation.

Typically, in electromagnetism, the electrical field inside a solid is described as the sum of two contributions

$$\mathbf{E} = \frac{1}{\varepsilon_0}(\mathbf{D} - \mathbf{P}), \tag{5.88}$$

where ε_0 is the permittivity of free space (vacuum). The electric displacement $\mathbf{D}$ can be viewed as an applied or external field. The polarization $\mathbf{P}$ is the response of the solid to the field. Since charges move away from charges with equal sign, the field due to the induced charges in a solid is opposite to the external field. Taking the divergence and using Gauss's law from Eq. (2.182) gives

$$\varepsilon_0 \nabla \cdot \mathbf{E} = \nabla \cdot \mathbf{D} - \nabla \cdot \mathbf{P} \quad \Rightarrow \quad \rho = \rho^{\text{ext}} + \rho^{\text{ind}}, \tag{5.89}$$

where we have used $\nabla \cdot \mathbf{D} = \rho^{\text{ext}}$ and $\nabla \cdot \mathbf{P} = -\rho^{\text{ind}}$, giving the free or external and induced charges, respectively. Within linear response theory, it is assumed that the polarization is directly proportional to the electric field inside the solid

$$\mathbf{P} = \varepsilon_0 \chi \mathbf{E}, \tag{5.90}$$

where χ is known as the electric susceptibility. χ is a property of the solid and can depend on the frequency of the electromagnetic field. Inserting Eq. (5.90) into (5.88) gives

$$\mathbf{E} = \frac{1}{\varepsilon_0}(\mathbf{D} - \varepsilon_0 \chi \mathbf{E}) \Rightarrow \mathbf{E} = \frac{1}{\varepsilon_0(1+\chi)}\mathbf{D} = \frac{\mathbf{D}}{\varepsilon_0 \varepsilon} \Rightarrow \rho = \frac{\rho^{\text{ext}}}{\varepsilon}, \tag{5.91}$$

where we have introduced the dielectric constant $\varepsilon = 1 + \chi$, also known as the relative permittivity. The dielectric constant allows us to directly relate the electric field and the total charge inside the solid to an external field.

We restrict the quantum-mechanical evaluation of ε to a solid with electrons described by their kinetic energy and the Coulomb interaction between them. From Eq. (4.73), the Coulomb interaction can be written as

$$\boldsymbol{H}_1 = \sum_{\mathbf{q}} U_{\mathbf{q}} \boldsymbol{\rho}_{\mathbf{q}}^{\dagger} \boldsymbol{\rho}_{\mathbf{q}}, \tag{5.92}$$

where the density operator is given by

$$\boldsymbol{\rho}_{\mathbf{q}}^{\dagger} = \sum_{\mathbf{k}} \mathbf{c}_{\mathbf{k}+\mathbf{q}}^{\dagger} \mathbf{c}_{\mathbf{k}} \text{ and } \boldsymbol{\rho}_{\mathbf{q}} = \sum_{\mathbf{k}} \mathbf{c}_{\mathbf{k}}^{\dagger} \mathbf{c}_{\mathbf{k}+\mathbf{q}} = \sum_{\mathbf{k}'} \mathbf{c}_{\mathbf{k}'-\mathbf{q}}^{\dagger} \mathbf{c}_{\mathbf{k}'} = \boldsymbol{\rho}_{-\mathbf{q}}^{\dagger}, \tag{5.93}$$

using the substitution $\mathbf{k} = \mathbf{k}' - \mathbf{q}$ in the equation on the right. In principle, the creation and annihilation operators do not commute if $\mathbf{q} = 0$. However, this term can be excluded from the summation since the static term cancels if a positively charge background is included, see Section 4.3.

The electrons in the solid interact via an interaction term $\boldsymbol{H}_{\text{ext}}(t)$ with a time-dependent external charge distribution $\rho^{\text{ext}}(t)$. The Schrödinger equation for such a system is given by

$$i\hbar \frac{\partial}{\partial t} |F(t)\rangle = (\boldsymbol{H} + \boldsymbol{H}_{\text{ext}}(t)) |F(t)\rangle. \tag{5.94}$$

The Hamiltonian for the system is $\boldsymbol{H}$. The time dependence of the wavefunction is split into the usual development under the Hamiltonian plus the effect of the perturbation described by a function $U(t)$,

$$|F(t)\rangle = e^{-\frac{i}{\hbar}\boldsymbol{H}t} U(t) |F\rangle. \tag{5.95}$$

For free electrons in a solid, $|F\rangle$ is taken as the Fermi sea, see Eq. (4.66). The state $|F(t)\rangle$ has excitations with respect to the Fermi sea created by the perturbation. Inserting this back into the Schrödinger equation gives

$$\left(\boldsymbol{H} + i\hbar \frac{\partial U(t)}{\partial t}\right) |F(t)\rangle = (\boldsymbol{H} + \boldsymbol{H}_{\text{ext}}(t)) |F(t)\rangle \Rightarrow i\hbar \frac{\partial U(t)}{\partial t} = \boldsymbol{H}_{\text{ext}}(t).$$

This leads to an effective equation of the perturbative part of the eigenfunction $|F(t)\rangle$. This result can be written in integral form as

$$U(t) \cong 1 - \frac{i}{\hbar} \int_0^t dt' \, \boldsymbol{H}_{\text{ext}}(t'). \tag{5.96}$$

Note that at $t = 0$, the time when the external field is switched on, this function needs to satisfy $U(0) = 1$, see Eq. (5.95). In the last step, the $U(0)$ result has been inserted, and higher terms have been neglected. This is known as linear response theory, which assumes that the effect of $\boldsymbol{H}_{\text{ext}}(t)$ is relatively small.

We now want to know how the charge distribution $\rho_{\mathbf{q}}$ has changed at a particular time $t > 0$. The expectation value of $\rho_{\mathbf{q}}$ in the time-dependent wavefunction is

$$\langle F(t)|\rho_{\mathbf{q}}|F(t)\rangle = \langle F|U^\dagger(t)e^{\frac{i}{\hbar}\boldsymbol{H}t}\rho_{\mathbf{q}}e^{-\frac{i}{\hbar}\boldsymbol{H}t}U(t)|F\rangle$$
$$= \langle F|\rho_{\mathbf{q}}(t)|F\rangle - \frac{i}{\hbar}\int_0^t dt'\ \langle F|\rho_{\mathbf{q}}(t)\boldsymbol{H}_{\text{ext}}(t') - \boldsymbol{H}_{\text{ext}}(t')\rho_{\mathbf{q}}(t)|F\rangle,$$

inserting the result from Eq. (5.96) and using $\rho_{\mathbf{q}}(t) \equiv e^{\frac{i}{\hbar}\boldsymbol{H}t}\rho_{\mathbf{q}}e^{-\frac{i}{\hbar}\boldsymbol{H}t}$. The first term is the expectation value of the charge in the ground state. This can be a very large number. We are more interested in the change that occurred due to the external perturbation, *i.e.* the induced charge. Assuming that the effect of the field on the charge distribution in the solid can be written as a Coulomb interaction $\boldsymbol{H}_{\text{ext}} = \sum_{\mathbf{q}} U_{\mathbf{q}} \boldsymbol{\rho}_{\mathbf{q}}^\dagger \boldsymbol{\rho}_{\mathbf{q}}^{\text{ext}}$ with a time-dependent external charge distribution $\rho_{\mathbf{q}}^{\text{ext}}(t')$, the above equation can be rewritten as

$$\langle F(t)|\rho_{\mathbf{q}}^{\text{ind}}|F(t)\rangle = -\frac{i}{\hbar}\int_0^t dt'\ \langle F|[\rho_{\mathbf{q}}(t), \boldsymbol{H}_{\text{ext}}(t')]|F\rangle$$
$$= -\frac{i}{\hbar}\int_0^t dt'\ \langle F|[\rho_{\mathbf{q}}(t), \rho_{\mathbf{q}}^\dagger(t')]|F\rangle U_{\mathbf{q}}\rho_{\mathbf{q}}^{\text{ext}}(t')$$
$$= \int_0^t dt'\ G_{\mathbf{q}}(t-t')U_{\mathbf{q}}\rho_{\mathbf{q}}^{\text{ext}}(t'), \tag{5.97}$$

using Eq. (4.73). This result is a particular form of the Kubo formula. From a materials point of view, the most interesting part is the Green's function,

$$G_{\mathbf{q}}(t) = -\frac{i}{\hbar}\langle F|[\rho_{\mathbf{q}}(t), \rho_{\mathbf{q}}^\dagger(0)]|F\rangle, \tag{5.98}$$

which tells us how the solid reacts to the external perturbation. It is a density–density correlation function. Therefore, at $t = 0$, a charge-conserving excitation is created with respect to the Fermi sea. This excitation is then removed at time t.

Let us first look at the response for free electrons when the Hamiltonian is given by the kinetic energy of the electrons, $\boldsymbol{H}_0 = \sum_{\mathbf{k}} \varepsilon_{\mathbf{k}} \mathbf{c}_{\mathbf{k}}^\dagger \mathbf{c}_{\mathbf{k}}$ with $\varepsilon_{\mathbf{k}} = \hbar^2\mathbf{k}^2/2m$. Since the expectation is for the Fermi sea, the electron–hole pair created by $\rho_{\mathbf{q}}^\dagger$ has to be removed by $\rho_{\mathbf{q}}(t)$ (a hole indicates an electron removed below the Fermi level). This gives

$$G_{\mathbf{q}}^0(t) = -\frac{i}{\hbar}\sum_{\mathbf{k}}\langle F|\mathbf{c}_{\mathbf{k}}^\dagger\mathbf{c}_{\mathbf{k}+\mathbf{q}}e^{-i\omega_{\mathbf{q}}t}\mathbf{c}_{\mathbf{k}+\mathbf{q}}^\dagger\mathbf{c}_{\mathbf{k}} - \mathbf{c}_{\mathbf{k}+\mathbf{q}}^\dagger\mathbf{c}_{\mathbf{k}}e^{i(-\omega_{\mathbf{q}})t}\mathbf{c}_{\mathbf{k}}^\dagger\mathbf{c}_{\mathbf{k}+\mathbf{q}}|F\rangle$$
$$= -\frac{i}{\hbar}\sum_{\mathbf{k}} e^{-i\omega_{\mathbf{q}}t}\left[(1-f_{\mathbf{k}+\mathbf{q}})f_{\mathbf{k}} - (1-f_{\mathbf{k}})f_{\mathbf{k}+\mathbf{q}}\right] = -\frac{i}{\hbar}e^{-i\omega_{\mathbf{q}}t}(f_{\mathbf{k}} - f_{\mathbf{k}+\mathbf{q}}),$$

with $\hbar\omega_{\mathbf{q}} = \varepsilon_{\mathbf{k}+\mathbf{q}} - \varepsilon_{\mathbf{k}}$. The result is the free propagator of an electron–hole pair, adjusted for the presence of the Fermi sea. Taking the Fourier transform gives

$$G_{\mathbf{q}}^0(\omega) = \sum_{\mathbf{k}}\frac{f_{\mathbf{k}} - f_{\mathbf{k}+\mathbf{q}}}{\hbar\omega - \hbar\omega_{\mathbf{q}} + i\eta} = \sum_{\mathbf{k}}\frac{f_{\mathbf{k}} - f_{\mathbf{k}+\mathbf{q}}}{\hbar\omega - \varepsilon_{\mathbf{k}+\mathbf{q}} + \varepsilon_{\mathbf{k}} + i\eta}. \tag{5.99}$$

Obviously, the electrons in the solid are not free and interact with each other via the Coulomb interaction in Eq. (4.73). Unfortunately, this problem is unsolvable without doing any approximations. The electron–hole pair created by the external field can create additional electron–hole pairs, which can create additional excitations. Excited electrons and holes can scatter, thereby creating more electron–hole pairs. A practical approach is known as the random-phase approximation. Here, the excited electron–hole pair returns to the Fermi sea, while exciting a new electron–hole pair. After many successive scatterings, an effective electron distribution is obtained that reflects the response to the external field at a particular wavevector $\mathbf{q}$. This successive scattering can be expressed in a series

$$G_{\mathbf{q}}(\omega) = G^0_{\mathbf{q}}(\omega) + G^0_{\mathbf{q}}(\omega)U_{\mathbf{q}}G^0_{\mathbf{q}}(\omega) + G^0_{\mathbf{q}}(\omega)U_{\mathbf{q}}G^0_{\mathbf{q}}(\omega)U_{\mathbf{q}}G^0_{\mathbf{q}}(\omega) + \cdots . \tag{5.100}$$

This series can be summed to give

$$G_{\mathbf{q}}(\omega) = \frac{G^0_{\mathbf{q}}(\omega)}{1 - G^0_{\mathbf{q}}(\omega)U_{\mathbf{q}}}. \tag{5.101}$$

The propagator for electron–hole pairs can be used to calculate the induced charge. The convolution in time in Eq. (5.97) turns into multiplication when looking at the frequency domain. The total charge is then

$$\begin{aligned}\rho_{\mathbf{q}}(\omega) &= \rho^{\text{ext}}_{\mathbf{q}}(\omega) + \rho^{\text{ind}}_{\mathbf{q}}(\omega) = \left(1 + G_{\mathbf{q}}(\omega)U_{\mathbf{q}}\right)\rho^{\text{ext}}_{\mathbf{q}}(\omega) \\ &= \frac{\rho^{\text{ext}}_{\mathbf{q}}(\omega)}{1 - G^0_{\mathbf{q}}(\omega)U_{\mathbf{q}}} = \frac{\rho^{\text{ext}}_{\mathbf{q}}(\omega)}{\varepsilon_{\mathbf{q}}(\omega)},\end{aligned} \tag{5.102}$$

where the dielectric function is given by

$$\varepsilon_{\mathbf{q}}(\omega) = 1 - G^0_{\mathbf{q}}(\omega)U_{\mathbf{q}}. \tag{5.103}$$

Let us first study the static response, *i.e.* the limit $\omega \to 0$. Additionally, we only look at the long-wavelength limit, *i.e.* $\mathbf{q} \to 0$,

$$\begin{aligned}G^0_{\mathbf{q}}(\omega) &= 2\sum_{\mathbf{k}} \frac{f_{\mathbf{k}} - f_{\mathbf{k}+\mathbf{q}}}{\varepsilon_{\mathbf{k}} - \varepsilon_{\mathbf{k}+\mathbf{q}}} \cong 2\sum_{\mathbf{k}} \frac{\frac{\partial f}{\partial \varepsilon_{\mathbf{k}}}(\varepsilon_{\mathbf{k}} - \varepsilon_{\mathbf{k}+\mathbf{q}})}{\varepsilon_{\mathbf{k}} - \varepsilon_{\mathbf{k}+\mathbf{q}}} = 2\sum_{\mathbf{k}} \frac{\partial f}{\partial \varepsilon_{\mathbf{k}}} \\ &= \int_0^\infty dE\, D(E)\frac{\partial f}{\partial E} \cong -\int_0^\infty dE\, D(E)\delta(E - E_F) = -D(E_F),\end{aligned}$$

where a factor 2 has been added to account for the spin degrees of freedom. This is then included in the free-electron density of states $D(E)$ from Eq. (4.68) when the summation over $\mathbf{k}$ is replaced by an integral over the energy. Note that at low temperatures, the Fermi–Dirac function is a downward step function at the chemical potential or Fermi energy. This becomes a negative δ function at the Fermi energy after taking the derivative. The dielectric function in Eq. (5.103) then becomes in the static limit

$$\varepsilon_{\mathbf{q}} \cong 1 + D(E_F)U_{\mathbf{q}}. \tag{5.104}$$

Using Eqs. (4.65), (4.68), and (4.74), the term in the denominator can also be written as

$$D(E_F)U_{\mathbf{q}} = \frac{1}{2\pi^2}\left(\frac{2m}{\hbar^2}\right)^{\frac{3}{2}}\sqrt{\frac{\hbar^2 k_F^2}{2m}}\frac{e^2}{\varepsilon_0 q^2} = \frac{me^2}{\pi^2\varepsilon_0\hbar^2}\frac{k_F}{q^2} = \frac{4}{\pi}\frac{k_F}{a_0 q^2} = \frac{\kappa_{\text{TF}}^2}{q^2},$$

where $a_0 = 4\pi\varepsilon_0\hbar^2/me^2$ is the Bohr radius, and $1/\kappa_{\text{TF}} = \sqrt{\pi a_0/4k_F}$ is the Thomas–Fermi screening length.

The potential in momentum space for a point charge now becomes

$$U_{\mathbf{q}} = \frac{e^2}{\varepsilon_{\mathbf{q}}\varepsilon_0 q^2} = \frac{e^2}{\left(1 + \frac{\kappa_{\text{TF}}^2}{q^2}\right)\varepsilon_0 q^2} = \frac{e^2}{\varepsilon_0}\frac{1}{q^2 + \kappa_{\text{TF}}^2}. \tag{5.105}$$

Transforming this potential back to real space gives a screened potential

$$U(\mathbf{r}) = \frac{1}{4\pi\varepsilon_0}\frac{e^2}{r}e^{-\kappa_{\text{TF}} r}. \tag{5.106}$$

The above results can be understood from a different perspective. Often, the chemical potential and the Fermi level are treated as equivalent quantities. However, in a system at equilibrium, the chemical potential is the same throughout the material. If this was not the case, the electrons would relax into the lower energy states. Let us define the Fermi level as the energy where the states go from occupied to unoccupied. In general, this is the same energy as the chemical potential. However, in the presence of an additional potential $U(\mathbf{r})$, the occupation can adjust such that $\mu(\mathbf{r}) = E_F(\mathbf{r}) - U(\mathbf{r})$ is constant. For example, if the potential is due to an additional electron, electrons will locally move away from the charge, thereby creating effectively a positive charge around the electron due to the fixed positive background. This screens the potential of the electron. However, the readjustment of the charge distributions costs energy. Equilibrium implies that the chemical potential remains constant throughout the material. Therefore,

$$\frac{dE_F}{dn}\Delta n(\mathbf{r}) - U^{\text{tot}}(\mathbf{r}) = 0 \;\Rightarrow\; \Delta n(\mathbf{r}) = \left.\frac{dn}{dE}\right|_{E_F} U^{\text{tot}}(\mathbf{r}) = D(E_F)U^{\text{tot}}(\mathbf{r}),$$

where $U^{\text{tot}}(\mathbf{r})$ is the potential energy due to the total charge $\rho(\mathbf{r})$. This is given by the Laplacian

$$\nabla^2 U^{\text{tot}}(\mathbf{r}) = -\frac{(-e)\rho(\mathbf{r})}{\varepsilon_0} \;\Rightarrow\; -q^2 U^{\text{tot}}_{\mathbf{q}} = \frac{e\rho_{\mathbf{q}}}{\varepsilon_0} \;\Rightarrow\; U^{\text{tot}}_{\mathbf{q}} = -\frac{e\rho_{\mathbf{q}}}{\varepsilon_0 q^2}.$$

The induced charge in momentum space is then

$$\rho^{\text{ind}}_{\mathbf{q}} = e\Delta n_{\mathbf{q}} = eD(E_F)U^{\text{tot}}_{\mathbf{q}}, = -D(E_F)\frac{e^2\rho_{\mathbf{q}}}{\varepsilon_0 q^2} = -D(E_F)U_{\mathbf{q}}\rho_{\mathbf{q}}. \tag{5.107}$$

The total charge is then given by

$$\rho_{\mathbf{q}} = \rho^{\text{ext}}_{\mathbf{q}} + \rho^{\text{ind}}_{\mathbf{q}} = \rho^{\text{ext}}_{\mathbf{q}} - D(E_F)U_{\mathbf{q}}\rho_{\mathbf{q}}. \;\Rightarrow\; \rho_{\mathbf{q}} = \frac{\rho^{\text{ext}}_{\mathbf{q}}}{1 + D(E_F)U_{\mathbf{q}}} = \frac{\rho^{\text{ext}}_{\mathbf{q}}}{\varepsilon_{\mathbf{q}}},$$

using Eq. (5.91). This reproduces the result in Eq. (5.104).

Conceptual summary. – The potential due to a charge inside a solid becomes screened due to the interactions with the charge density. It therefore becomes short ranged and has a similar dependence as the Yukawa potential in Eq. (4.153) describing the residual strong force mediated by massive π mesons. Effectively, one can view this as the photon field becoming massive. However, it does not really mean that photons have a mass. This is because we are dealing with an effective model that describes the effect of the entire solid in terms of a single effective potential. We are looking at the potential at a particular distance from the charge. However, the photon never reaches this point, since it interacted with other charges before that.

There are other problems where the effective model looks different from the microscopic model. In Section 3.10, we considered a local state coupled to a continuum. When considering the entire system, this can be described as a particle that is initially on the local state and then in the continuum states. However, when we remove the continuum states from the system, we are left with a particle on the local state with a finite lifetime. Another example is the rest mass of a particle. This problem can be viewed at the microscopic level as a particle continuously interacting with the Higgs field. However, when removing the Higgs field from the problem, we are left with a particle with a finite rest mass. As we shall see, something comparable happens when looking at the motion of photons in a solid. Photons move with less than the speed of light inside a solid (this explains why light refracts when entering a solid). However, this is again an effective model. When looking at a microscopic level, one has a photon propagating through a solid while interacting with the charge densities. However, when removing all the details of the solid, one is left with light traveling at less than the speed of light. Note that this works nicely for visible light, where the wavelength is much larger than the interatomic distances, but not for X-rays.

We also solved a different issue. In Section 4.3, we looked at the Coulomb interaction between electrons for a Fermi sphere. The end results were somewhat unphysical, since the effective electron mass diverged at the Fermi level. This was because the Coulomb energy was calculated for the Fermi sphere, which is the state lowest in energy for the kinetic energy. In this section, we allowed the electron density to respond to the presence of a charge. This makes the potential short ranged and also solves the problem of the diverging effective mass. However, what actually happens at a microscopic level is that charge is redistributed around the Fermi level to minimize the Coulomb interactions.

5.6 Plasmons in Solids

Introduction. – In Section 5.5, we looked at the response of a material within the random-phase approximation, *i.e.* the potential is allowed to excite electron–hole pairs with respect to the Fermi sphere. In the static limit, the solid effectively screens the potential of the external charge, leading to a short-ranged potential. In this section, we study the response for finite frequencies ω. This is of relevance for the interaction of solids and electromagnetic radiation, such as visible light. We study again the long-wavelength limit, *i.e.* $\mathbf{q} \to 0$. In real space, this implies that all electrons are moving in the same direction. It is a translational response of the entire electron system to the applied field. Note that the positive background of the nuclei is fixed.

Let us first approximate the free propagator for electron–hole pairs close to the Fermi level. This is done in order to make a connection to results that can be obtained from simple classical considerations. For small $\mathbf{q}$, the following approximations can be made:

$$\begin{aligned} \omega_{\mathbf{q}} &= \frac{1}{\hbar}(\varepsilon_{\mathbf{k}+\mathbf{q}} - \varepsilon_{\mathbf{k}}) = \frac{1}{\hbar}\nabla_{\mathbf{k}}\varepsilon_{\mathbf{k}} \cdot \mathbf{q} = \frac{\hbar\mathbf{k}}{m} \cdot \mathbf{q} \cong \mathbf{v}_F \cdot \mathbf{q}, \\ \frac{1}{\hbar}(f_{\mathbf{k}} - f_{\mathbf{k}+\mathbf{q}}) &= -\frac{1}{\hbar}\frac{\partial f_{\mathbf{k}}}{\partial \varepsilon_{\mathbf{k}}} \nabla_{\mathbf{k}}\varepsilon_{\mathbf{k}} \cdot \mathbf{q} \cong \delta(\varepsilon_{\mathbf{k}} - \mu)\mathbf{v}_F \cdot \mathbf{q}, \end{aligned} \tag{5.108}$$

where the zero-temperature limit has been taken, where the Fermi–Dirac function is a downward step function at the chemical potential μ. The free electron–hole propagator in Eq. (5.99) can then be evaluated

$$G^0_{\mathbf{q}}(\omega) \cong 2\int \frac{d\mathbf{k}}{(2\pi)^3} \frac{\mathbf{v}_F \cdot \mathbf{q}\, \delta(\varepsilon_{\mathbf{k}} - \mu)}{\omega - \mathbf{v}_F \cdot \mathbf{q} + i\eta} = \frac{2}{(2\pi)^2} \int_0^\pi d\theta\, \sin\theta\, \frac{k_F^2}{\hbar v_F} \frac{\dfrac{\mathbf{v}_F \cdot \mathbf{q}}{\omega + i\eta}}{1 - \dfrac{\mathbf{v}_F \cdot \mathbf{q}}{\omega + i\eta}},$$

where the factor 2 accounts for the spin; the integral over φ gives a factor 2π; and the integral over k only has a contribution at $k = k_F$ due to the δ function. The $\hbar v_F$ in the denominator comes from the change in argument in the δ function from energy to momentum $\delta(\varepsilon_{\mathbf{k}} - \mu) = \partial f(\varepsilon_{\mathbf{k}})/\partial \varepsilon_{\mathbf{k}} = (\partial f(\varepsilon_{\mathbf{k}})/\partial k)(\partial k/\partial \varepsilon_{\mathbf{k}}) = \delta(k - k_F)/(\hbar v_F)$. Expanding in the small term gives

$$G^0_{\mathbf{q}}(\omega) = \frac{1}{2\pi^2}\int_{-1}^{1} dx \frac{k_F^2}{\hbar v_F}\left\{\frac{v_F q x}{\omega + i\eta} + \left(\frac{v_F q x}{\omega + i\eta}\right)^2 + \left(\frac{v_F q x}{\omega + i\eta}\right)^3 + \cdots\right\}$$
$$= \frac{k_F^2}{2\pi^2 \hbar v_F}\left\{\frac{2}{3}\left(\frac{v_F q}{\omega + i\eta}\right)^2 + \frac{2}{5}\left(\frac{v_F q}{\omega + i\eta}\right)^4 + \cdots\right\}, \tag{5.109}$$

with $x = \cos\theta$. Note that all the odd terms in x cancel. This gives the final result:

$$G^0_{\mathbf{q}}(\omega) \cong \frac{k_F^3 q^2}{3\pi^2 m(\omega + i\eta)^2}\left\{1 + \frac{3}{5}\left(\frac{v_F q}{\omega + i\eta}\right)^2 + \cdots\right\} \cong \frac{nq^2}{m(\omega + i\eta)^2}, \tag{5.110}$$

with $v_F = \hbar k_F/m$. In the last step, only the leading term is retained, and use has been made of Eq. (4.65) expressing the result in the density of electrons n, *i.e.* $k_F = (3\pi^2 n)^{\frac{1}{3}}$. In the leading limit, the dielectric function in Eq. (5.103) becomes

$$\varepsilon(\omega) = 1 - G^0_{\mathbf{q}}(\omega)U_{\mathbf{q}} = 1 - \frac{\omega_p^2}{\omega^2} \quad \text{with} \quad \omega_p^2 = \frac{ne^2}{\varepsilon_0 m}, \tag{5.111}$$

using $U_{\mathbf{q}} = e^2/\varepsilon_0 q^2$ from Eq. (4.74). In this limit, the dielectric function is independent of $\mathbf{q}$. The frequency ω_p is known as the plasma frequency.

The collective plasmon mode can also be obtained within a semiclassical framework by considering the electrons as simple Lorentz oscillators. The electrons are then bound to the nucleus in harmonic potentials with frequency ω_0 or an effective spring constant $m\omega_0^2$. Newton's law then gives for the motion in an electric field $\mathbf{E}$

$$-e\mathbf{E} - m\omega_0^2\mathbf{r} - m\Gamma\mathbf{v} = m\mathbf{a} \quad \Rightarrow \quad \frac{d^2\mathbf{r}}{dt^2} + \Gamma\frac{d\mathbf{r}}{dt} + \omega_0^2\mathbf{r} = -\frac{e}{m}\mathbf{E}. \tag{5.112}$$

An additional damping term $-m\Gamma\mathbf{v}$ has been added to simulate the scattering of the electron. This scattering is generally related to the absorption of photons and therefore leads to evanescence of the electric field in the solid. The assumption is made that the electrons oscillate with the same frequency as the incoming field. The solutions are therefore of the form $\mathbf{r}\, e^{-i\omega t}$. The displacement is then given by

$$\mathbf{r} = \frac{e\mathbf{E}}{m}\frac{1}{\omega^2 - \omega_0^2 + i\Gamma\omega}. \tag{5.113}$$

The displacement of the electron with respect to the nucleus creates a dipole moment. The combined action of the dipoles per unit volume produces the polarization

$$\mathbf{P} = -en\mathbf{r} = -\frac{ne^2}{m}\frac{1}{\omega^2 - \omega_0^2 + i\Gamma\omega}\mathbf{E} = \varepsilon_0\chi(\omega)\mathbf{E}, \tag{5.114}$$

where n is the electron density. Again, the plasma frequency ω_p from Eq. (5.111) appears. This gives for the dielectric constant

$$\varepsilon = 1 + \chi = 1 - \frac{\omega_p^2}{\omega^2 - \omega_0^2 + i\Gamma\omega}. \tag{5.115}$$

This result reduces to Eq. (5.111) in the absence of a nuclear potential ($\omega_0 = 0$) and damping ($\Gamma = 0$). However, this simple model can also provide insights into how the dielectric response changes when the excitation energy is no longer zero.

The interaction with the solid affects the propagation of the photon. The effect of the medium in the propagation of the field can be included by replacing ε_0 with $\varepsilon\varepsilon_0$ in the wave equation in Eq. (2.65),

$$\nabla^2\mathbf{E} - \varepsilon\varepsilon_0\mu_0\frac{\partial^2\mathbf{E}}{\partial t^2} = \nabla^2\mathbf{E} - \frac{\hat{n}^2}{c^2}\frac{\partial^2\mathbf{E}}{\partial t^2} = 0, \tag{5.116}$$

using that $\varepsilon_0\mu_0 = 1/c^2$, where c is the speed of light. Due to the absorption of photon inside the solid, the dielectric function can be complex $\varepsilon = \varepsilon_1 + i\varepsilon_2$. This is commonly expressed in terms of a complex refractive index $\hat{n}$ that describes the change in the wavenumber q,

$$\hat{q} = \frac{\omega}{c}\hat{n} = q\hat{n} = q(n + i\kappa) \quad \Rightarrow \quad \varepsilon_1 = n^2 - \kappa^2, \ \ \varepsilon_2 = 2n\kappa, \tag{5.117}$$

with $q = \omega/c$, and where n is the refractive index and κ the extinction coefficient. Note that $\hat{n}$ and the related quantities n, κ, ε_1, and ε_2 are all functions of the frequency ω. These equations can be straightforwardly inverted to give

$$n = \sqrt{\frac{1}{2}\varepsilon_1 + \frac{1}{2}\sqrt{\varepsilon_1^2 + \varepsilon_2^2}} \quad \text{and} \quad \kappa = \sqrt{-\frac{1}{2}\varepsilon_1 + \frac{1}{2}\sqrt{\varepsilon_1^2 + \varepsilon_2^2}}. \tag{5.118}$$

Although the dielectric function is an important property, it is not always directly obvious how it relates to experimental properties. Let us look at an intuitive experiment such as reflectance, which measures how much electromagnetic radiation is reflected by the solid. Let us consider a wave traveling in the z-direction perpendicular to the surface of a solid. The electric field is perpendicular to the direction of propagation, which can be taken as the x-direction. The electric field is then

$$\mathbf{E} = Ee^{i\hat{q}z - i\omega t}\mathbf{e}_x, \tag{5.119}$$

where $\hat{q} = q, q\hat{n}$ in the vacuum and the solid, respectively. From Faraday's law, the magnetic field can be determined:

$$\nabla \times \mathbf{E} = -\mu_0\frac{\partial\mathbf{B}}{\partial t} \quad \Rightarrow \quad i\hat{q}E_x\mathbf{e}_y = i\mu_0\omega\mathbf{B} \quad \Rightarrow \quad \mathbf{B} = \frac{\hat{q}}{\mu_0\omega}E_i e^{i\hat{q}z - i\omega t}\mathbf{e}_y,$$

which means $\mathbf{B}$ is perpendicular to $\mathbf{E}$. At the surface, part of the wave is reflected, and part of it is transmitted. However, other Maxwell equations impose continuity of the electric field,

$$E_i + E_r = E_t, \tag{5.120}$$

where r and t stand for reflected and transmitted, respectively. For the magnetic field, we must satisfy

$$B_i - B_r = B_t \quad \Rightarrow \quad \frac{q}{\mu_0\omega}(E_i - E_r) = \frac{\hat{q}}{\mu_0\omega}E_t \quad \Rightarrow \quad E_i - E_r = \hat{n}E_t,$$

where we should take care that the direction of the magnetic field changes for the reflected light to maintain the right-handed rotation between the electric and magnetic fields for a wave traveling in the negative z-direction. Removing the transmitted wave from the continuity equations gives

$$(\hat{n} - 1)E_i + (1 + \hat{n})E_r = 0 \quad \Rightarrow \quad r = \frac{E_r}{E_i} = \frac{1 - \hat{n}}{1 + \hat{n}}. \tag{5.121}$$

The reflectance is then given by

$$R = r^* r = \left| \frac{1 - \hat{n}}{1 + \hat{n}} \right|^2 = \frac{(1 - n)^2 + \kappa^2}{(1 + n)^2 + \kappa^2}. \tag{5.122}$$

Using the frequency dependence of the dielectric function in Eq. (5.115) and its relation to n and κ in Eq. (5.118), the reflectance can be straightforwardly evaluated numerically.

Figure 5.1 shows the calculation of the index of refraction n, the extinction coefficient κ, and the reflectance R for the Lorentz model with $\omega_0 = 0$ (left) and $\frac{1}{2}\omega_p$ (right). For $\omega_0 = 0$, the spectral part of the dielectric function ε_2 diverges at $\omega \to 0$. Therefore, the excitation energy is zero, making this system a metal. The extinction coefficient is finite up to the plasma frequency ω_p, indicating that electromagnetic waves are not transmitted up to the energy $\hbar\omega_p$. This can also be seen from the reflectance, which is close to 1 up to the plasma frequency. A material that behaves close to a free-electron-like material is aluminum. With an electron configuration [Ne]$3s^2 3p^1$, there are three free electrons per atom leading to a density of 1.8×10^{29} electrons per m^3. The plasma frequency is then $\omega_p = 2.4 \times 10^{16}$ s^{-1}, giving $\hbar\omega_p = 15.8$ eV. More advanced calculations give an effective electron mass of $1.115m$, giving $\hbar\omega_p = 15.0$ eV. This is close to the experimentally observed energy of 14.7 eV. Since photons in the visible region have an energy of 1.66–3.27 eV, all visible light is reflected, explaining the mirror-like appearance of metals. The reflectance suddenly drops to zero at the plasma frequency, when the collective oscillation is no longer able to follow the frequency of the electromagnetic radiation. This does not necessarily mean that the material becomes transparent for these energies since additional absorption processes generally become accessible at these photon energies.

In semiconductors and insulators, there are no electronic excitations available with (close to) zero energy. For example, the lowest excitation energy in diamond is 5.47 eV. The behavior can be mimicked with the Lorentz oscillator by introducing a finite binding energy. The right side in Figure 5.1 shows an example for $\omega_0 = \frac{1}{2}\omega_p$. The spectral function ε_2 of the Lorentz oscillator shows a Lorentzian peak at ω_0. The extinction coefficient is close to zero up to ω_0. There are no transitions available for the photon to excite. The system becomes transparent up to an energy $\hbar\omega_0$. Since in diamond the energy gap is larger than the energy of photon in the visible region, diamond lets visible light through.

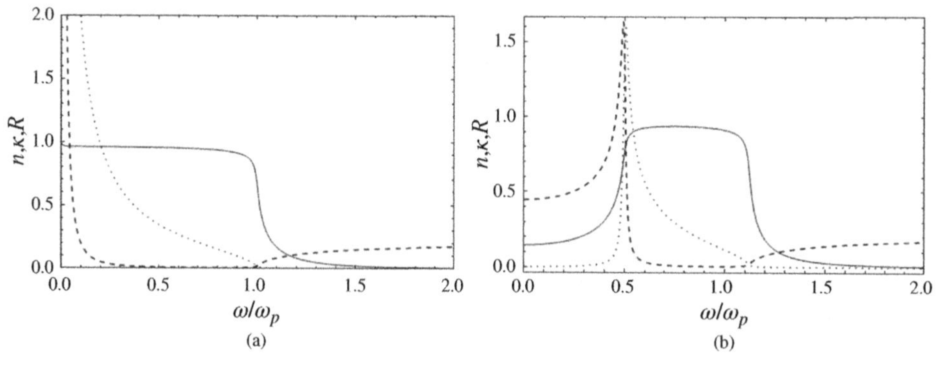

Figure 5.1 The index of refraction n (dashed), the extinction coefficient κ (dotted), and the reflectance R (solid) calculated for a Lorentz oscillator with oscillator frequency due to the nuclear potential $\omega_0 = 0$ (a) and $\frac{1}{2}\omega_p$ (b). The energy scale is relative to the plasma frequency ω_p. A broadening $\Gamma = 0.02\omega_p$ has been added.

However, note that the reflectance is not entirely zero, still allowing for partial reflection of the light.

Example 5.1 ***Index of Refraction***
Use Faraday's law to obtain Snell's law $n_1 \sin\theta_i = n_2 \sin\theta_t$ for the refraction of light between two mediums with a constant index of refraction and with $\kappa \cong 0$. This situation occurs in insulators for $\omega < \omega_0$, or, in other words, when the energy is below the insulating gap. The angles with the surface normal are θ_i and θ_t for the incoming and transmitted beams, respectively.

Solution
Faraday's law, see Eq. (2.182), becomes in the absence of a time-dependent magnetic field $\nabla \times \mathbf{E} = 0$. In integral form, this becomes $\oint \mathbf{E} \cdot d\mathbf{l} = 0$. For, a microscopic loop at the surface gives $(\mathbf{E}_i + \mathbf{E}_r - \mathbf{E}_t) \cdot d\mathbf{l} = 0$, where r indicates the reflected beam. Note that the electric field lies in the plane of the interface. The minus sign appears because the paths on each side of the interface are opposite. When the propagation of the electric field is perpendicular to the surface, which we take in the xy-plane, this reduces to Eq. (5.120). However, when the propagation is at an angle in the zx-plane, this becomes

$$E_i e^{i\hat{q}_{ix}x} + E_r e^{i\hat{q}_{rx}x} = E_t e^{i\hat{q}_{tx}x}, \tag{5.123}$$

for $z = 0$. Since this should be satisfied for any x, this implies that $\hat{q}_{ix} = \hat{q}_{rx} = \hat{q}_{tx}$. This gives

$$\hat{q}_{ix} = \hat{q}_{tx} \quad \Rightarrow \quad n_1 q \sin\theta_i = n_2 q \sin\theta_t, \tag{5.124}$$

which is Snell's law.

Conceptual summary. – In quantum mechanics, a material responds to the presence of an electric field by the excitation of electron–hole pairs. In the large wavelength limit, these results become comparable to a Lorentz oscillator. Essentially, all free electrons move together in response to the oscillating field. For low frequencies, the electrons are able to follow the electric field and reflect all electromagnetic radiation. This explains the mirror-like appearance for metals. However, for frequencies higher than the plasma frequency ω_p, the electrons are unable to reflect the incoming electromagnetic waves, and the material becomes transparent.

For semiconductors and insulators, there are no free electrons available, since there is a gap in the electronic structure. The solid can therefore not respond to the electromagnetic wave, and the material is transparent. A good example is diamond. The material can only respond when the energy of the photons is high enough to create electron–hole pairs. In many cases, the complex underlying physics can be taken into account by relatively simple models. A nice example is Snell's law where the dielectric function can be simplified to an effective change in the speed of light in different media.

Obviously, the electronic structure of solids is more complicated than described here, and more detailed calculations are necessary to explain, for example why silver reflects all visible light, but copper, which is right above it in the periodic table, has a reddish color. The answer lies in the position of the d bands below the Fermi level, which are inside and outside of the range of visible light for copper and silver, respectively.

The use of the word plasmon seems to indicate that the charge response can be viewed as a particle. In fact, the behavior of plasmons is comparable to that of a quantum harmonic oscillator with a frequency $\hbar\omega_p$. Plasmons can be excited in, for example electron energy-loss spectroscopy, where multiple plasmon excitations can be observed. The response to the electron introduced into the solid is very similar to that of the displaced harmonic oscillator treated in Eq. (3.3), and a Poisson distribution of plasmons is observed.

5.7 Superfluidity

Introduction. – In Section 4.2, it was shown that the composite-boson helium-4 forms a Bose–Einstein condensate. Liquid helium-4 has several unusual properties below the critical temperature. It behaves as mixture of two liquids, one component behaves like a regular liquid, and the other component is classified as a superfluid, *i.e.* it has zero viscosity, zero entropy, and a high thermal conductivity. It was already surmised by Lev Landau that having a Bose–Einstein condensate is not sufficient for superfluidity. In the condensate, the helium atoms have zero momentum, whereas flow implies a finite momentum. Flowing with zero viscosity means that no excitations are created in the superfluid phase. These excitations can occur by imperfections at the wall or other impurities.

An alternative way to look at the problem of zero viscosity is that an object or a particle can move through the superfluid without feeling any damping whatsoever. Let us consider the conditions to create elementary excitations in helium with a specific energy $\varepsilon_{\mathbf{k}}$ and momentum $\hbar\mathbf{k}$. To create such an excitation we require conservation of energy

$$\frac{1}{2}Mv^2 = \frac{1}{2}Mv'^2 + \varepsilon_{\mathbf{k}}, \tag{5.125}$$

where M is the mass of the moving body. Conservation of momentum gives

$$M\mathbf{v} = M\mathbf{v}' + \hbar\mathbf{k} \quad \Rightarrow \quad M\mathbf{v} - \hbar\mathbf{k} = M\mathbf{v}'. \tag{5.126}$$

Squaring and multiplying by $\frac{1}{2M}$ gives

$$\frac{1}{2}Mv^2 - \frac{1}{2}\hbar\mathbf{v}\cdot\mathbf{k} + \frac{1}{2M}\hbar^2\mathbf{k}^2 = \frac{1}{2}Mv'^2. \tag{5.127}$$

Subtracting this from the equation for conservation of energy gives the condition for the velocity

$$\frac{1}{2}\hbar\mathbf{v}\cdot\mathbf{k} - \frac{\hbar^2\mathbf{k}^2}{2M} = \varepsilon_{\mathbf{k}} \quad \Rightarrow \quad \hbar vk\cos\varphi = \varepsilon_{\mathbf{k}} + \frac{\hbar^2\mathbf{k}^2}{2M}, \tag{5.128}$$

where φ is the angle between $\mathbf{v}$ and $\mathbf{k}$. Let us take the limit $M \to \infty$, which can correspond to excitations created by interactions between the superfluid and the wall. We are looking for the minimum allowable velocity that creates no excitations. This occurs when $\mathbf{v}$ and $\mathbf{k}$ are parallel, *i.e.* $\varphi = 0$. We are therefore looking for a finite value of k where $\hbar vk > \varepsilon_{\mathbf{k}}$. If the helium atoms were free particles, the dispersion would be given by $\hbar^2k^2/2m$, where m is the mass of a helium-4 atom. The slope of a parabola $\hbar k/m$ approaches zero when $k \to 0$, so the line $\hbar vk$ is above the parabola for any finite v. Therefore, even though helium-4 is a Bose–Einstein condensate, this is not sufficient for superfluidity, since any motion causes excitations. The excitations imply dissipation, which means that the viscosity is finite. However, if the dispersion is linear $\varepsilon_{\mathbf{k}} = \hbar v_s k$, where v_s is the sound velocity of the superfluid, then no excitations are created in the superfluid for $v < v_s$, and the viscosity is zero.

Since for free particles, the eigenenergies are quadratic, additional interactions are needed to change the dispersion. A general Hamiltonian with a momentum-conserving interaction is given by

$$\boldsymbol{H} = \sum_{\mathbf{k}} \varepsilon_{\mathbf{k}} \boldsymbol{a}_{\mathbf{k}}^{\dagger} \boldsymbol{a}_{\mathbf{k}} + \sum_{\mathbf{k}\mathbf{k}'\mathbf{q}} U_{\mathbf{q}} \boldsymbol{a}_{\mathbf{k}+\mathbf{q}}^{\dagger} \boldsymbol{a}_{\mathbf{k}'-\mathbf{q}}^{\dagger} \boldsymbol{a}_{\mathbf{k}'} \boldsymbol{a}_{\mathbf{k}}. \tag{5.129}$$

Next, use is made of the two-fluid view of helium-4. It is assumed that the dominant interactions are with the larger Bose–Einstein condensate. Therefore, one pair of creation/annihilation operators is related to the condensate, giving an effective interaction

$$\sum_{\mathbf{k}} U_0 (\boldsymbol{a}_0^{\dagger} \boldsymbol{a}_{\mathbf{k}}^{\dagger} \boldsymbol{a}_{\mathbf{k}} \boldsymbol{a}_0 + \boldsymbol{a}_{\mathbf{k}}^{\dagger} \boldsymbol{a}_0^{\dagger} \boldsymbol{a}_0 \boldsymbol{a}_{\mathbf{k}} + \boldsymbol{a}_{\mathbf{k}}^{\dagger} \boldsymbol{a}_{-\mathbf{k}}^{\dagger} \boldsymbol{a}_0 \boldsymbol{a}_0 + \boldsymbol{a}_0^{\dagger} \boldsymbol{a}_0^{\dagger} \boldsymbol{a}_{\mathbf{k}} \boldsymbol{a}_{-\mathbf{k}}). \tag{5.130}$$

Since there are two fluids, there is no particle conservation in either fluid. There is a continuous interchange of helium atoms between the condensate and the superfluid. We assume now that all interactions with the condensate can be replaced by a scalar, *i.e.* $\boldsymbol{a}_0^{(\dagger)} |N_{\mathrm{BEC}}\rangle \cong \sqrt{N} |N_{\mathrm{BEC}}\rangle$, where N is a very large number. This leads to an effective single-boson Hamiltonian

$$\boldsymbol{H} = \sum_{\mathbf{k}} (\varepsilon_{\mathbf{k}} + 2NU_0) \boldsymbol{a}_{\mathbf{k}}^{\dagger} \boldsymbol{a}_{\mathbf{k}} + \sum_{\mathbf{k}} NU_0 (\boldsymbol{a}_{\mathbf{k}}^{\dagger} \boldsymbol{a}_{-\mathbf{k}}^{\dagger} + \boldsymbol{a}_{\mathbf{k}} \boldsymbol{a}_{-\mathbf{k}}). \tag{5.131}$$

This Hamiltonian looks very similar to Eq. (3.46), which was a quantum harmonic oscillator where the spring constant was changed. In analogy, we suggest a Bogoliubov transformation

$$\boldsymbol{b}_{\mathbf{k}}^{\dagger} = \cosh \frac{\alpha}{2} \boldsymbol{a}_{\mathbf{k}}^{\dagger} + \sinh \frac{\alpha}{2} \boldsymbol{a}_{-\mathbf{k}} \quad \text{with} \quad \tanh \alpha = \frac{2u_0 n}{\varepsilon_{\mathbf{k}} + 2u_0 n}, \tag{5.132}$$

using Eqs. (3.47) and (3.50) and $n = N/V$ and $u_0 = U_0 V$. The above Hamiltonian can then be written in diagonal form as

$$\boldsymbol{H} = \sum_{\mathbf{k}} E_{\mathbf{k}} \boldsymbol{b}_{\mathbf{k}}^{\dagger} \boldsymbol{b}_{\mathbf{k}} + \cdots \quad \text{with} \quad E_{\mathbf{k}} = \sqrt{\varepsilon_{\mathbf{k}}^2 + 4u_0 n \varepsilon_{\mathbf{k}}}, \tag{5.133}$$

using Eq. (3.52) and ignoring higher order terms and constant shifts.

So far, we have not yet established whether helium-4 is a superfluid. The result above can be extended to excitation energies with finite **k**. For small oscillations, this becomes

$$E_{\mathbf{k}} = \sqrt{\varepsilon_{\mathbf{k}}^2 + 4u_{\mathbf{k}} n \varepsilon_{\mathbf{k}}}. \tag{5.134}$$

We still need to introduce a model for the interaction. Let us take a hard-sphere potential given by

$$U(r) = \begin{cases} U & r < a_0 \\ 0 & r > a_0 \end{cases}. \tag{5.135}$$

This interaction takes into account the strong repulsive part around the He atom and neglects all the other r dependence. We now need to find the Fourier transform of the potential. Taking $\mathbf{k}$ along the z-axis gives for a general central potential $U(r) = u(r)/V$

$$\begin{aligned}
U_{\mathbf{k}} &= \frac{1}{V}\int_0^{2\pi} d\varphi \int_0^{\pi} d\theta \int_0^{\infty} drr^2 \sin\theta\, u(r) e^{ik\cos\theta} \\
&= 2\pi \int_{-1}^{1} d\cos\theta \int_0^{\infty} drr^2 u(r) e^{ikr\cos\theta} = 2\pi \int_0^{\infty} drr^2 u(r) \left[\frac{1}{ikr} e^{ikr\cos\theta}\right]_{-1}^{1} \\
&= 4\pi \int_0^{\infty} drr^2 u(r) \frac{1}{kr}\frac{e^{ikr} - e^{-ikr}}{2i} = 4\pi \int_0^{\infty} drr^2 u(r) \frac{\sin kr}{kr}. \qquad (5.136)
\end{aligned}$$

This integral simplifies when using our approximate potential

$$U_{\mathbf{k}} = 4\pi \frac{U}{V} \int_0^{a_0} dr\, r \frac{\sin kr}{k}. \qquad (5.137)$$

We can solve this using integration by parts. Taking $f_1 = r$ and $f_2' = \sin kr$ gives $f_1' = 1$ and $f_2 = -\frac{1}{k}\cos kr$. The integral then becomes

$$\begin{aligned}
u_{\mathbf{k}} = U_{\mathbf{k}} V &= \frac{4\pi U}{k^2}\left\{-[r\cos kr]_0^{a_0} + \int_0^{a_0} dr \cos kr\right\} \\
&= \frac{4\pi U}{k^2}\left\{-a_0 \cos ka_0 + \frac{1}{k}[\sin kr]_0^{a_0}\right\}, \qquad (5.138)
\end{aligned}$$

yielding

$$u_{\mathbf{k}} = \frac{4\pi U a_0}{k^2}\left(\frac{\sin ka_0}{ka_0} - \cos ka_0\right) \quad \Rightarrow \quad \bar{u}_{\bar{\mathbf{k}}} = \frac{u_{\mathbf{k}}}{\bar{\varepsilon}} = \frac{\bar{U}}{\bar{k}^2}\left(\frac{\sin \bar{k}}{\bar{k}} - \cos \bar{k}\right).$$

In the latter expression, the potential is expressed in the normalized quantities $\bar{k} = ka_0$ and $\bar{\varepsilon} = \hbar^2/2ma_0^2$, giving $\bar{U} = 4\pi(U/\bar{\varepsilon})(na_0^3)$. The normalized Fourier transform of the hard-sphere potential is plotted in Figure 5.2. The dispersion in normalized quantities is given by

$$\bar{E}_{\bar{\mathbf{k}}} = \frac{E_{\mathbf{k}}}{\bar{\varepsilon}} = \sqrt{\bar{k}^4 + 4\bar{k}^2 \bar{u}_{\bar{\mathbf{k}}}}, \qquad (5.139)$$

which is plotted in Figure 5.2. In the limit $\bar{k} \to 0$, this gives

$$\bar{u}_{\bar{\mathbf{k}}} \cong \frac{\bar{U}}{\bar{k}^2}\left(\frac{1}{\bar{k}}\left(\bar{k} - \frac{\bar{k}^3}{6}\right) - \left(1 - \frac{\bar{k}^2}{2}\right)\right) = \frac{\bar{U}}{3} \quad \Rightarrow \quad \bar{E}_{\bar{\mathbf{k}}} \cong \sqrt{\frac{4\bar{U}}{3}}\bar{k}.$$

This approximation is plotted in Figure 5.2. The inclusion of interactions gives the desired linear dispersion, so that superfluids can move without creating excitations.

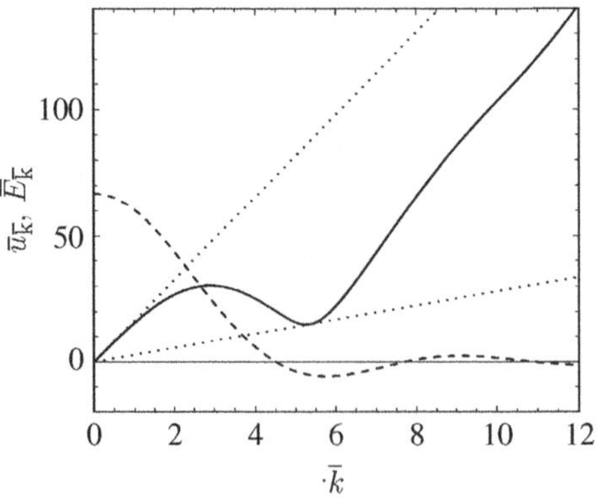

Figure 5.2 The excitation spectrum of helium-4 for a hard-sphere repulsive potential with radius a_0 and normalized strength $\bar{U} = 200$. The normalized potential $\bar{u}_{\bar{\mathbf{k}}}$ (dashed line) and dispersion relation $\bar{E}_{\bar{\mathbf{k}}}$ are plotted as a function of the normalized wavevector $\bar{k} = ka_0$. The dotted line shows the effective sound velocities without and with the inclusion of the roton part of the spectrum.

The actual maximum velocity of the superfluid is lower due to the presence of a minimum in the excitation energy related to the size of repulsive potential, see Figure 5.2. Excitations into this minimum are known as rotons.

Conceptual summary. – In this section, we have treated a final example of a collective phenomenon. Since helium-4 forms a Bose–Einstein condensate, one might naively assume that this is sufficient for superfluid behavior. However, the quadratic dispersion does not allow motion without the creation of excitations in the condensate leading to viscosity. Again, as we have seen before, it is interactions between particles, *i.e.* many-body effects, that play a crucial role in the understanding. Adding a hard-core repulsion between the helium atoms changes the dispersion and allows for superfluid behavior for velocities below an effective sound velocity.

Appendix A

Solutions to Problems

Problems for Section 1.1

1 For an infinitesimally small change in time dt, we have

$$dN(t) = N(t+dt) - N(t) = \sum_{k=0}^{\infty} \frac{1}{k!} \left((\alpha(t+dt))^k - (\alpha t)^k \right)$$
$$\cong \sum_{k=0}^{\infty} \frac{\alpha^k t^k}{k!} k \frac{dt}{t} = \alpha dt \sum_{k=1}^{\infty} \frac{(\alpha t)^{k-1}}{(k-1)!} = \alpha dt \sum_{k'=0}^{\infty} \frac{(\alpha t)^{k'}}{k'!},$$

with $k' = k - 1$. This can be rewritten as

$$\frac{dN(t)}{dt} = \alpha N(t). \tag{A.1}$$

This property is, of course, the hallmark of exponential growth where the change is directly proportional to the quantity itself and a growth rate α. Therefore, the exponential function satisfies this differential equation.

Problems for Section 1.2

1 The norm can be written as a product since

$$\begin{aligned}\hat{\mathbf{r}}\hat{\mathbf{r}} &= (\cos\varphi\ \mathbf{e}_x + \sin\varphi\ \mathbf{e}_y)(\cos\varphi\ \mathbf{e}_x + \sin\varphi\ \mathbf{e}_y) \\ &= \cos^2\varphi\ \mathbf{e}_x^2 + \sin^2\varphi\ \mathbf{e}_y^2 + \cos\varphi \sin\varphi(\mathbf{e}_x\mathbf{e}_y + \mathbf{e}_y\mathbf{e}_x) = 1,\end{aligned} \tag{A.2}$$

using $\mathbf{e}_x\mathbf{e}_y = -\mathbf{e}_y\mathbf{e}_x$ and $\mathbf{e}_i^2 = 1$, with $i = x, y$.

2 The vector quantities $\boldsymbol{i}_{yx}$ and $\mathbf{e}_i$, with $i = x, y$, anticommute, since

$$\boldsymbol{i}_{yx}\mathbf{e}_x = \mathbf{e}_y\mathbf{e}_x\mathbf{e}_x = -\mathbf{e}_x\mathbf{e}_y\mathbf{e}_x = -\mathbf{e}_x\boldsymbol{i}_{yx}, \tag{A.3}$$

$$\boldsymbol{i}_{yx}\mathbf{e}_y = \mathbf{e}_y\mathbf{e}_x\mathbf{e}_y = -\mathbf{e}_y\mathbf{e}_y\mathbf{e}_x = -\mathbf{e}_y\boldsymbol{i}_{yx}. \tag{A.4}$$

3 Using the expansion for the exponential, one obtains

$$e^{\boldsymbol{i}_{yx}\varphi}\mathbf{e}_i = \sum_{n=0}^{\infty} \frac{(\boldsymbol{i}_{yx}\varphi)^n}{n!}\mathbf{e}_i = \mathbf{e}_i \sum_{n=0}^{\infty} (-1)^n \frac{(\boldsymbol{i}_{yx}\varphi)^n}{n!} = \mathbf{e}_i e^{-\boldsymbol{i}_{yx}\varphi}, \tag{A.5}$$

where each anticommutation between $\boldsymbol{i}_{yx}$ and $\mathbf{e}_i$ gives a minus sign.

4 The product $\mathbf{r}'\mathbf{r}$ can be written as

$$\begin{aligned}\mathbf{r}'\mathbf{r} &= r'(\cos\varphi'\ \mathbf{e}_x + \sin\varphi'\ \mathbf{e}_y)r(\cos\varphi\ \mathbf{e}_x + \sin\varphi\ \mathbf{e}_y)\\ &= r'r(\cos\varphi'\cos\varphi\ \mathbf{e}_x^2 + \sin\varphi'\sin\varphi\ \mathbf{e}_y^2 + (\sin\varphi'\cos\varphi - \cos\varphi'\sin\varphi)\mathbf{e}_y\mathbf{e}_x)\\ &= r'r(\cos(\varphi'-\varphi) + \sin(\varphi'-\varphi)\mathbf{e}_y\mathbf{e}_x) = r'r e^{\imath_{yx}(\varphi'-\varphi)},\end{aligned}$$

using the angle-subtraction formulas. This can also be evaluated with complex exponentials

$$\mathbf{r}'\mathbf{r} = r'e^{\imath_{yx}\varphi'}\mathbf{e}_x r e^{\imath_{yx}\varphi}\mathbf{e}_x = r'r e^{\imath_{yx}\varphi'}e^{-\imath_{yx}\varphi}\mathbf{e}_x\mathbf{e}_x = r'r e^{\imath_{yx}(\varphi'-\varphi)}. \tag{A.6}$$

5 a) The product $\mathbf{e}_z\mathbf{e}_x$ gives

$$\mathbf{e}_z\mathbf{e}_x = \begin{pmatrix} 1 & 0 \\ 0 & -1 \end{pmatrix}\begin{pmatrix} 0 & 1 \\ 1 & 0 \end{pmatrix} = \begin{pmatrix} 0 & 1 \\ -1 & 0 \end{pmatrix}, \tag{A.7}$$

where $\mathbf{e}_x\mathbf{e}_z$

$$\mathbf{e}_x\mathbf{e}_z = \begin{pmatrix} 0 & 1 \\ 1 & 0 \end{pmatrix}\begin{pmatrix} 1 & 0 \\ 0 & -1 \end{pmatrix} = \begin{pmatrix} 0 & -1 \\ 1 & 0 \end{pmatrix}, \tag{A.8}$$

showing that $\mathbf{e}_x\mathbf{e}_z = -\mathbf{e}_z\mathbf{e}_x$.

b) Squaring the product $\mathbf{e}_z\mathbf{e}_x$ gives

$$(\mathbf{e}_z\mathbf{e}_x)^2 = \begin{pmatrix} 0 & 1 \\ -1 & 0 \end{pmatrix}\begin{pmatrix} 0 & 1 \\ -1 & 0 \end{pmatrix} = \begin{pmatrix} -1 & 0 \\ 0 & -1 \end{pmatrix} = -\mathbb{1}_2, \tag{A.9}$$

where $\mathbb{1}_2$ is a 2×2 identity matrix. This is the matrix equivalent of $(\mathbf{e}_z\mathbf{e}_x)^2 = -1$.

6 A scalar quantity remains unchanged under the transformation

$$e^{\imath_{yx}\frac{\varphi}{2}}1e^{-\imath_{yx}\frac{\varphi}{2}} = 1. \tag{A.10}$$

However, a vector is rotated over an angle φ,

$$e^{\imath_{yx}\frac{\varphi}{2}}\hat{\mathbf{r}}e^{-\imath_{yx}\frac{\varphi}{2}} = e^{\imath_{yx}\frac{\varphi}{2}}e^{\imath_{yx}\varphi'}\mathbf{e}_x e^{-\imath_{yx}\frac{\varphi}{2}} = e^{\imath_{yx}\frac{\varphi}{2}}e^{\imath_{yx}\varphi'}e^{\imath_{yx}\frac{\varphi}{2}}\mathbf{e}_x = e^{\imath_{yx}(\varphi+\varphi')}\mathbf{e}_x.$$

Problems for Section 1.3

1 Let us start with an arbitrary vector $\hat{\mathbf{r}}_1$. A new vector $\hat{\mathbf{r}}_{-1}$ rotated over 90° is obtained by

$$\hat{\mathbf{r}}_{-1} = \imath_{yx}\hat{\mathbf{r}}_1 = \frac{\partial\hat{\mathbf{r}}_1}{\partial\varphi} = -\sin\varphi\ \mathbf{e}_x + \cos\varphi\ \mathbf{e}_y. \tag{A.11}$$

Rotating again gives

$$\imath_{yx}\hat{\mathbf{r}}_{-1} = \frac{\partial\hat{\mathbf{r}}_{-1}}{\partial\varphi} = -\cos\varphi\ \mathbf{e}_x - \sin\varphi\ \mathbf{e}_y = -\mathbf{r}_1. \tag{A.12}$$

Therefore, any pair of orthogonal vectors transform into each other as $\imath_{yx}\hat{\mathbf{r}}_{\pm 1} = \pm\hat{\mathbf{r}}_{\mp 1}$.

2 Operating with $\hat{L}$ on the vector gives

$$\hat{L}\hat{\mathbf{r}} = -\boldsymbol{i}_{yx}\hat{i}_{yx}(\mathbf{e}_1 r^1 + \mathbf{e}_{-1}r^{-1}) = -\boldsymbol{i}_{yx}(\mathbf{e}_1(-i)r^1 + \mathbf{e}_{-1}ir^{-1})$$
$$= -((-i)\mathbf{e}_1(-i)r^1 + i\mathbf{e}_{-1}ir^{-1}) = \hat{\mathbf{r}}. \quad \text{(A.13)}$$

Since $r^{\pm 1}(-\varphi) = r^{\mp 1}(\varphi)$,

$$\hat{L}\hat{\mathbf{r}}(-\varphi) = -\boldsymbol{i}_{yx}\hat{i}_{yx}(\mathbf{e}_1 r^{-1} + \mathbf{e}_{-1}r^1) = -((-i)\mathbf{e}_1 ir^{-1} + i\mathbf{e}_{-1}(-i)r^1) = -\hat{\mathbf{r}}.$$

3 The suggested solution for the differential equation is the series $f(\varphi) = \sum_{n=0}^{\infty} c_n \varphi^n$. Its derivative is given by

$$\frac{df(\varphi)}{d\varphi} = \sum_{n=0}^{\infty} c_n n\varphi^{n-1} = \sum_{n'=0}^{\infty} c_{n'+1}(n'+1)\varphi^{n'}, \quad \text{(A.14)}$$

using $n' = n - 1$ or $n = n' + 1$. Inserting this into the differential equation gives

$$\sum_{n=1}^{\infty} \left(c_{n+1}(n+1) - ic_n\right)\varphi^n = 0 \quad \Rightarrow \quad c_{n+1} = \frac{i}{n+1}c_n, \quad \text{(A.15)}$$

removing the prime. The result on the right follows from the fact that φ^n are independent functions; therefore, the coefficients in front of them have to be zero. Shifting the index gives for the coefficient

$$c_n = \frac{i}{n}c_{n-1} = \frac{i^2}{n(n-1)}c_{n-2} = \frac{i^n}{n!}c_0. \quad \text{(A.16)}$$

The function is then

$$f(\varphi) = \sum_{n=0}^{\infty} c_n\varphi^n = c_0\sum_{n=0}^{\infty}\frac{i^n}{n!}\varphi^n = c_0 e^{i\varphi}, \quad \text{(A.17)}$$

using the definition of the exponential.

Problems for Section 1.4

1 a) The Fourier transform is given by

$$f_m = \langle m|f\rangle = \int_0^{2\pi} d\varphi\ \langle m|\varphi\rangle\langle\varphi|f\rangle = \frac{1}{\sqrt{2\pi}}\int_0^{2\pi} d\varphi\ e^{-im\varphi}f(\varphi).$$
$$= \frac{1}{\sqrt{2\pi}}\int_{\frac{\pi}{2}}^{\pi} d\varphi\ e^{-im\varphi} = \frac{1}{\sqrt{2\pi}}\left[\frac{e^{-im\varphi}}{-im}\right]_{\frac{\pi}{2}}^{\pi} = \frac{i}{m\sqrt{2\pi}}(e^{-im\pi} - e^{-im\frac{\pi}{2}}).$$

For $m = 0$, taking the limit gives $f_0 = \frac{1}{2}\sqrt{\frac{\pi}{2}}$.

b) The real part is plotted in Figure A.1. While the real part is even around $m = 0$, the imaginary part is odd.

c) Calculating the function back from f_m gives

$$f(\varphi) = \langle\varphi|f\rangle = \sum_{m=-m_{\text{max}}}^{m_{\text{max}}} \langle\varphi|m\rangle\langle m|f\rangle = \frac{1}{\sqrt{2\pi}}\sum_{m=-m_{\text{max}}}^{m_{\text{max}}} e^{im\varphi}f_m,$$

where different ranges of m are indicated. The function is plotted in Figure A.2 for different values of m_{max}.

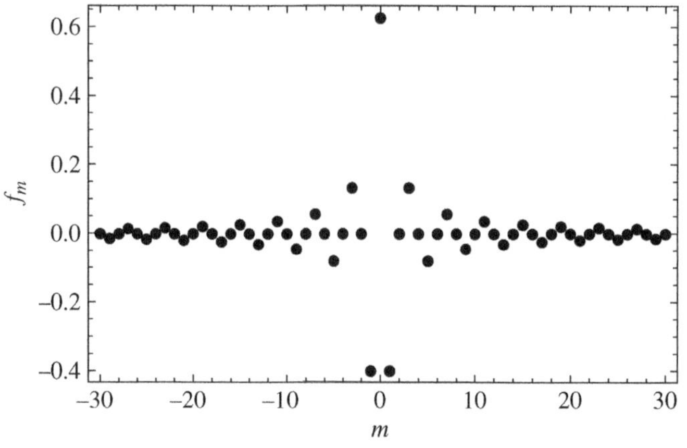

Figure A.1 The Fourier transform f_m of the square function in Eq. (1.97).

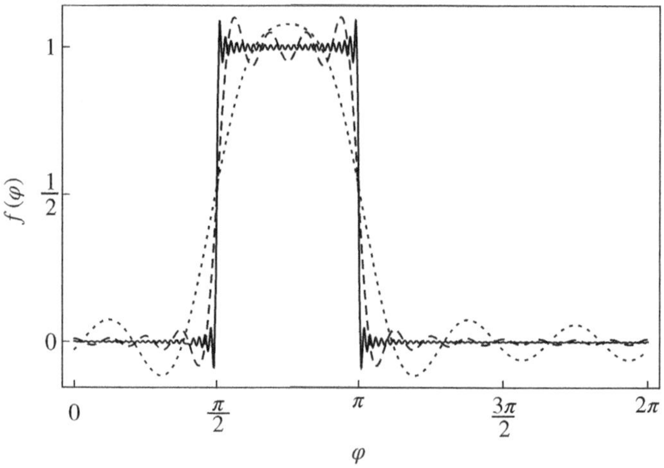

Figure A.2 The function $f(\varphi)$ calculated using the ranges $[-m_{\max}, m_{\max}]$, with $m_{\max} = 5$ (dotted), 15 (dashed), 100 (solid).

Problems for Section 1.5

1 For $\mathbf{e}_i$, we have

$$\begin{aligned}
\boldsymbol{i}\mathbf{e}_x &= \mathbf{e}_x\mathbf{e}_y\mathbf{e}_z\mathbf{e}_x = \mathbf{e}_x\mathbf{e}_x\mathbf{e}_y\mathbf{e}_z = \mathbf{e}_x\boldsymbol{i}, \\
\boldsymbol{i}\mathbf{e}_y &= \mathbf{e}_x\mathbf{e}_y\mathbf{e}_z\mathbf{e}_y = -\mathbf{e}_y\mathbf{e}_x\mathbf{e}_z\mathbf{e}_y = \mathbf{e}_y\mathbf{e}_x\mathbf{e}_y\mathbf{e}_z = \mathbf{e}_y\boldsymbol{i}, \\
\boldsymbol{i}\mathbf{e}_z &= \mathbf{e}_x\mathbf{e}_y\mathbf{e}_z\mathbf{e}_z = \mathbf{e}_z\mathbf{e}_x\mathbf{e}_y\mathbf{e}_z = \mathbf{e}_z\boldsymbol{i},
\end{aligned} \tag{A.18}$$

using the anticommutation of the unit vectors.

2 The product is given by

$$\begin{aligned}
(3\mathbf{e}_x + \mathbf{e}_y)(\mathbf{e}_x - 2\mathbf{e}_y) &= 3\mathbf{e}_x^2 - 2\mathbf{e}_y^2 - 6\mathbf{e}_x\mathbf{e}_y + \mathbf{e}_y\mathbf{e}_x \\
&= 1 - 7\mathbf{e}_x\mathbf{e}_y = 1 - 7\boldsymbol{i}\mathbf{e}_z.
\end{aligned} \tag{A.19}$$

The outer product is given by $-7\mathbf{e}_z$, indicating that the two vectors form a left-handed system. Note that, although both vectors lie in the xy-plane, restricting the results of a product to vectors leads to the introduction of a third dimension and an imaginary unit.

3 The area of a parallelogram can be viewed as a distorted rectangle with one side equal to r. The other side is the projection of r' onto a vector perpendicular to $\mathbf{r}$. This length is $r'\cos(90° - \alpha) = r'\sin\alpha$, where α is the angle between r and r'. The area is then $rr'\sin\alpha$, which is equal to the norm of the wedge product, see Eq. (1.119).

4 The order of the indices of the bivectors is important. In order for the bivectors to be consistent, they have to satisfy cyclic permutation, *i.e.* $x \to y \to z \to x$ or $yz \to zx \to xy \to yz$, see Eq. (1.108). Therefore, the bivector is more properly written as $4\boldsymbol{i}_{xy} - 3\boldsymbol{i}_{xz} = 4\boldsymbol{i}_{xy} + 3\boldsymbol{i}_{zx}$. This plane can also be defined by its normal $4\mathbf{e}_z + 3\mathbf{e}_y$.

Problems for Section 1.6

1 a) The spinors are effectively the square roots of the rotation axis. The spinors for the x-axis give half of the rotation from the z-axis to the positive and negative x-axis

$$\boldsymbol{e}_{x,\pm\frac{1}{2}} = e^{\pm\boldsymbol{i}_{xz}\frac{\pi}{4}} = \frac{1}{\sqrt{2}}(1 \pm \boldsymbol{i}_{xz}). \tag{A.20}$$

The spinors are defined with respect to the z-axis. Squaring them and working on the reference axis gives

$$\boldsymbol{e}_{x,\pm\frac{1}{2}}\boldsymbol{e}_{x,\pm\frac{1}{2}}\mathbf{e}_z = e^{\pm\boldsymbol{i}_{xz}\frac{\pi}{4}}e^{\pm\boldsymbol{i}_{xz}\frac{\pi}{4}}\mathbf{e}_z = e^{\pm\boldsymbol{i}_{xz}\frac{\pi}{2}}\mathbf{e}_z = \pm\boldsymbol{i}_{xz}\mathbf{e}_z = \pm\mathbf{e}_x,$$

where the sign corresponds to counterclockwise and clockwise rotations around the x-axis, respectively.

b) Operating with the rotation axis on the spinors gives

$$\begin{aligned}\mathbf{e}_x\boldsymbol{e}_{x,\pm\frac{1}{2}} &= \mathbf{e}_x\frac{1}{\sqrt{2}}(1 \pm \boldsymbol{i}_{xz}) = \frac{1}{\sqrt{2}}(\mathbf{e}_x \pm \mathbf{e}_z) = \pm\frac{1}{\sqrt{2}}(\mathbf{e}_z \pm \mathbf{e}_x) \\ &= \pm\mathbf{e}_{x,\pm\frac{1}{2}} = \pm\frac{1}{\sqrt{2}}(1 \pm \boldsymbol{i}_{xz})\mathbf{e}_z = \pm\boldsymbol{e}_{x,\pm\frac{1}{2}}\mathbf{e}_z,\end{aligned} \tag{A.21}$$

where $\mathbf{e}_z$ on the far right is the reference axis.

c) The imaginary unit in the yz-plane related to counterclockwise rotations around the x-axis is given by $-\boldsymbol{i}\mathbf{e}_x = -\mathbf{e}_x\mathbf{e}_y\mathbf{e}_z\mathbf{e}_x = -\mathbf{e}_y\mathbf{e}_z = \mathbf{e}_z\mathbf{e}_y = \boldsymbol{i}_{zy}$. Therefore, for its corresponding imaginary unit, Eq. (A.21) transforms into

$$\boldsymbol{i}_{zy}\boldsymbol{e}_{x,\pm\frac{1}{2}} = -\boldsymbol{i}\mathbf{e}_x\boldsymbol{e}_{x,\pm\frac{1}{2}} = \mp\boldsymbol{i}\mathbf{e}_{x,\pm\frac{1}{2}}. \tag{A.22}$$

2 Using only trigonometric functions, we obtain

$$\begin{aligned}\mathbf{r}_+ = \hat{\boldsymbol{r}}_\uparrow\hat{\boldsymbol{r}}_\uparrow\mathbf{e}_z &= \left(\cos\frac{\theta}{2} + \boldsymbol{i}_{x'z}\sin\frac{\theta}{2}\right)\left(\cos\frac{\theta}{2} + \boldsymbol{i}_{x'z}\sin\frac{\theta}{2}\right)\mathbf{e}_z \\ &= \left(\cos^2\frac{\theta}{2} - \sin^2\frac{\theta}{2} + \boldsymbol{i}_{x'z}2\sin\frac{\theta}{2}\cos\frac{\theta}{2}\right)\mathbf{e}_z = \cos\theta\,\mathbf{e}_z + \sin\theta\,\mathbf{e}_{x'} \\ &= \sin\theta(\cos\varphi\,\mathbf{e}_x + \sin\varphi\,\mathbf{e}_y) + \cos\theta\,\mathbf{e}_z,\end{aligned} \tag{A.23}$$

using the double-angle formulas. This is the unit vector indicating the rotation axis in spherical polar coordinates.

3 There are three possible imaginary units: $\boldsymbol{i}_{xz}$ does not work, since we are then left with $\boldsymbol{i}_{xz}\mathbf{e}_y = -\boldsymbol{i}$; $\boldsymbol{i}_{yx}$ only leaves one unit vector. We are therefore left with $\boldsymbol{i}_{zy}$. Substituting this for i gives

$$\mathbf{e}_{\pm 1} = \frac{1}{\sqrt{2}}(\mathbf{e}_x \pm \boldsymbol{i}_{zy}\mathbf{e}_y) = \frac{1}{\sqrt{2}}(\mathbf{e}_x \pm \mathbf{e}_z). \tag{A.24}$$

Operating with $\mathbf{e}_z$ gives

$$\mathbf{e}_z\mathbf{e}_{\pm 1} = \frac{1}{\sqrt{2}}(\mathbf{e}_z\mathbf{e}_x \pm 1) = \pm\frac{1}{\sqrt{2}}(1 \pm \boldsymbol{i}_{zx}) = \pm\frac{1}{\sqrt{2}}(\mathbf{e}_x \pm \mathbf{e}_z)\mathbf{e}_x = \pm\mathbf{e}_{\pm 1}\mathbf{e}_x.$$

Although the same spinor is returned, $\mathbf{e}_x$ appears as the reference axis of the coordinate system. Therefore, in order to find the rotation axis, the spinor squared has to be multiplied from the right by the reference axis. Using $\boldsymbol{e}_{\pm 1} = \mathbf{e}_{\pm 1}\mathbf{e}_x = (1 \pm \boldsymbol{i}_{zx})/\sqrt{2} = e^{\pm \boldsymbol{i}_{zx}\frac{\pi}{4}}$, we find

$$\boldsymbol{e}_{\pm 1}\boldsymbol{e}_{\pm 1}\mathbf{e}_x = e^{\pm \boldsymbol{i}_{zx}\frac{\pi}{4}}e^{\pm \boldsymbol{i}_{zx}\frac{\pi}{4}}\mathbf{e}_x = e^{\pm \boldsymbol{i}_{zx}\frac{\pi}{2}}\mathbf{e}_x = \pm\boldsymbol{i}_{zx}\mathbf{e}_x = \pm\mathbf{e}_z, \tag{A.25}$$

where $\pm$ corresponds to counterclockwise and clockwise rotations, respectively. If we are only interested in the transformation properties of the unit vectors, then there are a variety of ways to incorporate the same information: complex vectors or spinors, different reference axes, and so on. Although this is convenient, it also can make things very confusing.

Problems for Section 1.7

1 Written in terms of Pauli matrices, we have

$$3\mathbf{e}_x + \mathbf{e}_y = \begin{pmatrix} 0 & 3-i \\ 3+i & 0 \end{pmatrix}, \quad \mathbf{e}_x - 2\mathbf{e}_y = \begin{pmatrix} 0 & 1+2i \\ 1-2i & 0 \end{pmatrix}.$$

Multiplying gives

$$\begin{pmatrix} 0 & 3-i \\ 3+i & 0 \end{pmatrix}\begin{pmatrix} 0 & 1+2i \\ 1-2i & 0 \end{pmatrix} = \begin{pmatrix} 1-7i & 0 \\ 0 & 1+7i \end{pmatrix} = \mathbb{1}_2 - 7i\mathbf{e}_z.$$

2 The square of a vector $\mathbf{r}$ in terms of Pauli matrices is

$$\begin{aligned} \mathbf{rr} &= \begin{pmatrix} z & x-iy \\ x+iy & -z \end{pmatrix}\begin{pmatrix} z & x-iy \\ x+iy & -z \end{pmatrix} \\ &= \begin{pmatrix} z+x^2+y^2 & 0 \\ 0 & z+x^2+y^2 \end{pmatrix} = r^2\mathbb{1}_2. \end{aligned} \tag{A.26}$$

3 The imaginary unit $\boldsymbol{i} = \mathbf{e}_x\mathbf{e}_y\mathbf{e}_z$ is given by

$$\boldsymbol{i} = \mathbf{e}_x\begin{pmatrix} 0 & -i \\ i & 0 \end{pmatrix}\begin{pmatrix} 1 & 0 \\ 0 & -1 \end{pmatrix} = \begin{pmatrix} 0 & 1 \\ 1 & 0 \end{pmatrix}\begin{pmatrix} 0 & i \\ i & 0 \end{pmatrix} = \begin{pmatrix} i & 0 \\ 0 & i \end{pmatrix} = i\mathbb{1}_2.$$

Similarly, we can find $\mathbf{e}_z\mathbf{e}_y\mathbf{e}_x = -i\mathbb{1}_2 = -\boldsymbol{i}$. Using anticommutation, we find $\mathbf{e}_z\mathbf{e}_y\mathbf{e}_x = \mathbf{e}_y\mathbf{e}_x\mathbf{e}_z = -\mathbf{e}_x\mathbf{e}_y\mathbf{e}_z = -\boldsymbol{i}$.

4 The imaginary unit $\boldsymbol{i}_{zy}$ is given by

$$\boldsymbol{i}_{zy} = \mathbf{e}_z\mathbf{e}_y = \begin{pmatrix} 1 & 0 \\ 0 & -1 \end{pmatrix}\begin{pmatrix} 0 & -i \\ i & 0 \end{pmatrix} = \begin{pmatrix} 0 & -i \\ -i & 0 \end{pmatrix}. \tag{A.27}$$

Operating on the unit vectors gives

$$\boldsymbol{i}_{zy}\mathbf{e}_y = \begin{pmatrix} 0 & -i \\ -i & 0 \end{pmatrix}\begin{pmatrix} 0 & -i \\ i & 0 \end{pmatrix} = \begin{pmatrix} 1 & 0 \\ 0 & -1 \end{pmatrix} = \mathbf{e}_z,$$

$$\boldsymbol{i}_{zy}\mathbf{e}_z = \begin{pmatrix} 0 & -i \\ -i & 0 \end{pmatrix}\begin{pmatrix} 1 & 0 \\ 0 & -1 \end{pmatrix} = \begin{pmatrix} 0 & i \\ -i & 0 \end{pmatrix} = -\mathbf{e}_y. \tag{A.28}$$

Using anticommutation, we readily obtain $\boldsymbol{i}_{zy}\mathbf{e}_y = \mathbf{e}_z\mathbf{e}_y\mathbf{e}_y = \mathbf{e}_z$ and $\boldsymbol{i}_{zy}\mathbf{e}_z = \mathbf{e}_z\mathbf{e}_y\mathbf{e}_z = -\mathbf{e}_z\mathbf{e}_z\mathbf{e}_y = -\mathbf{e}_y$. This is obviously a cyclic permutation of the relations $\boldsymbol{i}_{yx}\mathbf{e}_x = \mathbf{e}_x$ and $\boldsymbol{i}_{yx}\mathbf{e}_y = -\mathbf{e}_y$ derived in Section 1.2. Although, for algebraic unit vectors, the results look similar, the 2×2 Pauli matrices look different for each direction, since they are defined for the zx-plane.

5 a) The multivector $\boldsymbol{H}$ can be split into a scalar E_0 and a vector $\mathbf{H}$,

$$\boldsymbol{H} = E_0 + \mathbf{H} = E_0 + \Delta\mathbf{e}_z + V\mathbf{e}_x = \begin{pmatrix} E_0 & 0 \\ 0 & E_0 \end{pmatrix} + \begin{pmatrix} \Delta & V \\ V & -\Delta \end{pmatrix},$$

with $\Delta = (\varepsilon_1 - \varepsilon_0)/2$, and $E_0 = (\varepsilon_1 + \varepsilon_0)/2$. Here, $\boldsymbol{H}$ and E_0 are used, anticipating the Hamiltonian and the energy, respectively.

b) The norm E of the multivector $\boldsymbol{H}$ is given by

$$\boldsymbol{H} = E \;\Rightarrow\; \mathbf{H} = E - E_0 \;\Rightarrow\; \mathbf{H}^2 = (E - E_0)^2 \;\Rightarrow\; E = E_0 \pm E',$$

where the norm of the vector is given by $E' = \sqrt{\mathbf{H}^2} = \sqrt{\Delta^2 + V^2}$. The result can be written as

$$E_\pm = E_0 \pm E' = E_0 \pm \sqrt{\Delta^2 + V^2} = \frac{\varepsilon_1 + \varepsilon_0}{2} \pm \sqrt{\left(\frac{\varepsilon_1 - \varepsilon_0}{2}\right)^2 + V^2}.$$

Obviously, these are simply the eigenvalues of the matrix in Eq. (1.168).

c) First, we align the z'-axis with the vector $\mathbf{H}$. The scalar component can be ignored, since it has no particular direction. The new z-axis is then

$$\mathbf{e}_{z'} = \cos\theta\;\mathbf{e}_z + \sin\theta\;\mathbf{e}_x, \quad \text{with} \quad \tan\theta = \frac{V}{\Delta}. \tag{A.29}$$

However, the axis itself is not an eigenvector of the vector $\mathbf{H}$. Using that $\cos\theta = \Delta/E'$ and $\sin\theta = V/E'$, one finds

$$\mathbf{H}\mathbf{e}_{z'} = E'\begin{pmatrix} \cos\theta & \sin\theta \\ \sin\theta & -\cos\theta \end{pmatrix}\begin{pmatrix} \cos\theta \\ \sin\theta \end{pmatrix} = E'\begin{pmatrix} 1 \\ 0 \end{pmatrix} = E'\mathbf{e}_z.$$

This corresponds to $\mathbf{r}\hat{\mathbf{r}}_+ = r\mathbf{e}_z$ or $\mathbf{r}\hat{\mathbf{r}}_+ = r$ from Eq. (1.121) but with a finite norm.

d) From Eq. (1.130), we know that the eigenvectors are given by the spinors,

$$\hat{\mathbf{r}}_\uparrow = \cos\frac{\theta}{2}\mathbf{e}_z + \sin\frac{\theta}{2}\mathbf{e}_x \quad \text{and} \quad \hat{\mathbf{r}}_\downarrow = -\sin\frac{\theta}{2}\mathbf{e}_z + \cos\frac{\theta}{2}\mathbf{e}_x, \tag{A.30}$$

using Eq. (1.128). For example, for the up spinor,

$$\mathbf{H}\hat{\mathbf{r}}_\uparrow = E'\begin{pmatrix} \cos\theta & \sin\theta \\ \sin\theta & -\cos\theta \end{pmatrix}\begin{pmatrix} \cos\frac{\theta}{2} \\ \sin\frac{\theta}{2} \end{pmatrix} = E'\begin{pmatrix} \cos\frac{\theta}{2} \\ \sin\frac{\theta}{2} \end{pmatrix} = E'\hat{\mathbf{r}}_\uparrow,$$

which can be straightforwardly evaluated using the angle addition formulas for trigonometric functions. Similarly, one finds $\mathbf{H}\hat{\mathbf{r}}_\downarrow = -E'\hat{\mathbf{r}}_\downarrow$.

e) The vector can also be written in the new z'-direction,

$$\mathbf{H} = E'\mathbf{e}_{z'} = \begin{pmatrix} E' & 0 \\ 0 & -E' \end{pmatrix}. \tag{A.31}$$

The choice of a different axis system does not affect the norm of the multivector $\boldsymbol{H} = \mathbf{H} + E_0 = E$. Expressing the spinors in the original basis gives

$$\mathbf{H} = E'(|\hat{\mathbf{r}}_\uparrow\rangle\langle\hat{\mathbf{r}}_\uparrow| - |\hat{\mathbf{r}}_\downarrow\rangle\langle\hat{\mathbf{r}}_\downarrow|) \tag{A.32}$$

$$\begin{aligned}
&= E'\left(\cos\frac{\theta}{2}|\mathbf{e}_z\rangle + \sin\frac{\theta}{2}|\mathbf{e}_x\rangle\right)\left(\cos\frac{\theta}{2}\langle\mathbf{e}_z| + \sin\frac{\theta}{2}\langle\mathbf{e}_x|\right) \\
&\quad - E'\left(-\sin\frac{\theta}{2}|\mathbf{e}_z\rangle + \cos\frac{\theta}{2}|\mathbf{e}_x\rangle\right)\left(-\sin\frac{\theta}{2}\langle\mathbf{e}_z| + \cos\frac{\theta}{2}\langle\mathbf{e}_x|\right) \\
&= E'\left(\cos^2\frac{\theta}{2} - \sin^2\frac{\theta}{2}\right)|\mathbf{e}_z\rangle\langle\mathbf{e}_z| + E'\left(\sin^2\frac{\theta}{2} - \cos^2\frac{\theta}{2}\right)|\mathbf{e}_x\rangle\langle\mathbf{e}_x| \\
&\quad + 2E'\sin\frac{\theta}{2}\cos\frac{\theta}{2}(|\mathbf{e}_z\rangle\langle\mathbf{e}_x| + |\mathbf{e}_x\rangle\langle\mathbf{e}_z|).
\end{aligned} \tag{A.33}$$

Using the double-angle formulas, one finds

$$\begin{aligned}
\mathbf{H} &= E'\cos\theta(|\mathbf{e}_z\rangle\langle\mathbf{e}_z| - |\mathbf{e}_x\rangle\langle\mathbf{e}_x|) + E'\sin\theta(|\mathbf{e}_z\rangle\langle\mathbf{e}_x| + |\mathbf{e}_x\rangle\langle\mathbf{e}_z|) \\
&= E'\cos\theta\,\mathbf{e}_z + E'\sin\theta\,\mathbf{e}_x = \Delta\mathbf{e}_z + V\mathbf{e}_x,
\end{aligned} \tag{A.34}$$

using $\cos\theta = \Delta/E'$ and $\sin\theta = V/E'$ from Eq. (A.29). The Pauli matrices can be written in bra-ket notation as $\mathbf{e}_z = |\mathbf{e}_z\rangle\langle\mathbf{e}_z| - |\mathbf{e}_x\rangle\langle\mathbf{e}_x|$ and $\mathbf{e}_x = |\mathbf{e}_z\rangle\langle\mathbf{e}_x| + |\mathbf{e}_x\rangle\langle\mathbf{e}_z|$. This is the vector in its original basis. Therefore, the spinors allow us to write the vector $\mathbf{H}$ in diagonal form. This is sometimes called a Bogoliubov transformation. Often, the half-angle trigonometric functions are expressed in terms of the parameters of the problem

$$\begin{aligned}
\cos^2\frac{\theta}{2} &= \frac{1}{2} + \frac{1}{2}\cos\theta \;\Rightarrow\; \cos\frac{\theta}{2} = \sqrt{\frac{1}{2} + \frac{\Delta}{2E'}}, \\
\sin^2\frac{\theta}{2} &= \frac{1}{2} - \frac{1}{2}\cos\theta \;\Rightarrow\; \sin\frac{\theta}{2} = \sqrt{\frac{1}{2} - \frac{\Delta}{2E'}},
\end{aligned} \tag{A.35}$$

using the double-angle formula for the cosine. Unfortunately, these expression somewhat obscure the half-angle nature of the coefficients.

Obviously, this problem can also be solved using standard techniques to obtain the eigenvalues and eigenvectors of a matrix.

Problems for Section 1.8

1 The $\boldsymbol{i}_{zy}$ corresponds to a rotation around the x-axis.

a) The eigenvalues are related to the norm n via $\boldsymbol{i}_{zy} = n$. Squaring gives $-1 = n^2$ or $n = \pm i$. The rotation is around the x-axis. The two directions of rotations are then $\pm\mathbf{e}_x = e^{\pm \boldsymbol{i}_{xz}\frac{\pi}{2}}\mathbf{e}_z$, where the z-direction is the reference axis for the Pauli matrices. The spinors are then $\boldsymbol{e}_{x,\pm\frac{1}{2}} = e^{\pm \boldsymbol{i}_{xz}\frac{\pi}{4}}$, so that $\boldsymbol{e}_{x,\pm\frac{1}{2}}\boldsymbol{e}_{x,\pm\frac{1}{2}}\mathbf{e}_z = \pm\mathbf{e}_x$. In ket notation, this is

$$|x, \pm\frac{1}{2}\rangle = \frac{1}{\sqrt{2}}\begin{pmatrix} 1 \\ \pm 1 \end{pmatrix}. \tag{A.36}$$

These are indeed eigenvectors, since

$$\boldsymbol{i}_{zy}|x, \pm\frac{1}{2}\rangle = \frac{1}{\sqrt{2}}\begin{pmatrix} 0 & -i \\ -i & 0 \end{pmatrix}\begin{pmatrix} 1 \\ \pm 1 \end{pmatrix} = \frac{1}{\sqrt{2}}\begin{pmatrix} \mp i \\ -i \end{pmatrix} = \mp\frac{i}{\sqrt{2}}\begin{pmatrix} 1 \\ \pm 1 \end{pmatrix},$$

using Eq. (1.170). Therefore, $\boldsymbol{i}_{zy}|x, \pm\frac{1}{2}\rangle = \mp i|x, \pm\frac{1}{2}\rangle$, giving the correct eigenvalues.

b) The eigenvectors are the same, since $\mathbf{e}_x = i\boldsymbol{i}_{zy}$. The eigenvalue equation is then $\mathbf{e}_x|x, \pm\frac{1}{2}\rangle = i\boldsymbol{i}_{zy}|x, \pm\frac{1}{2}\rangle = \pm|x, \pm\frac{1}{2}\rangle$. Note that the eigenvalue problems only differ by a factor i.

c) The ket-bra of the spinors is

$$|x, \pm\frac{1}{2}\rangle\langle x, \pm\frac{1}{2}| = \frac{1}{2}\begin{pmatrix} 1 \\ \pm 1 \end{pmatrix}(1 \pm 1) = \frac{1}{2}\begin{pmatrix} 1 & \pm 1 \\ \pm 1 & 1 \end{pmatrix}. \tag{A.37}$$

The identity matrix is

$$\mathbb{1}_2 = |x, \frac{1}{2}\rangle\langle x, \frac{1}{2}| + |x, -\frac{1}{2}\rangle\langle x, -\frac{1}{2}|. \tag{A.38}$$

The imaginary unit is given by

$$\boldsymbol{i}_{zy} = -i|x, \frac{1}{2}\rangle\langle x, \frac{1}{2}| + i|x, -\frac{1}{2}\rangle\langle x, -\frac{1}{2}|. \tag{A.39}$$

d) We want to demonstrate $-\boldsymbol{i}_{\hat{\mathbf{r}}}\hat{\mathbf{r}}_{\pm\frac{1}{2}} = \mp i\hat{\boldsymbol{r}}_{\pm\frac{1}{2}}$ for $\hat{\mathbf{r}} = \mathbf{e}_x$. Since $\boldsymbol{i}_{\underline{\mathbf{e}}_x} = i\boldsymbol{e}_x = \boldsymbol{i}_{yz} = -\boldsymbol{i}_{zy}$, this equation becomes $\boldsymbol{i}_{zy}\mathbf{e}_{x,\pm\frac{1}{2}} = \mp i\boldsymbol{e}_{x,\pm\frac{1}{2}}$ or, by multiplying with $\mathbf{e}_z$ from the right, $\boldsymbol{i}_{zy}\boldsymbol{e}_{x,\pm\frac{1}{2}} = \mp i\boldsymbol{e}_{x,\pm\frac{1}{2}}$, with $\boldsymbol{e}_{x,\pm\frac{1}{2}} = \mathbf{e}_{x,\pm\frac{1}{2}}\mathbf{e}_z$. This is the vector equivalent of the same equation in matrix form. Writing the spinor in terms of the Pauli matrices gives

$$\boldsymbol{e}_{x,\pm\frac{1}{2}} = e^{\pm \boldsymbol{i}_{xz}\frac{\pi}{4}} = \frac{1}{\sqrt{2}}(\mathbb{1}_2 \mp \boldsymbol{i}_{xz}) = \frac{1}{\sqrt{2}}\begin{pmatrix} 1 & \mp 1 \\ \pm 1 & 1 \end{pmatrix}. \tag{A.40}$$

This gives, on the one hand,

$$\boldsymbol{i}_{zy}\boldsymbol{e}_{x,\pm\frac{1}{2}} = \frac{1}{\sqrt{2}}\begin{pmatrix} 0 & -i \\ -i & 0 \end{pmatrix}\begin{pmatrix} 1 & \mp 1 \\ \pm 1 & 1 \end{pmatrix} = \frac{1}{\sqrt{2}}\begin{pmatrix} \mp i & -i \\ -i & \pm i \end{pmatrix}.$$

On the other hand, we obtain

$$\mp \boldsymbol{i}\mathbf{e}_{x,\pm\frac{1}{2}}\mathbf{e}_z = \mp i\frac{1}{\sqrt{2}}\begin{pmatrix} 1 & \mp 1 \\ \pm 1 & 1 \end{pmatrix}\begin{pmatrix} 1 & 0 \\ 0 & -1 \end{pmatrix} = \frac{1}{\sqrt{2}}\begin{pmatrix} \mp i & -i \\ -i & \pm i \end{pmatrix}.$$

e) The expectation values for $\mathbf{e}_x$ are

$$\langle x, \pm\frac{1}{2}|\mathbf{e}_x|x, \pm\frac{1}{2}\rangle = \frac{1}{2}(1, \pm 1)\begin{pmatrix} 0 & 1 \\ 1 & 0 \end{pmatrix}\begin{pmatrix} 1 \\ \pm 1 \end{pmatrix} = \frac{1}{2}(1, \pm 1)\begin{pmatrix} \pm 1 \\ 1 \end{pmatrix} = \pm 1.$$

The expectation values for y and z can be shown to be zero. This gives

$$\sum_i \langle x, \pm\frac{1}{2}|\mathbf{e}_i|x, \pm\frac{1}{2}\rangle \mathbf{e}_i = \pm\mathbf{e}_x, \tag{A.41}$$

which gives the rotation axis, where the $\pm$ sign indicates counterclockwise and clockwise rotations, respectively.

2 The unit vector in the $(1, 1, 1)$ direction is $\mathbf{e}_{111} = (\mathbf{e}_x + \mathbf{e}_y + \mathbf{e}_z)/\sqrt{3}$. The corresponding bivector is

$$\boldsymbol{i}_{111} = -\boldsymbol{i}\mathbf{e}_{111} = \frac{1}{\sqrt{3}}(\boldsymbol{i}_{yx} + \boldsymbol{i}_{zy} + \boldsymbol{i}_{xz}) = \frac{1}{\sqrt{3}}\begin{pmatrix} -i & -1-i \\ 1-i & i \end{pmatrix}.$$

Since $\mathbf{e}_z$ does not lie in the plane of rotation, we have to use an approach similar to Problem 6 in Section 1.2 to make sure that the projection onto the rotation axis is properly taken care of. The rotation over an angle θ is then given by $\boldsymbol{r}_{\frac{1}{2}}\mathbf{e}_z\boldsymbol{r}^*_{\frac{1}{2}}$, with

$$\boldsymbol{r}_{\frac{1}{2}} = e^{\boldsymbol{i}_{111}\frac{\theta}{2}} = \cos\frac{\theta}{2} + \boldsymbol{i}_{111}\sin\frac{\theta}{2}. \tag{A.42}$$

In terms of Pauli matrices, for $\theta = 120° = 2\pi/3$, this becomes, after some work,

$$\boldsymbol{r}_{\frac{1}{2}} = \cos\frac{\theta}{2}\mathbb{1}_2 + \sin\frac{\theta}{2}\boldsymbol{i}_{111} = \frac{1}{2}\begin{pmatrix} 1-i & -1-i \\ 1-i & 1+i \end{pmatrix}. \tag{A.43}$$

Rotating twice gives

$$\boldsymbol{r}_{\frac{1}{2}}\mathbf{e}_z\boldsymbol{r}^*_{\frac{1}{2}} = \begin{pmatrix} 0 & 1 \\ 1 & 0 \end{pmatrix} = \mathbf{e}_x \text{ and } \boldsymbol{r}_{\frac{1}{2}}\mathbf{e}_x\boldsymbol{r}^*_{\frac{1}{2}} = \begin{pmatrix} 0 & -i \\ i & 0 \end{pmatrix} = \mathbf{e}_y.$$

Obviously, we can continue to do this $\mathbf{e}_z \to \mathbf{e}_x \to \mathbf{e}_y \to \mathbf{e}_z$, etc.

Problems for Section 1.9

1 The rotation axis is $\hat{\mathbf{r}} = \cos 30°\mathbf{e}_z + \sin 30°\mathbf{e}_y = e^{\boldsymbol{i}_{yz}\frac{\pi}{6}}\mathbf{e}_z$. The spinors are the effective square root, and therefore, $\hat{\boldsymbol{r}}_\uparrow = e^{\boldsymbol{i}_{yz}\frac{\pi}{12}}$ or $\hat{\mathbf{r}}_\uparrow = \cos 15°\mathbf{e}_z + \sin 15°\mathbf{e}_y$. The spinor for clockwise rotations is rotated by 90° in the zy-plane, *i.e.* $\hat{\boldsymbol{r}}_\uparrow = \boldsymbol{i}_{yz}\hat{\boldsymbol{r}}_\uparrow$ or $\hat{\mathbf{r}}_\downarrow = -\sin 15°\mathbf{e}_z + \cos 15°\mathbf{e}_y$. The spinors are related to the axis of rotation

$$\hat{\mathbf{r}} = \cos 30°\mathbf{e}_z + \sin 30°\mathbf{e}_y = \begin{pmatrix} \frac{\sqrt{3}}{2} & -\frac{i}{2} \\ \frac{i}{2} & -\frac{\sqrt{3}}{2} \end{pmatrix}. \tag{A.44}$$

The spinors can also be obtained by numerically diagonalizing $\hat{\mathbf{r}}$, giving

$$|\hat{\mathbf{r}}_\uparrow\rangle = \begin{pmatrix} 0.965926 \\ 0.258819i \end{pmatrix} \quad \text{and} \quad |\hat{\mathbf{r}}_\downarrow\rangle = \begin{pmatrix} -0.258819 \\ 0.965926i \end{pmatrix}, \tag{A.45}$$

with eigenvalues ± 1, respectively. The numbers in the column vector correspond to $\cos 15^\circ$ and $\sin 15^\circ$. Some programs might give the result $i|\hat{\mathbf{r}}_\sigma\rangle$ for the eigenvectors. This is still correct since the real physical quantity is the direction of the rotation axis and an additional phase vector is irrelevant when calculating this axis via

$$\hat{\mathbf{r}}_\pm = \sum_{i=x,y,z} \langle \hat{\mathbf{r}}_{\pm\frac{1}{2}} | \mathbf{e}_i | \hat{\mathbf{r}}_{\pm\frac{1}{2}} \rangle \mathbf{e}_i = \pm(0.5\ \mathbf{e}_y + 0.866025\ \mathbf{e}_z). \tag{A.46}$$

2 a) Since $\theta = \pi/2$ and $\varphi = \pi/4$, Eq. (1.193) becomes

$$|\hat{\mathbf{r}}_\uparrow\rangle = \begin{pmatrix} \frac{1}{\sqrt{2}} \\ \frac{1}{2}(1+i) \end{pmatrix} \quad \text{and} \quad |\hat{\mathbf{r}}_\downarrow\rangle = \begin{pmatrix} -\frac{1}{\sqrt{2}} \\ \frac{1}{2}(1+i) \end{pmatrix}. \tag{A.47}$$

b) From Eq. (1.195), we find, after some laborious evaluations,

$$\hat{\mathbf{r}}_\pm = \sum_{i=x,y,z} \langle \hat{\mathbf{r}}_{\pm\frac{1}{2}} | \mathbf{e}_i | \hat{\mathbf{r}}_{\pm\frac{1}{2}} \rangle \mathbf{e}_i = \pm\frac{1}{\sqrt{2}}(\mathbf{e}_x + \mathbf{e}_y) = \pm\hat{\mathbf{r}}, \tag{A.48}$$

where $\hat{\mathbf{r}}_\pm$ correspond to counterclockwise and clockwise rotations around $\hat{\mathbf{r}}$, respectively.

c) The spinor gives half the rotation from the z-axis to the x'-axis, which is at 45° with the x-axis,

$$\begin{aligned} \hat{\boldsymbol{r}}_{\pm\frac{1}{2}} &= e^{\pm \boldsymbol{i}_{x'z}\frac{\pi}{4}} = \frac{1}{\sqrt{2}}(1 \pm \boldsymbol{i}_{x'z}) = \frac{1}{\sqrt{2}}(1 \pm e^{\boldsymbol{i}_{yx}\frac{\pi}{4}}\boldsymbol{i}_{xz}) \\ &= \frac{1}{\sqrt{2}} \pm \frac{1}{2}(\boldsymbol{i}_{xz} + \boldsymbol{i}_{yz}), \end{aligned} \tag{A.49}$$

which should be compared with Eq. (A.47). The change in sign for $\hat{\boldsymbol{r}}_{-\frac{1}{2}}$ with respect to Eqs. (A.47) and (1.134) is irrelevant for the rotation axis, which is the square of the spinor

$$\hat{\mathbf{r}}_\pm = \hat{\boldsymbol{r}}_{\pm\frac{1}{2}}\hat{\boldsymbol{r}}_{\pm\frac{1}{2}}\mathbf{e}_z = e^{\pm \boldsymbol{i}_{x'z}\frac{\pi}{2}}\mathbf{e}_z = \pm\boldsymbol{i}_{x'z}\mathbf{e}_z = \pm\mathbf{e}_{x'} = \pm\frac{1}{\sqrt{2}}(\mathbf{e}_x + \mathbf{e}_y).$$

However, the choice in Eq. (A.49) creates a left-handed system, which is not always desirable.

d) Writing Eq. (A.49) in Pauli matrices using Eq. (1.170) gives

$$\hat{\boldsymbol{r}}_{\pm\frac{1}{2}} = \frac{\mathbb{1}_2}{\sqrt{2}} \pm \frac{1}{2}(\boldsymbol{i}_{xz} + \boldsymbol{i}_{yz}) = \begin{pmatrix} \frac{1}{\sqrt{2}} & \pm\frac{1}{2}(-1+i) \\ \pm\frac{1}{2}(1+i) & \frac{1}{\sqrt{2}} \end{pmatrix}. \tag{A.50}$$

e) The rotation axis can be obtained again by squaring the spinors,

$$\hat{\mathbf{r}}_\pm = \hat{\boldsymbol{r}}_{\pm\frac{1}{2}}\hat{\boldsymbol{r}}_{\pm\frac{1}{2}}\mathbf{e}_z = \begin{pmatrix} 0 & \pm\frac{1}{\sqrt{2}}(1-i) \\ \pm\frac{1}{\sqrt{2}}(1+i) & 0 \end{pmatrix} = \pm\frac{1}{\sqrt{2}}(\mathbf{e}_x + \mathbf{e}_y).$$

Problems for Section 1.10

1 The function space for a periodic system in one dimension is given in Eq. (1.210). This can be expanded to three dimensions

$$\psi_{\mathbf{k}}(\mathbf{r}) = \langle \mathbf{r}|\mathbf{k}\rangle = \frac{1}{\sqrt{V}} e^{i\mathbf{k}\cdot\mathbf{r}}. \tag{A.51}$$

These functions are orthonormal, since

$$\langle \mathbf{k}'|\mathbf{k}\rangle = \int d\mathbf{r}\langle \mathbf{k}'|\mathbf{r}\rangle\langle \mathbf{r}|\mathbf{k}\rangle = \frac{1}{V}\int d\mathbf{r}\; e^{i(\mathbf{k}-\mathbf{k}')\cdot\mathbf{r}} = \delta_{\mathbf{k}',\mathbf{k}}. \tag{A.52}$$

In contrast to Eq. (1.221), the δ function is now a Kronecker δ function. Completeness follows from the orthogonality of space

$$\begin{aligned}\langle \mathbf{r}'|\mathbf{r}\rangle &= \delta(\mathbf{r}'-\mathbf{r}) = \frac{1}{(2\pi)^3}\int d\mathbf{k}\; e^{i\mathbf{k}\cdot(\mathbf{r}'-\mathbf{r})} \\ &\cong \frac{V}{(2\pi)^3}\sum_{\mathbf{k}} d\mathbf{k}\frac{1}{V}e^{i\mathbf{k}\cdot(\mathbf{r}'-\mathbf{r})} = \sum_{\mathbf{k}}\langle \mathbf{r}'|\mathbf{k}\rangle\langle \mathbf{k}|\mathbf{r}\rangle,\end{aligned} \tag{A.53}$$

using $d\mathbf{k} = dk_x dk_y dk_x = (2\pi/L)^3 = (2\pi)^3/V$. For the left- and right-hand side to be equal, we need $\sum_{\mathbf{k}}|\mathbf{k}\rangle\langle\mathbf{k}| = \mathbb{1}_\infty$, *i.e.* the function space is complete.

2 If the particle is in a box, then the functions have to go to zero at the edge of the box, *i.e.* $\varphi_{k_x}(x) = 0$ for $x = 0, L$. This can be done by taking linear combinations of the free-space solutions

$$\varphi \sim e^{ik_x x} \pm e^{-ik_x x} = A\cos k_x x, A\sin k_x x. \tag{A.54}$$

Only the second solution can be zero for $x = 0$ and $x = L$, giving wavenumbers $k_x = n\pi/L$. Normalizing gives

$$\langle k_x|k_x\rangle = \int_0^L dx\,\sin^2 k_x x = \int_0^L dx\,\left(\frac{1}{2}-\frac{1}{2}\cos 2k_x x\right) = \frac{L}{2}, \tag{A.55}$$

where the cosine term cancels. The normalized wavefunctions are then $\sqrt{2/L}\sin k_x x$. Note that the distance π/L between the k_x values is twice as small as for periodic boundary conditions, see Eq. (1.205). However, for a particle in a box, only positive k_x values are allowed, whereas for periodic boundary conditions, both positive and negative values are allowed. The density of states is therefore unchanged in the limit $L \to \infty$.

Problems for Section 1.11

1 The result can be obtained by writing the function in momentum space

$$\begin{aligned}\nabla\psi(\mathbf{r}) &= \frac{1}{(2\pi)^{\frac{3}{2}}}\nabla\int d\mathbf{k}\; e^{i\mathbf{k}\cdot\mathbf{r}}\psi(\mathbf{r}) = \frac{1}{(2\pi)^{\frac{3}{2}}}\int d\mathbf{k}\;\nabla e^{i\mathbf{k}\cdot\mathbf{r}}\psi(\mathbf{r}) \\ &= \frac{1}{(2\pi)^{\frac{3}{2}}}\int d\mathbf{k}\; e^{i\mathbf{k}\cdot\mathbf{r}}\left(i\mathbf{k}\psi(\mathbf{r})\right).\end{aligned} \tag{A.56}$$

2 A Fourier transform is not necessarily done in Cartesian coordinates. For problems that involve atoms, for example the scattering of radiation, it is more convenient to perform the Fourier transform in spherical polar coordinates. Let us take the angle between $\mathbf{r}$ and $\mathbf{k}$ to be θ,

$$\begin{aligned} n(\mathbf{k}) &= \int d\mathbf{r}\, e^{-i\mathbf{k}\cdot\mathbf{r}} n(\mathbf{r}) = \int_0^\infty dr \int_0^{2\pi} d\varphi \int_0^\pi d\theta\, r^2 \sin\theta n(r) e^{-ikr\cos\theta} \\ &= 2\pi \int_0^\infty dr r^2 n(r) \frac{e^{ikr} - e^{-ikr}}{ikr} = 4\pi \int_0^\infty dr r^2 n(r) \frac{\sin kr}{kr}. \end{aligned} \tag{A.57}$$

3 The nonunitary Fourier transform of the convolution is

$$\begin{aligned} A(\mathbf{k}) &= \int d\mathbf{r}\, e^{-i\mathbf{k}\cdot\mathbf{r}} A(\mathbf{r}) = \int d\mathbf{r}\, e^{-i\mathbf{k}\cdot\mathbf{r}} \left(\int d\mathbf{r}'\, G(\mathbf{r}-\mathbf{r}') F(\mathbf{r}') \right) \\ &= \int d\mathbf{r}'\, F(\mathbf{r}') \left(\int d\mathbf{r}\, e^{-i\mathbf{k}\cdot\mathbf{r}}\, G(\mathbf{r}-\mathbf{r}') \right). \end{aligned} \tag{A.58}$$

Making the substitution $\mathbf{s} = \mathbf{r} - \mathbf{r}'$ or $\mathbf{r} = \mathbf{s} + \mathbf{r}'$, this becomes

$$A(\mathbf{k}) = \left(\int d\mathbf{r}'\, F(\mathbf{r}') e^{-i\mathbf{k}\cdot\mathbf{r}'} \right) \left(\int d\mathbf{s} e^{-i\mathbf{k}\cdot\mathbf{s}}\, G(\mathbf{s}) \right) = G(\mathbf{k}) F(\mathbf{k}), \tag{A.59}$$

using $d\mathbf{r} = d\mathbf{s}$, since $\mathbf{r}'$ is constant while integrating over $\mathbf{r}$.

Problems for Section 1.12

1 The vectors given in the problem are shown in Figure A.3a.

a) The vector $\mathbf{a}^3$ pointing toward the center of the cube is the shortest vector. However, it transforms into itself under cyclic permutation. Let us take the vector $\mathbf{a}^1 = \frac{a}{2}(-\mathbf{e}_x + \mathbf{e}_y + \mathbf{e}_z)$ pointing toward an atom at the center of a neighboring cube. Cyclic permutation gives

$$\mathbf{a}^2 = \frac{a}{2}(\mathbf{e}_x - \mathbf{e}_y + \mathbf{e}_z), \quad \mathbf{a}^3 = \frac{a}{2}(\mathbf{e}_x + \mathbf{e}_y - \mathbf{e}_z). \tag{A.60}$$

The vectors are shown in Figure A.3b.

b) We first determine the volume of the parallelepiped. This can be done by considering the number of atoms in the cube. In addition to the atom at the center, each atom at the corners is 1/8 in the cube. The total number of atoms is then $1 + 8(1/8) = 2$. The volume of the primitive cell is $V = a^3/2$. The first reciprocal lattice vector is

$$\mathbf{a}_1 = \frac{\mathbf{a}^2 \times \mathbf{a}^3}{V} = \frac{2}{a^3}\left(\frac{a}{2}\right)^2 (\mathbf{e}_x - \mathbf{e}_y + \mathbf{e}_z) \times (\mathbf{e}_x + \mathbf{e}_y - \mathbf{e}_z) = \frac{1}{a}(\mathbf{e}_y + \mathbf{e}_z).$$

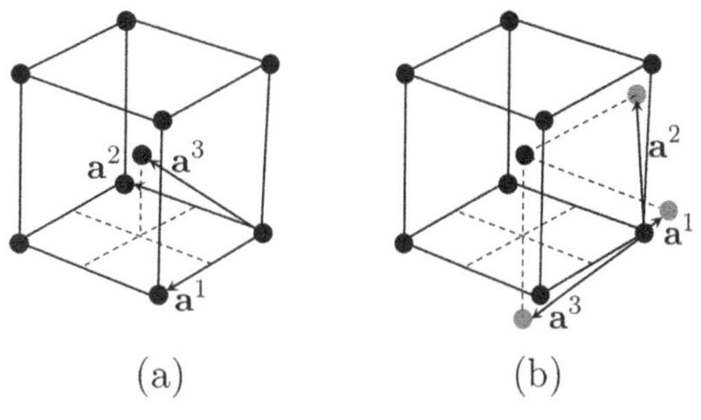

Figure A.3 Two different sets of bcc primitive unit vectors.

The other two can be obtained via cyclic permutation

$$\mathbf{a}_2 = \frac{1}{a}(\mathbf{e}_z + \mathbf{e}_x) \quad \text{and} \quad \mathbf{a}_3 = \frac{1}{a}(\mathbf{e}_x + \mathbf{e}_y). \tag{A.61}$$

Plotting this lattice leads to a face-centered cubic lattice, *i.e.* a simple cubic lattice with additional atoms at the center of the sides.

Problems for Section 1.13

1 The lattice vectors for a triangular lattice are given by $\mathbf{a}^1 = a\mathbf{e}_x$ and $\mathbf{a}^2 = \frac{1}{2}a\mathbf{e}_x + \frac{1}{2}\sqrt{3}a\mathbf{e}_y$ (note that other orientations and choices are possible). In matrix form, using Eq. (1.288), this can be written as

$$\mathrm{U} = \begin{pmatrix} \mathbf{a}^1 \\ \mathbf{a}^2 \end{pmatrix} = \begin{pmatrix} a & 0 \\ \frac{1}{2}a & \frac{1}{2}\sqrt{3}a \end{pmatrix}. \tag{A.62}$$

Taking the inverse (this can be done numerically) gives

$$\mathrm{U}^{-1} = \begin{pmatrix} \mathbf{a}_1 & \mathbf{a}_2 \end{pmatrix} = \frac{1}{a}\begin{pmatrix} 1 & 0 \\ -\frac{1}{\sqrt{3}} & \frac{2}{\sqrt{3}} \end{pmatrix}. \tag{A.63}$$

Plotting, see Figure A.4, shows that this is still a triangular lattice but rotated in order to satisfy $\mathbf{a}^i \cdot \mathbf{a}_j = \delta_{ij}$. The distance between the nearest-neighbor points in reciprocal space is $1/a\sqrt{3}$.

2 The unitary transformation from a Cartesian basis to the primitive vectors of the bcc lattice is given by Eq. (A.60). In matrix form, this is

$$\mathrm{U} = \begin{pmatrix} \mathbf{a}^1 \\ \mathbf{a}^2 \\ \mathbf{a}^3 \end{pmatrix} = \frac{a}{2}\begin{pmatrix} -1 & 1 & 1 \\ 1 & -1 & 1 \\ 1 & 1 & -1 \end{pmatrix}, \tag{A.64}$$

see Eq. (1.288). The inverse can be obtained numerically and is

$$\mathrm{U}^{-1} = \begin{pmatrix} \mathbf{a}_1 & \mathbf{a}_2 & \mathbf{a}_3 \end{pmatrix} = \frac{1}{a}\begin{pmatrix} 0 & 1 & 1 \\ 1 & 0 & 1 \\ 1 & 1 & 0 \end{pmatrix}, \tag{A.65}$$

which is a face-centered cubic (fcc) lattice. Note that if we started out with an fcc lattice in real space, its reciprocal would be bcc.

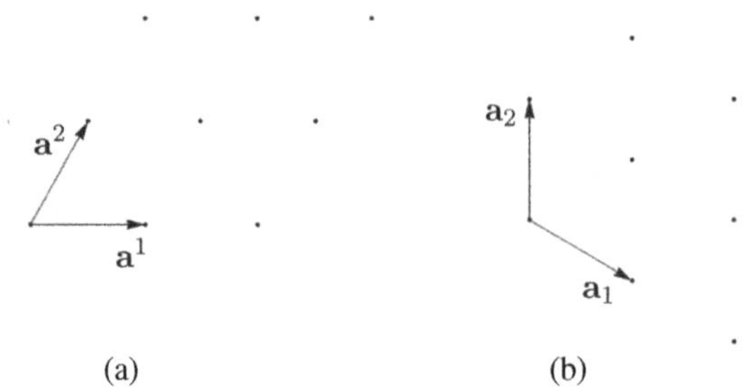

Figure A.4 A triangular lattice (a) and its reciprocal (b).

Problems for Section 1.14

1 In order to connect different m values, we need to change the up spinors into down spinors and vice versa. However, we want to keep the number of spinors constant. We therefore suggest $\mathbf{L}_+ = \boldsymbol{a}_\uparrow^\dagger \boldsymbol{a}_\downarrow$ and $\mathbf{L}_- = \boldsymbol{a}_\downarrow^\dagger \boldsymbol{a}_\uparrow$. For example, $\mathbf{L}_- e_1 = \boldsymbol{a}_\downarrow^\dagger \boldsymbol{a}_\uparrow \boldsymbol{a}_\uparrow^\dagger \boldsymbol{a}_\uparrow^\dagger/\sqrt{2} = \boldsymbol{a}_\downarrow^\dagger 2\boldsymbol{a}_\uparrow^\dagger/\sqrt{2} = \sqrt{2}e_0$ and $\mathbf{L}_- e_0 = \boldsymbol{a}_\downarrow^\dagger \boldsymbol{a}_\uparrow \boldsymbol{a}_\uparrow^\dagger \boldsymbol{a}_\downarrow^\dagger = \boldsymbol{a}_\downarrow^\dagger \boldsymbol{a}_\downarrow^\dagger = \sqrt{2}\boldsymbol{e}_{-1}$. Note that it is not possible to connect $\boldsymbol{e}_1$ and $\boldsymbol{e}_{-1}$ with $\mathbf{L}_\pm$.

2 Both $\boldsymbol{e}_x$ and $\boldsymbol{e}_y$ are linear combinations of $\boldsymbol{e}_{\pm 1}$. We do not want to make transitions to $\boldsymbol{e}_0$, so $\mathbf{L}_\pm$ from the preceding problem does not work. $\mathbf{L}_z$ is related to $\mathbf{e}_z$, see Eq. (1.330). We know from Section 1.6 that a rotation in the xy plane is related to $\boldsymbol{i}_{yx} = -\boldsymbol{i}\boldsymbol{e}_z$, which is the imaginary unit for rotations in the xy-plane. So, let us try $-\boldsymbol{i}\mathbf{L}_z$. Using Eq. (1.348), we find

$$
\begin{aligned}
-\boldsymbol{i}\mathbf{L}_z \boldsymbol{e}_x &= -\boldsymbol{i}\mathbf{L}_z \frac{1}{\sqrt{2}}(-e_1 + \boldsymbol{e}_{-1}) = \frac{i}{\sqrt{2}}(e_1 - (-\boldsymbol{e}_{-1})) = \boldsymbol{e}_y, \\
-\boldsymbol{i}\mathbf{L}_z \boldsymbol{e}_y &= -\boldsymbol{i}\mathbf{L}_z \frac{i}{\sqrt{2}}(e_1 + \boldsymbol{e}_{-1}) = \frac{1}{\sqrt{2}}(e_1 + (-\boldsymbol{e}_{-1})) = -\boldsymbol{e}_x.
\end{aligned} \tag{A.66}
$$

This is exactly what we would expect from $\boldsymbol{i}_{yx}$. Note that $\mathbf{L}_z$ also transforms the unit vectors into each other but with complex coefficients.

Problems for Section 1.15

1 For $l = 2$ and $m = 1$, only the $i = j = 1$ and $k = 0$ term remains in the summation

$$
C_1^2(\hat{\boldsymbol{r}}) = \sqrt{3!1!}\frac{1}{1!1!0!}\cos\theta\left(-\frac{\sin\theta e^{i\varphi}}{2}\right) = -\sqrt{\frac{3}{2}}\cos\theta\sin\theta e^{i\varphi}.
$$

For $l = 2$ and $m = -2$, only the $i = j = 0$ and $k = 2$ term remains in the summation

$$
C_{-2}^2(\hat{\boldsymbol{r}}) = \sqrt{0!4!}\frac{1}{0!0!2!}\left(\frac{\sin\theta e^{-i\varphi}}{2}\right)^2 = \frac{1}{2}\sqrt{\frac{3}{2}}\sin^2\theta e^{-2i\varphi}
$$

2 The only term that effectively changes is

$$
\left(-\frac{\sin\theta e^{-i\varphi}}{2}\right)^j\left(\frac{\sin\theta e^{i\varphi}}{2}\right)^k = (-1)^{k-j}\left(\frac{\sin\theta e^{-i\varphi}}{2}\right)^{k'}\left(-\frac{\sin\theta e^{i\varphi}}{2}\right)^{j'},
$$

using $j' = k$ and $k' = j$, giving $j' - k' = -(j - k) = -m$ using $j - k = m$. This directly gives $(C_m^l(\hat{\boldsymbol{r}}))^* = (-1)^m C_{-m}^l(\hat{\boldsymbol{r}})$.

Problems for Section 1.16

1 The imaginary unit related to $\mathbf{L}_z$ is

$$
\mathbf{L}_z = \begin{pmatrix} 1 & 0 & 0 \\ 0 & 0 & 0 \\ 0 & 0 & -1 \end{pmatrix} \quad \Rightarrow \quad \boldsymbol{I}_{yx} = -\boldsymbol{i}\mathbf{L}_z = \begin{pmatrix} -i & 0 & 0 \\ 0 & 0 & 0 \\ 0 & 0 & i \end{pmatrix}. \tag{A.67}
$$

Additionally, we can define the identity for the x and y (or $m = \pm 1$) directions

$$\mathbb{1}_{xy} = \boldsymbol{I}_{yx}^4 = \begin{pmatrix} 1 & 0 & 0 \\ 0 & 0 & 0 \\ 0 & 0 & 1 \end{pmatrix}. \tag{A.68}$$

Using $\boldsymbol{I}_{yx}^2 = -\mathbb{1}_{xy}$, $\boldsymbol{I}_{yx}^3 = -\boldsymbol{I}_{yx}$, etc., the exponential for rotation can be expanded as follows:

$$\begin{aligned} \mathrm{R}_{\varphi}^{z} &= e^{-i\mathbf{L}_z\varphi} = e^{\boldsymbol{I}_{yx}\varphi} = \sum_{n=0}^{\infty} \frac{(\boldsymbol{I}_{yx}\varphi)^n}{n!} \\ &= \mathbb{1}_3 + \mathbb{1}_{xy}\left(-\frac{\varphi^2}{2!} + \frac{\varphi^4}{4!} - \frac{\varphi^6}{6!} + \cdots\right) + \boldsymbol{I}_{yx}\left(\varphi - \frac{\varphi^3}{3!} + \cdots\right). \end{aligned} \tag{A.69}$$

Note that $\boldsymbol{I}_{yx}^0 = \mathbb{1}_3$ cannot be expressed in $\boldsymbol{I}_{yx}$ and $\mathbb{1}_{xy}$, since it includes the identity in the z-direction. Using the Taylor series for the sine and cosine gives

$$\mathrm{R}_{\varphi}^{z} = \mathbb{1}_3 + \mathbb{1}_{xy}(\cos\varphi - 1) + \boldsymbol{i}_{yx}\sin\varphi = \begin{pmatrix} e^{-i\varphi} & 0 & 0 \\ 0 & 1 & 0 \\ 0 & 0 & e^{i\varphi} \end{pmatrix}.$$

This matrix is diagonal with matrix elements given by $(\mathrm{R}_{\varphi}^{z})_{mm} = e^{-im\varphi}$. The rotation of a spherical harmonic over an angle φ_0 around the z-axis is given by

$$\begin{aligned} \langle \boldsymbol{r}|\mathrm{R}_{\varphi_0}^{z}|lm\rangle &= e^{-im\varphi_0}\langle \boldsymbol{r}|lm\rangle = e^{-im\varphi_0}Y_{lm}(\boldsymbol{r}) \\ &= Y_{lm}(\theta, \varphi - \varphi_0) = Y_{lm}(\mathrm{R}_{-\varphi_0}^{z}\boldsymbol{r}), \end{aligned} \tag{A.70}$$

using that the φ dependence of $Y_{lm}(\theta,\varphi) \equiv Y_{lm}(\boldsymbol{r})$ is $e^{im\varphi}$. This is indeed a rotation over φ_0 around the z-axis.

2 a) The matrix elements can be evaluated using Eq. (1.385), for example for the state $|lm\rangle$, with $l = 1$ and $m = 0$,

$$\mathbf{L}_1|10\rangle = -\frac{1}{\sqrt{2}}\boldsymbol{a}_{\uparrow}^{\dagger}\boldsymbol{a}_{\downarrow}|1;1\rangle = -\frac{1}{\sqrt{2}}\sqrt{2\times 1}|2;0\rangle = -|11\rangle, \tag{A.71}$$

using the notation $|n_{\uparrow};n_{\downarrow}\rangle$ to indicate the number of spinors. This leads to the following matrices for $\mathbf{L}_{\pm 1}$:

$$\mathbf{L}_1 = \begin{pmatrix} 0 & -1 & 0 \\ 0 & 0 & -1 \\ 0 & 0 & 0 \end{pmatrix} \quad \text{and} \quad \mathbf{L}_{-1} = \begin{pmatrix} 0 & 0 & 0 \\ 1 & 0 & 0 \\ 0 & 1 & 0 \end{pmatrix}, \tag{A.72}$$

where the basis functions are in the order $m = 1, 0, -1$. This gives for $\mathbf{L}_x$, see Eq. (1.385),

$$\mathbf{L}_x = \frac{1}{\sqrt{2}}(-\mathbf{L}_1 + \mathbf{L}_{-1}) = \frac{1}{\sqrt{2}}\begin{pmatrix} 0 & 1 & 0 \\ 1 & 0 & 1 \\ 0 & 1 & 0 \end{pmatrix}. \tag{A.73}$$

b) Numerically, the eigenvectors are found to be

$$|11'\rangle = \frac{1}{2}\begin{pmatrix} 1 \\ \sqrt{2} \\ 1 \end{pmatrix}, \quad |10'\rangle = \frac{1}{\sqrt{2}}\begin{pmatrix} -1 \\ 0 \\ 1 \end{pmatrix}, \quad |1,-1'\rangle = \frac{1}{2}\begin{pmatrix} 1 \\ -\sqrt{2} \\ 1 \end{pmatrix}.$$

The eigenvalues of $|1m'\rangle$ are $m' = 1, 0, -1$.

c) Using Eqs. (1.317) and (1.318), the spinors for the x-axis are

$$\hat{r}_\uparrow^\dagger = \frac{1}{\sqrt{2}}(a_\uparrow^\dagger + a_\downarrow^\dagger) \quad \text{and} \quad \hat{r}_\downarrow^\dagger = \frac{1}{\sqrt{2}}(-a_\uparrow^\dagger + a_\downarrow^\dagger), \tag{A.74}$$

taking $\theta = \frac{\pi}{2}$ and $\varphi = 0$.

d) In the new spinor basis, the eigenvectors can be written in the same fashion as Eq. (1.342). This gives

$$\begin{aligned} e_{1'} &= \frac{1}{\sqrt{2}}\hat{r}_\uparrow^\dagger \hat{r}_\uparrow^\dagger = \frac{1}{\sqrt{2}}\frac{1}{2}\left(a_\uparrow^\dagger + a_\downarrow^\dagger\right)\left(a_\uparrow^\dagger + a_\downarrow^\dagger\right) \\ &= \frac{1}{2\sqrt{2}}\left(a_\uparrow^\dagger a_\uparrow^\dagger + 2a_\uparrow^\dagger a_\downarrow^\dagger + a_\downarrow^\dagger a_\downarrow^\dagger\right) = \frac{1}{2}(e_1 + \sqrt{2}e_0 + e_{-1}) \end{aligned} \tag{A.75}$$

and

$$\begin{aligned} e_{0'} &= \hat{r}_\uparrow^\dagger \hat{r}_\downarrow^\dagger = \frac{1}{2}\left(a_\uparrow^\dagger + a_\downarrow^\dagger\right)\left(-a_\uparrow^\dagger + a_\downarrow^\dagger\right) \\ &= \frac{1}{2}\left(-a_\uparrow^\dagger a_\uparrow^\dagger + a_\downarrow^\dagger a_\downarrow^\dagger\right) = \frac{1}{\sqrt{2}}(-e_1 + e_{-1}). \end{aligned} \tag{A.76}$$

Similarly, we obtain $e_{-1'} = \frac{1}{2}(e_1 - \sqrt{2}e_0 + e_{-1})$. This reproduces the eigenvectors found numerically.

e) In term of the new spinors, we have $a_{\pm\frac{1}{2}}^\dagger = (\hat{r}_\uparrow^\dagger \mp \hat{r}_\downarrow^\dagger)/\sqrt{2}$. The operator can then be rewritten as

$$\begin{aligned} \mathbf{L}_x &= \frac{1}{\sqrt{2}}(-\mathbf{L}_1 + \mathbf{L}_{-1}) = \frac{1}{2}\left(a_\uparrow^\dagger a_\downarrow + a_\downarrow^\dagger a_\uparrow\right) \\ &= \frac{1}{4}\left(\left(\hat{r}_\uparrow^\dagger - \hat{r}_\downarrow^\dagger\right)\left(\hat{r}_\uparrow + \hat{r}_\downarrow\right) + \left(\hat{r}_\uparrow^\dagger + \hat{r}_\downarrow^\dagger\right)\left(\hat{r}_\uparrow - \hat{r}_\downarrow\right)\right) \\ &= \frac{1}{2}\left(\hat{r}_\uparrow^\dagger \hat{r}_\uparrow - \hat{r}_\downarrow^\dagger \hat{r}_\downarrow\right) = \mathbf{L}_{0'} = \mathbf{L}_{z'}, \end{aligned} \tag{A.77}$$

using Eq. (1.385). This operator is diagonal and gives eigenvalues $m' = 1, 0, -1$, as expected.

Problems for Section 1.17

1 a) The gradient in spherical polar coordinates is given by

$$\nabla f = \frac{\partial f}{\partial r}e_r + \frac{1}{r}\frac{\partial f}{\partial \theta}e_\theta + \frac{1}{r\sin\theta}\frac{\partial f}{\partial \varphi}e_\varphi. \tag{A.78}$$

The angular momentum is given by the outer product of the position and the momentum

$$\hat{\mathbf{L}} = \mathbf{r} \times \mathbf{p} = -i\hbar \begin{vmatrix} e_r & e_\theta & e_\varphi \\ r & 0 & 0 \\ \frac{\partial}{\partial r} & \frac{1}{r}\frac{\partial}{\partial \theta} & \frac{1}{r\sin\theta}\frac{\partial}{\partial \varphi} \end{vmatrix} = -i\hbar\left(-e_\theta \frac{1}{\sin\theta}\frac{\partial}{\partial\varphi} + e_\varphi \frac{\partial}{\partial\theta}\right).$$

The unit vectors in a spherical polar coordinate system can be expressed in terms of Cartesian coordinates as

$$e_\theta = e_x \cos\theta\cos\varphi + e_y \cos\theta\sin\varphi - e_z \sin\theta, \tag{A.79}$$

$$e_\varphi = -e_x \sin\varphi + e_y \cos\varphi. \tag{A.80}$$

This gives

$$\hat{L}_x = i\hbar\left(\sin\varphi\frac{\partial}{\partial\theta} + \cot\theta\cos\varphi\frac{\partial}{\partial\varphi}\right),$$
$$\hat{L}_y = i\hbar\left(-\cos\varphi\frac{\partial}{\partial\theta} + \cot\theta\sin\varphi\frac{\partial}{\partial\varphi}\right),$$
$$\hat{L}_z = -i\hbar\frac{\partial}{\partial\varphi}. \tag{A.81}$$

We take $\hbar \equiv 1$ in the following. The step operators are then

$$\hat{L}_{\pm 1} = \mp\frac{1}{\sqrt{2}}(\hat{L}_x \pm i\hat{L}_y) = \frac{e^{\pm i\varphi}}{\sqrt{2}}\left[-\frac{\partial}{\partial\theta} \mp i\cot\theta\frac{\partial}{\partial\varphi}\right]. \tag{A.82}$$

b) We can use a separation of variables for the spherical harmonic $Y_{lm}(\theta,\varphi) = \Theta_{lm}(\theta)e^{im\varphi}$, where it is straightforward to show that the φ-dependent part is a complex exponential. Since $m = l = 1$ is the maximum value, stepping up should give zero,

$$\hat{L}_1 Y_{11}(\theta,\varphi) = \hat{L}_1\Theta_{11}(\theta)e^{i\varphi} = \frac{e^{2i\varphi}}{\sqrt{2}}\left(-\frac{d\Theta_{11}(\theta)}{d\theta} + \cot\theta\Theta_{11}(\theta)\right) = 0.$$

It is easy to check that the solution is $\Theta_{11}(\theta) = \sin\theta$. The normalization of the function of φ is $\sqrt{2\pi}$. The integral over θ gives

$$\int_0^\pi d\theta\sin\theta\sin^2\theta = -\int_0^\pi(1-\cos^2\theta)d\cos\theta = \left[\frac{1}{3}\cos^3\theta - \cos\theta\right]_0^\pi = \frac{4}{3}.$$

The normalized wavefunction is therefore

$$Y_{11}(\theta,\varphi) = -\sqrt{\frac{3}{8\pi}}\sin\theta e^{i\varphi}, \tag{A.83}$$

taking into account the Condon–Shortley phase.

c) Stepping down gives

$$\hat{L}_{-1}Y_{11} = -\frac{1}{\sqrt{2}}\sqrt{\frac{3}{8\pi}}e^{-i\varphi}\left(-\frac{\partial}{\partial\theta} + i\cot\theta\frac{\partial}{\partial\varphi}\right)\sin\theta e^{i\varphi}$$
$$= \frac{1}{\sqrt{2}}\sqrt{\frac{3}{8\pi}}e^{-i\varphi}(\cos\theta + \cot\theta\sin\theta)e^{i\varphi} = \sqrt{\frac{3}{4\pi}}\cos\theta = Y_{10}.$$

It is straightforward to show that this function is indeed normalized. However, note that the same transformation properties are contained in $\mathbf{L}_{-1}|11\rangle = \frac{1}{\sqrt{2}}\boldsymbol{a}_\downarrow^\dagger\boldsymbol{a}_\uparrow|2;0\rangle = \frac{1}{\sqrt{2}}\sqrt{2\times 1}|1;1\rangle = |10\rangle$.

Problems for Section 2.1

1 Let us take, for example

$$\boldsymbol{i}\mathbf{e}_1' = \mathbf{e}_0'\mathbf{e}_1'\mathbf{e}_2'\mathbf{e}_3'\mathbf{e}_1' = \mathbf{e}_0'\mathbf{e}_1'\mathbf{e}_1'\mathbf{e}_2'\mathbf{e}_3' = -\mathbf{e}_1'\mathbf{e}_0'\mathbf{e}_1'\mathbf{e}_2'\mathbf{e}_3' = -\mathbf{e}_1'\boldsymbol{i}. \tag{A.84}$$

Therefore, they anticommute. This can be shown to be the case for all $\mathbf{e}_\mu'$. For the regular Cartesian unit vectors, we can write for $\mu = 1$,

$$\boldsymbol{i}\mathbf{e}_1 = \boldsymbol{i}\mathbf{e}_1'\mathbf{e}_0' = -\mathbf{e}_1'\boldsymbol{i}\mathbf{e}_0' = \mathbf{e}_1'\mathbf{e}_0'\boldsymbol{i} = \mathbf{e}_1\boldsymbol{i}, \tag{A.85}$$

using the result above.

2 Multiplying gives

$$\begin{aligned}
\mathbf{p}_1'\mathbf{p}_2' &= (E_1\mathbf{e}_0' + p_{1x}\mathbf{e}_x' + p_{1y}\mathbf{e}_y' + p_{1z}\mathbf{e}_z')(E_2\mathbf{e}_0' + p_{2x}\mathbf{e}_x' + p_{2y}\mathbf{e}_y' + p_{2z}\mathbf{e}_z') \\
&= (E_1E_2 - p_{1x}p_{2x} - p_{1y}p_{2y} - p_{1z}p_{2z}) + (E_1p_{2x} - p_{1x}E_2)\mathbf{e}_0'\mathbf{e}_x' \\
&\quad + (E_1p_{2y} - p_{1y}E_2)\mathbf{e}_0'\mathbf{e}_y' + (E_1p_{2z} - p_{1z}E_2)\mathbf{e}_0'\mathbf{e}_z' \\
&\quad + (p_{1x}p_{2y} - p_{1y}p_{2x})\mathbf{e}_x'\mathbf{e}_y' + (p_{1y}p_{2z} - p_{1z}p_{2y})\mathbf{e}_y'\mathbf{e}_z' \\
&\quad + (p_{1z}p_{2x} - p_{1x}p_{2z})\mathbf{e}_z'\mathbf{e}_x' \\
&= \mathbf{p}_1' \cdot \mathbf{p}_2' + \mathbf{e}_0'(E_1\mathbf{p}_2' - \mathbf{p}_1'E_2) + \mathbf{p}_2' \wedge \mathbf{p}_1'.
\end{aligned} \tag{A.86}$$

When $\mathbf{p}_2' = \mathbf{p}_1'$, the only term left is $\mathbf{p}_1'\mathbf{p}_1' = \mathbf{p}_1' \cdot \mathbf{p}_1'$.

3 The highest grade multivector given is the pseudoscalar $\boldsymbol{i} = \mathbf{e}_0'\mathbf{e}_x'\mathbf{e}_y'\mathbf{e}_z'$. Trying to raise this to a higher grade gives, for example,

$$\boldsymbol{i}\mathbf{e}_1' = \mathbf{e}_0'\mathbf{e}_1'\mathbf{e}_2'\mathbf{e}_3'\mathbf{e}_1' = \mathbf{e}_0'\mathbf{e}_1'\mathbf{e}_1'\mathbf{e}_2'\mathbf{e}_3' = -\mathbf{e}_0'\mathbf{e}_2'\mathbf{e}_3', \tag{A.87}$$

which is a multivector of grade 3.

Problems for Section 2.2

1 The commutation is given by

$$e^{\hat{\mathbf{p}}\alpha}\mathbf{e}_0' = \sum_{n=0}^{\infty}\frac{(\hat{\mathbf{p}}\alpha)^n}{n!}\mathbf{e}_0' = \sum_{n=0}^{\infty}\mathbf{e}_0'\frac{(-\hat{\mathbf{p}}\alpha)^n}{n!} = \mathbf{e}_0'e^{-\hat{\mathbf{p}}\alpha}, \tag{A.88}$$

where each commutation between $\hat{\mathbf{p}}$ and $\mathbf{e}_0'$ gives a negative sign.

2 The product is given by

$$\begin{aligned}
\hat{\mathbf{p}}_1'\hat{\mathbf{p}}_2' &= (\cosh\alpha_1\mathbf{e}_0' + \sinh\alpha_1\hat{\mathbf{p}}')(\cosh\alpha_2\mathbf{e}_0' + \sinh\alpha_2\hat{\mathbf{p}}') \\
&= \cosh\alpha_1\cosh\alpha_2 - \sinh\alpha_1\sinh\alpha_2 \\
&\quad + (\sinh\alpha_1\cosh\alpha_2 - \cosh\alpha_1\sinh\alpha_2)\hat{\mathbf{p}}'\mathbf{e}_0' \\
&= \cosh(\alpha_1 - \alpha_2) + \sinh(\alpha_1 - \alpha_2)\hat{\mathbf{p}}'\mathbf{e}_0',
\end{aligned} \tag{A.89}$$

using the hyperbolic angle subtraction formulas. This problem can be simplified by recognizing that the four-momenta can be written in terms of hyperbolic functions

$$\hat{\mathbf{p}}_1'\hat{\mathbf{p}}_2' = e^{\hat{\mathbf{p}}\alpha_1}\mathbf{e}_0'e^{\hat{\mathbf{p}}\alpha_2}\mathbf{e}_0' = e^{\hat{\mathbf{p}}\alpha_1}e^{-\hat{\mathbf{p}}\alpha_2}\mathbf{e}_0'\mathbf{e}_0' = e^{\hat{\mathbf{p}}(\alpha_1-\alpha_2)}, \tag{A.90}$$

which reproduces the above result.

3 We are interested how the wavelength changes when the source is moving away. It takes a period T to move the distance of the wavelength λ. Using Eq. (2.48),

$$\begin{aligned}
\lambda' &= \gamma(\lambda + vT) = \gamma\lambda\left(1 + v\frac{T}{\lambda}\right) = \gamma\lambda\left(1 + v\frac{k}{\omega}\right) = \gamma\lambda\left(1 + \frac{v}{c}\right) \\
&= \lambda\frac{1 + \frac{v}{c}}{\sqrt{1 - \frac{v^2}{c^2}}} = \lambda\sqrt{\frac{1 + \frac{v}{c}}{1 - \frac{v}{c}}}.
\end{aligned} \tag{A.91}$$

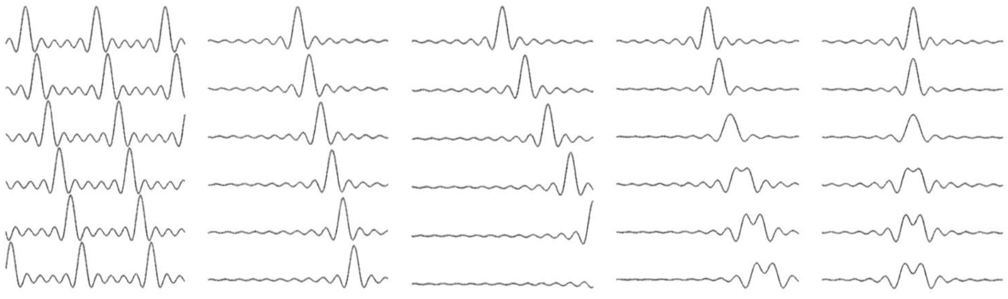

Figure A.5 The function $f(x, t)$ with $a = 5$, $b = 0$ for $\Delta k = 0.2$ and $(a, b) = (5, 0), (10, 0), (5, 1), (0, 1)$ for $\Delta k = 0.02$ (from left to right) and for $t = 0, 1, 2, 3, 4, 5$ (from top to bottom). The different time steps are shifted in the y-direction. The plot range is from $-40 \le x \le 40$.

Note that the speed of light $\lambda/T = 2\pi/T/(2\pi/\lambda) = \omega/k = c$ is constant in any frame. When the light source is moving away ($v > 0$), the wavelength becomes longer. This is known as a redshift. When the source is moving toward the observer ($v < 0$), the wavelength becomes shorter, giving a blueshift.

Problems for Section 2.3

1 The functions for the different conditions are plotted in Figure (A.5). From Problem 1 in Section 1.4, we know that the Fourier transform of a block wave is a sinc function. Since the Fourier transform is now a block wave, the expected function in real space is a sinc function. Some other comments on the plot: For $\Delta k = 0.2$, there is a recurrence of the sinc function. Since we discretized the momentum space, the real space is now periodic. The period is given by $L = 2\pi/\Delta k$, which is about $L = 31.4$ for $\Delta k = 0.2$. Since often one is trying to calculate a continuous block function via discretization, this would imply that Δk is not small enough. For $\Delta k = 0.02$, $L = 314$, we have a single sinc function within the plot range, which moves in time in the positive x-direction. The group velocity is given by $v_g = \partial\omega_k/\partial k = 5, 10$ for $\omega_k = 5k, 10k$. Note that the wave packet simply shifts with respect to time. Adding a quadratic component to the dispersion leads to distortions in the wavepacket. Removing the linear term, $\omega_k = k^2$ removes the motion of the wave packet since $v_g = \left.\frac{\partial\omega_k}{\partial k}\right|_{k=0} = 2k|_{k=0} = 0$. The wave packet does spread out. A quadratic dispersion with a finite group velocity is given by $\omega_k = (k - k_0)^2$ (not shown).

Problems for Section 2.4

1 a) The bivector between the time and momentum directions in the Dirac notation is $\hat{\mathbf{p}}\mathbf{e}'_0 = \hat{\mathbf{p}}'$. However, note that this is an imaginary unit since $\hat{\mathbf{p}}'^2 = -1$. The four-momentum can be written in exponential form as

$$\hat{\mathbf{p}} = Ee^{\hat{\mathbf{p}}'\delta}\mathbf{e}'_0 = \mathbf{e}'_0 E\cos\delta + \hat{\mathbf{p}}E\sin\delta = mc^2\mathbf{e}'_0 + \mathbf{p}, \tag{A.92}$$

with

$$E^2 = m^2c^4 + \mathbf{p}^2c^2 \quad \text{and} \quad \cos\delta = \frac{mc^2}{E}. \tag{A.93}$$

Note that this is a regular rotation and not a hyperbolic transformation. The spinors are then

$$\hat{\mathbf{p}}^+ = e^{\hat{\mathbf{p}}'\frac{\delta}{2}} \quad \text{and} \quad \hat{\mathbf{p}}^- = \hat{\mathbf{p}}' e^{\hat{\mathbf{p}}'\frac{\delta}{2}}. \tag{A.94}$$

b) Acting with the four-momentum on the spinors gives

$$\begin{aligned}
\hat{\mathbf{p}}\hat{\mathbf{p}}^+ &= E e^{\hat{\mathbf{p}}'\delta}\mathbf{e}_0' e^{\hat{\mathbf{p}}'\frac{\delta}{2}} = E e^{\hat{\mathbf{p}}'\delta} e^{-\hat{\mathbf{p}}'\frac{\delta}{2}}\mathbf{e}_0' = E e^{\hat{\mathbf{p}}'\frac{\delta}{2}}\mathbf{e}_0' = E\hat{\mathbf{p}}^+\mathbf{e}_0', \\
\hat{\mathbf{p}}\hat{\mathbf{p}}^- &= E e^{\hat{\mathbf{p}}'\delta}\mathbf{e}_0'\hat{\mathbf{p}}' e^{\hat{\mathbf{p}}'\frac{\delta}{2}} = -E\hat{\mathbf{p}}' e^{\hat{\mathbf{p}}'\delta} e^{-\hat{\mathbf{p}}'\frac{\delta}{2}}\mathbf{e}_0' = -E\hat{\mathbf{p}}' e^{\hat{\mathbf{p}}'\frac{\delta}{2}}\mathbf{e}_0' = -E\hat{\mathbf{p}}^-.
\end{aligned}$$

The result $\hat{\mathbf{p}}\hat{\mathbf{p}}^\pm = \pm E\hat{\mathbf{p}}^\pm$ should be compared with Eq. (2.87).

c) Squaring the spinors gives

$$\begin{aligned}
E\hat{\mathbf{p}}^+\hat{\mathbf{p}}^+\mathbf{e}_0' &= E e^{\hat{\mathbf{p}}'\frac{\delta}{2}} e^{\hat{\mathbf{p}}'\frac{\delta}{2}}\mathbf{e}_0' = E e^{\hat{\mathbf{p}}'\delta}\mathbf{e}_0' = \hat{\mathbf{p}}, \\
E\hat{\mathbf{p}}^-\hat{\mathbf{p}}^-\mathbf{e}_0' &= E\hat{\mathbf{p}}' e^{\hat{\mathbf{p}}'\frac{\delta}{2}}\hat{\mathbf{p}}' e^{\hat{\mathbf{p}}'\frac{\delta}{2}}\mathbf{e}_0' = E\hat{\mathbf{p}}'^2 e^{\hat{\mathbf{p}}'\delta}\mathbf{e}_0' = -\hat{\mathbf{p}}.
\end{aligned} \tag{A.95}$$

Since in the Dirac basis, the four-momentum is written as a rotation, the result is comparable to Eq. (1.127).

Problems for Section 2.5

1 The transition between particles and antiparticles is given by the real unit **1**, see Eqs. (2.91) and (2.115), which converts between the basis and its dual in Minkowski space,

$$|\hat{\mathbf{p}}_\uparrow^-\rangle = \mathbf{1}|\hat{\mathbf{p}}_\uparrow^+\rangle = \mathbf{1}\mathbf{e}_0' = \begin{pmatrix} 0 & 0 & 1 & 0 \\ 0 & 0 & 0 & 1 \\ 1 & 0 & 0 & 0 \\ 0 & 1 & 0 & 0 \end{pmatrix}\begin{pmatrix} 1 \\ 0 \\ 0 \\ 0 \end{pmatrix} = \begin{pmatrix} 0 \\ 0 \\ 1 \\ 0 \end{pmatrix} = \underline{\mathbf{e}}_0'. \tag{A.96}$$

The other way to change between the different Dirac spinors is by multiplying with the momentum, see Eq. (2.83), which is $\hat{\mathbf{p}} = \mathbf{e}_z$, for a particle at rest (z being the reference axis),

$$|\hat{\mathbf{p}}_\uparrow^-\rangle = \mathbf{e}_z\mathbf{e}_0' = \begin{pmatrix} 0 & 0 & 1 & 0 \\ 0 & 0 & 0 & -1 \\ 1 & 0 & 0 & 0 \\ 0 & -1 & 0 & 0 \end{pmatrix}\begin{pmatrix} 1 \\ 0 \\ 0 \\ 0 \end{pmatrix} = \begin{pmatrix} 0 \\ 0 \\ 1 \\ 0 \end{pmatrix} = \mathbf{e}_z'. \tag{A.97}$$

The transition between the up and the down spins is determined by

$$|\hat{\mathbf{p}}_\downarrow^+\rangle = \boldsymbol{i}_{xz}\mathbf{e}_0' = \mathbf{e}_x\mathbf{e}_z\mathbf{e}_0' = \begin{pmatrix} 0 & -1 & 0 & 0 \\ 1 & 0 & 0 & 0 \\ 0 & 0 & 0 & -1 \\ 0 & 0 & 1 & 0 \end{pmatrix}\begin{pmatrix} 1 \\ 0 \\ 0 \\ 0 \end{pmatrix} = \begin{pmatrix} 0 \\ 1 \\ 0 \\ 0 \end{pmatrix},$$

using Eq. (2.116). The unit spinor with the 1 in the fourth row can be obtained by $|\hat{\mathbf{p}}_\downarrow^-\rangle = \mathbf{1}|\hat{\mathbf{p}}_\downarrow^+\rangle = \boldsymbol{i}_{xz}|\hat{\mathbf{p}}_\uparrow^-\rangle$. This gives the different spinors for a particle at rest.

2 The spinor in 4×4 matrix form can be expressed as

$$\begin{aligned}
\hat{\mathbf{p}}_\uparrow^+ &= e^{\hat{\mathbf{p}}\frac{\alpha}{2}} = \cosh\frac{\alpha}{2} + \hat{\mathbf{p}}\sinh\frac{\alpha}{2} \\
&= \cosh\frac{\alpha}{2} + \sinh\frac{\alpha}{2}(\sin\theta\cos\varphi\mathbf{e}_x + \sin\theta\sin\varphi\mathbf{e}_y + \cos\theta\mathbf{e}_z)
\end{aligned}$$

$$= \begin{pmatrix} c & 0 & sp_z & sp_{-1} \\ 0 & c & sp_1 & -sp_z \\ sp_z & sp_{-1} & c & 0 \\ sp_1 & -sp_z & 0 & c \end{pmatrix} = (|\hat{\mathbf{p}}_\uparrow^+\rangle\ |\hat{\mathbf{p}}_\downarrow^+\rangle\ |\hat{\mathbf{p}}_\uparrow^-\rangle\ |\hat{\mathbf{p}}_\downarrow^-\rangle), \tag{A.98}$$

using $\hat{p}_z = \cos\theta$, $p_{\pm 1} = e^{\pm i\varphi}\sin\theta$, and the shorthands $c = \cosh\frac{\alpha}{2}$ and $s = \sinh\frac{\alpha}{2}$. Note that the columns in the matrix become the spinors when multiplying by the rest spinors from the previous problem.

The matrix for $\hat{\mathbf{p}}$ is the same as that of $\hat{\mathbf{p}}^+$ except that $\alpha/2$ is replaced by α. After multiplying the 4×4 matrix on the spinors (this is preferably done using analytical mathematical codes), one finds

$$\hat{\mathbf{p}}|\hat{\mathbf{p}}_\sigma^\pm\rangle = \pm|\hat{\mathbf{p}}_\sigma^\pm\rangle, \tag{A.99}$$

with $\sigma = \uparrow, \downarrow$. This reproduces the results in Eq. (2.97).

The four-momenta can be calculated using Eq. (2.121),

$$\mathsf{p}' = mc\sum_{\mu=0}^{3}\langle\hat{\mathbf{p}}_{\pm\frac{1}{2}}^\pm|\mathbf{e}'_\mu|\hat{\mathbf{p}}_{\pm\frac{1}{2}}^\pm\rangle\mathbf{e}'^\mu. \tag{A.100}$$

For $\langle\hat{\mathbf{p}}_{\pm\frac{1}{2}}^\pm|$, one has to take the transpose and the conjugate. For example,

$$\langle\hat{\mathbf{p}}_\uparrow^+| = \left(\cosh\frac{\alpha}{2}, 0, -\sinh\frac{\alpha}{2}\cos\theta, -\sinh\frac{\alpha}{2}e^{-i\varphi}\sin\theta\right), \tag{A.101}$$

where $|\hat{\mathbf{p}}^+\rangle$ is given by the first column in Eq. (A.98). After some more hard work, the expectation values are given by

$$\begin{aligned} \mathsf{p}_0 &= mc\left\langle\hat{\mathbf{p}}_{\pm\frac{1}{2}}^\pm|\mathbf{e}'_0|\hat{\mathbf{p}}_{\pm\frac{1}{2}}^\pm\right\rangle = mc\cosh\alpha, \\ \mathsf{p}_x &= mc\left\langle\hat{\mathbf{p}}_{\pm\frac{1}{2}}^\pm|\mathbf{e}'_x|\hat{\mathbf{p}}_{\pm\frac{1}{2}}^\pm\right\rangle = -mc\sinh\alpha\sin\theta\cos\varphi, \\ \mathsf{p}_y &= mc\left\langle\hat{\mathbf{p}}_{\pm\frac{1}{2}}^\pm|\mathbf{e}'_y|\hat{\mathbf{p}}_{\pm\frac{1}{2}}^\pm\right\rangle = -mc\sinh\alpha\sin\theta\sin\varphi, \\ \mathsf{p}_z &= mc\left\langle\hat{\mathbf{p}}_{\pm\frac{1}{2}}^\pm|\mathbf{e}'_z|\hat{\mathbf{p}}_{\pm\frac{1}{2}}^\pm\right\rangle = -mc\sinh\alpha\cos\theta, \end{aligned} \tag{A.102}$$

which are indeed the components of the covariant four-momentum. This is the matrix expression of Eq. (2.98).

3 The matrix for $\hat{\mathbf{p}}_\uparrow^+$ is given in Eq. (A.98). Matrix multiplication and using hyperbolic double-angle formulas (or letting analytical mathematical software do the work) gives

$$\hat{\mathbf{p}}_\uparrow^+\hat{\mathbf{p}}_\uparrow^+\mathbf{e}'_0 = \begin{pmatrix} c & 0 & -sp_z & -sp_{-1} \\ 0 & c & -sp_1 & sp_z \\ sp_z & sp_{-1} & -c & 0 \\ sp_1 & -sp_z & 0 & -c \end{pmatrix}$$

$$= \cosh\alpha\ \mathbf{e}'_0 + \sinh\alpha(\sin\theta\cos\varphi\mathbf{e}'_x + \sin\theta\sin\varphi\mathbf{e}'_y + \cos\theta\mathbf{e}'_z) = \hat{\mathbf{p}}, \tag{A.103}$$

with the shorthands $c = \cosh\alpha$ and $s = \sinh\alpha$. However, note that $\hat{\mathbf{p}}_\uparrow^+\hat{\mathbf{p}}_\uparrow^+ = e^{\hat{\mathbf{p}}\frac{\alpha}{2}}e^{\hat{\mathbf{p}}\frac{\alpha}{2}} = e^{\hat{\mathbf{p}}\alpha}$ does exactly the same thing as the matrix multiplications. The down spinor can be obtained using

$$\hat{\mathbf{p}}_{\uparrow}^{-} = \hat{\mathbf{p}}\hat{\mathbf{p}}_{\uparrow}^{+} = (p_x\mathbf{e}_x + p_y\mathbf{e}_y + p_z\mathbf{e}_z)\hat{\mathbf{p}}_{\uparrow}^{+}$$

$$= \begin{pmatrix} 0 & 0 & p_z & p_{-1} \\ 0 & 0 & -p_1 & p_z \\ p_z & p_{-1} & 0 & 0 \\ p_1 & -p_z & 0 & 0 \end{pmatrix} \hat{\mathbf{p}}_{\uparrow}^{+} = \begin{pmatrix} s & 0 & cp_z & cp_{-1} \\ 0 & s & cp_1 & -cp_z \\ cp_z & cp_{-1} & s & 0 \\ cp_1 & -cp_z & 0 & s \end{pmatrix}. \tag{A.104}$$

Note that this operation is the same as $\hat{\mathbf{p}}_{\uparrow}^{-} = \hat{\mathbf{p}}\hat{\mathbf{p}}_{\uparrow}^{+} = \hat{\mathbf{p}}(c + \hat{\mathbf{p}}s) = (s + \hat{\mathbf{p}}c)$. After some more matrix multiplications, we find $\hat{\mathbf{p}}_{\uparrow}^{-}\hat{\mathbf{p}}_{\uparrow}^{-}\mathbf{e}_0' = \hat{\mathbf{p}}$.

Now that we have found all the matrices, we can additionally multiply them to find $\hat{\mathbf{p}}\hat{\mathbf{p}}_{\uparrow}^{\pm} = \pm\hat{\mathbf{p}}^{\pm}\mathbf{e}_0'$. Instead of matrix multiplication, this can also be obtained using $\hat{\mathbf{p}}\hat{\mathbf{p}}_{\uparrow}^{+} = e^{\hat{\mathbf{p}}\alpha}\mathbf{e}_0' e^{\hat{\mathbf{p}}\frac{\alpha}{2}} = e^{\hat{\mathbf{p}}\alpha}e^{-\hat{\mathbf{p}}\frac{\alpha}{2}}\mathbf{e}_0' = e^{\hat{\mathbf{p}}\frac{\alpha}{2}}\mathbf{e}_0' = \hat{\mathbf{p}}^{+}\mathbf{e}_0'$ and $\hat{\mathbf{p}}\hat{\mathbf{p}}_{\uparrow}^{-} = e^{\hat{\mathbf{p}}\alpha}\mathbf{e}_0'\hat{\mathbf{p}}e^{\hat{\mathbf{p}}\frac{\alpha}{2}} = -\hat{\mathbf{p}}e^{\hat{\mathbf{p}}\alpha}e^{-\hat{\mathbf{p}}\frac{\alpha}{2}}\mathbf{e}_0' = -\hat{\mathbf{p}}e^{\hat{\mathbf{p}}\frac{\alpha}{2}}\mathbf{e}_0' = -\hat{\mathbf{p}}^{-}\mathbf{e}_0'$, using the fact that $\hat{\mathbf{p}}$ commutes with itself and anticommutes with $\mathbf{e}_0'$.

Problems for Section 2.6

1 The electric field of a large plate capacitor is given by $\mathbf{E}_i = \frac{\sigma}{\varepsilon_0}\mathbf{e}_i$, where the electric field is perpendicular to the plate, and σ is the surface charge density. For E_x, the length contraction is in the x-direction, which changes the distance x between the plates. However, since the electric field does not depend on the distance $E_x' = E_x$. For E_y, the contraction changes the area of the plate from $A = l_x l_z$ to $A' = (l_x/\gamma)l_z = A/\gamma$, with $\gamma = 1/\sqrt{1 - v^2/c^2}$. This changes the charge density to $\sigma' = Q/A' = \gamma Q/A = \gamma\sigma$, where Q is the charge on the plate. The electric field then becomes $E_y' = \gamma\sigma/\varepsilon_0 = \gamma E_y$. This is consistent with the Lorentz boost.

2 The original electromagnetic field tensor is $\mathsf{F} = \boldsymbol{i}\mathbf{B}$. Since the Lorentz boost is perpendicular to the magnetic field, the boosted tensor is

$$\begin{aligned} \mathsf{F}' &= e^{\mathbf{e}_x\alpha}\mathsf{F} = \boldsymbol{i}\cosh\alpha\mathbf{B} + \boldsymbol{i}\sinh\alpha(B_y\mathbf{e}_x\mathbf{e}_y + B_z\mathbf{e}_x\mathbf{e}_z) \\ &= \boldsymbol{i}\gamma\mathbf{B} + \frac{\gamma v}{c}(-B_y\mathbf{e}_z + B_z\mathbf{e}_y) = \frac{\mathbf{E}'}{c} + \boldsymbol{i}\mathbf{B}', \end{aligned} \tag{A.105}$$

giving

$$\mathbf{E}' = -\gamma\mathbf{v} \times \mathbf{B} \quad \text{and} \quad \mathbf{B}' = \gamma\mathbf{B}, \tag{A.106}$$

with $\mathbf{v} = v\mathbf{e}_x$.

3 The electric field causes a motion in the z-direction. The magnetic field in the x-direction forces rotations in the yz-plane. The motion is therefore restricted to the yz-plane, *i.e.* $\mathbf{v} = v_y\mathbf{e}_y + v_z\mathbf{e}_z$. The force due to $\mathbf{B} = B\mathbf{e}_x$ is then proportional to $\mathbf{v} \times \mathbf{B} = -\boldsymbol{i}\mathbf{v}\mathbf{B} = -\boldsymbol{i}v_yB\mathbf{e}_y\mathbf{e}_x - \boldsymbol{i}v_zB\mathbf{e}_z\mathbf{e}_x = -v_yB\mathbf{e}_z + v_zB\mathbf{e}_y$. Newton's equation of motion is then

$$\mathbf{F} = q(E\mathbf{e}_z + v_zB\mathbf{e}_y - v_yB\mathbf{e}_z) = m\mathbf{a} = m(a_y\mathbf{e}_y + a_z\mathbf{e}_z). \tag{A.107}$$

Introducing the cyclotron frequency $\omega_c = qB/m$, this leads to two coupled differential equations:

$$\frac{d^2y}{dt^2} = \omega_c\frac{dz}{dt}, \quad \frac{d^2z}{dt^2} = \omega_c\left(\frac{E}{B} - \frac{dy}{dt}\right). \tag{A.108}$$

Differentiating gives

$$\frac{d^3y}{dt^3} = \omega_c \frac{d^2z}{dt^2} = -\omega_c^2\left(\frac{dy}{dt} - \frac{E}{B}\right) \quad \Rightarrow \quad \frac{d^2u}{dt^2} = -\omega_c^2 u, \tag{A.109}$$

with $u = y - (E/B)t$ and $du/dt = dy/dt - E/B$. This gives $u = a\cos\omega_c t + b\sin\omega_c t$ or

$$y = a\cos\omega_c t + b\sin\omega_c t + \frac{E}{B}t = \frac{E}{B}\left(\frac{1}{\omega_c}\sin\omega_c t + t\right), \tag{A.110}$$

since the boundary conditions give $a = 0$ and $\omega_c(-a + b) + E/B = 0$. Equation (A.108) then gives

$$z = \frac{E}{\omega_c B}(1 - \cos\omega_c t). \tag{A.111}$$

Using $\cos^2\omega_c t + \sin^2\omega_c t = 1$, this can also be rewritten as $(y - R\omega_c t)^2 + (z - R)^2 = R^2$. This is the equation of a circle with radius $R = E/B\omega_c$ whose origin $R\omega_c t\mathbf{e}_y + R\mathbf{e}_z$ moves at a constant velocity $v = \omega_c R = E/B$ in the y direction. The particle therefore moves as a spot on the circumference of a circle rolling in the y-direction with velocity v. Contrary to the expectation, the overall motion of the particle is not in the direction of the electric field.

Problems for Section 2.7

1 In the static limit, there are no currents, so $\mathbf{J}'(\mathbf{r}') = c\rho(\mathbf{r})\mathbf{e}'_0$. For a localized charge, $\mathbf{J}'(\mathbf{r}') = ce\delta(\mathbf{r})\mathbf{e}'_0$, the Fourier transform is $\mathbf{J}'(\mathbf{q}') = ce\mathbf{e}'_0$. Additionally, for a static problem, the time dependence is removed, and the charge density only depends on the position $\mathbf{r}$. In the absence of time, there is no frequency dependence, which reduces the Green's function in Eq. (2.167) to $G^0(\mathbf{q}') = -\mu_0/\mathbf{q}^2$. Equation (2.166) then becomes

$$\mathbf{A}'(\mathbf{q}') = \frac{V(\mathbf{q})}{c}\mathbf{e}'_0 = -G^0(\mathbf{q}')\mathbf{J}'(\mathbf{q}') = \frac{\mu_0}{\mathbf{q}^2}c\rho(\mathbf{q})\mathbf{e}'_0 = \frac{e}{c\varepsilon_0\mathbf{q}^2}\mathbf{e}'_0,$$

using $\rho(\mathbf{q}) = e$ and $\mu_0\varepsilon_0 = 1/c^2$. This gives for the potential

$$V(\mathbf{q}) = \frac{e}{\varepsilon_0\mathbf{q}^2} \quad \Rightarrow \quad \mathbf{q}^2 V(\mathbf{q}) = \frac{e}{\varepsilon_0} \quad \Rightarrow \quad \nabla^2 V(\mathbf{r}) = -\frac{e\delta(\mathbf{r})}{\varepsilon_0},$$

which is Poisson's equation. The potential is then

$$V(\mathbf{q}) = \frac{e}{\varepsilon_0\mathbf{q}^2} \quad \Rightarrow \quad V(\mathbf{r}) = \frac{1}{4\pi\varepsilon_0}\frac{e}{r}, \tag{A.112}$$

where the Fourier transform of the $1/r$ potential is given in Eq. (4.74).

Problems for Section 2.8

1 For $x \leq 0$, the wavefunction is $\psi(x) = e^{ikx} + Re^{-ikx}$. The wavefunction is not normalized since we are mainly interested in the amplitudes of the reflected (R) and transmitted (T) wavefunctions relative to the incoming wavefunction. For $x > 0$, the solution traveling in the positive x-direction is $\psi(x) = Te^{iqx}$, with

$$q^2 = \frac{2m}{\hbar^2}(E - U) = k^2 - \kappa^2, \quad \text{with} \quad \kappa^2 = \frac{2m}{\hbar^2}U. \tag{A.113}$$

If $U > E$, the value of q becomes complex, leading to a damped wave $\psi(x) = e^{-qx}$.

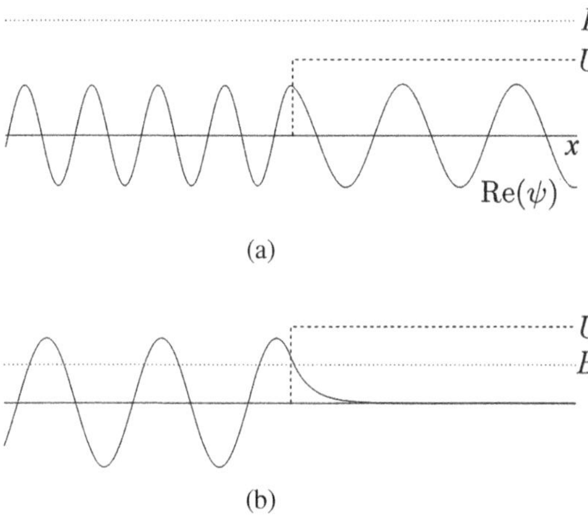

Figure A.6 The real part of the wavefunction $\psi(x)$ as a function of x for different energies. (a) For $E > U$, the particle passes the potential step, causing a change in wavevector. (b) For $E < U$, the wavefunction is evanescent inside the potential step.

The condition that the wavefunction is continuous implies that $1 + R = T$. The continuity of the derivatives gives $ik(1 - R) = iqT$. The coefficients can be straightforwardly solved, giving

$$R = \frac{k-q}{k+q} \quad \text{and} \quad T = \frac{2k}{k+q}, \tag{A.114}$$

where q can be real or complex. When $E > U$, the particle can pass the potential barrier. However, its wavevector decreases, leading to an increase in wavelength, see Figure A.6a. On the other hand, when $E < U$, the particle cannot pass the potential step. Since the barrier height is not infinite, the particle's wavefunction can still penetrate the region with the higher potential, see Figure A.6b. The square of the reflection coefficient is given by

$$|R|^2 = \frac{(k-q)^2}{(k+q)^2} \ (q \text{ is real}), \quad |R|^2 = \frac{k+i|q|}{k-i|q|}\frac{k-i|q|}{k+i|q|} = 1 \ (q \text{ is complex}).$$

We see that if $q \cong k$, which occurs for $E \gg U$, the reflection coefficient goes to zero, and the transmission $|T|^2 = 4k^2/(k+q)^2 \to 1$. Therefore, the particle barely notices the potential step. On the other hand, when $E < U$, q becomes complex, and the reflection coefficient $|R|^2 = 1$. However, $|T|^2$ is not equal to zero. This is because the wavefunction penetrates the barrier, although it decays exponentially.

2 The wavefunction can be split into three parts:

$$\begin{aligned}
\psi(x) &= e^{ikx} + Re^{-ikx} \ \text{ for } \ x < 0, \\
\psi(x) &= Ae^{iqx} + Be^{-iqx} \ \text{ for } \ 0 \le x \le L, \\
\psi(x) &= Te^{ikx} \ \text{ for } \ x > L,
\end{aligned} \tag{A.115}$$

where q is given in Eq. (A.113). The different coefficients can be found by imposing continuity of the wavefunction and its derivative at $x = 0$ and $x = L$, leading to the following set of equations:

$$1 + R = A + B, \quad k(1 - R) = q(A - B), \tag{A.116}$$

$$Ae^{iqL} + Be^{-iqL} = Te^{ikL}, \quad q(Ae^{iqL} - Be^{-iqL}) = kTe^{ikL}. \tag{A.117}$$

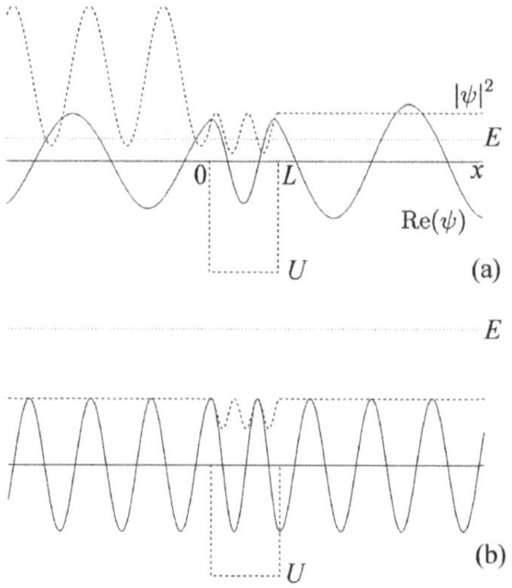

Figure A.7 (a) Typical behavior of the wavefunction $\psi(x)$ when crossing a negative potential well that has a value $U < 0$ from $0 \leq x \leq L$ for two different energy values. The solid line gives the real part of $\psi(x)$; the dotted line gives the square $|\psi(x)|^2$. (b) The wavefunction $\psi(x)$ for a value of q where there is no reflection.

After some cumbersome algebra, we obtain the following expressions for the reflection and the transmission:

$$R = iN(q^2 - k^2)\sin qL \text{ and } T = 2Ne^{-ikL}kq,$$
$$\text{with } N = (2kq\cos qL - i(k^2 + q^2)\sin qL)^{-1} \tag{A.118}$$

Figure A.7a shows the typical behavior of a running wave passing over a potential well with $U < 0$. For $0 \leq x \leq L$, the momentum q increases, see Eq. (A.113), and the wavelength decreases. When looking at $|\psi|^2$, we see that for $x > L$, the transmitted wave is again a running wave with

$$|\psi(x)|^2 = |T|^2 = \left(1 + \left(\frac{k^2 - q^2}{2kq}\right)^2 \sin^2 qL\right)^{-1}. \tag{A.119}$$

For $x < 0$, the incoming and the reflected waves interfere, leading to the (partial) formation of a standing wave. From Eq. (A.119), we see that the transmission becomes 100% when $\sin qL = 0$ or $q = n\pi/L$. At the same time, the reflection becomes zero, see Eq. (A.118). This situation is shown in Figure A.7b. In this case, there are running waves on both sides of the potential well.

For $E < 0$, there are bound states. Symmetry tells us that the bound states are symmetric around the center of the potential well at $x = L/2$. The nonnormalized even and odd solutions are $\psi_+(x) = \cos q(x - \frac{L}{2})$ and $\psi_-(x) = \sin q(x - \frac{L}{2})$ for $0 \leq x \leq L$. Since the potential well is not infinitely deep, the wavefunctions penetrate the potential barrier. Since the bound states exceed the potential well, the value of q is not given by $q = n\pi/L$ as for an infinite potential well. Since $E < 0$, the solutions are damped outside the potential well, and we can write $\psi_\pm(x) = Ce^{\kappa x}$ for $x < 0$ and $\psi_\pm(x) = \pm Ce^{-\kappa(x-L)}$ for $x > L$, where symmetry imposes that the solutions on the left and right sides of the potential well are even and odd with respect to $x = L/2$. The coefficient C follows from the continuity of the wavefunction and its first derivative. For example, for the even solution, we obtain at $x = 0$

$$C = \cos\left(-q_+\frac{L}{2}\right) \text{ and } \kappa_+ C = -q_+ \sin\left(-q_+\frac{L}{2}\right), \tag{A.120}$$

and for the odd solution

$$C = \sin\left(-q_-\frac{L}{2}\right) \quad \text{and} \quad \kappa\_C = q_- \cos\left(-q_-\frac{L}{2}\right), \tag{A.121}$$

giving the relations $\kappa_+ = q_+ \tan q_+ \frac{L}{2}$ and $\kappa_- = -q_- \cot q_- \frac{L}{2}$. We prefer to express the solutions in terms of the energy $\kappa^2 = -2mE/\hbar^2$ $q^2 = 2m(E-U)/\hbar^2$. This allows us to rewrite the condition for $\kappa_\pm$ into an equation for the value of q,

$$\frac{1}{q_+}\sqrt{\frac{2m}{\hbar^2}|U| - q_+^2} = \tan q_+\frac{L}{2}, \quad \frac{1}{q_-}\sqrt{\frac{2m}{\hbar^2}|U| - q_-^2} = -\cot q_-\frac{L}{2}.$$

The tangent diverges for $q_+L/2 = (2n'+1)\pi/2$ or $q_+ = (2n'+1)\pi/L$, with $n' = 0, 1, 2, \ldots$ If $\hbar^2 q_+^2/2m \ll |U|$, the function on the left-hand side crosses the tangent approximately at the q values where the tangent diverges. For the cotangent, the corresponding q_- values are $q_- = 2n'\pi/L$. Therefore, solutions are obtained for $q = n\pi/L$, with $n = 1, 2, 3, \ldots$ When the energy increases, the nth value for q becomes smaller but stays in the region $(n-1)\pi/L < q < n\pi/L$. Since the argument of the square root has to be larger than zero, there is a finite number of bound states, or, equivalently, there are no bound states for $E > 0$. The cutoff for the momentum is given by $q^2 = 2mU/\hbar^2$. Once we have determined the value of $q_\pm$, we can obtain the values of $\kappa_\pm$ and $C_\pm$. Some typical bound states are shown in Figure A.8.

3 The solution for the potential well directly allows us to write down that of the potential barrier when $E < U$ and $U > 0$. In this case, damping of the wavefunction occurs in the region $0 \leq x \leq L$. Defining

$$\kappa = \frac{q}{i} = \sqrt{\frac{2m}{\hbar^2}(U-E)}, \tag{A.122}$$

and noting that $\cos i\kappa L = \cosh \kappa L$ and $\sin i\kappa L = i \sinh \kappa L$, we can rewrite Eq. (A.118) as

$$R = N(\kappa^2 + k^2)\sinh \kappa L \quad \text{and} \quad T = 2iNe^{-ikL}k\kappa,$$
$$\text{with} \quad N = (2ik\kappa \cosh \kappa L + (k^2 - \kappa^2)\sinh \kappa L)^{-1}. \tag{A.123}$$

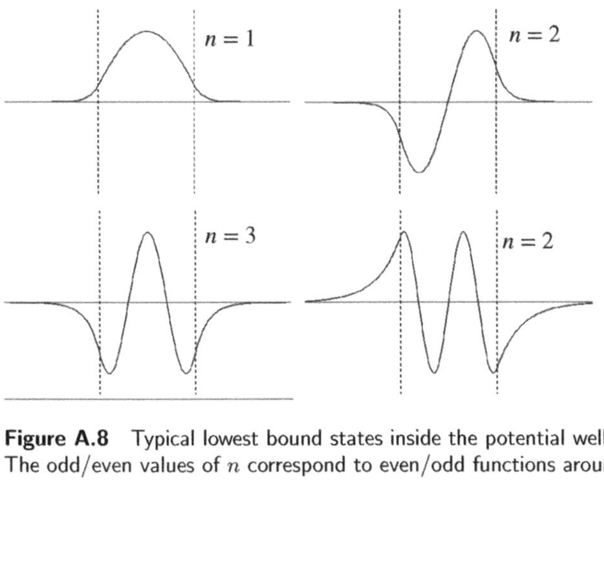

Figure A.8 Typical lowest bound states inside the potential well (indicated by the dashed line). The odd/even values of n correspond to even/odd functions around $x = L/2$, respectively.

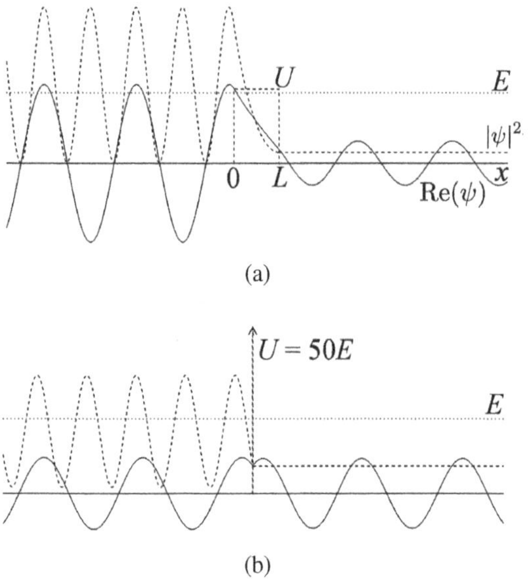

Figure A.9 Typical behavior of the wavefunction $\psi(x)$ when crossing a potential barrier that has a value $U > 0$ from $0 \leq x \leq L$ for two different types of barriers. The solid line gives the real part of $\psi(x)$; the dotted line gives the square $|\psi(x)|^2$. (a) Typical damped behavior through the potential barrier for $E < U$. (b) Phase shift behavior in the limit that $L \to 0$.

A typical situation is shown in Figure A.9a, where only a small portion of the particle's wavefunction passes through the barrier. In analogy to Eq. (A.119), the transmission coefficient can be written as

$$|T|^2 = \left(1 + \left(\frac{k^2+\kappa^2}{2k\kappa}\right)^2 \sinh^2\kappa L\right)^{-1} \cong \left(\frac{2k\kappa}{k^2+\kappa^2}\right)^2 e^{-2\kappa L},$$

where the last term on the right-hand side gives the result in the limit $\kappa L \gg 1$. As expected, the probability of tunneling through a thick barrier is exponentially suppressed. However, we see that the particle can tunnel through the barrier if κ is small compared to $1/L$. From Eq. (A.122), we see that this occurs when the energy E is close to the barrier height U.

We can also study the results in the limit that the width of the potential barrier L is much less than the wavelength $2\pi/k$ of the particle. The hyperbolic functions can then be approximated by $\cosh \kappa L \cong 1$ and $\sinh \kappa L \cong \kappa L$. The transmission coefficient can then be approximated by

$$T \cong \frac{e^{-ikL}}{1 - i\dfrac{k^2-\kappa^2}{2k}L} = \frac{e^{-ikL}}{1 + i\dfrac{mUL}{\hbar^2 k}}. \tag{A.124}$$

In the limit $L \to 0$, we have

$$T = \frac{1}{1+i\alpha} = \frac{1-i\alpha}{1+\alpha^2} = \frac{e^{i\chi}}{\sqrt{1+\alpha^2}}, \tag{A.125}$$

with $\alpha = mU'/\hbar^2 k$ (with $U' = UL$), and χ is a phase shift at the potential barrier. This solution is the same as that for a δ-function potential barrier $U(x) = U'\delta(x)$. From Figure A.9b, we see that the behavior at the δ-function step appears to be closely described by a phase shift. The phase shift in this limit is determined by Eq. (A.125): $\chi = -\arctan\alpha = -\arctan mU/\hbar^2 k$. The wavefunction therefore continues after the barrier with a phase $e^{ikx+i\chi}$, where $\chi < 0$.

4 The Schrödinger equation for bound states with $E < 0$ becomes

$$-\frac{\hbar^2}{2m}\frac{d^2\psi}{dx^2} = -|E|\psi \quad \Rightarrow \quad \psi(x) = Ae^{\kappa x} + Be^{-\kappa x}, \tag{A.126}$$

with $\kappa = \sqrt{2m|E|}/\hbar$. When removing the divergent solutions, we have $\psi(x) = Ae^{\kappa x}$ for $x < 0$ and $Be^{-\kappa x}$ for $x > 0$. Although the wavefunction is continuous, giving $A = B$, the derivative is not since the potential goes to infinity. Integrating the Schrödinger equation around $x = 0$ gives

$$-\frac{\hbar^2}{2m}\int_{-\epsilon}^{\epsilon} dx\, \frac{d^2\psi(x)}{dx^2} + \int_{-\epsilon}^{\epsilon} dx\, U(x)\psi(x) = E\int_{-\epsilon}^{\epsilon} dx\, \psi(x),$$

where we take the limit $\epsilon \to 0$. The first integral gives the derivative $d\psi/dx$. The integral on the right-hand side goes to zero if the integration interval $\epsilon \to 0$. So, we find for the change in derivative

$$\left.\frac{d\psi(x)}{dx}\right|_{\epsilon} - \left.\frac{d\psi(x)}{dx}\right|_{-\epsilon} = -\frac{2mU_0}{\hbar^2}\int_{-\epsilon}^{\epsilon} dx\, \delta(x)\psi(x) = -\frac{2mU_0}{\hbar^2}\psi(0).$$

For the decaying exponential, the limit $\epsilon \to 0$ gives

$$-A\kappa e^{-\kappa x}\big|_{\epsilon} - A\kappa e^{\kappa x}\big|_{-\epsilon} = -2\kappa A = -\frac{2mU_0}{\hbar^2}A. \tag{A.127}$$

This gives $\kappa = mU_0/\hbar^2$. The energy of the bound state is given by $E = -\hbar^2\kappa^2/2m = -mU_0^2/2\hbar^2$. The wavefunction is normalized by $2A^2\int_0^\infty dx e^{-2\kappa x} = A^2/\kappa = 1$, giving $A = \sqrt{\kappa} = \sqrt{mU_0}/\hbar$. The δ-function well has, regardless of the strength U_0, a single bound state,

$$\psi(x) = \sqrt{\kappa}\exp\left(-\kappa|x|\right), \quad E = -\frac{\hbar^2\kappa^2}{2m}. \tag{A.128}$$

Problems for Section 2.9

1 For $\mathbf{B} = B\mathbf{e}_z$, the spin aligns itself in the z-direction. Using Eq. (2.213),

$$H' = \mu_B(\mathbf{B}\cdot\mathbf{L} \pm B) = \mu_B B(\mathbf{L}_z \pm 1). \tag{A.129}$$

Working on the spherical harmonics gives

$$H'|lm\rangle = \mu_B B(\mathbf{L}_z \pm 1)|lm\rangle = \mu_B B(m \pm 1)|lm\rangle. \tag{A.130}$$

The splittings for $l = 1$ are $2, 1, 0, 0, -1, -2$ in units $\mu_B B$ for $2\uparrow, 0\uparrow, 1\downarrow, -1, \uparrow, 0\downarrow, -1\downarrow$. For typical atomic problems, this splitting is rather small, since $\mu_B = 5.788 \times 10^{-5}$ eV/T, and a large magnetic field is of the order of 1–10 T. The Zeeman splitting is therefore often smaller or comparable in size to other interactions, such as the spin–orbit interaction.

Problems for Section 2.11

1 The spin–orbit interaction can be written as $\mathbf{L}\cdot\mathbf{S} = -\mathbf{L}_1\mathbf{S}_{-1} + \mathbf{L}_0\mathbf{S}_0 - \mathbf{L}_{-1}\mathbf{S}_1$, taking the strength of the coupling $\zeta = 1$. The diagonal matrix elements are $\langle m\sigma|\mathbf{L}_0\mathbf{S}_0|m\sigma\rangle = m\sigma$. The values of the nonzero off-diagonal matrix elements are, using Eqs. (1.371) and (1.372) and $\mathbf{L}_{\pm 1} = \mp\mathbf{L}_\pm/\sqrt{2}$, $-\langle 0\uparrow|\mathbf{L}_{-1}\mathbf{S}_1|1\downarrow\rangle =$

$\frac{1}{2}\sqrt{(1+1)(1-1+1)}\sqrt{(\frac{1}{2}-(-\frac{1}{2}))(\frac{1}{2}+(-\frac{1}{2})+1)} = \frac{1}{\sqrt{2}}$ in units of $\hbar^2$. The matrix is then

$$
\boldsymbol{H}''_{\mathrm{SOI}} = \begin{pmatrix} \frac{1}{2} & & & & & \\ & -\frac{1}{2} & \frac{1}{\sqrt{2}} & & & \\ & \frac{1}{\sqrt{2}} & 0 & & & \\ & & & 0 & \frac{1}{\sqrt{2}} & \\ & & & \frac{1}{\sqrt{2}} & -\frac{1}{2} & \\ & & & & & \frac{1}{2} \end{pmatrix} \begin{matrix} 1\uparrow \\ 1\downarrow \\ 0\uparrow \\ 0\downarrow \\ -1\uparrow \\ -1\downarrow \end{matrix}, \tag{A.131}
$$

where only nonzero matrix elements are indicated. This matrix can be solved numerically or by recognizing that it consists of 1×1 and 2×2 matrices. The eigenvalues are $\frac{1}{2}$ $(4\times)$, with eigenvectors $|\frac{3}{2}, \pm\frac{3}{2}\rangle = |\pm 1, \pm\frac{1}{2}\rangle$ and $|\frac{3}{2}, \pm\frac{1}{2}\rangle = \sqrt{\frac{1}{3}}|\pm 1, \mp\frac{1}{2}\rangle + \sqrt{\frac{2}{3}}|0, \pm\frac{1}{2}\rangle$, and -1 $(2\times)$, with eigenvectors $|\frac{1}{2}, \pm\frac{1}{2}\rangle = -\sqrt{\frac{2}{3}}|\pm 1, \mp\frac{1}{2}\rangle + \sqrt{\frac{1}{3}}|0, \pm\frac{1}{2}\rangle$. The eigenvalues are consistent with Eq. (2.245). The eigenvectors are the same as those found using step operators in the examples in Section 2.11.

Problems for Section 2.12

1 a) The time dependence contains three terms:

$$
\frac{d\langle \hat{M}\rangle}{dt} = \int d\mathbf{r}\, \frac{\partial \psi^*}{\partial t}\hat{M}\psi + \int d\mathbf{r}\, \psi^* \hat{M}\frac{\partial \psi}{\partial t} + \int d\mathbf{r}\, \psi^* \frac{\partial \hat{M}}{\partial t}\psi. \tag{A.132}
$$

The time dependence of the wavefunction is given by the Schrödinger equation,

$$
\hat{H}\psi = i\hbar\frac{\partial \psi}{\partial t} \quad \Rightarrow \quad \hat{H}\psi^* = -i\hbar\frac{\partial \psi^*}{\partial t}, \tag{A.133}
$$

where $\hat{H}^* = \hat{H}$, since the Hamiltonian is Hermitian. This gives

$$
\frac{d\langle \hat{M}\rangle}{dt} = \frac{i}{\hbar}\langle[\hat{H}, \hat{M}]\rangle + \left\langle \frac{\partial \hat{M}}{\partial t} \right\rangle. \tag{A.134}
$$

This is known as the Ehrenfest theorem.

b) Since the position operator is time independent, the Ehrenfest theorem reduces to

$$
\frac{d\langle \hat{\mathbf{r}}\rangle}{dt} = \frac{i}{\hbar}\langle[\hat{H}, \hat{\mathbf{r}}]\rangle = \frac{i}{\hbar}\langle[\frac{\hat{\mathbf{p}}^2}{2m}, \hat{\mathbf{r}}]\rangle. \tag{A.135}
$$

Since operators commute with themselves, the potential, being a function of the position, commutes with the position operator. The commutator can be evaluated using

$$
[\hat{\mathbf{p}}^2, \hat{\mathbf{r}}] = \hat{\mathbf{p}}[\hat{\mathbf{p}}, \hat{\mathbf{r}}] + [\hat{\mathbf{p}}, \hat{\mathbf{r}}]\hat{\mathbf{p}} = -2i\hbar\hat{\mathbf{p}}. \tag{A.136}
$$

This gives for the time dependence

$$
\frac{d\langle \hat{\mathbf{r}}\rangle}{dt} = \frac{\langle \hat{\mathbf{p}}\rangle}{m}. \tag{A.137}
$$

c) For the momentum, the Ehrenfest theorem gives

$$\frac{d\langle \hat{\mathbf{p}} \rangle}{dt} = \frac{i}{\hbar}\langle [\hat{H}, \hat{\mathbf{p}}] \rangle. \tag{A.138}$$

Since any operator commutes with itself, the kinetic energy drops out, and we are left with the commutator of the potential energy,

$$[\hat{H}, \hat{\mathbf{p}}] = (U(\mathbf{r})\hat{\mathbf{p}} - \hat{\mathbf{p}}U(\mathbf{r})) = i\hbar\nabla U(\mathbf{r}), \tag{A.139}$$

using $\hat{\mathbf{p}} = -i\hbar\nabla$. The equation of motion then becomes

$$\frac{d\langle \hat{\mathbf{p}} \rangle}{dt} = -\langle \nabla U \rangle. \tag{A.140}$$

d) The gradient of the potential can be related to the force $\mathbf{F} = -\nabla U$. This gives

$$\frac{d\langle \hat{\mathbf{p}} \rangle}{dt} = \langle \mathbf{F} \rangle. \tag{A.141}$$

Combining with the equation of the position, we find

$$\frac{d^2\langle \mathbf{r} \rangle}{dt^2} = \langle \mathbf{F} \rangle, \tag{A.142}$$

which is Newton's equation of motion.

2 $d\mathbf{S}/dt$ can be obtained from Heisenberg's equation of motion

$$\frac{d\mathbf{S}}{dt} = \frac{1}{i\hbar}[\mathbf{S}, \hat{H}] = -\frac{ge}{2im\hbar}[\mathbf{S}, \mathbf{S}\cdot\mathbf{B}]. \tag{A.143}$$

We can split the commutator into its components

$$[\mathbf{S}, \mathbf{S}\cdot\mathbf{B}] = \sum_{i=x,y,z} e_i[S_i, \mathbf{S}\cdot\mathbf{B}]. \tag{A.144}$$

The commutators can be evaluated for each of the components. For example, for the x component, we have

$$\begin{aligned}[S_x, \mathbf{S}\cdot\mathbf{B}] &= [S_x, S_x]B_x + [S_x, S_y]B_y + [S_x, S_z]B_z \\ &= i\hbar(S_zB_y - S_yB_z) = -i\hbar(\mathbf{S}\times\mathbf{B})_x,\end{aligned} \tag{A.145}$$

using the commutation relation for the angular momentum. The other components follow directly from cyclic permutation, and we obtain

$$[\mathbf{S}, \mathbf{S}\cdot\mathbf{B}] = -i\hbar\mathbf{S}\times\mathbf{B} \quad \Rightarrow \quad \frac{d\mathbf{S}}{dt} = \frac{ge}{2m}\mathbf{S}\times\mathbf{B}. \tag{A.146}$$

(b) For $\mathbf{B}$ in the y-direction, $dS_y/dt = 0$. The other two components are given by

$$\frac{dS_x}{dt} = -\omega_B S_z \quad \text{and} \quad \frac{dS_z}{dt} = \omega_B S_x, \tag{A.147}$$

with $\omega_B = geB/2m \cong eB/m$. These equations can be combined to give

$$\frac{d^2S_z}{dt^2} = -\omega_B^2 S_z. \tag{A.148}$$

The solution is given by

$$S_z(t) = a_z \cos\omega_B t + b_z \sin\omega_B t. \tag{A.149}$$

We have a similar solution for S_x. At $t = 0$, we have $S_z(0) = a_z$. The coefficient b_z can be related to $S_x(0)$ using $dS_z/dt = \omega_B S_x$,

$$-a_z\omega_B \sin\omega_B t + b_z\omega_B \cos\omega_B t = \omega_B(a_x \cos\omega_B t + b_x \sin\omega_B t)$$
$$\Rightarrow\ b_z = a_x = S_x(0) \text{ for } t = 0. \tag{A.150}$$

This gives for S_z and S_x:

$$S_z(t) = S_z(0)\cos\omega_B t + S_x(0)\sin\omega_B t, \tag{A.151}$$

$$S_x(t) = \frac{1}{\omega_B}\frac{dS_z}{dt} = S_x(0)\cos\omega_B t - S_z(0)\sin\omega_B t. \tag{A.152}$$

The y component remains unchanged, *i.e.* $S_y(t) = S_y(0)$. The spin therefore precesses around the magnetic field.

Problems for Section 3.1

1 Applying the step-up operator on $\varphi_0(\xi)$ gives

$$\hat{a}^\dagger\varphi_0(\xi) = \frac{1}{\sqrt{2}}\left(\xi - \frac{d}{d\xi}\right)\frac{1}{\pi^{\frac{1}{4}}}e^{-\frac{\xi^2}{2}} = \frac{1}{\sqrt{2}\pi^{\frac{1}{4}}}(\xi - (-\xi))e^{-\frac{\xi^2}{2}} = \frac{1}{\sqrt{2}\pi^{\frac{1}{4}}}2\xi e^{-\frac{\xi^2}{2}}.$$

For $\varphi_2(\xi)$, we have

$$\begin{aligned}\varphi_2(\xi) &= \frac{1}{\sqrt{2}}(\hat{a}^\dagger)^2\varphi_0(\xi) = \frac{1}{\sqrt{2}}\hat{a}^\dagger(\hat{a}^\dagger\varphi_0(\xi)) = \frac{1}{\sqrt{2}}\hat{a}^\dagger\varphi_1(\xi)\\ &= \frac{1}{2}\left(\xi - \frac{d}{d\xi}\right)\sqrt{2}\frac{1}{\pi^{\frac{1}{4}}}\xi e^{-\frac{\xi^2}{2}} = \frac{1}{\sqrt{2\sqrt{\pi}}}\left(\xi^2 e^{-\frac{\xi^2}{2}} - e^{-\frac{\xi^2}{2}} - \xi e^{-\frac{\xi^2}{2}}(-\xi)\right)\\ &= \frac{1}{2\sqrt{2\sqrt{\pi}}}\left(4\xi^2 - 2\right)e^{-\frac{\xi^2}{2}}.\end{aligned} \tag{A.153}$$

2 Inserting $\varphi_(\xi)$ into Eq. (3.3) and performing the derivatives gives

$$\frac{d^2H_n(\xi)}{d\xi^2} - 2\xi\frac{dH_n(\xi)}{d\xi} + (\lambda_n - 1)H_n(\xi) = 0, \tag{A.154}$$

$\lambda_n \equiv 2E_n/\hbar\omega_0$. For $n = 0$, we have $H_0(\xi) = c_0$, where c_0 is a constant. Inserting in Eq. (A.154) gives

$$(\lambda_0 - 1)H_0(\xi) = (\lambda_0 - 1)c_0 = 0\ \Rightarrow\ \lambda_0 = 1\ \Rightarrow\ E_0 = \frac{\lambda_0}{2}\hbar\omega_0 = \frac{1}{2}\hbar\omega_0.$$

Note that there is freedom to choose c_0. In fact, there are different choices for the normalization of the Hermite polynomials. The physicist's convention is to take $c_n = 2^n$ for a particular n. This gives $c_0 = 1$ and $H_0(\xi) = 1$. For $H_1(\xi) = c_1\xi + c_0$, we have

$$-2\xi c_1 + (\lambda_1 - 1)(c_1\xi + c_0) = 0\ \Rightarrow\ (\lambda_1 - 3)c_1\xi + (\lambda_1 - 1)c_0 = 0.$$

The first term gives $\lambda_1 = 3$ or $E_1 = \frac{3}{2}\hbar\omega_0$. However, in that case, $c_0 = 0$ for the second term to disappear. In fact, the second term is related to $H_0(\xi)$. So, we obtain $H_1(\xi) = c_1\xi = 2\xi$. Note that the eigenenergies are not continuous but separated by $\hbar\omega_0$.

For $H_2(\xi) = c_2\xi^2 + c_1\xi + c_0$, we have $dH_2(\xi)/d\xi = 2c_2\xi + c_1$ and $d^2H_2(\xi)/d\xi^2 = 2c_2$. Inserting gives

$$2c_2 - 2\xi(2c_2\xi + c_1) + (\lambda_2 - 1)(c_2\xi^2 + c_1\xi + c_0) = 0. \tag{A.155}$$

Rearranging into independent terms gives

$$(-4 + (\lambda_2 - 1))\,c_2\xi^2 + (-2 + (\lambda_2 - 1))\,c_1\xi + (2c_2 + (\lambda_2 - 1)c_0) = 0.$$

Since we want to keep the ξ^2 term, this implies that $\lambda_2 = 5$ or $E_2 = \frac{5}{2}\hbar\omega_0$. This differs again by $\hbar\omega_0$ from E_1. However, since $-2 + \lambda_2 - 1 = 2$, this means $c_1 = 0$. The third term gives the relation $c_0 = -c_2/2$. Therefore, the polynomial can be written as $H_2(\xi) = a(2\xi^2 - 1) = 4\xi^2 - 2$ following convention. This agrees with the solutions for the preceding problem.

3 For $n = 3$, we have

$$|3\rangle = \frac{1}{\sqrt{3!}}(\boldsymbol{a}^\dagger)^3|0\rangle = \frac{1}{\sqrt{6}}(\boldsymbol{a}^\dagger)^3|0\rangle. \tag{A.156}$$

Normalization follows from

$$\langle 3|3\rangle = \frac{1}{3!}\langle 0|\boldsymbol{a}^3(\boldsymbol{a}^\dagger)^3|0\rangle = \frac{1}{2!}\langle 0|\boldsymbol{a}^2(\boldsymbol{a}^\dagger)^2|0\rangle = \langle 0|\boldsymbol{a}(\boldsymbol{a}^\dagger)|0\rangle = 1. \tag{A.157}$$

We find the following results: $\boldsymbol{a}^\dagger|3\rangle = \sqrt{3+1}|4\rangle = 2|4\rangle$, $\boldsymbol{a}|3\rangle = \sqrt{3}|2\rangle$, and $\boldsymbol{a}^\dagger\boldsymbol{a}|3\rangle = \sqrt{3}\boldsymbol{a}^\dagger|2\rangle = \sqrt{3}\sqrt{3}|2\rangle = 3|2\rangle$. The expectation value of $\boldsymbol{\xi}$ is

$$\frac{1}{\sqrt{2}}\langle 3|\boldsymbol{a} + \boldsymbol{a}^\dagger|3\rangle = \frac{1}{\sqrt{2}}\left(\sqrt{3}\langle 3|2\rangle + \sqrt{4}\langle 3|4\rangle\right) = 0, \tag{A.158}$$

and of $\boldsymbol{\xi}^2$

$$\begin{aligned}\frac{1}{2}\langle 3|(\boldsymbol{a} + \boldsymbol{a}^\dagger)(\boldsymbol{a} + \boldsymbol{a}^\dagger)|3\rangle &= \frac{1}{2}\langle 3|(\boldsymbol{a}^\dagger\boldsymbol{a}^\dagger + \boldsymbol{a}^\dagger\boldsymbol{a} + \boldsymbol{a}\boldsymbol{a}^\dagger + \boldsymbol{a}\boldsymbol{a})|3\rangle \\ &= \frac{1}{2}(\sqrt{5\times 4}\langle 3|5\rangle + \sqrt{3\times 3}\langle 3|3\rangle + \sqrt{4\times 4}\langle 3|3\rangle + \sqrt{2\times 3}\langle 3|1\rangle) = \frac{7}{2}.\end{aligned}$$

Problems for Section 3.2

1 The interaction depends on the distance $x_1 - x_2$. It is therefore convenient to define the following new coordinates:

$$x_\pm = \frac{1}{\sqrt{2}}(x_1 \pm x_2) \quad \text{and} \quad \hat{p}_\pm = \frac{1}{\sqrt{2}}(\hat{p}_1 \pm \hat{p}_2). \tag{A.159}$$

Alternatively, we can view this as separating the motion of the center of mass and the relative motion of the two oscillators. The original coordinates can be expressed in terms of the rotated coordinates as

$$x_{1/2} = \frac{1}{\sqrt{2}}(x_+ \pm x_-) \quad \text{and} \quad \hat{p}_{1/2} = \frac{1}{\sqrt{2}}(\hat{p}_+ \pm \hat{p}_-). \tag{A.160}$$

The Hamiltonian for two quantum harmonic oscillators coupled via U then becomes

$$\frac{\hat{p}_+^2}{2m} + \frac{1}{2}m\omega_0^2x_+^2 + \frac{\hat{p}_-^2}{2m} + \frac{1}{2}m(\omega_0^2 + 2\omega_1^2)x_-^2. \tag{A.161}$$

These are two independent harmonic oscillators. The energies are therefore given by

$$E_{n_+n_-} = \left(n_+ + \frac{1}{2}\right)\hbar\omega_+ + \left(n_- + \frac{1}{2}\right)\hbar\omega_-, \quad \text{with } \omega_\pm = \omega_0, \sqrt{\omega_0^2 + 2\omega_1^2}.$$

Note that the solutions are the same as the classical problem in Eq. (1.166). Since the oscillators in the rotated system are independent, we can use a separation of variables $\varphi_{n_+n_-}(\xi_+,\xi_-) = \varphi_{n_+}(\xi_+)\varphi_{n_-}(\xi_-)$, introducing the reduced variables $\xi_\pm = \sqrt{m\omega_\pm/\hbar}x_\pm$. This gives a total wavefunction

$$\varphi_{n_+n_-}(\xi_+,\xi_-) = \frac{1}{\sqrt{2^{n_-+n_+}\pi n_+!n_-!}} H_{n_+}(\xi_+)H_{n_-}(\xi_-)e^{-\frac{1}{2}(\xi_+^2+\xi_-^2)}.$$

2 The displaced wavefunction can be expanded in a series as

$$\varphi_0(x-x_0) = 1 - x_0\frac{d\varphi_0}{dx} + \frac{x_0^2}{2!}\frac{d^2\varphi_0}{dx^2} + \cdots = e^{-x_0\frac{d}{dx}}\varphi_0(x) = e^{-\frac{i}{\hbar}x_0\hat{p}_x}\varphi_0(x),$$

with $\hat{p}_x = -i\hbar d/dx$. We can write the momentum operator in terms of step operators as $\hat{p}_x = i\sqrt{m\hbar\omega_0/2}(\hat{a}^\dagger - \hat{a})$. Inserting this gives

$$\varphi_0(x-x_0) = e^{x_0\sqrt{\frac{m\omega_0}{2\hbar}}(\hat{a}^\dagger-\hat{a})}\varphi_0(x) = e^{\sqrt{g}(\hat{a}^\dagger-\hat{a})}\varphi_0(x), \tag{A.162}$$

where $g \equiv \frac{1}{2}m\omega_0^2x_0^2/\hbar\omega_0$. The Baker–Campbell–Hausdorff formula for this problem can be written as

$$e^{X+Y} = e^{-\frac{1}{2}[X,Y]}e^Ye^X \quad\Rightarrow\quad e^{\sqrt{g}(\boldsymbol{a}^\dagger-\boldsymbol{a})} = e^{-\frac{g}{2}}e^{\sqrt{g}\boldsymbol{a}^\dagger}e^{-\sqrt{g}\boldsymbol{a}}, \tag{A.163}$$

using $-\frac{1}{2}[\sqrt{g}\boldsymbol{a}^\dagger, -\sqrt{g}\boldsymbol{a}] = -\frac{g}{2}[\boldsymbol{a},\boldsymbol{a}^\dagger] = -g/2$. The switch is made from the operators $\hat{a}^{(\dagger)}$ to the vector operators $\boldsymbol{a}^{(\dagger)}$, which more easily describe the transformation properties of the eigenfunctions. Applying this to the state $|0\rangle$ gives

$$|g\rangle = e^{-\frac{g}{2}}e^{\sqrt{g}\boldsymbol{a}^\dagger}e^{-\sqrt{g}\boldsymbol{a}}|0\rangle = e^{-\frac{g}{2}}\left(1 + \sqrt{g}\boldsymbol{a}^\dagger + \frac{(\sqrt{g}\boldsymbol{a}^\dagger)^2}{2!} + \cdots\right)|0\rangle.$$

Since $e^{-\sqrt{g}\boldsymbol{a}} = (1 - \sqrt{g}\boldsymbol{a} + \cdots)|0\rangle = |0\rangle$, with $\boldsymbol{a}|0\rangle = 0$. Using the expression of the eigenfunctions $|n\rangle = (\boldsymbol{a}^\dagger)^n|0\rangle/\sqrt{n!}$, we can write

$$|g\rangle = e^{-\frac{g}{2}}\left(|0\rangle + \sqrt{g}|1\rangle + \frac{(\sqrt{g})^2}{\sqrt{2!}}|2\rangle + \cdots\right)|0\rangle = e^{-\frac{g}{2}}\sum_{n=0}^{\infty}\frac{g^{\frac{n}{2}}}{\sqrt{n!}}|n\rangle,$$

which reproduces Eq. (3.39).

Problems for Section 3.3

1 Rodrigues formula for $n_r = 2$ gives

$$\begin{aligned} L_2^\nu(\rho) &= \frac{\rho^{-\nu}}{2!}\left(\frac{d}{d\rho} - 1\right)^2\rho^{2+\nu} = \frac{1}{2\rho^\nu}\left(\frac{d}{d\rho} - 1\right)\left((2+\nu)\rho^{1+\nu} - \rho^{2+\nu}\right) \\ &= \frac{1}{2\rho^\nu}\left((2+\nu)(1+\nu)\rho^\nu - (2+\nu)\rho^{1+\nu} - (2+\nu)\rho^{1+\nu} + \rho^{2+\nu}\right) \\ &= \frac{1}{2}\left((2+3\nu+\nu^2) - (4+2\nu)\rho + \rho^2\right). \end{aligned} \tag{A.164}$$

Problems for Section 3.4

1 Using Eq. (3.98), we find for $m_j = \pm\frac{3}{2}$

$$\begin{aligned} \langle\mathbf{r}|\frac{3}{2},\pm\frac{3}{2}\rangle &= \sqrt{\frac{0!}{3!}}(\xi \pm \boldsymbol{i}\eta)^3 L_0^3(\xi^2+\eta^2) \\ &= \frac{1}{\sqrt{6}}\xi^3 \pm \boldsymbol{i}\sqrt{\frac{3}{2}}\xi^2\eta - \sqrt{\frac{3}{2}}\xi\eta^2 \mp \frac{\boldsymbol{i}}{\sqrt{6}}\eta^3, \end{aligned} \tag{A.165}$$

with $L_0^3(\xi^2+\eta^2)=1$. For $m_j=\pm\frac{1}{2}$, the eigenfunctions are

$$\langle \mathbf{r}|\frac{3}{2},\pm\frac{1}{2}\rangle = -\sqrt{\frac{1!}{2!}}(\xi \pm i\eta)^1 L_1^1(\xi^2+\eta^2)$$
$$= \frac{1}{\sqrt{2}}\xi^3 - \sqrt{2}\xi \pm \frac{i}{\sqrt{2}}\xi^2\eta + \frac{1}{\sqrt{2}}\xi\eta^2 \mp i\sqrt{2}\eta \pm \frac{i}{\sqrt{2}}\eta^3,$$

with $L_1^1(\xi^2+\eta^2)=2-\xi^2-\eta^2$. This reproduces the results obtained with the Hermite polynomials in Eq. (3.111).

Problems for Section 3.5

1 a) The higher order correction can be rewritten as

$$-\frac{\hat{\mathbf{p}}^4}{8m^3c^2} = -\frac{1}{2mc^2}\left(\frac{\hat{\mathbf{p}}^2}{2m}\right)^2, \quad \text{with} \quad \frac{\hat{\mathbf{p}}^2}{2m}\varphi_{nlm} = (E_n - U)\varphi_{nlm}.$$

Using $U(r) = -e^2/4\pi\varepsilon_0 r = -2Ra_0/r$, where R is the Rydberg constant and a_0 is the Bohr radius, the expectation value can be written as

$$-\frac{1}{2mc^2}\left\langle \frac{\hat{\mathbf{p}}^2}{2m}\frac{\hat{\mathbf{p}}^2}{2m}\right\rangle = -\frac{1}{2mc^2}(E_n^2 - 2E_n\langle U\rangle + \langle U^2\rangle)$$
$$= -\frac{1}{2mc^2}\left(E_n^2 - 2E_n(-2Ra_0)\left\langle\frac{1}{r}\right\rangle + (-2Ra_0)^2\left\langle\frac{1}{r^2}\right\rangle\right)$$
$$= -\frac{E_n^2}{2mc^2}\left(1 - 4 + \frac{4n}{l+\frac{1}{2}}\right) = E_n\frac{\alpha^2}{n^2}\left(\frac{n}{l+\frac{1}{2}} - \frac{3}{4}\right),$$

with $E_n = -R/n^2$ and using the expectation values from the problem. The final result is rewritten in terms of the fine-structure constant

$$\alpha^2 = \frac{|U(a_0)|}{mc^2} = \frac{2R}{mc^2} = 2\frac{me^4}{2(4\pi\varepsilon_0)^2\hbar^2}\frac{1}{mc^2} = \left(\frac{e^2}{4\pi\varepsilon_0\,\hbar c}\right)^2, \tag{A.166}$$

giving $E_n/mc^2 = -R/n^2mc^2 = -\alpha^2/2n^2$. The fine-structure constant can also be written as

$$\frac{|U(\lambda_C)|}{mc^2} = \frac{e^2}{4\pi\varepsilon_0}\frac{mc}{h}\frac{1}{mc^2} = \frac{\alpha}{2\pi}, \quad \text{with } \alpha \cong \frac{1}{137}, \tag{A.167}$$

where $\lambda_C = h/mc$ is the Compton wavelength. The fine-structure constant therefore gives the ratio of the Coulomb energy at Compton wavelength $\lambda_C = 2.426\times10^{-12}$ m and the rest energy of the electron, and α^2 gives the ratio of the Coulomb at the Bohr radius $a_0 = 5.291\times10^{-11}$ m and the rest energy. This also implies that $\alpha = \lambda/2\pi a_0$.

b) Using Eq. (2.234), the expectation value is given by

$$\langle H''_{\text{Darwin}}\rangle = \frac{e^2\hbar^2}{8\varepsilon_0 m^2c^2}\langle\delta(\mathbf{r})\rangle = \frac{e^2\hbar^2}{8\varepsilon_0 m^2c^2}|\varphi_{nlm}(0)|^2. \tag{A.168}$$

The value of the wavefunction at the nucleus is only finite for $l=0$, giving $\varphi_{nlm}(0) = \sqrt{(2/na_0)^3(n-1)!/(2nn!)}L_{n-1}^1(0)/\sqrt{4\pi} = 2/\sqrt{4\pi(na_0)^3}$ using Eq. (3.128) and $L_{n-1}^1(0) = n$. This gives

$$\langle H''_{\text{Darwin}}\rangle = \frac{e^2\hbar^2}{8\varepsilon_0 m^2c^2a_0}\frac{1}{(a_0n^2)^2}n = \frac{2E_n^2}{mc^2}n = -E_n\frac{\alpha^2}{n^2}n.$$

c)

$$\langle H''_{\text{SOI}}\rangle = \frac{1}{2m^2c^2}\langle\frac{dU}{rdr}\mathbf{L}\cdot\mathbf{S}\rangle = \frac{1}{2m^2c^2}\frac{e^2}{4\pi\varepsilon_0}\left\langle\frac{1}{r^3}\right\rangle\langle\mathbf{L}\cdot\mathbf{S}\rangle.$$

Using the results from Section 2.11 and $\langle 1/r^3\rangle = 1/(na_0)^3 l(l+\frac{1}{2})(l+1)$, we obtain, after some work,

$$\langle H''_{\text{SOI}}\rangle = -E_n\frac{\alpha^2}{n^2}\frac{n}{2}\frac{j(j+1)-l(l+1)-\frac{3}{4}}{l(l+\frac{1}{2})(l+1)}. \tag{A.169}$$

d) With some effort, the results can be combined to give

$$E_{nj} = E_n\left(1+\frac{\alpha^2}{n^2}\left(\frac{n}{j+\frac{1}{2}}-\frac{3}{4}\right)\right). \tag{A.170}$$

The levels for $n=2$ are $2s_{\frac{1}{2}}$, $2p_{\frac{1}{2}}$, and $2p_{\frac{3}{2}}$. Since the energy only depends on j, the levels $2s_{\frac{1}{2}}$ and $2p_{\frac{1}{2}}$ are degenerate and 4.5×10^{-5} eV lower in energy compared to $2p_{\frac{3}{2}}$.

Problems for Section 3.6

1 The integrated intensity can be written as

$$I_q = \int_{-\infty}^{\infty} I_q(\omega)d\hbar\omega = \sum_f |\langle f|\boldsymbol{T}^1_q|i\rangle|^2 = \langle i|(\boldsymbol{T}^1_q)^*\boldsymbol{T}^1_q|i\rangle = -\langle i|\boldsymbol{T}^1_{-q}\boldsymbol{T}^1_q|i\rangle.$$

The difference in intensities is given by

$$I_1 - I_{-1} = \langle i|\boldsymbol{T}^1_1\boldsymbol{T}^1_{-1} - \boldsymbol{T}^1_{-1}\boldsymbol{T}^1_1|i\rangle.$$

The initial and final states are p^6d^n and p^5d^{n+1} configurations, respectively. Therefore, the electron removed from the p level has to be put back again when looking at an expectation value for the initial state. This gives, for example for $\boldsymbol{T}^1_1\boldsymbol{T}^1_{-1}$

$$\begin{pmatrix} \sqrt{6} & 0 & 0 \\ 0 & \sqrt{3} & 0 \\ 0 & 0 & 1 \\ 0 & 0 & 0 \\ 0 & 0 & 0 \end{pmatrix}\begin{pmatrix} \sqrt{6} & 0 & 0 & 0 & 0 \\ 0 & \sqrt{3} & 0 & 0 & 0 \\ 0 & 0 & 1 & 0 & 0 \end{pmatrix} = \begin{pmatrix} 6 & 0 & 0 & 0 & 0 \\ 0 & 3 & 0 & 0 & 0 \\ 0 & 0 & 1 & 0 & 0 \\ 0 & 0 & 0 & 0 & 0 \\ 0 & 0 & 0 & 0 & 0 \end{pmatrix},$$

see Section 3.7. The difference is then given by

$$\boldsymbol{T}^1_1\boldsymbol{T}^1_{-1} - \boldsymbol{T}^1_{-1}\boldsymbol{T}^1_1 = 3\begin{pmatrix} 2 & 0 & 0 & 0 & 0 \\ 0 & 1 & 0 & 0 & 0 \\ 0 & 0 & 0 & 0 & 0 \\ 0 & 0 & 0 & -1 & 0 \\ 0 & 0 & 0 & 0 & -2 \end{pmatrix} = 3\boldsymbol{L}_z. \tag{A.171}$$

The integrated intensity is then given by

$$I_1 - I_{-1} = 3\langle i|\boldsymbol{L}_z|i\rangle. \tag{A.172}$$

This is known as a sum rule.

Problems for Section 3.7

1 The matrix can be written as

$$\boldsymbol{H} = \begin{pmatrix} 0 & 0 & -t_s & -t_s \\ 0 & 0 & -t_p & t_p \\ -t_s & -t_p & 0 & 0 \\ -t_s & t_p & 0 & 0 \end{pmatrix}. \tag{A.173}$$

Solving gives eigenenergies $\mp\sqrt{2}t_s$ and $\mp\sqrt{2}t_p$ for $(|s\rangle \pm |s^+\rangle)/\sqrt{2}$ and $(|p_x\rangle \pm |s^-\rangle)/\sqrt{2}$, respectively, with $|s^\pm\rangle = (|s_1\rangle \pm |s_2\rangle)/\sqrt{2}$. To obtain bonding in one direction, hybrids can be made $|sp^\pm\rangle = (|s\rangle \pm |p_x\rangle)/\sqrt{2}$,

$$\boldsymbol{U} = \begin{pmatrix} \frac{1}{\sqrt{2}} & \frac{1}{\sqrt{2}} & 0 & 0 \\ \frac{1}{\sqrt{2}} & -\frac{1}{\sqrt{2}} & 0 & 0 \\ 0 & 0 & 1 & 0 \\ 0 & 0 & 0 & 1 \end{pmatrix}. \tag{A.174}$$

The Hamiltonian in this basis is

$$\boldsymbol{H'} = \boldsymbol{U}^\dagger \boldsymbol{H} \boldsymbol{U} = \begin{pmatrix} 0 & 0 & -t^+ & -t^- \\ 0 & 0 & -t^- & -t^+ \\ -t^+ & -t^- & 0 & 0 \\ -t^- & -t^+ & 0 & 0 \end{pmatrix}, \tag{A.175}$$

where $\boldsymbol{U}^\dagger = \boldsymbol{U}$, and $t^\pm = (t_s \pm t_p)/\sqrt{2}$. In the limit $t = t_s \cong t_p$, the matrix elements are $t^\pm \cong \sqrt{2}t, 0$, and the hybrids $|sp^\pm\rangle$ couple predominantly in the positive and negative x directions, respectively.

2 a) The periodicity of the lattice is $2a$. The reciprocal lattice vectors are therefore

$$K = 0, \pm\frac{2\pi}{2a}, \pm\frac{4\pi}{2a}, \dots = 0, \pm\frac{\pi}{a}, \pm\frac{2\pi}{a}, \dots \tag{A.176}$$

The Bragg condition in one dimension gives

$$\mathbf{k}\cdot\mathbf{K} = \frac{1}{2}K^2 \quad \Rightarrow \quad k = \frac{1}{2}K = \pm\frac{\pi}{2a}, \tag{A.177}$$

which determines the first Brillouin zone $[-\frac{\pi}{2a}, \frac{\pi}{2a}]$.

b) The matrix element is given by

$$\varepsilon_k = \sum_{\delta=\pm 1} \langle p_{x,i+\delta}|H|s_i\rangle e^{ik\delta a} = t(e^{ika} - e^{-ika}) = 2it\sin ka.$$

Note that the matrix element changes sign in the negative direction due to the sign change of the lobes of the p_x orbital. The matrix is given by

$$H = \begin{pmatrix} 0 & \varepsilon_k \\ \varepsilon_k^* & \Delta \end{pmatrix} = \begin{pmatrix} 0 & 2it\sin ka \\ -2it\sin ka & \Delta \end{pmatrix}. \tag{A.178}$$

The eigenvalues are

$$E_k^\pm = \frac{\Delta}{2} \pm \frac{1}{2}\sqrt{\Delta^2 + 4|\varepsilon_k|^2}. \tag{A.179}$$

c) See Figure A.10.

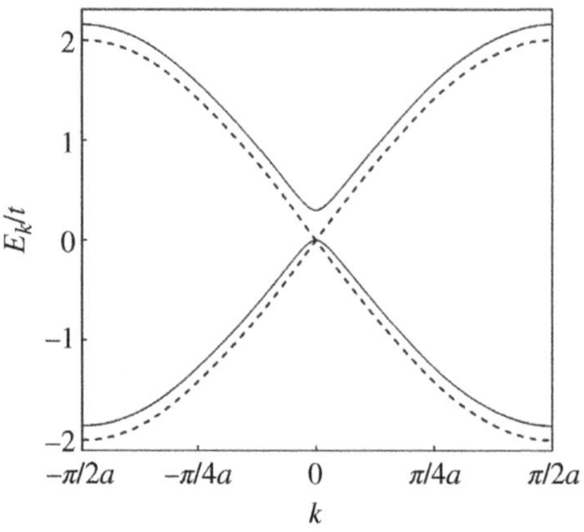

Figure A.10 The tight-binding dispersion for a chain of alternating s and p_x orbitals. The dashed and solid lines give the results for $\Delta/t = 0, 0.3$, respectively.

d) The s and p_x orbitals are arranged as

$$\cdots(-+)(+)(+-)(-)(-+)(+)\cdots, \tag{A.180}$$

where only the signs of the nodes are indicated. Note that the minima are at $k = \pm\pi/2a$. Therefore, the orbitals change sign in each unit cell.

3 The distortion leads to an increase in the periodicity to $2a$. The Brillouin zone then decreases to $[-\frac{\pi}{2a}, \frac{\pi}{2a}]$. There are two inequivalent sites inside a unit cell. The hopping between is given by

$$\varepsilon_{\mathbf{k}} = \sum_{\delta} t_{\delta} e^{-i\mathbf{k}\cdot\boldsymbol{\delta}} = (t + \Delta t)e^{ika'} + (t - \Delta t)e^{-ik(2a-a')}. \tag{A.181}$$

Note that the distance to the next equivalent atom is $a' + (2a - a') = 2a$. Effectively, the atoms are forming dimers in the chain for $a' < a$. The matrix is then

$$\boldsymbol{H} = \begin{pmatrix} 0 & \varepsilon_{\mathbf{k}} \\ \varepsilon^*_{\mathbf{k}} & 0 \end{pmatrix}, \tag{A.182}$$

where the off-diagonal matrix element is complex. After some work, the eigenenergies are found to be

$$E^{\pm}_{\mathbf{k}} = \pm\sqrt{|\varepsilon_{\mathbf{k}}|^2} = \pm\sqrt{2(t^2 + (\Delta t)^2) + 2(t^2 - (\Delta t)^2)\cos 2ka}.$$

For $\Delta t = 0$, this becomes $E^{\pm}_{\mathbf{k}} = \pm t\sqrt{2 + 2\cos 2ka} = \pm 2t\cos ka$. This reproduces the eigenenergies in Eq. (3.166), except that we are still working in the Brillouin zone $[-\frac{\pi}{2a}, \frac{\pi}{2a}]$. This assumes that the two sites are unequal, so we obtain two bands in the Brillouin zone. At the edge of the Brillouin zone, $k = \pi/2a$, giving $E^{\pm}_{\mathbf{k}} = \pm 2\Delta t$. Therefore, the Peierls distortion leads to a gap in the dispersion, see Figure A.11.

Problems for Section 3.8

1 From Eq. (3.182), we know that the eigenfunctions can be written as

$$\psi_{n\mathbf{k}}(\mathbf{r}) = \sum_{\mathbf{K}} c^n_{\mathbf{k}-\mathbf{K}} e^{i(\mathbf{k}-\mathbf{K})\cdot\mathbf{r}}, \tag{A.183}$$

Figure A.11 The band structure for a one-dimensional chain with a Peierls distortion. The dashed and solid lines give the results for $\Delta t/t = 0, 0.2$, respectively.

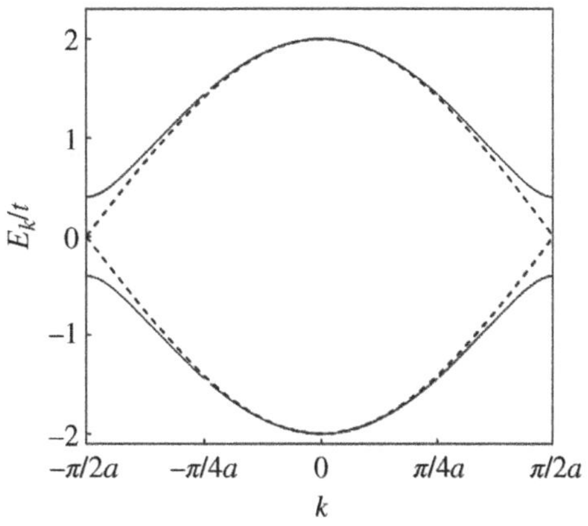

where n labels the different eigenstates for a particular $\mathbf{k}$. Separating the part related to $\mathbf{k}$ gives

$$\psi_{n\mathbf{k}}(\mathbf{r}) = e^{i\mathbf{k}\cdot\mathbf{r}} u_{n\mathbf{k}}(\mathbf{r}), \quad \text{with } u_{n\mathbf{k}}(\mathbf{r}) = \sum_{\mathbf{K}} c^n_{\mathbf{k}-\mathbf{K}} e^{-i\mathbf{K}\cdot\mathbf{r}}. \tag{A.184}$$

This shows that the eigenfunction can be expressed in a plane-wave part $e^{i\mathbf{k}\cdot\mathbf{r}}$ and a function $u_{n\mathbf{k}}(\mathbf{r})$. The latter has the periodicity of the lattice

$$u_{n\mathbf{k}}(\mathbf{r}+\mathbf{R}) = \sum_{\mathbf{K}} c^n_{\mathbf{k}-\mathbf{K}} e^{-i\mathbf{K}\cdot(\mathbf{r}+\mathbf{R})} = u_{n\mathbf{k}}(\mathbf{r}), \tag{A.185}$$

since $e^{-i\mathbf{K}\cdot\mathbf{R}} = 1$. In free space, translation over a vector $\boldsymbol{\tau}$ gives $\psi_{\mathbf{k},\text{free}}(\mathbf{r}+\boldsymbol{\tau}) = e^{i\mathbf{k}\cdot(\mathbf{r}+\boldsymbol{\tau})}/\sqrt{V} = e^{i\mathbf{k}\cdot\boldsymbol{\tau}}\psi_{\mathbf{k},\text{free}}(\mathbf{r})$. Therefore, the only change in the wavefunction is a phase factor. In a solid, there is no longer translational symmetry for any vector $\boldsymbol{\tau}$, but only for the lattice vectors $\mathbf{R}$. However, translation over a lattice vector should only change the wavefunction by a phase factor. For nearly free electron wavefunctions, we have

$$\psi_{n\mathbf{k}}(\mathbf{r}+\mathbf{R}) = e^{i\mathbf{k}\cdot(\mathbf{r}+\mathbf{R})} u_{n\mathbf{k}}(\mathbf{r}+\mathbf{R}) = e^{i\mathbf{k}\cdot\mathbf{R}} e^{i\mathbf{k}\cdot\mathbf{r}} u_{n\mathbf{k}}(\mathbf{r}) = e^{i\mathbf{k}\cdot\mathbf{R}} \psi_{n\mathbf{k}}(\mathbf{r}),$$

using the periodicity of $u_{n\mathbf{k}}(\mathbf{r})$. Therefore, the translation over $\mathbf{R}$ only changes the eigenfunction by a phase factor.

For tight-binding wavefunctions, we find

$$\begin{aligned}\varphi_{\mathbf{k}}(\mathbf{r}+\mathbf{R}') &= \frac{1}{\sqrt{N}} \sum_{\mathbf{R}} e^{i\mathbf{k}\cdot\mathbf{R}} \varphi(\mathbf{r}+\mathbf{R}'-\mathbf{R}) \\ &= \frac{1}{\sqrt{N}} e^{i\mathbf{k}\cdot\mathbf{R}'} \sum_{\mathbf{R}-\mathbf{R}'} e^{i\mathbf{k}\cdot(\mathbf{R}-\mathbf{R}')} \varphi(\mathbf{r}+\mathbf{R}'-\mathbf{R}) = e^{i\mathbf{k}\cdot\mathbf{R}'} \varphi_{\mathbf{k}}(\mathbf{r}).\end{aligned}$$

Therefore, Bloch's theorem also holds in this case.

2 a) The reciprocal lattice vectors are given by $\mathbf{K} = 2\pi(n_x\mathbf{e}_x + n_y\mathbf{e}_y)/a$. The first Brillouin zone is determined by the von Laue conditions in Eq. (1.275). For the smallest reciprocal lattice vectors, this gives

$$\mathbf{k}\cdot\left(\pm\mathbf{e}_x\frac{2\pi}{a}\right) = \frac{1}{2}\left(\frac{2\pi}{a}\right)^2 \quad \Rightarrow \quad k_x = \pm\frac{\pi}{a}. \tag{A.186}$$

Since there is no restriction on k_y, all $\mathbf{k} = \pm\frac{\pi}{a}\mathbf{e}_x + k_y\mathbf{e}_y$ satisfy this condition. Similarly, for the smallest reciprocal lattice vector in the y-direction it gives another line that satisfies the von Laue conditions $\mathbf{k} = k_x\mathbf{e}_x \pm \frac{\pi}{a}\mathbf{e}_y$. Therefore, the unique $\mathbf{k}$ values lie in a square around $\mathbf{k} = 0$.

b) The free electron states inside the first Brillouin zone can be obtained by centering free-electron bands at different $\mathbf{K}$ values,

$$\varepsilon_{\mathbf{k}-\mathbf{K}} = \frac{\hbar^2}{2m}(\mathbf{k} - \mathbf{K})^2. \tag{A.187}$$

One then plots the values inside the first Brillouin zone, as was done in Figure 3.4. The band indicated by 1 in Figure A.12 is centered around $\mathbf{K} = 0$. The bands $2_{\text{I-IV}}$ are related to $\mathbf{K} = \frac{2\pi}{a}\mathbf{e}_x, \frac{2\pi}{a}\mathbf{e}_y, -\frac{2\pi}{a}\mathbf{e}_x, -\frac{2\pi}{a}\mathbf{e}_y$. The band 3_I is related to $\mathbf{K} = \frac{2\pi}{a}(\mathbf{e}_x + \mathbf{e}_y)$. Effectively, the eigenenergies still form a parabola, but all the states have been folded back inside the first Brillouin zone.

c) The amount of reciprocal space occupied by one $\mathbf{k}$ point is $A_0 = (2\pi/L)^2$. For a free-electron model where the energy is given by $\varepsilon_k = \hbar^2k^2/2m$, the lines of constant energy are circles. The number of k-points inside the circle as a function of the energy is

$$N(E) = \frac{\pi k^2}{A_0} = \pi k^2\left(\frac{L}{2\pi}\right)^2 = \frac{A}{4\pi}k^2 = \frac{A}{4\pi}\frac{2m}{\hbar^2}E, \tag{A.188}$$

where A is the area (*i.e.* the size) of the system. The number of states per area $n(E)$ is then

$$n(E) = \frac{N(E)}{A} = \frac{1}{4\pi}\frac{2m}{\hbar^2}E \;\Rightarrow\; D(E) = \frac{dn(E)}{dE} = \frac{1}{4\pi}\frac{2m}{\hbar^2}.$$

The density of states $D(E)$ giving the change in the number of states with energy is then a constant. The density of states allows us to calculate up to what energy the eigenstates are filled for a certain number ν of electrons per site, see Figure A.12.

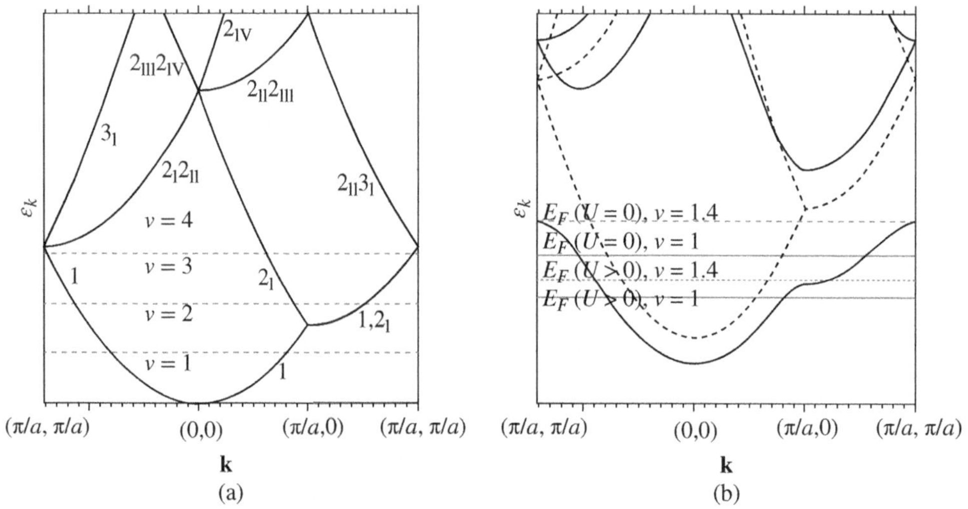

Figure A.12 (a) The free-electron bands in two dimensions along particular directions indicated in reciprocal space. The dashed lines show up to what energy the states are filled for different numbers ν of electrons per site. (b) The effect of including a rather large U on the lowest bands.

The constant density of states is the reason why the distance in energy is directly proportional to ν, as shown in Figure A.12. Note that the density of states for free electrons is not constant in one and three dimensions.

d) Figure A.12 shows how the bands are affected by a relatively large U_K. The associated density of states (not shown) is also strongly changed. First, there is a gap in the energy levels. Secondly, some bands are severely flattened, leading to a strong increase in the effective mass. Note, also, that for a filling of $\nu = 1.4$ electrons per site, the band crosses the Fermi level between $(0,0)$ and $(\pi/a, 0)$ for $U = 0$ and between $(\pi/a, 0)$ and $(\pi/a, \pi/a)$ for $U > 0$.

Problems for Section 3.9

1 In the limit $W \to 0$, the unperturbed Green's function in Eq. (3.204) can be approximated. Using that

$$\frac{E+W}{E-W} = \frac{1+\frac{W}{E}}{1-\frac{W}{E}} \cong \left(1+\frac{W}{E}\right)^2 \cong 1+\frac{2W}{E}, \tag{A.189}$$

keeping only the lowest order W/E. Note that since the majority of the spectral weight is at $E \cong U$, this also corresponds to the $W/|U| \ll 1$ limit. The Green's function is then

$$G^0(E) = \frac{1}{2W}\log\left(1+\frac{2W}{E}\right) \cong \frac{1}{E}. \tag{A.190}$$

The total Green's function in Eq. (3.203) is then

$$G_{00}(E) = \frac{1}{(G^0(E))^{-1} - U} = \frac{1}{E-U}. \tag{A.191}$$

In this limit, all band effects are gone, and the particle is simply stuck on the site with the attractive potential $-|U|$.

2 This problem is solved by numerically diagonalizing the matrix in Eq. (4.103) for $W = 1$, $U = -4$, and $N = 1000$. The width of the square density of states is $2W = 2$. Analytical approaches are possible but mathematically complex. The results are plotted in Figure A.13. The vertical dotted lines give the unperturbed energies $\varepsilon_i = -W + 2W(i-1)/(N-1)$ close to $N/2$. For $N = 1000$, the states are separated by 0.002 (using $W = 1$). The dots show the eigenenergies with $U = -4$. The density of states shift by 0.004 to higher energy, which has been removed for clarity of the figure. First of all, the eigenstates close to $N/2$ appear shifted. This can be understood as follows. The local potential U effectively removes a state from the density of states, leading to a bound state at $E_1 \cong U = -4$. This now leaves $N-1$ levels to describe the density of states. The removal of this state barely affects the width of the density of states, which is 1.99948 instead of 2, for $U \neq 0$ and $U = 0$, respectively. This is much less than the energy separation $2W/(N-1) = 0.002$ between the levels. The bottom and top of the band are therefore barely affected, *i.e.* $|E_2\rangle \cong |\varepsilon_1\rangle$ and $|E_N\rangle \cong |\varepsilon_N\rangle$. However, since there is one level less in the band, the spacing between the levels is slightly different, $\Delta\varepsilon = 2W/(N-1) = 0.002002$ and $\Delta E = 2W/(N-2) = 0.002004$ for $U = 0$ and $|U| \neq 0$, respectively. Although this

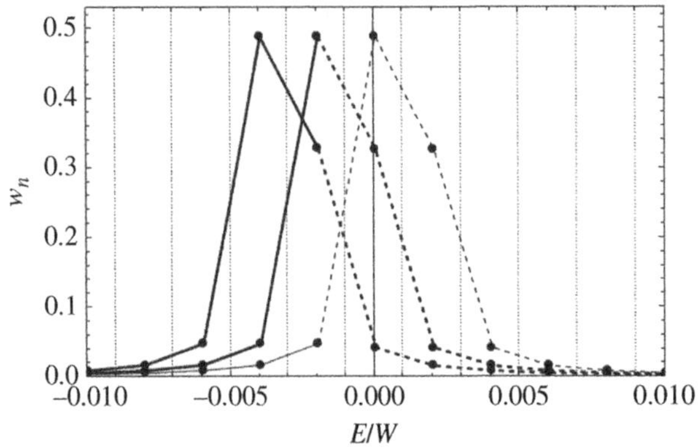

Figure A.13 The eigenstates close to the center of the band from $-W$ to W with $W = 1$ and $U = -4$. The states are obtained by diagonalizing numerically the matrix in Eq. (4.103) for $N = 1000$. The vertical dotted lines indicate the eigenenergies for $U = 0$. The dots indicate the energy positions for finite U. The three curves indicate the overlap $|\langle E_n|\varepsilon_i\rangle|^2$, where $\varepsilon_i = -W + 2W(i-1)/(N-1)$ are the unperturbed states, and E_n are the energies for $U = -4$. For left to right, the curves are for $i = N/2 - 1, N/2, N/2 + 1$.

is small, at the center of the band this amounts to $\frac{1}{2}N(\Delta E - \Delta\varepsilon) = 0.001$, which is half the splitting between the levels. This is why E_n is shifted by approximately $\Delta\varepsilon/2$ close to the center of the band, see Figure A.13.

This also affects the overlap $\langle E_n|\varepsilon_i\rangle$. Close to the center of the band $|E_n\rangle \cong a|\varepsilon_{n-1}\rangle + b|\varepsilon_n\rangle$. Therefore, an electron initially in $|\varepsilon_i\rangle$ predominantly scatters into $|E_i\rangle$ and $|E_{i+1}\rangle$ if the potential is switched on suddenly. This can be seen from the weights $|\langle E_n|\varepsilon_i\rangle|^2$ plotted in Figure A.13. Now let us assume that the states are filled up to $i = N/2$. For the lowest final state, the states $|E_n\rangle$ are also filled up to $n = N/2$. Let us look at the $i = N/2$ (the central curve in Figure A.13). This state scatters into $|E_{\frac{N}{2}}\rangle$, which is part of the lowest final state (indicated by the solid line). However, it also scatters into $|E_{\frac{N}{2}+1}\rangle$, which is not part of the lowest final state. Numerically, 41% scatters into states that are not part of the lowest perturbed state. However, this is only one state out of many states, so the effect is rather small. However, in more realistic systems, there are many states with the same energy (in this case $E = 0$), and the probability of a transition from the lowest initial state to the lowest final state goes to zero. This is called Anderson's orthogonality catastrophe. However, this weight has to go somewhere. This system generally responds to the sudden appearance of a local potential by creating electron–hole pairs. These are excitations where an electron is taken from the occupied states $n \leq N/2$ and put into the unoccupied states $n > N/2$ with respect to the lowest state (all states up to $n = N/2$ filled for a half-filled band).

Problems for Section 3.10

1 The Green's function $G_{11}(E)$ can be written as

$$G_{11} = G_1^0 + G_1^0 V_{12} G_{21}. \tag{A.192}$$

A particle can therefore stay in state 1 with the unperturbed Green's function G_1^0, where $G_i^0 = (E - \varepsilon_i)^{-1}$. It is implicitly assumed that E contains a small positive imaginary part for convergence. Alternatively, the particle can hop to state 2 via V_{12} and then find a way back again to 1. The full Green's function G_{21} can be written in terms of Green's functions and hopping matrix elements as

$$G_{21} = G_2^0 V_{21} G_{11} + G_2^0 V_{23} G_{31}, \quad \text{with } G_{31} = G_3^0 V_{32} G_{21}. \tag{A.193}$$

From state 2, there are two options: hop back to 1 or hop to 3. From state 3, the particle can only hop back to 2. Inserting G_{31} back into G_{21} and solving for G_{21} gives

$$G_{21} = \frac{1}{(G_2^0)^{-1} - V_{23} G_3^0 V_{32}} V_{21} G_{11}, \tag{A.194}$$

where $V_{ij} = V_{ji}$. Inserting this into Eq. (A.195) and solving for G_{11} gives

$$\begin{aligned} G_{11} &= \frac{1}{(G_1^0)^{-1} - V_{12} \dfrac{1}{(G_2^0)^{-1} - V_{23} G_3^0 V_{32}} V_{21}} \\ &= \frac{1}{E - \varepsilon_1 - V_{12} \dfrac{1}{E - \varepsilon_2 - V_{23} \dfrac{1}{E - \varepsilon_3} V_{32}} V_{21}}. \end{aligned} \tag{A.195}$$

Note that for an infinite number of states with $V_{ij} = V_{i,i\pm 1}\delta_{j,i\pm 1}$, this would lead to a continued fraction expansion. The spectral function is plotted in Figure A.14.

The Green's function is directly related to a spectral function through Fermi's golden rule, see Eq. (2.283). The eigenstates can also be obtained by solving the determinant of the following matrix:

$$\boldsymbol{H} = \begin{pmatrix} \varepsilon_1 & V_{12} & 0 \\ V_{21} & \varepsilon_2 & V_{23} \\ 0 & V_{32} & \varepsilon_3 \end{pmatrix} \quad \Rightarrow \quad |\boldsymbol{H} - E\mathbb{1}_3| = 0, \tag{A.196}$$

for the basis $|i\rangle$, with $i = 1, 2, 3$. For the given values of the parameters, the eigenenergies can be obtained numerically, giving $E_n = -1.145$, 0.476, and 3.669. The spectral function is given by

$$I(E) = -\frac{1}{\pi} \text{ Im } [G_{11}(E)] = \sum_{n=1}^{3} |\langle E_n|1\rangle|^2 \delta(E - E_n). \tag{A.197}$$

The spectral weights for the eigenstates $|E_n\rangle = \sum_{i=1}^{3} a_{ni}|i\rangle$, where a_{ni} is a coefficient, are $|\langle E_n|1\rangle|^2 = a_{n1}^2 = 0.352$, 0.618, and 0.029. Broadening the δ functions in Eq. (A.197) produces the same plot as Figure A.14.

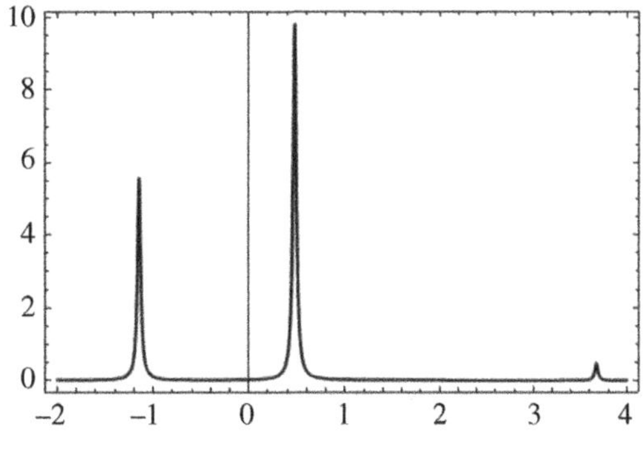

Figure A.14 The imaginary part of the Green's function in Eq. (A.195), *i.e.* $-\text{Im}[G_{11}(E)]/\pi$ for the parameters given in the problem. The energy E contains a small imaginary part $0.02i$ that broadens the poles.

Problems for Section 4.1

1 Since k_2 and k_4 are different between the bra and ket, the matrix element must include these quantum numbers

$$\begin{aligned}\langle k_1k_3k_4|\boldsymbol{H}_1|k_1k_2k_3\rangle &= \langle 0|\mathbf{c}_{k_1}\mathbf{c}_{k_3}\mathbf{c}_{k_4}\boldsymbol{H}_1\mathbf{c}^{\dagger}_{k_3}\mathbf{c}^{\dagger}_{k_2}\mathbf{c}^{\dagger}_{k_1}|0\rangle \\ &= -\langle 0|\mathbf{c}_{k_3}\mathbf{c}_{k_1}\mathbf{c}_{k_4}\boldsymbol{H}_1\mathbf{c}^{\dagger}_{k_2}\mathbf{c}^{\dagger}_{k_1}\mathbf{c}^{\dagger}_{k_3}|0\rangle + \langle 0|\mathbf{c}_{k_1}\mathbf{c}_{k_3}\mathbf{c}_{k_4}\boldsymbol{H}_1\mathbf{c}^{\dagger}_{k_3}\mathbf{c}^{\dagger}_{k_2}\mathbf{c}^{\dagger}_{k_1}|0\rangle \\ &= -\langle k_1k_4|\boldsymbol{H}_1|k_1k_2\rangle + \langle k_3k_4|\boldsymbol{H}_1|k_2k_3\rangle = -H_{k_1k_4k_2k_1} + H_{k_3k_4k_3k_2}.\end{aligned}$$

Note that two matrix elements are obtained for a two-particle interaction, whereas there is only one matrix element for a one-particle interaction.

2 a) The Green's function under the interaction $\hat{H}' = (e/m)\hat{\mathbf{p}}\cdot\mathbf{A}$ is given by

$$G_{\mathbf{k},\mathbf{k}}(E) = G^0_{\mathbf{k}}(E) + G^0_{\mathbf{k}}(E)\sum_{\mathbf{k}'\mathbf{q}\alpha} V_{\mathbf{k},\mathbf{k}'\mathbf{q}\alpha}G_{\mathbf{k}'\mathbf{q}\alpha,\mathbf{k}}(E), \tag{A.198}$$

where $G^0_{\mathbf{k}}(E) = (E-\varepsilon^0_{\mathbf{k}})^{-1}$ is the free-electron propagator. The matrix element $V_{\mathbf{k},\mathbf{k}'\mathbf{q}\alpha} = \langle\mathbf{k}|\boldsymbol{H}'|\mathbf{k}'\mathbf{q}\alpha\rangle$ scatters the electron from $\mathbf{k}$ to $\mathbf{k}'$ under the emission of a photon with momentum $\mathbf{q}$ and polarization vector $\boldsymbol{\epsilon}_{\mathbf{q}\alpha}$. However, for the lowest order we want this state to scatter back to the state with only an electron with wavevector $\mathbf{k}$. This gives

$$G_{\mathbf{k},\mathbf{k}}(E) = G^0_{\mathbf{k}}(E) + G^0_{\mathbf{k}}(E)\sum_{\mathbf{k}'\mathbf{q}\alpha} V_{\mathbf{k},\mathbf{k}'\mathbf{q}\alpha}G^0_{\mathbf{k}'\mathbf{q}\alpha}(E)V_{\mathbf{k}'\mathbf{q}\alpha,\mathbf{k}}G_{\mathbf{k},\mathbf{k}}(E),$$

where $G^0_{\mathbf{k}'\mathbf{q}\alpha} = (E-\varepsilon^0_{\mathbf{k}'}-\hbar\omega_{\mathbf{q}})^{-1}$ is the free propagator for an electron and a photon. This gives the Green's function

$$G_{\mathbf{k},\mathbf{k}}(E) = \frac{1}{(G^0_{\mathbf{k}}(E))^{-1} - \sum_{\mathbf{q}\alpha}|V_{\mathbf{k},\mathbf{k}'\mathbf{q}\alpha}|^2 G^0_{\mathbf{k}'\mathbf{q}\alpha}(E)}. \tag{A.199}$$

b) Using Eq. (4.16), the matrix element is

$$\begin{aligned}V_{\mathbf{k}'\mathbf{q}\alpha,\mathbf{k}} &= \frac{e}{m_0}\sqrt{\frac{\hbar}{2\varepsilon_0\omega_{\mathbf{q}}V}}\langle\mathbf{k}'\mathbf{q}\alpha|\hat{\mathbf{p}}\cdot\boldsymbol{\epsilon}_{\mathbf{q}\alpha}\boldsymbol{a}^{\dagger}_{\mathbf{q}\alpha}|\mathbf{k}\rangle \\ &= \frac{e}{m_0}\sqrt{\frac{\hbar^3}{2\varepsilon_0\omega_{\mathbf{q}}V}}\mathbf{k}\cdot\boldsymbol{\epsilon}_{\mathbf{q}\alpha}\delta_{\mathbf{k}',\mathbf{k}},\end{aligned}$$

using linear polarization vectors and $\hat{\mathbf{p}}|\mathbf{k}\rangle = \hbar\mathbf{k}|\mathbf{k}\rangle$. Note that the dipole approximation $e^{i\mathbf{q}\cdot\mathbf{r}} \cong 1$ changes the momentum conservation from $\delta_{\mathbf{k}'+\mathbf{q},\mathbf{k}}$ to $\delta_{\mathbf{k}',\mathbf{k}}$. The matrix element squared is

$$|V_{\mathbf{k}'\mathbf{q}\alpha,\mathbf{k}}|^2 = \frac{\hbar^3e^2}{2\varepsilon_0\omega_{\mathbf{q}}m_0^2}(\mathbf{k}\cdot\boldsymbol{\epsilon}_{\mathbf{q}\alpha})^2\delta_{\mathbf{k}',\mathbf{k}}. \tag{A.200}$$

Using that the sum over three orthogonal unit vectors equals an identity matrix, *i.e.* $\sum_i|\mathbf{e}_i\rangle\langle\mathbf{e}_i| = \mathbb{1}_3$ or $\sum_\alpha|\boldsymbol{\epsilon}_{\mathbf{q}\alpha}\rangle\langle\boldsymbol{\epsilon}_{\mathbf{q}\alpha}| + |\hat{\mathbf{q}}\rangle\langle\hat{\mathbf{q}}| = \mathbb{1}_3$, the angular dependence can be rewritten as

$$\sum_\alpha (\hat{\mathbf{k}} \cdot \boldsymbol{\epsilon}_{\mathbf{q}\alpha})^2 = \sum_\alpha \langle \hat{\mathbf{k}} | \boldsymbol{\epsilon}_{\mathbf{q}\alpha} \rangle \langle \boldsymbol{\epsilon}_{\mathbf{q}\alpha} | \hat{\mathbf{k}} \rangle = \langle \hat{\mathbf{k}} | (\mathbb{1}_3 - |\hat{\mathbf{q}}\rangle\langle\hat{\mathbf{q}}|) | \hat{\mathbf{k}} \rangle$$
$$= 1 - |\langle \hat{\mathbf{k}} | \hat{\mathbf{q}} \rangle|^2 = 1 - \cos^2\theta, \tag{A.201}$$

where the hat indicates a unit vector, and θ is the angle between the wavevectors of the electron and the photon.

c) The second term in the denominator in Eq. (A.199) causes an energy shift

$$\Delta E = \sum_{\mathbf{k}'\mathbf{q}} |V_{\mathbf{k},\mathbf{k}'\mathbf{q}\alpha}|^2 G^0_{\mathbf{k}\mathbf{q}\alpha}(\varepsilon^0_{\mathbf{k}}) = \sum_{\mathbf{k}'\mathbf{q}\alpha} \frac{e^2\hbar^3\mathbf{k}^2}{2\varepsilon_0 m_0^2 V \omega_{\mathbf{q}}} \delta_{\mathbf{k}',\mathbf{k}} \frac{1-\cos^2\theta}{\varepsilon^0_{\mathbf{k}} - \varepsilon^0_{\mathbf{k}'} - \hbar\omega_{\mathbf{q}}}$$
$$= \frac{e^2\hbar^3\mathbf{k}^2}{2\varepsilon_0 m_0^2 V} \frac{V}{(2\pi)^3} \int d\mathbf{q} \frac{1-\cos^2\theta}{-\hbar\omega_{\mathbf{q}}^2}.$$

With $\int d\mathbf{q} = \int d\varphi d\theta dq\, q^2 \sin\theta = (1/\hbar^2 c^3) \int d\varphi d(\cos\theta) d(\hbar\omega_{\mathbf{q}})\, \hbar\omega_{\mathbf{q}}^2$, the integral over the angular coordinates gives $8\pi/3$, leading to

$$\Delta E = -\frac{8\pi}{3} \frac{e^2\hbar\mathbf{k}^2}{16\pi^3\varepsilon_0 m_0^2 c^3} \int_0^\infty d(\hbar\omega) = -\frac{2}{3\pi}\left(\frac{e^2}{4\pi\varepsilon_0\hbar c}\right)\frac{(\hbar\mathbf{k})^2}{m_0^2c^2}\int_0^\infty d(\hbar\omega)$$
$$= -\frac{(\hbar\mathbf{k})^2}{2m_0}\frac{4\alpha}{3\pi}\frac{1}{m_0c^2}\int_0^\infty d(\hbar\omega),$$

using the definition of the fine-structure constant in Eq. (A.166). Note that the integral in this expression diverges. This is the self-energy of the electron.

d) The major problem in the evaluation lies in the use of the dipole approximation. More accurate calculations can be done using quantum electrodynamics. However, these calculations become rather technical. The bottom line is that at some point the wavelength of the light becomes smaller than the "size" of the electron, and the interaction will start to cancel out. This happens when $mc^2 \cong hc/\lambda$ of $\lambda_C = h/mc$, which is the Compton wavelength. So, instead of having an integrand that goes to infinity, there is a finite cutoff $E_{\text{cutoff}} \sim mc^2$. The energy of the electron is then

$$\varepsilon_{\mathbf{k}} = \frac{(\hbar\mathbf{k})^2}{2m} = \varepsilon^0_{\mathbf{k}} - \Delta E = \frac{(\hbar\mathbf{k})^2}{2m_0}\left(1 - \frac{4\alpha}{3\pi}\frac{E_{\text{cutoff}}}{m_0c^2}\right).$$

Since we cannot switch off the self-energy, m is the actual value that is measured in an experiment, whereas m_0 is the bare mass. Taking $E_{\text{cutoff}} = \kappa m_0 c^2$, the experimental mass in terms of the bare mass is

$$\frac{1}{m} = \frac{1}{m_0}(1 - a\alpha) \quad \Rightarrow \quad m = \frac{m_0}{1 - a\alpha} \cong m_0(1 + a\alpha),$$

where $a = 4\kappa/3\pi$ is close to unity.

e) The electromagnetic radiation interacts, to lowest order, only with the bare particle, giving an interaction

$$\frac{g^0 e}{2m_0} \cong \frac{g^0 e}{2m(1 - a\alpha)} = \frac{ge}{2m}, \quad \text{with } g \cong g^0(1 + a\alpha), \tag{A.202}$$

where $g^0 = 2$, see Eq. (2.224). A lengthy calculation shows that $a = 1/2\pi$, which gives

$$g \cong 2(1 + \frac{\alpha}{2\pi}) = 2.002322. \tag{A.203}$$

This is close to the experimental value of $g = 2.002319$.

Problems for Section 4.2

1 Indicating the spin states as $\uparrow$ and $\downarrow$, the possible configurations can be written as

$$(\uparrow + \downarrow)^N = \sum_k \begin{pmatrix} N \\ k \end{pmatrix} \uparrow^{N-k}\downarrow^k, \quad \text{with} \quad g_k = \begin{pmatrix} N \\ k \end{pmatrix} = \frac{N!}{(N-k)!k!},$$

where g_k gives the number of microstates. Using that $N! \sim N^N$ and $N_{\pm\frac{1}{2}} = \frac{1}{2}(N \pm M)$, the magnetization is given by $\overline{M} = (N_{\frac{1}{2}} - N_{-\frac{1}{2}})/N = M/N$. The logarithm of $g_k \rightarrow g_M$ can then be written as

$$\begin{aligned}\ln g_M &= \ln \frac{N!}{(\frac{1}{2}(N-M))!(\frac{1}{2}(N+M))!} \\ &\cong N \ln N - \frac{1}{2}(N-M)\ln\frac{1}{2}(N-M) - \frac{1}{2}(N+M)\ln\frac{1}{2}(N+M) \\ &= -\frac{N}{2}(1-\overline{M})\ln\frac{1}{2}(1-\overline{M}) - \frac{N}{2}(1+\overline{M})\ln\frac{1}{2}(1+\overline{M}),\end{aligned}$$

where $-N \leq M \leq N$ and $-1 \leq \overline{M} \leq 1$. Using that

$$\begin{aligned}-\frac{N}{2}(1 \pm \overline{M})\ln\left(\frac{1}{2}(1 \pm \overline{M})\right) &\cong -\frac{N}{2}(1 \pm \overline{M})\left(\pm\overline{M} - \frac{1}{2}\overline{M}^2 - \ln 2\right) \\ &= N\left(\mp\frac{1}{2}\overline{M} - \frac{1}{4}\overline{M}^2 + \frac{1}{2}(1 \pm \overline{M})\ln 2\right).\end{aligned}$$

The probability of finding a state with a particular M is proportional to

$$g_M \sim e^{-\frac{1}{2}N\overline{M}^2} = e^{-\frac{M^2}{2N}}, \tag{A.204}$$

where the terms with opposite signs cancel. Therefore, the probability of randomly finding all the spins in the same direction is $e^{-\frac{N}{2}}$. Even for a relatively small number of spins, this number is very small, *e.g.* for $N = 100$, $e^{-100/2} \cong 2 \times 10^{-22}$. Fluctuations around $\overline{M} = 0$ with a probability of the order of $e^{-\frac{M^2}{2N}} = e^{-1}$ have a magnitude of $M \sim \sqrt{N}$ or $\overline{M} = 1/\sqrt{N}$. The relative fluctuations in magnetization from $\overline{M} = 0$ are again very small for macroscopic number of spins. Despite this low probability, we will see that in the presence of interactions, finite magnetization can be found.

Problems for Section 4.3

1 a) The local Green's function splits into two. If the site is empty, the energy is ε_0; if the site contains a down spin, the energy is $\varepsilon_0 + U$. Each state has a probability of $\frac{1}{2}$. The local Green's function is then

$$G^0(E) = \frac{1}{2}\frac{1}{E + \dfrac{U}{2}} + \frac{1}{2}\frac{1}{E - \dfrac{U}{2}} = \frac{E}{E^2 - \left(\dfrac{U}{2}\right)^2}, \tag{A.205}$$

using $\varepsilon_0 = -U/2$ and $\varepsilon_0 + U = U/2$. It is implicitly assumed that E includes a small imaginary part $i\eta$ for convergence.

b) The full-Green's function is given by

$$G_{\mathbf{R}\mathbf{R}'}(E) = G^0(E) + G^0(E)\sum_{\mathbf{R}''} t_{\mathbf{R}\mathbf{R}''} G_{\mathbf{R}''\mathbf{R}'}(E). \tag{A.206}$$

This series closes on itself when taking the Fourier transform

$$G_{\mathbf{RR}'}(E) = \langle \mathbf{R}' - \mathbf{R}|G\rangle = \frac{1}{N}\sum_{\mathbf{k}} \langle \mathbf{R}' - \mathbf{R}|\mathbf{k}\rangle\langle \mathbf{k}|G\rangle$$
$$= \frac{1}{N}\sum_{\mathbf{k}} e^{i\mathbf{k}\cdot(\mathbf{R}'-\mathbf{R})} G_{\mathbf{k}}(E), \qquad \text{(A.207)}$$

where, due to translation symmetry, the Green's function only depends on the difference $\mathbf{R}' - \mathbf{R}$. The total Green's function then becomes

$$\frac{1}{N}\sum_{\mathbf{k}} e^{i\mathbf{k}\cdot(\mathbf{R}'-\mathbf{R})} G_{\mathbf{k}} = G^0 + G^0 \frac{1}{N}\sum_{\mathbf{k}\mathbf{R}''} t_{\mathbf{RR}''} e^{i\mathbf{k}\cdot(\mathbf{R}'-\mathbf{R}'')} G_{\mathbf{k}}$$
$$= \frac{1}{N}\sum_{\mathbf{k}} e^{i\mathbf{k}\cdot(\mathbf{R}'-\mathbf{R})} \left(G^0 + G^0 \sum_{\boldsymbol{\delta}} e^{i\mathbf{k}\cdot\boldsymbol{\delta}} t\, G_{\mathbf{k}} \right),$$

using $t_{\mathbf{RR}''} = t\delta_{\mathbf{R}'',\mathbf{R}+\boldsymbol{\delta}}$. Using $\varepsilon_{\mathbf{k}} = \sum_{\boldsymbol{\delta}} e^{i\mathbf{k}\cdot\boldsymbol{\delta}} t$, this gives

$$G_{\mathbf{k}} = G^0 + G^0 \varepsilon_{\mathbf{k}} G_{\mathbf{k}} \quad \Rightarrow \quad G_{\mathbf{k}}(E) = \frac{1}{(G^0(E))^{-1} - \varepsilon_{\mathbf{k}}}. \qquad \text{(A.208)}$$

Using Eq. (A.205), the final Green's function is then

$$G_{\mathbf{k}}(E) = \frac{1}{E - \varepsilon_{\mathbf{k}} - \left(\frac{U}{2}\right)^2 \frac{1}{E}} \qquad \text{(A.209)}$$

c) The Green's function has the same shape as Eq. (3.221). It is therefore equivalent to the matrix

$$H = \begin{pmatrix} \varepsilon_{\mathbf{k}} & \frac{U}{2} \\ \frac{U}{2} & 0 \end{pmatrix}, \quad \text{with } E_{\pm} = \frac{\varepsilon_{\mathbf{k}}}{2} \pm \frac{1}{2}\sqrt{\varepsilon_{\mathbf{k}}^2 + U^2}. \qquad \text{(A.210)}$$

The eigenvectors are $\cos\theta_{\mathbf{k}}|\mathbf{k}\rangle - \sin\theta_{\mathbf{k}}|0\rangle$ and $\sin\theta_{\mathbf{k}}|\mathbf{k}\rangle + \cos\theta_{\mathbf{k}}|0\rangle$, with $\tan 2\theta_{\mathbf{k}} = U/\varepsilon_{\mathbf{k}}$, where $|0\rangle$ is a nondispersive local state. The weights are given by the inner product with $|\mathbf{k}\rangle$, giving $\cos^2\theta_{\mathbf{k}}$ and $\sin^2\theta_{\mathbf{k}}$ (the eigenvectors interchange for $\varepsilon_{\mathbf{k}} > 0$ and $\varepsilon_{\mathbf{k}} < 0$).

d) For a one-dimensional system, the dispersion is $\varepsilon_k = -2t\cos ka$, see Eq. (3.166). The eigenenergies and weights are plotted in Figure A.15.

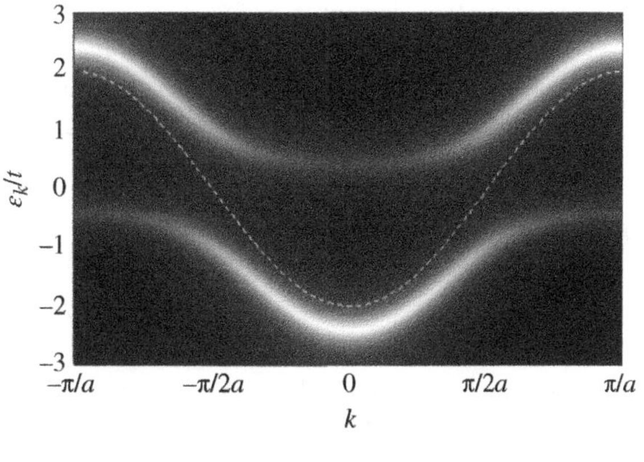

Figure A.15 The eigenenergies and their weights for a Hubbard model with $U = 2t$ in one dimension. The dashed line shows the dispersion for $U = 0$.

Problems for Section 4.4

1 We are looking for the matrix elements of $\langle 1m'|\boldsymbol{C}_q^k R^k|1m\rangle$. First, the possible values of k giving a nonzero matrix element are $k = 0, 1, 2$. The $k = 0$ term is finite. However, the associated matrix is $\mathbb{1}_3$ and does not split the p orbitals. Since the potential has inversion symmetry, the $k = 1$ terms are zero. This leaves $k = 2$. The potential remains the same under a 90° rotation around the z-axis. The same should be true for the spherical harmonics. The φ dependence is given by $e^{iq\varphi}$. Rotating over $\frac{\pi}{2}$ gives $e^{iq(\varphi+\frac{\pi}{2})}$. This only remains unchanged for $q = 0$. Therefore, we are left with $\boldsymbol{C}_0^k$. However, for the situation $a_i = a$, the potential should also remain unchanged under a rotation over 90° around the x and y axes and $\boldsymbol{C}_0^k$ does not. Therefore, apart from $k = 0$, there is no additional potential. Alternatively, we can also look at the orbitals. The p_i orbitals simply rotate into each other under the symmetry operation of the potential (inversion symmetry, rotation over 90° and 180° around the $\mathbf{e}_i$ axis, rotation around 120° around the (111) direction, etc.). Therefore, the orbitals are all equivalent for this potential.

This changes for the situation $a_x = a_y = a$ and $a_z = a'$, when the potential does not have the symmetry operations of 90° rotations around the x and y directions or rotations around 120° around the (111) direction. Therefore, the symmetry operations of the system cannot tell us whether the p_z orbital is the same as the p_x and p_y orbitals. For the potential due to the neighboring point charges, we cannot discard the $\boldsymbol{C}_0^k$ term. The matrix associated with the spherical harmonic C_0^k is $\boldsymbol{W}_0^2 = \boldsymbol{W}_{3z^2-r^2}^2$, see Eq. (4.94). Therefore, the $m = 0$ (p_z) orbital will have a different energy than the $m = \pm 1$ (related to the p_x and p_y orbitals). Symmetry therefore tells you how the orbitals split. However, it does not tell you by how much, since we do not know the details of the potential.

2 Following the reasoning from the preceding problem, for $a_i = a$, all matrix elements are zero for $k = 1, 2, 3$, due to inversion symmetry and the 90° rotations around x, y, and z. However, for $k = 4$, $e^{iq(\varphi+\frac{\pi}{2})} = 1$ for $q = 0$ and also for $q = \pm 4$. Along the z-axis, only the $q = 0$ term remains. For the x and y directions, we have both the $q = 0$ and $q = \pm 4$ terms. Due to the additional term it is now possible to satisfy the condition that the potential has to be equal along all axes. This gives a relation between the coefficients for $C_0^4(\hat{\mathbf{r}})$ and $C_{\pm 4}^4(\hat{\mathbf{r}})$, which we will not derive. C_0^4 only gives rise to diagonal matrix elements; $C_{\pm 4}^4$ only couples $m = \mp 2 \to \pm 2$. This explains the nonzero matrix elements.

The 5×5 matrix is effectively a 3×3 matrix with diagonal matrix elements -4Δ, 6Δ, and -4Δ and a 2×2 matrix with off-diagonal matrix elements. The latter is easily solved, giving eigenenergies $\Delta(1 \pm 5) = -4\Delta, 6\Delta$. We therefore have the following eigenenergies $-4\Delta(3\times), 6\Delta(2\times)$:

Problems for Section 4.5

1 For a p^3 configuration, there are a total of $\binom{6}{3} = 6 \times 5 \times 4/3 \times 2 = 20$ configurations, which are given in Table A.1. Let us now see what different LS terms are in these 20 configurations. The configuration $|1\uparrow; 1\downarrow; 0\uparrow\rangle$ must belong to an LS term with

Table A.1 Different configurations for p^3

p^3	$M_S = \frac{3}{2}$	$\frac{1}{2}$
$M = 2$		$\|1\uparrow;1\downarrow;0\uparrow\rangle$
1		$\|1\uparrow;0\uparrow;0\downarrow\rangle$, $\|1\uparrow;1\downarrow;-1\uparrow\rangle$
0	$\|1\uparrow;0\uparrow;-1\uparrow\rangle$	$\|1\downarrow;0\uparrow;-1\uparrow\rangle$, $\|1\uparrow;0\downarrow;-1\uparrow\rangle$, $\|1\uparrow;0\uparrow;-1\downarrow\rangle$
-1		$\|0\uparrow;0\downarrow;-1\uparrow\rangle$, $\|1\uparrow;-1\uparrow;-1\downarrow\rangle$
-2		$\|0\uparrow;-1\uparrow;-1\downarrow\rangle$

Only positive M_S values are indicated.

$L = 2$ and $S = \frac{1}{2}$ since there are no configurations with higher L and S that could give rise to this term. This term is denoted as a ^2D. The configuration $|1\uparrow;0\uparrow;-1\uparrow\rangle$ can only come from an LS term with $S = \frac{3}{2}$ and $L = 0$, therefore a ^4S. The first M and M_S terms for which we see more than one configuration are $M = 1$ and $M_S = \frac{1}{2}$. The LS states are linear combinations of the configurations $|1\uparrow;0\uparrow;0\downarrow\rangle$, $|1\uparrow;1\downarrow;-1\uparrow\rangle$. One of these combinations is part of the ^2D multiplet. The other combination belongs to the LS term with $L = 1$ and $S = \frac{1}{2}$, *i.e.* ^2P. For $M = 0$ and $M_S = \frac{1}{2}$, there are three configurations. However, ^2D, ^2P, and ^4S all have an $M = 0$ and $M_S = \frac{1}{2}$ component, so that it accounts for all three configurations. The total degeneracy is $5\times 2 + 3\times 2 + 1\times 4 = 20$, so we have indeed found all the LS terms.

The diagonal energy for a two-particle interaction in a three-particle configuration is given by all the possible pairs. Therefore, for the ^2D

$$\begin{aligned}&\langle 1\uparrow;1\downarrow;0\uparrow|\boldsymbol{H}_1|1\uparrow;1\downarrow;0\uparrow\rangle\\&=\langle 1\uparrow;1\downarrow|\boldsymbol{H}_1|1\uparrow;1\downarrow\rangle+\langle 1\uparrow;0\uparrow|\boldsymbol{H}_1|1\uparrow;0\uparrow\rangle+\langle 1\downarrow;0\uparrow|\boldsymbol{H}_1|1\downarrow;0\uparrow\rangle\\&=F^0+\frac{1}{25}F^2+F^0-\frac{1}{5}F^2+F^0-\frac{2}{25}F^2=3F^0-\frac{6}{25}F^2,\end{aligned} \tag{A.211}$$

where Eqs. (4.115) and (4.116) have been used. For the last matrix element, only the direct part in Eq. (4.116) has been used; the exchange part is zero due to the opposite spin of the electrons. Note that the 3 in front of the monopole term F^0 just counts the number of different pairs.

For the ^4S, we have

$$\begin{aligned}&\langle 1\uparrow;0\uparrow;-1\uparrow|\boldsymbol{H}_1|1\uparrow;0\uparrow;-1\uparrow\rangle=\langle 1\uparrow;0\uparrow|\boldsymbol{H}_1|1\uparrow;0\uparrow\rangle\\&\quad+\langle 1\uparrow;-1\uparrow|\boldsymbol{H}_1|1\uparrow;-1\uparrow\rangle+\langle 0\uparrow;-1\uparrow|\boldsymbol{H}_1|0\uparrow;-1\uparrow\rangle.\end{aligned}$$

This contains some matrix elements that we have not yet evaluated, for example

$$\begin{aligned}&\langle 1\uparrow;-1\uparrow|\boldsymbol{H}_1|1\uparrow;-1\uparrow\rangle=F^0+W^2_{3z^2-r^2,11}W^2_{3z^2-r^2,-1,-1}\frac{F^2}{25}\\&\quad-\sum_{q=x^2-y^2,xy}W^2_{q,1,-1}W^2_{q,-1,1}\frac{F^2}{25}\\&=F^0+\frac{F^2}{25}(1\times 1-(-i\sqrt{3})(i\sqrt{3})-\sqrt{3}\sqrt{3})=F^0-\frac{1}{5}F^2,\end{aligned} \tag{A.212}$$

giving only the nonzero matrix elements of the tensors from Section 4.4. The radial matrix elements are the same for the direct and exchange parts. The other matrix element is left to the reader. The total matrix element is then

$$\langle 1\uparrow;0\uparrow;-1\uparrow|\boldsymbol{H}_1|1\uparrow;0\uparrow;-1\uparrow\rangle = 3\left(F^0-\frac{1}{5}F^2\right) = 3F^0-\frac{3}{5}F^2.$$

All the matrix elements have the same value, since they are all matrix elements between electrons with spin up in different p orbitals.

For the ^2P term, we have to make use of the diagonal sum rule, which says that the trace of a matrix (*i.e.* the sum over the diagonal matrix elements) remains unchanged under a unitary transformation. In other words, the sum of the diagonal matrix elements is equal to the sum of the eigenenergies. We have two configurations of $M=1$ and $M_S=\frac{1}{2}$. The matrix element for the first is

$$\begin{aligned}\langle 1\uparrow;0\uparrow;0\downarrow|\boldsymbol{H}_1|1\uparrow;0\uparrow;0\downarrow\rangle &= \langle 1\uparrow;0\uparrow|\boldsymbol{H}_1|1\uparrow;0\uparrow\rangle\\ &+\langle 1\uparrow;0\downarrow|\boldsymbol{H}_1|1\uparrow;0\downarrow\rangle+\langle 0\uparrow;0\downarrow|\boldsymbol{H}_1|0\uparrow;0\downarrow\rangle\\ &= F^0-\frac{1}{5}F^2+F^0-\frac{2}{25}F^2+F^0+\frac{4}{25}F^2 = 3F^0-\frac{3}{25}F^2,\end{aligned}$$

using the results from Eqs. (4.116) and (A.211). The final matrix element is $\langle 0\uparrow;0\downarrow|\boldsymbol{H}_1|0\uparrow;0\downarrow\rangle = F^0+(W^2_{3z^2-r^2,00})^2F^2/25 = F^0+(-2)^2F^2/25$. The other matrix element is

$$\begin{aligned}\langle 1\uparrow;1\downarrow;-1\uparrow|\boldsymbol{H}_1|1\uparrow;1\downarrow;-1\uparrow\rangle &= \langle 1\uparrow;1\downarrow|\boldsymbol{H}_1|1\uparrow;1\downarrow\rangle\\ &+\langle 1\uparrow;-1\uparrow|\boldsymbol{H}_1|1\uparrow;-1\uparrow\rangle+\langle 1\downarrow;-1\uparrow|\boldsymbol{H}_1|1\downarrow;-1\uparrow\rangle\\ &= F^0+\frac{1}{25}F^2+F^0-\frac{5}{25}F^2+F^0+\frac{1}{25}F^2 = 3F^0-\frac{3}{25}F^2.\end{aligned}$$

According to the diagonal sum rule, the sum of the matrix elements of the configurations equals the sum of the eigenenergies

$$2\left(3F^0-\frac{3}{25}F^2\right) = E(^2D)+E(^2P) = 3F^0-\frac{3}{25}F^2+E(^2P),$$

giving $E(^2P)=3F^0$.

Therefore, for the p^3 configuration, we find that the ^4S state is the lowest eigenstate, in agreement with Hund's rule that says that S should be maximum. Pauli's principle forces electrons with the same spin into different orbitals, which lowers their Coulomb repulsion.

2 a) In the limit of a weak spin–orbit interaction, L and S are still considered good quantum numbers and can be coupled to a total J. This gives $J=L+S, L+S-1,\ldots,|L-S|$. This gives $J=2$ for ^1D, $J=0$ for ^1S, and $J=2,1,0$ for ^3P. The energies can be calculated using

$$\mathbf{L}\cdot\mathbf{S} = \frac{1}{2}\left(J(J+1)-L(L+1)-S(S+1)\right). \tag{A.213}$$

This gives 0 for ^1D and ^1S and $\zeta,-\zeta,-2\zeta$ for ^3P.

b) When there is no Coulomb interaction, the states are given by $j=\frac{3}{2},\frac{1}{2}$. The possible combinations are $\frac{3}{2}\frac{3}{2}$, $\frac{3}{2}\frac{1}{2}$, and $\frac{1}{2}\frac{1}{2}$ with energies ζ, $-\zeta/2$, and -2ζ, respectively.

c) The configuration of the $M=2$ state is $|1\uparrow;1\downarrow\rangle$. The diagonal energy is $\langle 1\uparrow;1\downarrow|L_zS_z|1\uparrow;1\downarrow\rangle = 1\times\frac{1}{2}+1\times(-\frac{1}{2})=0$. Its energy could be changed by the off-diagonal terms. However, the state $L_-S_+|1\uparrow;1\downarrow\rangle\sim|1\uparrow;0\uparrow\rangle$ is part of ^3P which we assumed to be far removed in energy. Therefore, there is no change in energy due to the spin–orbit interaction.

Problems for Section 4.6

1 a) Since the u and d quarks play a comparable role to the $\uparrow$ and $\downarrow$ spins, the eigenvectors for the triplet are given by $|uu\rangle$, $(|ud\rangle + |du\rangle)/\sqrt{2}$, and $|dd\rangle$. For the singlet, we have $(|ud\rangle - |du\rangle)/\sqrt{2}$.

b) Adding up and down quarks to the two-quark effective $S = 0$ state gives

$$\frac{1}{\sqrt{2}}(|ud\rangle - |du\rangle)|u\rangle \text{ and } \frac{1}{\sqrt{2}}(|ud\rangle - |du\rangle)|d\rangle. \tag{A.214}$$

c) There is only one way to create effective $S_z = \pm\frac{3}{2}$, namely $|uuu\rangle$ and $|ddd\rangle$, respectively. The normalized functions when stepping down/up each of the quarks are

$$\frac{1}{\sqrt{3}}(|duu\rangle + |udu\rangle + |uud\rangle) \text{ and } \frac{1}{\sqrt{3}}(|udd\rangle + |dud\rangle + |ddu\rangle).$$

We will see in the following section that these correspond to the Δ baryons.

d) There are a total of $2^3 = 8$ possible configurations, so there should be eight eigenvectors. So far, we have found six, so there are two remaining eigenstates. This must again be an $S = \frac{1}{2}$ state. These eigenvectors must be orthogonal to the previously found eigenvectors, giving

$$\frac{1}{\sqrt{6}}(|duu\rangle + |udu\rangle - 2|uud\rangle) \text{ and } \frac{1}{\sqrt{6}}(|udd\rangle + |dud\rangle - 2|ddu\rangle).$$

These are the wavefunctions for the proton and the neutron.

2 The Gell-Mann Matrix $\boldsymbol{\lambda}_8$ restricted to up and down quarks gives a diagonal term

$$\boldsymbol{\lambda}_8 = \frac{1}{\sqrt{3}}(\boldsymbol{a}_u^\dagger \boldsymbol{a}_u + \boldsymbol{a}_d^\dagger \boldsymbol{a}_d) = \frac{1}{\sqrt{3}}(\boldsymbol{a}_u^\dagger \underline{\boldsymbol{a}}_u^\dagger - \boldsymbol{a}_d^\dagger \underline{\boldsymbol{a}}_d^\dagger) = \sqrt{\frac{2}{3}}\pi^0, \tag{A.215}$$

using the fact that the wavefunction is the same as that of the π^0 meson, see Eq. (4.128). The $\boldsymbol{\lambda}_1$ and $\boldsymbol{\lambda}_2$ are

$$\begin{aligned} \boldsymbol{\lambda}_1 &= \boldsymbol{a}_u^\dagger \boldsymbol{a}_d + \boldsymbol{a}_d^\dagger \boldsymbol{a}_u = -\boldsymbol{a}_u^\dagger \underline{\boldsymbol{a}}_d^\dagger + \boldsymbol{a}_d^\dagger \underline{\boldsymbol{a}}_u^\dagger, \\ \boldsymbol{\lambda}_2 &= i(-\boldsymbol{a}_u^\dagger \boldsymbol{a}_d + \boldsymbol{a}_d^\dagger \boldsymbol{a}_u) = i(\boldsymbol{a}_u^\dagger \underline{\boldsymbol{a}}_d^\dagger + \boldsymbol{a}_d^\dagger \underline{\boldsymbol{a}}_u^\dagger). \end{aligned} \tag{A.216}$$

We can make a unitary transformation of these two operators

$$\begin{aligned} -\frac{1}{\sqrt{2}}(\boldsymbol{\lambda}_1 + i\boldsymbol{\lambda}_2) &= -\sqrt{2}\boldsymbol{a}_u^\dagger \boldsymbol{a}_d = \sqrt{2}\boldsymbol{a}_u^\dagger \underline{\boldsymbol{a}}_d^\dagger = \sqrt{2}\pi^+, \\ \frac{1}{\sqrt{2}}(\boldsymbol{\lambda}_1 - i\boldsymbol{\lambda}_2) &= \sqrt{2}\boldsymbol{a}_d^\dagger \boldsymbol{a}_u = \sqrt{2}\boldsymbol{a}_d^\dagger \underline{\boldsymbol{a}}_u^\dagger = \sqrt{2}\pi^-, \end{aligned} \tag{A.217}$$

which correspond to the $\pi^\pm$ mesons, see Figure 4.2.

Problems for Section 4.7

1 Squaring the total spin $\mathbf{S} = \mathbf{s}_1 + \mathbf{s}_2 + \mathbf{s}_3$ gives

$$\mathbf{S}^2 = (\mathbf{s}_1 + \mathbf{s}_2 + \mathbf{s}_3)^2 = \mathbf{s}_1^2 + \mathbf{s}_2^2 + \mathbf{s}_3^2 + 2\sum_{i<j} \mathbf{s}_i \cdot \mathbf{s}_j. \tag{A.218}$$

The spin–spin coupling is then

$$\sum_{i<j} \mathbf{s}_i \cdot \mathbf{s}_j = \frac{1}{2}\left(S(S+1) - 3\frac{1}{2}(\frac{1}{2}+1)\right) = \frac{1}{2}S(S+1) - \frac{9}{8} = \pm\frac{3}{4}$$

for $S = \frac{3}{2}, \frac{1}{2}$, respectively. The masses are then

$$M \cong 3m \pm \frac{3A}{4m^2} = 1.23, 0.93 \text{ GeV} \quad \text{for} \quad \Delta^+,\ p \tag{A.219}$$

or 2.2×10^{-27} and 1.65×10^{-27} kg, respectively. The results are somewhat off due to the simplification $m = m_u = m_d$ and the approximate value of m. The Δ^+ baryon has a lifetime of about 5.6×10^{-24} s.

2 The proton wavefunction is given by

$$\begin{aligned}|p\uparrow\rangle = \frac{1}{\sqrt{18}} & (2|u\uparrow u\uparrow d\downarrow\rangle - |u\uparrow u\downarrow d\uparrow\rangle - |u\downarrow u\uparrow d\uparrow\rangle + 2|u\uparrow d\downarrow u\uparrow\rangle \\ & - |u\uparrow d\uparrow u\downarrow\rangle - |u\downarrow d\uparrow u\uparrow\rangle + 2|d\downarrow u\uparrow u\uparrow\rangle - |d\uparrow u\uparrow u\downarrow\rangle \\ & - |d\uparrow u\downarrow u\uparrow\rangle).\end{aligned} \tag{A.220}$$

The magnetic moment of the proton is then given by

$$\begin{aligned}\mu_p &= \langle p\uparrow|\mu_1 + \mu_2 + \mu_3|p\uparrow\rangle \\ &= 3 \times \frac{1}{18}((4\mu_u + 4\mu_u - 4\mu_d) + (-\mu_u + \mu_u + \mu_d) + (\mu_u - \mu_u + \mu_d)) \\ &= \frac{1}{3}(4\mu_u - \mu_d) = \mu_N.\end{aligned} \tag{A.221}$$

For a neutron, one can interchange u and d. This gives

$$\mu_n = \frac{1}{3}(4\mu_d - \mu_e) = -\frac{2}{3}\mu_N. \tag{A.222}$$

Therefore, the expected ratio is $\mu_p/\mu_n = -\frac{3}{2}$. This is close to the experimentally observed value of -1.46. However, the absolute values are quite a bit off, $\mu_p = 2.79\mu_N$ and $\mu_p = -1.91\mu_N$, and require a more detailed calculation.

Problems for Section 5.1

1 The interaction between the free electron and the large spin can be viewed as an effective magnetic field,

$$\boldsymbol{H} = \begin{pmatrix} -B & 0 & t & 0 \\ 0 & B & 0 & t \\ t & 0 & -B\cos\theta & -Be^{-i\varphi}\sin\theta \\ 0 & t & -Be^{i\varphi}\sin\theta & B\cos\theta \end{pmatrix} \begin{matrix} 1\uparrow \\ 1\downarrow \\ 2\uparrow \\ 2\downarrow \end{matrix}, \tag{A.223}$$

where the different configurations of sites 1 and 2 are given on the right. Note that the hopping does not change the spin. The direction of the spin at site 1 is taken as a reference, so θ and φ are relatively to the direction of the spin at site 1. The interactions with the spin are $-2\mathbf{B} \cdot \mathbf{S}$, with $\mathbf{B}_1 = B\mathbf{e}_z$. Solving the matrix (one can use analytical mathematical software) gives for the two lowest eigenenergies

$$E_-^{\mp} = -\sqrt{B^2 + t^2 \pm 2Bt\cos\frac{\theta}{2}} \cong -B \mp t\cos\frac{\theta}{2}. \tag{A.224}$$

In the limit $B \gg t$, we can view this as the hopping of a free electron with its spin parallel to the large spin. The effective hopping matrix element is $t \cos \frac{\theta}{2}$, and bonding–antibonding states are found. The reduction in the matrix element is the same as that found in the Stern–Gerlach experiment, where the probabilities were calculated when the direction of a magnetic field is changed, see Eq. (2.229). The lowest energy is found for $\theta = 0$. An extended system will therefore have a tendency to become a ferromagnet. This ferromagnetic coupling is called double exchange. There is no gain from the hopping when the spins are antiparallel, *i.e.* $\theta = \pi$. The other two eigenenergies are, approximately, $E_{\pm}^{\mp} \cong B \mp t \cos \frac{\theta}{2}$, which is the electron hopping with its spin antiparallel to the large spins.

Problems for Section 5.2

1 Since the number of states is preserved, $D_s(E)dE = D_n(\xi)d\xi$. The density of states in the superconducting state $D_s(E)$ is then

$$D_s(E) = D_n(\xi)\frac{d\xi}{dE} \cong D_n(0)\frac{d}{dE}\sqrt{E^2 - \Delta^2} = \frac{ED_n(0)}{\sqrt{E^2 - \Delta^2}}, \tag{A.225}$$

using that $E_{\mathbf{k}} = \sqrt{\xi_{\mathbf{k}} + \Delta_{\mathbf{k}}}$ and taking the density of states in the normal state $D_n(0)$. Note that $D_s(E) = 0$ for $E < \Delta$.

Problems for Section 5.3

1 In order to remove the Higgs field from Eq. (5.66), the interaction is squared, and the expectation value for the Higgs field is taken:

$$\begin{aligned} \langle\phi_0|\boldsymbol{H}_{\mathrm{EW}}\boldsymbol{H}_{\mathrm{EW}}|\phi_0\rangle = (0, \phi_0) &\begin{pmatrix} e\mathbf{A}' + G\cos 2\theta\ \mathbf{Z}' & \sqrt{2}g'\mathbf{W}'_- \\ \sqrt{2}g'\mathbf{W}'_+ & G\mathbf{Z}' \end{pmatrix} \\ &\times \begin{pmatrix} e\mathbf{A}' + G\cos 2\theta\ \mathbf{Z}' & \sqrt{2}g'\mathbf{W}'_- \\ \sqrt{2}g'\mathbf{W}'_+ & G\mathbf{Z}' \end{pmatrix} \begin{pmatrix} 0 \\ \phi_0 \end{pmatrix} \\ &= 2g'^2\phi_0^2\mathbf{W}'_- \cdot \mathbf{W}'_+ + G^2\phi_0^2\mathbf{Z}' \cdot \mathbf{Z}'. \end{aligned} \tag{A.226}$$

For free space, we can view this interaction as an effective Klein–Gordan equation $\hbar^2\mathbf{q} - m^2c^2 = 0$, see Eq. (4.150),

$$\begin{aligned} &(\hbar^2\mathbf{q}'^2_{W_+} - g'^2\phi_0^2)\mathbf{W}'^{\dagger}_+ \cdot \mathbf{W}'_+ + (\hbar^2\mathbf{q}'^2_{W_-} - g'^2\phi_0^2)\mathbf{W}'^{\dagger}_- \cdot \mathbf{W}'_- \\ &\quad + (\hbar^2\mathbf{q}'^2_Z - G^2\phi_0^2)\mathbf{Z}' \cdot \mathbf{Z}' = 0. \end{aligned} \tag{A.227}$$

The ratio of the masses is then

$$\frac{M^2_{W_\pm}c^2}{M_Z^2c^2} = \frac{g'^2\phi_0^2}{G\phi_0^2} = \cos^2\theta \quad \Rightarrow \quad \frac{M_W}{M_Z} = \cos\theta = \cos 28^\circ = 0.88,$$

using $g' = G\cos\theta$, see Eq. (2.293). The experimental masses are $M_{W_\pm} = 80.2$ and $M_Z = 91.19\,\mathrm{GeV}$, giving a ratio of 0.88.

Problems for Section 5.4

1 a) Replacing $-J \to J$ and changing the spin operators gives

$$H = J\sum_{i\delta} \mathbf{S}_{i+\delta} \cdot \mathbf{S}_i = J\sum_{i\delta} \left(-\mathbf{S}_{i+\delta,0}\mathbf{S}_{i0} + \mathbf{S}_{i+\delta,1}\mathbf{S}_{i1} + \mathbf{S}_{i+\delta,-1}\mathbf{S}_{i,-1}\right),$$

where the summation over i only goes over one of the sublattices.

b)

$$H_{\text{eff}} = JS\sum_{i\delta} \left(a_i^\dagger a_i + a_{i+\delta}^\dagger a_{i+\delta} + a_{i+\delta}^\dagger a_i^\dagger + a_{i+\delta} a_i\right),$$

leaving out the shift in energy. The diagonal terms give the increase in energy when the spin is changed from $M_S = S$ to $M_S = S - 1$ on either sublattice. The motion of the spin between the sublattices is given by the last two terms. However, note that this is a result of two step-up or two step-down operators.

c) In momentum space, we have

$$H_{\text{eff}} = \frac{zJS}{2}\sum_k \left(2a_k^\dagger a_k + \gamma_k(a_k^\dagger a_k^\dagger + a_k a_k)\right),$$

where $z = 2$ is the number of nearest neighbors, and $\gamma_{\mathbf{k}} = \frac{1}{z}\sum_\delta e^{ik\delta} = \cos k$, with $z = 2$, for a one-dimensional system, and $k \in [-\pi, \pi]$.

d) The Hamiltonian is similar to the quadratic perturbation treated in Section 3.3. The energy is given by

$$\hbar\omega_k = JS\sqrt{2^2 - (2\gamma_k)^2} = 2JS\sqrt{1 - \gamma_k^2} = 2JS\sqrt{1 - \cos^2 k} = 2JS\sin k.$$

Note that the dispersion in the limit $k \to 0$ is $\sim k$, whereas for a ferromagnetic system, we had found a k^2 dependence.

Index

Geometric Quantum Mechanics, First Edition. Michel van Veenendaal.

f

g

h

n

o

p

t

u

v

Printed and bound by CPI Group (UK) Ltd, Croydon, CR0 4YY
27/10/2024
14580678-0003